SHIP RESISTANCE AND PROPULSION

Ship Resistance and Propulsion is dedicated to providing a comprehensive and modern scientific approach to evaluating ship resistance and propulsion. The study of propulsive power enables the size and mass of the propulsion engines to be established and estimates made of the fuel consumption and likely operating costs. This book, written by experts in the field, includes the latest developments from applied research, including those in experimental and CFD techniques, and provides guidance for the practical estimation of ship propulsive power for a range of ship types. This text includes sufficient published standard series data for hull resistance and propeller performance to enable practitioners to make ship power predictions based on material and data contained within the book. A large number of fully worked examples are included to illustrate applications of the data and powering methodologies; these include cargo and container ships, tankers and bulk carriers, ferries, warships, patrol craft, work boats, planing craft and yachts. The book is aimed at a broad readership including practising naval architects and marine engineers, sea-going officers, small craft designers and undergraduate and postgraduate degree students. It should also appeal to others involved in transportation, transport efficiency and eco-logistics, who need to carry out reliable estimates of ship power requirements.

Anthony F. Molland is Emeritus Professor of Ship Design at the University of Southampton in the United Kingdom. For many years, Professor Molland has extensively researched and published papers on ship design and ship hydrodynamics including propellers and ship resistance components, ship rudders and control surfaces. He also acts as a consultant to industry in these subject areas and has gained international recognition through presentations at conferences and membership on committees of the International Towing Tank Conference (ITTC). Professor Molland is the co-author of *Marine Rudders and Control Surfaces* (2007) and editor of *The Maritime Engineering Reference Book* (2008).

Stephen R. Turnock is Professor of Maritime Fluid Dynamics at the University of Southampton in the United Kingdom. Professor Turnock lectures on many subjects, including ship resistance and propulsion, powercraft performance, marine renewable energy and applications of CFD. His research encompasses both experimental and theoretical work on energy efficiency of shipping, performance sport, underwater systems and renewable energy devices, together with the application of CFD for the design of propulsion systems and control surfaces. He acts as a consultant to industry in these subject areas, and as a member of the committees of the International Towing Tank Conference (ITTC) and International Ship and Offshore Structures Congress (ISSC). Professor Turnock is the co-author of *Marine Rudders and Control Surfaces* (2007).

Dominic A. Hudson is Senior Lecturer in Ship Science at the University of Southampton in the United Kingdom. Dr. Hudson lectures on ship resistance and propulsion, powercraft performance and design, recreational and high-speed craft and ship design. His research interests are in all areas of ship hydrodynamics, including experimental and theoretical work on ship resistance components, seakeeping and manoeuvring, together with ship design for minimum energy consumption. He is a member of the 26th International Towing Tank Conference (ITTC) specialist committee on high-speed craft and was a member of the 17th International Ship and Offshore Structures Congress (ISSC) committee on sailing yacht design.

Ship Resistance and Propulsion

PRACTICAL ESTIMATION OF SHIP PROPULSIVE POWER

Anthony F. Molland
University of Southampton

Stephen R. Turnock
University of Southampton

Dominic A. Hudson
University of Southampton

CAMBRIDGE
UNIVERSITY PRESS

CAMBRIDGE UNIVERSITY PRESS
Cambridge, New York, Melbourne, Madrid, Cape Town,
Singapore, São Paulo, Delhi, Tokyo, Mexico City

Cambridge University Press
32 Avenue of the Americas, New York, NY 10013-2473, USA

www.cambridge.org
Information on this title: www.cambridge.org/9780521760522

First published 2011

Printed in the United States of America

A catalog record for this publication is available from the British Library.

Library of Congress Cataloging in Publication data

Molland, Anthony F.
Ship resistance and propulsion : practical estimation of ship propulsive power /
Anthony F. Molland, Stephen R. Turnock, Dominic A. Hudson.
    p.   cm.
Includes bibliographical references and index.
ISBN 978-0-521-76052-2 (hardback)
1. Ship resistance.   2. Ship resistance – Mathematical models.
3. Ship propulsion.   4. Ship propulsion – Mathematical models.
I. Turnock, Stephen R.   II. Hudson, Dominic A.   III. Title.
VM751.M65   2011
623.8′12–dc22   2011002620

ISBN 978-0-521-76052-2 Hardback

Contents

Preface

New ship types and applications continue to be developed in response to economic, societal and technical factors, including changes in operational speeds and fluctuations in fuel costs. These changes in ship design all depend on reliable estimates of ship propulsive power. There is a growing need to minimise power, fuel consumption and operating costs driven by environmental concerns and from an economic perspective. The International Maritime Organisation (IMO) is leading the shipping sector in efforts to reduce emissions such as NOx, SOx and CO_2 through the development of legislation and operational guidelines.

The estimation of ship propulsive power is fundamental to the process of designing and operating a ship. Knowledge of the propulsive power enables the size and mass of the propulsion engines to be established and estimates made of the fuel consumption and likely operating costs. The methods whereby ship resistance and propulsion are evaluated will never be an exact science, but require a combination of analysis, experiments, computations and empiricism. This book provides an up-to-date detailed appraisal of the data sources, methods and techniques for establishing propulsive power.

Notwithstanding the quantity of commercial software available for estimating ship resistance and designing propellers, it is our contention that rigorous and robust engineering design requires that engineers have the ability to carry out these calculations from first principles. This provides a transparent view of the calculation process and a deeper understanding as to how the final answer is obtained. An objective of this book is to include enough published standard series data for hull resistance and propeller performance to enable practitioners to make ship power predictions based on material and data contained within the book. A large number of fully worked examples are included to illustrate applications of the data and powering methodologies; these include cargo and container ships, tankers and bulk carriers, ferries, warships, patrol craft, work boats, planing craft and yachts.

The book is aimed at a broad readership, including practising professional naval architects and marine engineers and undergraduate and postgraduate degree students. It should also be of use to other science and engineering students and professionals with interests in the marine field.

The book is arranged in 17 chapters. The first 10 chapters broadly cover resistance, with Chapter 10 providing both sources of resistance data and useable data. Chapters 11 to 16 cover propellers and propulsion, with Chapter 16 providing both sources of propeller data and useable data. Chapter 17 includes a number of worked example applications. For the reader requiring more information on basic fluid mechanics, Appendix A1 provides a background to the physics of fluid flow. Appendix A2 derives a wave resistance formula and Appendices A3 and A4 contain tabulated resistance and propeller data. References are provided at the end of each chapter to facilitate readers' access to the original sources of data and information and further depth of study when necessary.

Proceedings, conference reports and standard procedures of the International Towing Tank Conference (ITTC) are referred to frequently. These provide an invaluable source of reviews and developments of ship resistance and propulsion. The proceedings and procedures are freely available through the website of the Society of Naval Architects and Marine Engineers (SNAME), which kindly hosts the ITTC website, http://ittc.sname.org. The University of Southampton Ship Science Reports, referenced in the book, can be obtained free from www.eprints .soton.ac.uk.

The authors acknowledge the help and support of their colleagues at the University of Southampton. Thanks must also be conveyed to national and international colleagues for their continued support over the years. Particular acknowledgement should also be made to the many undergraduate and postgraduate students who, over many years, have contributed to a better understanding of the subject through research and project and assignment work.

Many of the basic sections of the book are based on notes of lectures on ship resistance and propulsion delivered at the University of Southampton. In this context, particular thanks are due to Dr. John Wellicome, who assembled and delivered many of the original versions of the notes from the foundation of the Ship Science degree programme in Southampton in 1968.

Finally, the authors wish especially to thank their respective families for their practical help and support.

Anthony F. Molland
Stephen R. Turnock
Dominic A. Hudson
Southampton 2011

Nomenclature

A	Wetted surface area, thin ship theory (m^2)
A_0	Propeller disc area $[\pi D^2/4]$
A_D	Propeller developed blade area ratio, or developed blade area (m^2)
A_E	Propeller expanded blade area ratio
A_P	Projected bottom planing area of planing hull (m^2) or projected area of propeller blade (m^2)
A_T	Transverse frontal area of hull and superstructure above water (m^2)
A_X	Midship section area (m^2)
b	Breadth of catamaran demihull (m), or mean chine beam of planing craft (m)
B	Breadth of monohull or overall breadth of catamaran (m)
B_{pa}	Mean breadth over chines $[= A_P/L_P]$ (m)
B_{px}	Maximum breadth over chines (m)
B_{WL}	Breadth on waterline (m)
c	Section chord (m)
C_A	Model-ship correlation allowance coefficient
C_B	Block coefficient
C_{Dair}	Coefficient of air resistance $[R_{\mathrm{air}}/\frac{1}{2}\rho_a A_T V^2]$
C_f	Local coefficient of frictional resistance
C_F	Coefficient of frictional resistance $[R_F/\frac{1}{2}\rho_W S V^2]$
C_L	Lift coefficient
C_M	Midship coefficient $[A_X/(B \times T)]$
C_P	Prismatic coefficient $[\nabla/(L \times A_X)]$ or pressure coefficient
C_R	Coefficient of residuary resistance $[R_R/\frac{1}{2}\rho S V^2]$
C_S	Wetted surface coefficient $[S/\sqrt{\nabla \cdot L}]$
C_T	Coefficient of total resistance $[R_T/\frac{1}{2}\rho S V^2]$
C_V	Coefficient of viscous resistance $[R_V/\frac{1}{2}\rho S V^2]$
C_W	Coefficient of wave resistance $[R_W/\frac{1}{2}\rho S V^2]$
C_{WP}	Coefficient of wave pattern resistance $[R_{WP}/\frac{1}{2}\rho S V^2]$
D	Propeller diameter (m)

D_{air}	Aerodynamic drag, horizontal (planing craft) (N)
D_{APP}	Appendage resistance (N)
D_F	Planing hull frictional resistance, parallel to keel (N)
Demihull	One of the hulls which make up the catamaran
E	Energy in wave front
F_H	Hydrostatic pressure acting at centre of pressure of planing hull (N)
F_P	Pressure force over wetted surface of planing hull (N)
Fr	Froude number $[V/\sqrt{g \cdot L}]$
Fr_h	Depth Froude number $[V/\sqrt{g \cdot h}]$
Fr_∇	Volume Froude number $[V/\sqrt{g \cdot \nabla^{1/3}}]$
Fx	Yacht sail longitudinal force (N)
Fy	Yacht sail transverse force (N)
g	Acceleration due to gravity (m/s^2)
G	Gap between catamaran hulls (m)
GM	Metacentric height (m)
h	Water depth (m)
H	Wave height (m)
H_T	Transom immersion (m)
i_E	Half angle of entrance of waterline (deg.), see also $\frac{1}{2}\,\alpha_E$
J	Propeller advance coefficient (V_A/nD)
k	Wave number
K_T	Propeller thrust coefficient ($T/\rho n^2 D^4$)
K_Q	Propeller torque coefficient $Q/\rho n^2 D^5$)
L	Length of ship (m)
L_{air}	Aerodynamic lift, vertically upwards (planing craft) (N)
L_{APP}	Appendage lift (N)
L_{BP}	Length of ship between perpendiculars (m)
l_c	Wetted length of chine, planing craft (m)
LCB	Longitudinal centre of buoyancy (%L forward or aft of amidships)
LCG	Longitudinal centre of gravity (%L forward or aft of amidships)
L_f	Length of ship (ft)
l_K	Wetted length of keel, planing craft (m)
l_m	Mean wetted length, planing craft [$= (l_K + l_c)/2$]
L_{OA}	Length of ship overall (m)
lp	Distance of centre of pressure from transom (planing craft)(m)
L_P	Projected chine length of planing hull (m)
L_{PS}	Length between pressure sources
L_{WL}	Length on waterline (m)
$L/\nabla^{1/3}$	Length–displacement ratio
n	Propeller rate of revolution (rps)
N	Propeller rate of revolution (rpm), or normal bottom pressure load on planing craft (N)
P	Propeller pitch (m)
P_{AT}	Atmospheric pressure (N/m^2)
P/D	Propeller pitch ratio

P_D	Delivered power (kW)
P_E	Effective power (kW)
P_L	Local pressure (N/m^2)
P_S	Installed power (kW)
P_V	Vapour pressure (N/m^2)
Q	Propeller torque (Nm)
R_{air}	Air resistance (N)
R_{app}	Appendage resistance (N)
Re	Reynolds Number ($\rho VL/\mu$ or VL/ν)
R_F	Frictional resistance (N)
R_{Fh}	Frictional resistance of yacht hull (N)
R_{Ind}	Induced resistance of yacht (N)
rps	Revolutions per second
rpm	Revolutions per minute
R_R	Residuary resistance (N)
R_{Rh}	Residuary resistance of yacht hull (N)
R_{RK}	Residuary resistance of yacht keel (N)
R_T	Total hull resistance (N)
R_V	Viscous resistance (N)
R_{VK}	Viscous resistance of yacht keel (N)
R_{VR}	Viscous resistance of yacht rudder (N)
R_W	Wave resistance (N)
R_{WP}	Wave pattern resistance (N)
S	Wetted surface area (m^2)
S_{APP}	Wetted area of appendage (m^2)
S_C	Wetted surface area of yacht canoe body (m^2) or separation between catamaran demihull centrelines (m)
sfc	Specific fuel consumption
S_P	Propeller/hull interaction on planing craft (N)
t	Thrust deduction factor, or thickness of section (m)
T	Draught (m), or propeller thrust (N), or wave period (secs)
T_C	Draught of yacht canoe body (m)
U	Speed (m/s)
V	Speed (m/s)
Va	Wake speed ($V_S(1 - w_T)$) (m/s)
V_A	Relative or apparent wind velocity (m/s)
V_K	Ship speed (knots)
$V_K/\sqrt{L_f}$	Speed length ratio (knots and feet)
V_R	Reference velocity (m/s)
V_S	Ship speed (m/s)
W	Channel width (m)
w_T	Wake fraction
Z	Number of blades of propeller
$(1+k)$	Form-factor, monohull
$(1+\beta k)$	Form factor, catamaran
$\frac{1}{2}\alpha_E$	Half angle of entrance of waterline (deg.), see also i_E

β	Viscous resistance interference factor, or appendage scaling factor, or deadrise angle of planing hull (deg.) or angle of relative or apparent wind (deg.)
δ	Boundary layer thickness (m)
ε	Angle of propeller thrust line to heel (deg.)
η_D	Propulsive coefficient ($\eta_0 \eta_H \eta_R$)
η_O	Open water efficiency ($JK_T/2\pi\,K_Q$)
η_H	Hull efficiency $(1-t)/(1-w_T)$
η_R	Relative rotative efficiency
η_T	Transmission efficiency
γ	Surface tension (N/m), or wave height decay coefficient, or course angle of yacht (deg.), or wave number
ϕ	Heel angle (deg.), or hydrodynamic pitch angle (deg.)
λ	Leeway angle (deg.)
μ	Dynamic viscosity (g/ms)
ν	Kinematic viscosity (μ/ρ) (m²/s)
ρ	Density of water (kg/m³)
ρ_a	Density of air (kg/m³)
σ	Cavitation number, or source strength, or allowable stress (N/m²)
τ	Wave resistance interference factor (catamaran resistance/monohull resistance), or trim angle of planing hull (deg.)
τ_c	Thrust/unit area, cavitation (N/m²)
τ_R	Residuary resistance interference factor (catamaran resistance/monohull resistance)
τ_W	Surface or wall shear stress (N/m²)
θ	Wave angle (deg.)
ζ	Wave elevation (m)
∇	Ship displacement volume (m³)
∇_C	Displacement volume of yacht canoe body (m³)
Δ	Ship displacement mass ($\nabla\rho$) (tonnes), or displacement force ($\nabla\rho g$) (N)

Conversion of Units

1 m = 3.28 ft	1 ft = 12 in.
1 in. = 25.4 mm	1 km = 1000 m
1 kg = 2.205 lb	1 tonne = 1000 kg
1 ton = 2240 lb	1 lb = 4.45 N
1 lbs/in.² = 6895 N/m²	1 bar = 14.7 lbs/in.²
1 mile = 5280 ft	1 nautical mile (Nm) = 6078 ft
1 mile/hr = 1.61 km/hr	1 knot = 1 Nm/hr
$Fr = 0.2974\, V_K/\sqrt{L_f}$	1 knot = 0.5144 m/s
1 HP = 0.7457 kW	1 UK gal = 4.546 litres

Abbreviations

ABS	American Bureau of Shipping
AEW	Admiralty Experiment Works (UK)
AFS	Antifouling systems on ships
AHR	Average hull roughness
AP	After perpendicular
ARC	Aeronautical Research Council (UK)
ATTC	American Towing Tank Conference
BDC	Bottom dead centre
BEM	Boundary element method
BEMT	Blade element-momentum theory
BMEP	Brake mean effective pressure
BMT	British Maritime Technology
BN	Beaufort Number
BSRA	British Ship Research Association
BTTP	British Towing Tank Panel
CAD	Computer-aided design
CCD	Charge-coupled device
CFD	Computational fluid dynamics
CG	Centre of gravity
CLR	Centre of lateral resistance
CODAG	Combined diesel and gas
CP	Controllable pitch (propeller)
CSR	Continuous service rating
DES	Detached eddy simulation
DNS	Direct numerical simulation
DNV	Det Norske Veritas
DSYHS	Delft systematic yacht hull series
DTMB	David Taylor Model Basin
EFD	Experimental fluid dynamics
FEA	Finite element analysis
FP	Forward perpendicular, or fixed pitch (propeller)
FRP	Fibre-reinforced plastic
FV	Finite volume

GL	Germanischer Lloyd
GPS	Global Positioning System
HP	Horsepower
HSVA	Hamburg Ship Model Basin
IESS	Institute of Engineers and Shipbuilders in Scotland
IMarE	Institute of Marine Engineers (became IMarEST from 2001)
IMarEST	Institute of Marine Engineering, Science and Technology
IMechE	Institution of Mechanical Engineers
IMO	International Maritime Organisation
INSEAN	Instituto di Architectura Navale (Rome)
ISO	International Standards Organisation
ITTC	International Towing Tank Conference
JASNAOE	Japan Society of Naval Architects and Ocean Engineers
LCG	Longitudinal centre of gravity
LDA	Laser Doppler anemometry
LDV	Laser Doppler velocimetry
LE	Leading edge of foil or fin
LES	Large eddy simulation
LR	Lloyd's Register of Shipping
MAA	Mean apparent amplitude
MARIN	Maritime Research Institute of the Netherlands (formerly NSMB)
MCR	Maximum continuous rating
MEMS	Microelectromechanical systems
NACA	National Advisory Council for Aeronautics (USA)
NECIES	North East Coast Institution of Engineers and Shipbuilders
NPL	National Physical Laboratory (UK)
NSMB	The Netherlands Ship Model Basin (later to become MARIN)
NTUA	National Technical University of Athens
ORC	Offshore Racing Congress
P	Port
PIV	Particle image velocimetry
QPC	Quasi propulsive coefficient
RANS	Reynolds Averaged Navier–Stokes
RB	Round back (section)
RINA	Royal Institution of Naval Architects
ROF	Rise of floor
rpm	Revolutions per minute
rps	Revolutions per second
S	Starboard
SAC	Sectional area curve
SCF	Ship correlation factor
SG	Specific gravity
SNAJ	Society of Naval Architects of Japan (later to become JASNAOE)
SNAK	Society of Naval Architects of Korea
SNAME	Society of Naval Architects and Marine Engineers (USA)
SP	Self-propulsion
SSPA	Statens Skeppsprovingansalt, Götaborg, Sweden

STG	Schiffbautechnische Gesellschaft, Hamburg
TBT	Tributyltin
TDC	Top dead centre
TDW	Tons deadweight
TE	Trailing edge of foil or fin
TEU	Twenty foot equivalent unit [container]
UTS	Ultimate tensile stress
VCB	Vertical centre of buoyancy
VLCC	Very large crude carrier
VPP	Velocity prediction program
VWS	Versuchsanstalt für Wasserbau und Schiffbau Berlin (Berlin Model Basin)
WUMTIA	Wolfson Unit for Marine Technology and Industrial Aerodynamics, University of Southampton

Figure Acknowledgements

The authors acknowledge with thanks the assistance given by the following companies and publishers in permitting the reproduction of illustrations and tables from their publications:

Figures 8.5, 10.7, 10.9, 10.10, 10.12, 12.24, 14.17, 14.18, 14.19, 14.20, 14.21, 14.22, 14.23, 14.24, 14.30, 15.4, 15.14, 15.15, 15.17, 16.1, 16.2 and Tables A3.13, A3.14, A3.15, A4.3 reprinted courtesy of The Society of Naval Architects and Marine Engineers (SNAME), New York.

Figures 3.28, 3.29, 4.4, 4.5, 4.6, 7.6, 7.10, 7.16, 7.28, 8.4, 10.2, 10.3, 10.4, 10.5, 10.13, 10.14, 10.20, 10.21, 12.26, 14.15, 16.7, 16.8, 16.9, 16.10, 16.15, 16.16, 16.17, 16.19, 16.26 and Tables A3.1, A3.6, A3.24, A3.25, A3.26, A4.4, A4.5 reprinted courtesy of The Royal Institution of Naval Architects (RINA), London.

Figures 10.11, 16.24 and Tables A3.2, A3.3, A3.4, A3.5, A3.12, A3.23, A4.1, A4.2, A4.6 reprinted courtesy of IOS Press BV, Amsterdam.

Figures 4.8, 8.3, 8.12, 8.13, 16.3, 16.4, 16.5, 16.6, 16.13 reprinted courtesy of MARIN, Wageningen.

Figure 10.23 and Tables 10.13, 10.14, 10.15, 10.16, 10.17 reprinted courtesy of The HISWA Symposium Foundation, Amsterdam.

Figures A1.1, A1.7, A1.8 and Sections A1.1–A1.7 reprinted courtesy of Elsevier Ltd., Oxford.

Figure 10.22 reprinted courtesy of The Japan Society of Naval Architects and Ocean Engineers (JASNAOE), Tokyo (formerly The Society of Naval Architects of Japan (SNAJ), Tokyo).

Figure 3.10 reprinted courtesy of WUMTIA, University of Southampton and Dubois Naval Architects Ltd., Lymington.

Figure 7.3 reprinted courtesy of WUMTIA, University of Southampton.

Figure 12.29 reprinted courtesy of The University of Newcastle upon Tyne.

Figures 12.31, 13.9, 15.17 reprinted courtesy of The North East Coast Institution of Engineers and Shipbuilders (NECIES), Newcastle upon Tyne.

Table A3.7 reprinted courtesy of *Ship Technology Research*, Hamburg.

Table A3.27 reprinted courtesy of STG, Hamburg.

Figure 10.6 reprinted courtesy of BMT Group Ltd., Teddington.

Figure 12.25 reprinted courtesy of The Institution of Mechanical Engineers (IMechE), London.

Figures 16.27, 16.28 and Tables 16.8, 16.9 reprinted courtesy of The Offshore Racing Congress (ORC).

1 Introduction

The estimation of ship propulsive power is fundamental to the process of designing and operating a ship. A knowledge of the propulsive power enables the size and mass of the propulsion engines to be established and estimates made of the fuel consumption and operating costs. The estimation of power entails the use of experimental techniques, numerical methods and theoretical analysis for the various aspects of the powering problem. The requirement for this stems from the need to determine the correct match between the installed power and the ship hull form during the design process. An understanding of ship resistance and propulsion derives from the fundamental behaviour of fluid flow. The complexity inherent in ship hydrodynamic design arises from the challenges of scaling from practical model sizes and the unsteady flow interactions between the viscous ship boundary layer, the generated free-surface wave system and a propulsor operating in a spatially varying inflow.

History

Up to the early 1860s, little was really understood about ship resistance and many of the ideas on powering at that time were erroneous. Propeller design was very much a question of trial and error. The power installed in ships was often wrong and it was clear that there was a need for a method of estimating the power to be installed in order to attain a certain speed.

In 1870, W. Froude initiated an investigation into ship resistance with the use of models. He noted that the wave configurations around geometrically similar forms were similar if compared at corresponding speeds, that is, speeds proportional to the square root of the model length. He propounded that the total resistance could be divided into skin friction resistance and residuary, mainly wavemaking, resistance. He derived estimates of frictional resistance from a series of measurements on planks of different lengths and with different surface finishes [1.1], [1.2]. Specific residuary resistance, or resistance per ton displacement, would remain constant at corresponding speeds between model and ship. His proposal was initially not well received, but gained favour after full-scale tests had been carried out. HMS *Greyhound* (100 ft) was towed by a larger vessel and the results showed a substantial level of agreement with the model predictions [1.3]. Model tests had been vindicated and

1

the way opened for the realistic prediction of ship power. In a 1877 paper, Froude gave a detailed explanation of wavemaking resistance which lent further support to his methodology [1.4].

In the 1860s, propeller design was hampered by a lack of understanding of negative, or apparent, slip; naval architects were not fully aware of the effect of wake. Early propeller theories were developed to enhance the propeller design process, including the momentum theory of Rankine [1.5] in 1865, the blade element theory of Froude [1.6] in 1878 and the actuator disc theory of Froude [1.7] in 1889. In 1910, Luke [1.8] published the first of three important papers on wake, allowing more realistic estimates of wake to be made for propeller design purposes. Cavitation was not known as such at this time, although several investigators, including Reynolds [1.9], were attempting to describe its presence in various ways. Barnaby [1.10] goes some way to describing cavitation, including the experience of Parsons with *Turbinia*. During this period, propeller blade area was based simply on thrust loading, without a basic understanding of cavitation.

By the 1890s the full potential of model resistance tests had been realised. Routine testing was being carried out for specific ships and tests were also being carried out on series of models. A notable early contribution to this is the work of Taylor [1.11], [1.12] which was closely followed by Baker [1.13].

The next era saw a steady stream of model resistance tests, including the study of the effects of changes in hull parameters, the effects of shallow water and to challenge the suitability and correctness of the Froude friction values [1.14]. There was an increasing interest in the performance of ships in rough water. Several investigations were carried out to determine the influence of waves on motions and added resistance, both at model scale and from full-scale ship measurements [1.15].

Since about the 1960s there have been many developments in propulsor types. These include various enhancements to the basic marine propeller such as tip fins, varying degrees of sweep, changes in section design to suit specific purposes and the addition of ducts. Contra-rotating propellers have been revisited, cycloidal propellers have found new applications, waterjets have been introduced and podded units have been developed. Propulsion-enhancing devices have been proposed and introduced including propeller boss cap fins, upstream preswirl fins or ducts, twisted rudders and fins on rudders. It can of course be noted that these devices are generally at their most efficient in particular specific applications.

From about the start of the 1980s, the potential future of computational fluid dynamics (CFD) was fully realised. This would include the modelling of the flow around the hull and the derivation of viscous resistance and free-surface waves. This generated the need for high quality benchmark data for the physical components of resistance necessary for the validation of the CFD. Much of the earlier data of the 1970s were revisited and new benchmark data developed, in particular, for viscous and wave drag. Much of the gathering of such data has been coordinated by the International Towing Tank Conference (ITTC). Typical examples of the application of CFD to hull form development and resistance prediction are given in [1.16] and [1.17].

Propeller theories had continued to be developed in order to improve the propeller design process. Starting from the work of Rankine, Froude and Perring, these included blade element-momentum theories, such as Burrill [1.18] in 1944, and

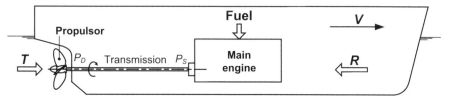

Figure 1.1. Overall concept of energy conversion.

Lerbs [1.19] in 1952 using a development of the lifting line and lifting surface methods where vorticity is distributed over the blade. Vortex lattice methods, boundary element, or panel, methods and their application to propellers began in the 1980s. The 1990s saw the application of CFD and Reynolds-Averaged Navier–Stokes (RANS) solvers applied to propeller design and, bringing us to the current period, CFD modelling of the combined hull and propeller [1.20].

Powering: Overall Concept

The overall concept of the powering system may be seen as converting the energy of the fuel into useful thrust (T) to match the ship resistance (R) at the required speed (V), Figure 1.1. It is seen that the overall efficiency of the propulsion system will depend on:

Fuel type, properties and quality.
The efficiency of the engine in converting the fuel energy into useful transmittable power.
The efficiency of the propulsor in converting the power (usually rotational) into useful thrust (T).

The following chapters concentrate on the performance of the hull and propulsor, considering, for a given situation, how resistance (R) and thrust (T) may be estimated and then how resistance may be minimised and thrust maximised. Accounts of the properties and performance of engines are summarised separately.

The main components of powering may be summarised as the effective power P_E to tow the vessel in calm water, where $P_E = R \times V$ and the propulsive efficiency η, leading to the propulsive (or delivered) power P_D, defined as: $P_D = P_E/\eta$. This is the traditional breakdown and allows the assessment of the individual components to be made and potential improvements to be investigated.

Improvements in Efficiency

The factors that drive research and investigation into improving the overall efficiency of the propulsion of ships are both economic and environmental. The main economic drivers amount to the construction costs, disposal costs, ship speed and, in particular, fuel costs. These need to be combined in such a way that the shipowner makes an adequate rate of return on the investment. The main environmental drivers amount to emissions, pollution, noise, antifoulings and wave wash.

The emissions from ships include NOx, SOx and CO_2, a greenhouse gas. Whilst NOx and SOx mainly affect coastal regions, carbon dioxide (CO_2) emissions have a

Table 1.1. *Potential savings in resistance and propulsive efficiency*

RESISTANCE (a) Hull resistance	Principal dimensions: main hull form parameters, U- or V-shape sections Local detail: bulbous bows, vortex generators Frictional resistance: WSA, surface finish, coatings
(b) Appendages	Bilge keels, shaft brackets, rudders: careful design
(c) Air drag	Design and fairing of superstructures Stowage of containers
PROPULSIVE EFFICIENCY (d) Propeller	Choice of main dimensions: $D, P/D, BAR$, optimum diameter, rpm. Local detail: section shape, tip fins, twist, tip rake, skew etc. Surface finish

(e) Propeller–hull interaction	Main effects: local hull shape, U, V or 'circular' forms
	    [resistance vs. propulsion]
	Changes in wake, thrust deduction, hull efficiency
	Design of appendages: such as shaft brackets and rudders
	Local detail: such as pre- and postswirl fins, upstream duct,
	    twisted rudders

global climatic impact and a concentrated effort is being made worldwide towards their reduction. The International Maritime Organisation (IMO) is co-ordinating efforts in the marine field, and the possibilities of CO_2 Emissions Control and an Emissions Trading Scheme are under consideration.

The likely extension of a carbon dioxide based emissions control mechanism to international shipping will influence the selection of propulsion system components together with ship particulars. Fuel costs have always provided an economic imperative to improve propulsive efficiency. The relative importance of fuel costs to overall operational costs influences the selection of design parameters such as dimensions, speed and trading pattern. Economic and environmental pressures thus combine to create a situation which demands a detailed appraisal of the estimation of ship propulsive power and the choice of suitable machinery. There are, however, some possible technical changes that will decrease emissions, but which may not be economically viable. Many of the auxiliary powering devices using renewable energy sources, and enhanced hull coatings, are likely to come into this category. On the basis that emissions trading for ships may be introduced in the future, all means of improvement in powering and reduction in greenhouse gas emissions should be explored and assessed, even if such improvements may not be directly economically viable.

The principal areas where improvements might be expected to be made at the design stage are listed in Table 1.1. It is divided into sections concerned first with resistance and then propulsive efficiency, but noting that the two are closely related in terms of hull form, wake fraction and propeller–hull interaction. It is seen that there is a wide range of potential areas for improving propulsive efficiency.

Power reductions can also be achieved through changes and improvements in operational procedures, such as running at a reduced speed, weather routeing, running at optimum trim, using hydrodynamically efficient hull coatings, hull/propeller cleaning and roll stabilisation. Auxiliary propulsion devices may also be employed, including wind assist devices such as sails, rotors, kites and wind turbines, wave propulsion devices and solar energy.

The following chapters describe the basic components of ship powering and how they can be estimated in a practical manner in the early stages of a ship design. The early chapters describe fundamental principles and the estimation of the basic components of resistance, together with influences such as shallow water, fouling and rough weather. The efficiency of various propulsors is described including the propeller, ducted propeller, supercavitating propeller, surface piercing and podded propellers and waterjets. Attention is paid to their design and off design cases and how improvements in efficiency may be made. Databases of hull resistance and propeller performance are included in Chapters 10 and 16. Worked examples of the overall power estimate using both the resistance and propulsion data are described in Chapter 17.

References are provided at the end of each chapter. Further more detailed accounts of particular subject areas may be found in the publications referenced and in the more specialised texts such as [1.21] to [1.29].

REFERENCES (CHAPTER 1)

1.1 Froude, W. Experiments on the surface-friction experienced by a plane moving through water, *42nd Report of the British Association for the Advancement of Science*, Brighton, 1872.

1.2 Froude, W. Report to the Lords Commissioners of the Admiralty on experiments for the determination of the frictional resistance of water on a surface, under various conditions, performed at Chelston Cross, under the Authority of their Lordships, *44th Report by the British Association for the Advancement of Science*, Belfast, 1874.

1.3 Froude, W. On experiments with HMS *Greyhound. Transactions of the Royal Institution of Naval Architects*, Vol. 15, 1874, pp. 36–73.

1.4 Froude, W. Experiments upon the effect produced on the wave-making resistance of ships by length of parallel middle body. *Transactions of the Royal Institution of Naval Architects*, Vol. 18, 1877, pp. 77–97.

1.5 Rankine, W.J. On the mechanical principles of the action of propellers. *Transactions of the Royal Institution of Naval Architects*, Vol. 6, 1865, pp. 13–35.

1.6 Froude, W. On the elementary relation between pitch, slip and propulsive efficiency. *Transactions of the Royal Institution of Naval Architects*, Vol. 19, 1878, pp. 47–65.

1.7 Froude, R.E. On the part played in propulsion by differences in fluid pressure. *Transactions of the Royal Institution of Naval Architects*, Vol. 30, 1889, pp. 390–405.

1.8 Luke, W.J. Experimental investigation on wake and thrust deduction values. *Transactions of the Royal Institution of Naval Architects*, Vol. 52, 1910, pp. 43–57.

1.9 Reynolds, O. The causes of the racing of the engines of screw steamers investigated theoretically and by experiment. *Transactions of the Royal Institution of Naval Architects*, Vol. 14, 1873, pp. 56–67.

1.10 Barnaby, S.W. Some further notes on cavitation. *Transactions of the Royal Institution of Naval Architects*, Vol. 53, 1911, pp. 219–232.

1.11 Taylor, D.W. The influence of midship section shape upon the resistance of ships. *Transactions of the Society of Naval Architects and Marine Engineers*, Vol. 16, 1908.

1.12 Taylor, D.W. *The Speed and Power of Ships*. U.S. Government Printing Office, Washington, DC, 1943.

1.13 Baker, G.S. Methodical experiments with mercantile ship forms. *Transactions of the Royal Institution of Naval Architects*, Vol. 55, 1913, pp. 162–180.

1.14 Stanton, T.E. The law of comparison for surface friction and eddy-making resistance in fluids. *Transactions of the Royal Institution of Naval Architects*, Vol. 54, 1912, pp. 48–57.

1.15 Kent, J.L. The effect of wind and waves on the propulsion of ships. *Transactions of the Royal Institution of Naval Architects*, Vol. 66, 1924, pp. 188–213.

1.16 Valkhof, H.H., Hoekstra, M. and Andersen, J.E. Model tests and CFD in hull form optimisation. *Transactions of the Society of Naval Architects and Marine Engineers*, Vol. 106, 1998, pp. 391–412.

1.17 Huan, J.C. and Huang, T.T. Surface ship total resistance prediction based on a nonlinear free surface potential flow solver and a Reynolds-averaged Navier-Stokes viscous correction. *Journal of Ship Research*, Vol. 51, 2007, pp. 47–64.

1.18 Burrill, L.C. Calculation of marine propeller performance characteristics. *Transactions of the North East Coast Institution of Engineers and Shipbuilders*, Vol. 60, 1944.

1.19 Lerbs, H.W. Moderately loaded propellers with a finite number of blades and an arbitrary distribution of circulation. *Transactions of the Society of Naval Architects and Marine Engineers*, Vol. 60, 1952, pp. 73–123.

1.20 Turnock, S.R., Phillips, A.B. and Furlong, M. URANS simulations of static drift and dynamic manoeuvres of the KVLCC2 tanker, *Proceedings of the SIM-MAN International Manoeuvring Workshop*, Copenhagen, April 2008.

1.21 Lewis, E.V. (ed.) *Principles of Naval Architecture*. The Society of Naval Architects and Marine Engineers, New York, 1988.

1.22 Harvald, S.A. *Resistance and Propulsion of Ships*. Wiley Interscience, New York, 1983.

1.23 Breslin, J.P. and Andersen, P. *Hydrodynamics of Ship Propellers*. Cambridge Ocean Technology Series, Cambridge University Press, Cambridge, UK, 1996.

1.24 Carlton, J.S. *Marine Propellers and Propulsion*. 2nd Edition, Butterworth-Heinemann, Oxford, UK, 2007.

1.25 Bose, N. *Marine Powering Predictions and Propulsors*. The Society of Naval Architects and Marine Engineers, New York, 2008.

1.26 Faltinsen, O.M. *Hydrodynamics of High-Speed Marine Vehicles*. Cambridge University Press, Cambridge, UK, 2005.

1.27 Bertram, V. *Practical Ship Hydrodynamics*. Butterworth-Heinemann, Oxford, UK, 2000.

1.28 Kerwin, J.E. and Hadler, J.B. Principles of Naval Architecture: Propulsion. The Society of Naval Architects and Marine Engineers, New York, 2010.

1.29 Larsson, L. and Raven, H.C. Principles of Naval Architecture: Ship Resistance and Flow. The Society of Naval Architects and Marine Engineers, New York, 2010.

2 Propulsive Power

2.1 Components of Propulsive Power

During the course of designing a ship it is necessary to estimate the power required to propel the ship at a particular speed. This allows estimates to be made of:

(a) Machinery masses, which are a function of the installed power, and
(b) The expected fuel consumption and tank capacities.

The power estimate for a new design is obtained by comparison with an existing similar vessel or from model tests. In either case it is necessary to derive a power estimate for one size of craft from the power requirement of a different size of craft. That is, it is necessary to be able to *scale* powering estimates.

The different components of the powering problem scale in different ways and it is therefore necessary to estimate each component separately and apply the correct scaling laws to each.

One fundamental division in conventional powering methods is to distinguish between the *effective power* required to drive the ship and the *power delivered* to the propulsion unit(s). The power delivered to the propulsion unit exceeds the effective power by virtue of the efficiency of the propulsion unit being less than 100%.

The main components considered when establishing the ship power comprise the ship resistance to motion, the propeller open water efficiency and the hull–propeller interaction efficiency, and these are summarised in Figure 2.1.

Ship power predictions are made either by

(1) Model experiments and extrapolation, or
(2) Use of standard series data (hull resistance series and propeller series), or
(3) Theoretical (e.g. components of resistance and propeller design).
(4) A mixture of (1) and (2) or (1), (2) and (3).

2.2 Propulsion Systems

When making power estimates it is necessary to have an understanding of the performance characteristics of the chosen propulsion system, as these determine the operation and overall efficiency of the propulsion unit.

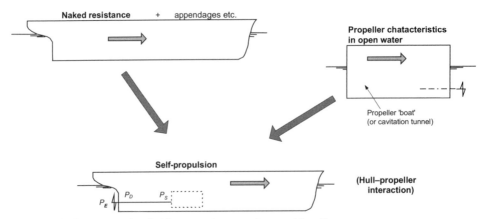

Figure 2.1. Components of ship powering – main considerations.

A fundamental requirement of any ship propulsion system is the efficient conversion of the power (P) available from the main propulsion engine(s) [prime mover] into useful thrust (T) to propel the ship at the required speed (V), Figure 2.2.

There are several forms of main propulsion engines including:

Diesel.
Gas turbine.
Steam turbine.
Electric.
(And variants / combinations of these).

and various propulsors (generally variants of a propeller) which convert the power into useful thrust, including:

Propeller, fixed pitch (FP).
Propeller, controllable pitch (CP).
Ducted propeller.
Waterjet.
Azimuthing podded units.
(And variants of these).

Each type of propulsion engine and propulsor has its own advantages and disadvantages, and applications and limitations, including such fundamental attributes as size, cost and efficiency. All of the these propulsion options are in current use and the choice of a particular propulsion engine and propulsor will depend on the ship type and its design and operational requirements. Propulsors and propulsion machinery are described in Chapters 11 and 13.

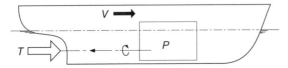

Figure 2.2. Conversion of power to thrust.

The overall assessment of the marine propulsion system for a particular vessel will therefore require:

(1) A knowledge of the required thrust (T) at a speed (V), and its conversion into required power (P),
(2) A knowledge and assessment of the physical properties and efficiencies of the available propulsion engines,
(3) The assessment of the various propulsors and engine-propulsor layouts.

2.3 Definitions

(1) Effective power (P_E) = power required to tow the ship at the required speed
= total resistance × ship speed
= $R_T \times V_S$
(2) Thrust power (P_T) = propeller thrust × speed past propeller
= $T \times V_a$
(3) Delivered power (P_D) = power required to be delivered to the propulsion unit (at the tailshaft)
(4) Quasi-propulsive coefficient (QPC) (η_D) = $\dfrac{\text{effective power}}{\text{delivered power}} = \dfrac{P_E}{P_D}$.

The total installed power will exceed the delivered power by the amount of power lost in the transmission system (shafting and gearing losses), and by a design power margin to allow for roughness, fouling and weather, i.e.

(5) Transmission Efficiency (η_T) = $\dfrac{\text{delivered power}}{\text{power required at engine}}$, hence,
(6) Installed power (P_I) = $\dfrac{P_E}{\eta_D} \times \dfrac{1}{\eta_T}$ + margin (roughness, fouling and weather)

The powering problem is thus separated into three parts:

(1) The estimation of effective power
(2) The estimation of QPC (η_D)
(3) The estimation of required power margins

The estimation of the effective power requirement involves the estimation of the total resistance or drag of the ship made up of:

1. Main hull naked resistance.
2. Resistance of appendages such as shafting, shaft brackets, rudders, fin stabilisers and bilge keels.
3. Air resistance of the hull above water.

The QPC depends primarily upon the efficiency of the propulsion device, but also depends on the interaction of the propulsion device and the hull. Propulsor types and their performance characteristics are described in Chapters 11, 12 and 16.

The required power margin for fouling and weather will depend on the areas of operation and likely sea conditions and will typically be between 15% and 30% of installed power. Power margins are described in Chapter 3.

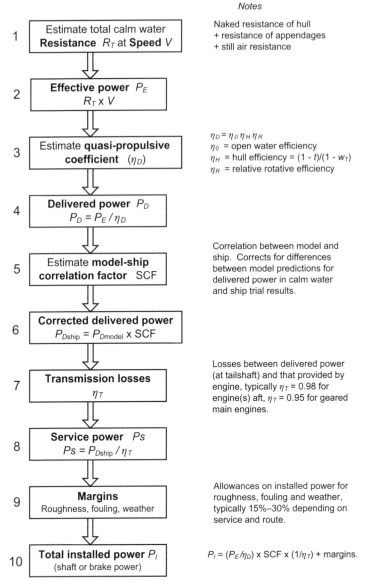

Figure 2.3. Components of the ship power estimate.

The overall components of the ship power estimate are summarised in Section 2.4.

2.4 Components of the Ship Power Estimate

The various components of the ship power estimate and the stages in the powering process are summarised in Figure 2.3.

The total calm water resistance is made up of the hull naked resistance, together with the resistance of appendages and the air resistance.

The propeller quasi-propulsive coefficient (QPC), or η_D, is made up of the open water, hull and relative rotative efficiencies. The hull efficiency is derived as $(1-t)/(1-w_T)$, where t is the thrust deduction factor and w_T is the wake fraction.

For clarity, the model-ship correlation allowance is included as a single-ship correlation factor, SCF, applied to the overall delivered power. Current practice recommends more detailed corrections to individual components of the resistance estimate and to the components of propeller efficiency. This is discussed in Chapter 5.

Transmission losses, η_T, between the engine and tailshaft/propeller are typically about $\eta_T = 0.98$ for direct drive engines aft, and $\eta_T = 0.95$ for transmission via a gearbox.

The margins in stage 9 account for the increase in resistance, hence power, due to roughness, fouling and weather. They are derived in a scientific manner for the purpose of installing propulsion machinery with an adequate reserve of power. This stage should not be seen as adding a margin to allow for uncertainty in the earlier stages of the power estimate.

The total installed power, P_I, will typically relate to the MCR (maximum continuous rating) or CSR (continuous service rating) of the main propulsion engine, depending on the practice of the ship operator.

3 Components of Hull Resistance

3.1 Physical Components of Main Hull Resistance

3.1.1 Physical Components

An understanding of the components of ship resistance and their behaviour is important as they are used in scaling the resistance of one ship to that of another size or, more commonly, scaling resistance from tests at model size to full size. Such resistance estimates are subsequently used in estimating the required propulsive power.

Observation of a ship moving through water indicates two features of the flow, Figure 3.1, namely that there is a wave pattern moving with the hull and there is a region of turbulent flow building up along the length of the hull and extending as a wake behind the hull.

Both of these features of the flow absorb energy from the hull and, hence, constitute a resistance force on the hull. This resistance force is transmitted to the hull as a distribution of pressure and shear forces over the hull; the shear stress arises because of the viscous property of the water.

This leads to the first possible physical breakdown of resistance which considers the *forces acting*:

(1) Frictional resistance

The fore and aft components of the tangential shear forces τ acting on each element of the hull surface, Figure 3.2, can be summed over the hull to produce the total shear resistance or *frictional resistance*.

(2) Pressure resistance

The fore and aft components of the pressure force P acting on each element of hull surface, Figure 3.2, can be summed over the hull to produce a *total pressure resistance*.

The frictional drag arises purely because of the viscosity, but the pressure drag is due in part to viscous effects and to hull wavemaking.

An alternative physical breakdown of resistance considers *energy dissipation*.

(3) Total viscous resistance

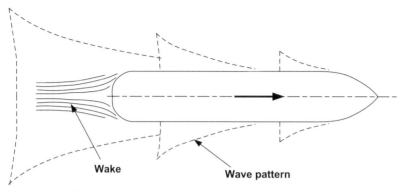

Figure 3.1. Waves and wake.

Bernoulli's theorem (see Appendix A1.5) states that $\frac{P}{g} + \frac{V^2}{2g} + h = H$ and, in the absence of viscous forces, H is constant throughout the flow. By means of a Pitôt tube, local total head can be measured. Since losses in total head are due to viscous forces, it is possible to measure the *total viscous resistance* by measuring the total head loss in the wake behind the hull, Figure 3.3.

This resistance will include the skin frictional resistance and part of the pressure resistance force, since the total head losses in the flow along the hull due to viscous forces result in a pressure loss over the afterbody which gives rise to a resistance due to pressure forces.

(4) Total wave resistance

The wave pattern created by the hull can be measured and analysed into its component waves. The energy required to sustain each wave component can be estimated and, hence, the *total wave resistance* component obtained.

Thus, by physical measurement it is possible to identify the following methods of breaking down the total resistance of a hull:

1. Pressure resistance + frictional resistance
2. Viscous resistance + remainder
3. Wave resistance + remainder

These three can be combined to give a final resistance breakdown as:

$$\text{Total resistance} = \quad \text{Frictional resistance}$$
$$+ \text{Viscous pressure resistance}$$
$$+ \text{Wave resistance}$$

The experimental methods used to derive the individual components of resistance are described in Chapter 7.

Figure 3.2. Frictional and pressure forces.

Figure 3.3. Measurement of total viscous resistance.

It should also be noted that each of the resistance components obeys a different set of scaling laws and the problem of scaling is made more complex because of interaction between these components.

A summary of these basic hydrodynamic components of ship resistance is shown in Figure 3.4. When considering the *forces acting*, the total resistance is made up of the sum of the tangential shear and normal pressure forces acting on the wetted surface of the vessel, as shown in Figure 3.2 and at the top of Figure 3.4. When considering energy dissipation, the total resistance is made up of the sum of the energy dissipated in the wake and the energy used in the creation of waves, as shown in Figure 3.1 and at the bottom of Figure 3.4.

Figure 3.5 shows a more detailed breakdown of the basic resistance components together with other contributing components, including wave breaking, spray, transom and induced resistance. The total skin friction in Figure 3.5 has been divided into two-dimensional flat plate friction and three-dimensional effects. This is used to illustrate the breakdown in respect to some model-to-ship extrapolation methods, discussed in Chapter 4, which use flat plate friction data.

Wave breaking and spray can be important in high-speed craft and, in the case of the catamaran, significant wave breaking may occur between the hulls at particular speeds. Wave breaking and spray should form part of the total wavemaking

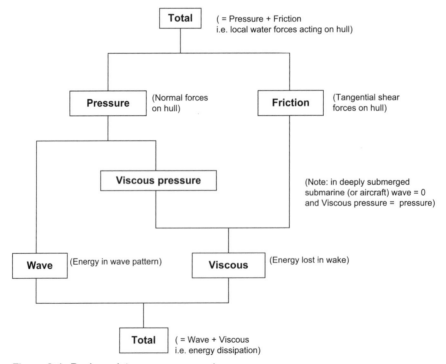

Figure 3.4. Basic resistance components.

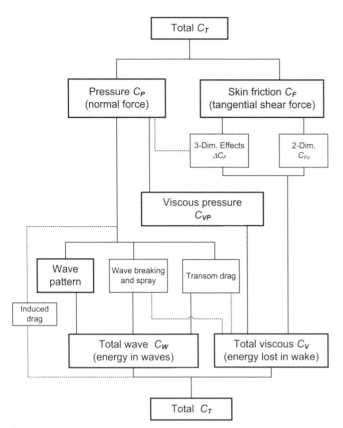

Figure 3.5. Detailed resistance components.

resistance, but, in practice, this energy will normally be lost in the wake; the dotted line in Figure 3.5 illustrates this effect.

The transom stern, used on most high-speed vessels, is included as a pressure drag component. It is likely that the large low-pressure area directly behind the transom, which causes the transom to be at atmospheric pressure rather than stagnation pressure, causes waves and wave breaking and spray which are not fully transmitted to the far field. Again, this energy is likely to be lost in the wake, as illustrated by the dotted line in Figure 3.5.

Induced drag will be generated in the case of yachts, resulting from the lift produced by keels and rudders. Catamarans can also create induced drag because of the asymmetric nature of the flow between and over their hulls and the resulting production of lift or sideforce on the individual hulls. An investigation reported in [3.1] indicates that the influence of induced drag for catamarans is likely to be very small. Multihulls, such as catamarans or trimarans, will also have wave resistance interaction between the hulls, which may be favourable or unfavourable, depending on ship speed and separation of the hulls.

Lackenby [3.2] provides further useful and detailed discussions of the components of ship resistance and their interdependence, whilst [3.3 and 3.4] pay particular attention to the resistance components of catamarans.

The following comments are made on the resistance components of some high-speed craft and sailing vessels.

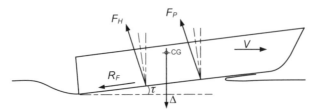

Figure 3.6. Planing craft forces.

3.1.1.1 Planing Craft

The basic forces acting are shown in Figure 3.6 where, for a trim angle τ, F_P is the pressure force over the wetted surface, F_H is the hydrostatic force acting at the centre of pressure of the hull and R_F is the skin friction resistance. Trim τ has an important influence on drag and, for efficient planing, τ is small. As the speed of planing is increased, the wetted length and consequently the wedge volume decrease rapidly, lift becomes mainly dynamic and $F_H \ll F_P$. A reasonable proportion of buoyant reaction should be maintained, for example in the interests of seakeeping.

The resistance components may be summarised as

$$R_T = R_F + R_W + R_I, \tag{3.1}$$

where R_I is the drag resulting from the inclination of the pressure force F_P to the vertical. At high speed, wavemaking resistance R_W becomes small. Spray resistance may be important, depending on hull shape and the use of spray rails, according to Savitsky *et al.* [3.5]. The physics of planing and the forces acting are described in some detail in [3.6] and [3.7]. The estimation of the resistance of planing craft is described in Chapter 10.

3.1.1.2 Sailing Vessels

The sailing vessel has the same basic resistance components as a displacement or semi-displacement craft, together with extra components. The fundamental extra component incurred by a sailing vessel is the induced drag resulting from the lift produced by the keel(s) and rudder(s) when moving at a yaw angle. The production of lift is fundamental to resisting the sideforce(s) produced by the sails, Figure 11.12. Some consider the resistance due to heel a separate resistance, to be added to the upright resistance. Further information on sailing vessels may be obtained from [3.8] and [3.9]. The estimation of the resistance of sailing craft is outlined in Chapter 10.

3.1.1.3 Hovercraft and Hydrofoils

Hovercraft (air cushion vehicles) and hydrofoil craft have resistance components that are different from those of displacement and semi-displacement ships and require separate treatment. Outline summaries of their components are given as follows:

(1) Air cushion vehicles (including surface effect ships or sidewall hovercraft).

The components of resistance for air cushion vehicles include

(a) Aerodynamic (or profile) drag of the above-water vehicle
(b) Inlet momentum drag due to the ingestion of air through the lift fan, where the air must acquire the craft speed
(c) Drag due to trim

The trim drag is the resultant force of two physical effects: 1) the wave drag due to the pressure in the cushion creating a wave pattern and 2) outlet momentum effects due to a variable air gap at the base of the cushion and consequent non-uniform air outflow. The air gap is usually larger at the stern than at the bow and thus the outflow momentum creates a forward thrust.

(d) Other resistance components include sidewall drag (if present) for surface effect ships, water appendage drag (if any) and intermittent water contact and spray generation

For hovercraft with no sidewalls, the intermittent water contact and spray generation drag are usually estimated as that drag not accounted for by (a)–(c).

The total power estimate for air cushion vehicles will consist of the propulsive power required to overcome the resistance components (a)–(d) *and* the lift fan power required to sustain the cushion pressure necessary to support the craft weight at the required (design) air gap. The basic physics, design and performance characteristics of hovercraft are described in some detail in [3.6], [3.7], [3.10] and [3.12].

(2) Hydrofoil-supported craft

Hydrofoil-supported craft experience the same resistance components on their hulls as conventional semi-displacement and planing hulls at lower speeds and as they progress to being supported by the foils. In addition to the hull resistance, there is the resistance due to the foil support system. This consists of the drag of the non-lifting components such as vertical support struts, antiventilation fences, rudders and propeller shafting and the drag of the lifting foils. The lifting foil drag comprises the profile drag of the foil section, the induced drag caused by generation of lift and the wavemaking drag of the foil beneath the free surface. The lift generated by a foil in proximity to the free surface is reduced from that of a deeply immersed foil because of wavemaking, flow curvature and a reduction in onset flow speed. The induced drag is increased relative to a deeply submerged foil as a result of the free surface increasing the downwash. The basic physics, design and performance characteristics of hydrofoil craft are described in some detail in [3.6], [3.7] and [3.11].

3.1.2 Momentum Analysis of Flow Around Hull

3.1.2.1 Basic Considerations
The resistance of the hull is clearly related to the momentum changes taking place in the flow. An analysis of these momentum changes provides a precise definition of what is meant by each resistance component in terms of *energy dissipation*.

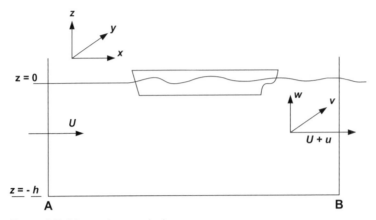

Figure 3.7. Momentum analysis.

Consider a model held in a stream of speed U in a rectangular channel of breadth b and depth h, Figure 3.7. The momentum changes in the fluid passing through the 'control box' from plane A to plane B downstream can be related to the forces on the control planes and the model.

Let the free-surface elevation be $z = \zeta(x, y)$ where ζ is taken as small, and let the disturbance to the flow have a velocity $q = (u, v, w)$. For continuity of flow, flow through A = flow through B,

$$U \cdot b \cdot h = \int_{-b/2}^{b/2} \int_{-h}^{\varsigma_B} (U + u) dz \, dy \tag{3.2}$$

where $\varsigma_B = \varsigma(x_B, y)$. The momentum flowing out through B in unit time is

$$M_B = \rho \int_{-b/2}^{b/2} \int_{-h}^{\varsigma_B} (U + u)^2 dz \, dy. \tag{3.3}$$

The momentum flowing in through A in unit time is

$$M_A = \rho U^2 \cdot b \cdot h.$$

Substituting for $U \cdot b \cdot h$ from Equation (3.2),

$$M_A = \rho \int_{-b/2}^{b/2} \int_{-h}^{\varsigma_B} U(U + u) dz \, dy. \tag{3.4}$$

Hence, the rate of change of momentum of fluid flowing through the control box is $M_B - M_A$
i.e.

$$M_B - M_A = \rho \int_{-b/2}^{b/2} \int_{-h}^{\varsigma_B} u(U + u) dz \, dy. \tag{3.5}$$

This rate of change of momentum can be equated to the forces on the fluid in the control box and, neglecting friction on the walls, these are R (hull resistance), F_A

(pressure force on plane A) and F_B (pressure force on plane B). Therefore, $M_B - M_A = -R + F_A - F_B$.

Bernoulli's equation can be used to derive expressions for the pressures at A and B, hence, for forces F_A and F_B

$$H = \frac{P_A}{\rho} + \frac{1}{2}U^2 + gz = \frac{P_B}{\rho} + \frac{1}{2}[(U+u)^2 + v^2 + w^2] + gz_B + \frac{\Delta P}{\rho}, \quad (3.6)$$

where ΔP is the loss of pressure in the boundary layer and $\Delta P/\rho$ is the corresponding loss in total head.

If atmospheric pressure is taken as zero, then ahead of the model, $P_A = 0$ on the free surface where $z = 0$ and the constant term is $H = \frac{1}{2}U^2$. Hence,

$$P_B = -\rho \left\{ gz_B + \frac{\Delta P}{\rho} + \frac{1}{2}\left[2Uu + u^2 + v^2 + w^2\right] \right\}. \quad (3.7)$$

On the upstream control plane,

$$F_A = \int_{-b/2}^{b/2} \int_{-h}^{0} P_A dz\,dy = -\rho g \int_{-b/2}^{b/2} \int_{-h}^{0} z\,dz\,dy = \frac{1}{2}\rho g \int_{-b/2}^{b/2} h^2 dy = \frac{1}{2}\rho gbh^2. \quad (3.8)$$

On the downstream control plane, using Equation (3.7),

$$F_B = \int_{-b/2}^{b/2} \int_{-h}^{\varsigma_B} P_B dz\,dy = -\rho \int_{-b/2}^{b/2} \int_{-h}^{\varsigma_B} \left\{ gz_B + \frac{\Delta P}{\rho} + \frac{1}{2}[2Uu + u^2 + v^2 + w^2] \right\} dz\,dy$$

$$= \frac{1}{2}\rho g \int_{-b/2}^{b/2} \left(h^2 - \varsigma_B^2 \right) dy - \int_{-b/2}^{b/2} \int_{-h}^{\varsigma_B} \Delta P\,dz\,dy - \frac{\rho}{2} \int_{-b/2}^{b/2} \int_{-h}^{\varsigma_B} [2Uu + u^2 + v^2 + w^2] dz\,dy.$$

$$(3.9)$$

The resistance force $R = F_A - F_B - (M_B - M_A)$.

Substituting for the various terms from Equations (3.5), (3.8) and (3.9),

$$R = \left\{ \frac{1}{2}\rho g \int_{-b/2}^{b/2} \varsigma_B^2\,dy + \frac{1}{2}\rho \int_{-b/2}^{b/2} \int_{-h}^{\varsigma_B} (v^2 + w^2 - u^2)dz\,dy \right\} + \int_{-b/2}^{b/2} \int_{-h}^{\varsigma_B} \Delta P dz\,dy. \quad (3.10)$$

In this equation, the first two terms may be broadly associated with wave pattern drag, although the perturbation velocities v, w and u, which are due mainly to wave orbital velocities, are also due partly to induced velocities arising from the viscous shear in the boundary layer. The third term in the equation is due to viscous drag.

The use of the first two terms in the analysis of wave pattern measurements, and in formulating a wave resistance theory, are described in Chapters 7 and 9.

3.1.2.2 Identification of Induced Drag

A *ficticious* velocity component u' may be defined by the following equation:

$$\frac{p_B}{\rho} + \frac{1}{2}[(U+u')^2 + v^2 + w^2] + gz_B = \frac{1}{2}U^2, \quad (3.11)$$

where u' is the equivalent velocity component required for no head loss in the boundary layer. This equation can be compared with Equation (3.6). u' can be calculated from Δp since, by comparing Equations (3.6) and (3.11),

$$\tfrac{1}{2}\rho\,(U+u')^2 = \tfrac{1}{2}\rho\,(U+u)^2 + \Delta p,$$

then

$$R = \frac{1}{2}\rho g \int_{-b/2}^{b/2} \zeta_B^2\,dy + \frac{1}{2}\rho \int_{-b/2}^{b/2} \int_{-h}^{\zeta_B} (v^2 + w^2 - u'^2)\,dz\,dy$$
$$+ \iint_{\text{wake}} \left\{ \Delta p + \frac{1}{2}\rho(u'^2 - u^2) \right\} dz\,dy. \qquad (3.12)$$

The integrand of the last term $\{\Delta p + \tfrac{1}{2}\rho(u'^2 - u^2)\}$ is different from zero only inside the wake region for which $\Delta p \neq 0$. In order to separate *induced drag* from wave resistance, the velocity components (u_I, v_I, w_I) of the wave orbit motion can be introduced. [The components (u, v, w) include both wave orbit *and* induced velocities.] The velocity components (u_I, v_I, w_I) can be calculated by measuring the free-surface wave pattern, and applying linearised potential theory.

It should be noted that, from measurements of wave elevation ζ and perturbation velocities u, v, w over plane B, the wave resistance could be determined. However, measurements of subsurface velocities are difficult to make, so linearised potential theory is used, in effect, to deduce these velocities from the more conveniently measured surface wave pattern ζ. This is discussed in Chapter 7 and Appendix A2. Recent developments in PIV techniques would allow subsurface velocities to be measured; see Chapter 7.

Substituting (u_I, v_I, w_I) into the last Equation (3.12) for R,

$$R = R_W + R_V + R_I,$$

where R_W is the wave pattern resistance

$$R_W = \frac{1}{2}\rho g \int_{-b/2}^{b/2} \zeta_B^2\,dy + \frac{1}{2}\rho \int_{-b/2}^{b/2} \int_{-h}^{\zeta_B} (v_I^2 + w_I^2 - u_I^2)\,dz\,dy, \qquad (3.13)$$

R_V is the total viscous resistance

$$R_V = \iint_{\text{wake}} \left\{ \Delta p + \frac{1}{2}\rho(u'^2 - u^2) \right\} dz\,dy \qquad (3.14)$$

and R_I is the induced resistance

$$R_I = \frac{1}{2}\rho \int_{-b/2}^{b/2} \int_{-h}^{\zeta_B} (v^2 - v_I^2 + w^2 - w_I^2 - u'^2 + u_I^2)\,dz\,dy. \qquad (3.15)$$

For normal ship forms, R_I is expected to be small.

3.1.3 Systems of Coefficients Used in Ship Powering

Two principal forms of presentation of resistance data are in current use. These are the ITTC form of coefficients, which were based mainly on those already in use in the aeronautical field, and the Froude coefficients [3.13].

3.1.3.1 ITTC Coefficients

The resistance coefficient is

$$C_T = \frac{R_T}{1/2\rho \, SV^2},\qquad(3.16)$$

where

$$R_T = \text{total resistance force,}$$
$$S = \text{wetted surface area of hull, } V = \text{ship speed,}$$
$$C_T = \text{total resistance coefficient or, for various components,}$$
$$C_F = \text{frictional resistance coefficient (flat plate),}$$
$$C_V = \text{total viscous drag coefficient, including form allowance, and}$$
$$C_W = \text{wave resistance coefficient.}$$

The speed parameter is Froude number

$$Fr = \frac{V}{\sqrt{gL}},$$

and the hull form parameters include the wetted area coefficient

$$C_S = \frac{S}{\sqrt{\nabla L}}$$

and the slenderness coefficient

$$C_\nabla = \frac{\nabla}{L^3},$$

where

$$\nabla = \text{immersed volume, } L = \text{ship length.}$$

3.1.3.2 Froude Coefficients

A basic design requirement is to have the least power for given displacement Δ and speed V. Froude chose a resistance coefficient representing a 'resistance per ton displacement' (i.e. R/Δ). He used a circular notation to represent non-dimensional coefficients, describing them as Ⓚ, Ⓛ, Ⓢ, Ⓜ and Ⓒ.

The speed parameters include

$$Ⓚ = \frac{V}{\sqrt{\dfrac{g\nabla^{1/3}}{4\pi}}} \quad \text{and} \quad Ⓛ = \frac{V}{\sqrt{\dfrac{gL}{4\pi}}} \propto Fr$$

(4π was introduced for wave speed considerations). The hull form parameters include the wetted area coefficient

$$Ⓢ = \frac{S}{\nabla^{2/3}},$$

and the slenderness coefficient

$$\text{\textcircled{M}} = \frac{L}{\nabla^{1/3}}$$

(now generally referred to as the length-displacement ratio). The resistance coefficient is

$$\text{\textcircled{C}} = \frac{1000\,R}{\Delta\,\text{\textcircled{K}}^2}, \tag{3.17}$$

i.e. resistance per ton R/Δ (R = resistance force and Δ = displacement force, same units).

The 1000 was introduced to create a more convenient value for $\text{\textcircled{C}}$. The subscript for each component is as for C_T, C_F, etc., i.e. $\text{\textcircled{C}}_T$, $\text{\textcircled{C}}_F$. The presentation R/Δ was also used by Taylor in the United States in his original presentation of the Taylor standard series [3.14].

Some other useful relationships include the following: in imperial units,

$$\text{\textcircled{C}} = \frac{427.1\,P_E}{\Delta^{2/3}V^3} \tag{3.18}$$

($P_E = hp$, Δ = tons, V = knots), and in metric units,

$$\text{\textcircled{C}} = \frac{579.8\,P_E}{\Delta^{2/3}V^3} \tag{3.19}$$

($P_E = kW$, Δ = tonnes, V = knots, and using 1 knot = 0.5144 m/s).

The following relationships between the ITTC and Froude coefficients allow conversion between the two presentations:

$$C_s = S/(\nabla.L)^{1/2},$$
$$\text{\textcircled{S}} = S/\nabla^{2/3},$$
$$\text{\textcircled{M}} = L/\nabla^{1/3},$$
$$Cs = \text{\textcircled{S}}/\text{\textcircled{M}}^{1/2} \text{ and}$$
$$C_T = (8 \times \pi/1000) \times \text{\textcircled{C}}/\text{\textcircled{S}}.$$

It should be noted that for fixed Δ and V, the smallest $\text{\textcircled{C}}$ implies least P_E, Equation (3.19), but since S can change for a fixed Δ, this is *not* true for C_T, Equation (3.16).

It is for such reasons that other forms of presentation have been proposed. These have been summarised by Lackenby [3.15] and Telfer [3.16]. Lackenby and Telfer argue that, *from a design point of view*, resistance per unit of displacement (i.e. R/Δ), plotted to a suitable base, should be used in order to rank alternative hull shapes correctly. In this case, taking R/Δ as the criterion of performance, and using only displacement and length as characteristics of ship size, the permissible systems of presentation are

(1) $\dfrac{R}{\Delta^{2/3}V^2}$ (e.g. $\text{\textcircled{C}}$) on $\dfrac{V}{\Delta^{1/6}}$ (e.g. $\text{\textcircled{K}}$)

(2) $\dfrac{RL}{\Delta V^2}$ on Fr

(3) $\dfrac{R}{\Delta}$ on either $\dfrac{V}{\Delta^{1/6}}$ or Fr.

Consistent units must be used in all cases in order to preserve the non-dimensional nature of the coefficients.

For skin friction resistance, Lackenby points out that the most useful characteristic of ship size to be included is wetted surface area, which is one of the primary variables affecting this resistance.

A coefficient of the form $C_F = R_F/1/2\,\rho SV^2$ is then acceptable. In which case, the other components of resistance also have to be in this form. This is the form of the ITTC presentation, described earlier, which has been in common use for many years.

It is also important to be able to understand and apply the circular coefficient notation, because many useful data have been published in $\copyright - Fr$ format, such as the BSRA series of resistance tests discussed in Chapter 10.

Doust [3.17] uses the resistance coefficient $RL/\Delta V^2$ in his regression analysis of trawler resistance data. Sabit [3.18] also uses this coefficient in his regression analysis of the BSRA series resistance data, which is discussed in Chapter 10. It is useful to note that this coefficient is related to $\copyright$ as follows:

$$\frac{RL}{\Delta V^2} = 2.4938\ \copyright \times \left[\frac{L}{\nabla^{1/3}}\right]. \qquad (3.20)$$

3.1.4 Measurement of Model Total Resistance

3.1.4.1 Displacement Ships

The model resistance to motion is measured in a test tank, also termed a towing tank. The first tank to be used solely for such tests was established by William Froude in 1871 [3.19]. The model is attached to a moving carriage and towed down the tank at a set constant speed (V), and the model resistance (R) is measured, Figure 3.8. The towing force will normally be in line with the propeller shafting in order to minimise unwanted trim moments during a run. The model is normally free to trim and to rise/sink vertically and the amount of sinkage and trim during a run is measured. Typical resistance test measurements, as described in ITTC (2002)

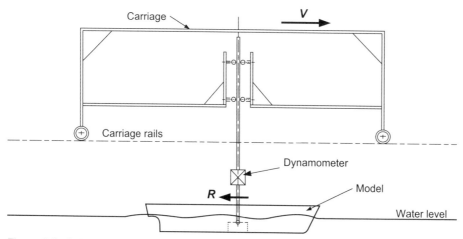

Figure 3.8. Schematic layout of model towing test.

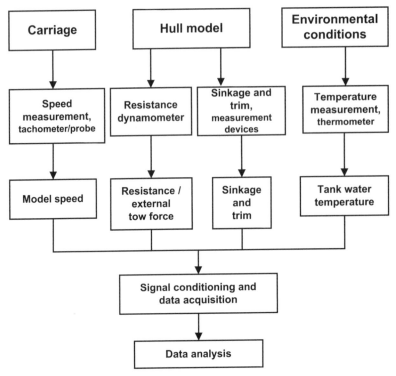

Figure 3.9. Resistance test measurements.

[3.20], are shown in Figure 3.9. An example of a model undergoing a resistance test is shown in Figure 3.10.

Model speed is measured either from the carriage wheel speed (speed over ground), from the time taken for the carriage to travel over a known distance (speed over ground) or by a Pitôt-static tube or water speed meter attached to the carriage ahead of the model (speed through water). If the model speed remains constant during the course of a run and the tank water does not develop any significant drift during the test programme, then all of these methods are equally satisfactory.

Figure 3.10. Model resistance test. Photograph courtesy of WUMTIA and Dubois Naval Architects Ltd.

The total hydrodynamic resistance to motion (R) is measured by a dynamometer. The dynamometer may be mechanical, using a spring balance and counterbalance weights, or electromechanical, where the displacement of flexures is measured by a linear voltmeter or inductor, or where the flexures have strain gauges attached. In all cases, a calibration procedure of measured output against an applied calibration force will take place before and after the experiments. Two-component (resistance and sideforce) or three-component (resistance, sideforce and yaw moment [torque]) dynamometers may be used in the case of yawed tests for assessing manoeuvring performance and for the testing of yacht models.

Model test tanks vary in size from about 60 m × 3.7 m × 2 m water depth up to 300 m × 12 m × 3 m water depth, with carriage speeds ranging from 3 m/s to 15 m/s. A 'beach' is normally incorporated at the end of the tank (and sometimes down the sides) in order to absorb the waves created by the model and to minimise wave reflections back down the tank. This also helps to minimise the settling time between runs.

Circulating water channels are also employed for resistance and other tests. In this case, in Figure 3.8, the model is static and the water is circulated at speed V. Such channels often have glass side and bottom windows, allowing good flow visualisation studies. They are also useful when multiple measurements need to be carried out, such as hull surface pressure or skin friction measurements (see Chapter 7). Circulating water channels need a lot of power to circulate the water, compared say with a wind tunnel, given the density of water is about 1000 times that of air.

For many years, models were made from paraffin wax. Current materials used for models include wood, high-density closed-cell foam and fibre reinforced plastic (FRP). The models, or plugs for plastic models, will normally be shaped using a multiple-axis cutting machine. Each model material has its merits, depending on producibility, accuracy, weight and cost. Model size may vary from about 1.6 m in a 60-m tank up to 9 m in a 300-m tank.

The flow over the fore end of the model may be laminar, whilst turbulent flow would be expected on the full-scale ship. Turbulence stimulators will normally be incorporated near the fore end of the model in order to stimulate turbulent flow. Turbulence stimulators may be in the form of sand strips, trip wires or trip studs located about 5% aft of the fore end of the model. Trip studs will typically be of 3 mm diameter and 2.5 mm height and spaced at 25 mm intervals. Trip wires will typically be of 0.90 mm diameter. For further details see [3.20], [3.21], [3.22] and [3.23]. Corrections for the parasitic drag of the turbulence stimulators will normally be carried out; a detailed investigation of such corrections is contained in Appendix A of [3.24]. One common correction is to assume that the deficit in resistance due to laminar flow ahead of the trip wire or studs balances the additional parasitic drag of the wire or studs.

In order to minimise scale effect, the model size should be as large as possible without incurring significant interference (blockage) effects from the walls and tank floor. A typical assumption is that the model cross-sectional area should not be more than 0.5% of the tank cross-sectional area. Blockage speed corrections may be applied if necessary. Typical blockage corrections include those proposed by Hughes [3.25] and Scott [3.26], [3.27]. The correction proposed by Hughes (in its approximate form) is as follows:

$$\frac{\Delta V}{V} = \frac{m}{1 - m - Fr_h^2},\qquad(3.21)$$

where ΔV is the correction to speed, Fr_h is the depth Froude number.

$$Fr_h = \frac{V}{\sqrt{gh}},$$

where h is the tank water depth and m is the mean blockage

$$m = \frac{a}{A_T} = \frac{\nabla}{L_M A_T},$$

where L_M and ∇ are model length and displacement, and A_T is the tank section area.

The temperature of the test water will be measured in the course of the experiments and appropriate corrections made to the resistance data. Viscosity values for water are a function of water temperature. As a result, Re and, hence, C_F vary with water temperature in a manner which can be calculated from published viscosity values (see Tables A1.1 and A1.2 in Appendix A1). Standard practice is to correct all model results to, and predict ship performance for, 15°C (59°F), i.e.

$$C_{T(15)} = C_{T(T)} + (C_{F(15)} - C_{F(T)}).\qquad(3.22)$$

The ITTC recommended procedure for the standard resistance test is described in ITTC 2002 [3.20]. Uncertainty analysis of the results should take place, involving the accuracy of the model and the measurements of resistance and speed. A background to uncertainty analysis is given in [3.28], and recommended procedures are described in [3.20].

After the various corrections are made (e.g. to speed and temperature), the model resistance test results will normally be presented in terms of the total resistance coefficient $C_T (= R/\frac{1}{2}\rho S V^2)$ against Froude number Fr, where S is the static wetted area of the model. Extrapolation of the model results to full scale is described in Chapter 4.

3.1.4.2 High-Speed Craft and Sailing Vessels

High-speed craft and sailing yachts develop changes in running attitude when under way. Compared with a conventional displacement hull, this leads to a number of extra topics and measurements to be considered in the course of a resistance test. The changes and measurements required are reviewed in [3.29]. Because of the higher speeds involved, such craft may also be subject to shallow water effects and corrections may be required, as discussed in Chapters 5 and 6. The typical requirements for testing high-speed craft, compared with displacement hulls, are outlined as follows:

(i) Semi-displacement craft

High-speed semi-displacement craft develop changes in running trim and wetted surface area when under way. The semi-displacement craft is normally tested free to heave (vertical motion) and trim, and the heave/sinkage and trim are measured

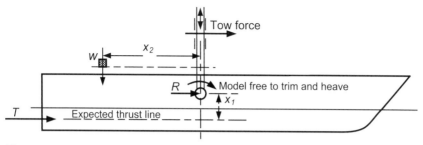

Figure 3.11. Trim compensation for offset tow line.

during the course of a test run. Running wetted surface area may be measured during a run, for example, by noting the wave profile against a grid on the hull (or by a photograph) and applying the new girths (up to the wave profile) to the body plan. There are conflicting opinions as to whether static or running wetted area should be used in the analysis. Appendix B in [3.24] examines this problem in some detail and concludes that, for examination of the physics, the running wetted area should be used, whilst for practical powering purposes, the use of the static wetted area is satisfactory. It can be noted that, for this reason, standard series test data for semi-displacement craft, such as those for the NPL Series and Series 64, are presented in terms of static wetted area.

Since the high-speed semi-displacement craft is sensitive to trim, the position and direction of the tow force has to be considered carefully. The tow force should be located at the longitudinal centre of gravity (LCG) and in the line of the expected thrust line, otherwise erroneous trim changes can occur. If, for practical reasons, the tow force is not in line with the required thrust line, then a compensating moment can be applied, shown schematically in Figure 3.11. If the tow line is offset from the thrust line by a distance x_1, then a compensating moment ($w \times x_2$) can be applied, where ($w \times x_2$) = ($R \times x_1$). This process leads to an effective shift in the LCG. w will normally be part of the (movable) ballast in the model, and the lever x_2 can be changed as necessary to allow for the change in R with change in speed. Such corrections will also be applied as necessary to inclined shaft/thrust lines.

(ii) Planing craft

A planing craft will normally be run free to heave and trim. Such craft incur significant changes in trim with speed, and the position and direction of the tow line is important. Like the semi-displacement craft, compensating moments may have to be applied, Figure 3.11. A friction moment correction may also be applied to allow for the difference in friction coefficients between model and ship (model is too large). If R_{Fm} is the model frictional resistance corresponding to C_{Fm} and R_{Fms} is the model frictional resistance corresponding to C_{Fs}, then, assuming the friction drag acts at half draught, a counterbalance moment can be used to counteract the force ($R_{Fm} - R_{Fms}$). This correction is analogous to the skin friction correction in the model self-propulsion experiment, Chapter 8.

Care has to be taken with the location of turbulence stimulation on planing craft, where the wetted length varies with speed. An alternative is to use struts or wires in the water upstream of the model [3.29].

Air resistance can be significant in high-speed model tests and corrections to the resistance data may be necessary. The actual air speed under the carriage should be measured with the model removed. Some tanks include the superstructure and then make suitable corrections based on airflow speed and suitable drag coefficients. The air resistance can also cause trimming moments, which should be corrected by an effective shift of LCG.

Planing craft incur significant changes in wetted surface area with change in speed. Accurate measurement of the running wetted area and estimation of the frictional resistance is fundamental to the data analysis and extrapolation process. Methods of measuring the running wetted surface area include noting the position of the fore end of the wetted area on the centreline and at the side chines, using underwater photography, or using a clear bottom on the model. An alternative is to apply the running draught and trim to the hydrostatic information on wetted area, although this approach tends not to be very accurate. The spray and spray root at the leading edge of the wetted area can lead to difficulties in differentiating between spray and the solid water in contact with the hull.

For high-speed craft, appendage drag normally represents a larger proportion of total resistance than for conventional displacement hulls. If high-speed craft are tested without appendages, then the estimated trim moments caused by the append-ages should be compensated by an effective change in LCG.

Captive tests on planing craft have been employed. For fully captive tests, the model is fixed in heave and trim whilst, for partially captive tests, the model is tested free to heave over a range of fixed trims [3.29]. Heave and trim moment are meas-ured, together with lift and drag. Required values will be obtained through interpol-ation of the test data in the postanalysis process.

Renilson [3.30] provides a useful review of the problems associated with meas-uring the hydrodynamic performance of high-speed craft.

(iii) Sailing craft

A yacht model will normally be tested in a semi-captive arrangement, where it is free to heave and trim, but fixed in heel and yaw. A special dynamometer is required that is capable of measuring resistance, sideforce, heave, trim, roll moment and yaw moment. Measurement of the moments allows the centre of lateral resistance (CLR) to be determined. A typical test programme entails a matrix of tests covering a range of heel and yaw angles over a range of speeds. Small negative angles of heel and yaw will also be tested to check for any asymmetry. Typical test procedures, dynamomet-ers and model requirements are described by Claughton et al. [3.8].

In the case of a yacht, the position and line of the tow force is particularly important because the actual position, when under sail, is at the centre of effort of the sails. This leads to trim and heel moments and a downward component of force. As the model tow fitting will be at or in the model, these moments and force will need to be compensated for during the model test. An alternative approach that has been used is to apply the model tow force at the estimated position of the centre of effort of the sails, with the model set at a predetermined yaw angle. This is a more elegant approach, although it does require more complex model arrange-ments [3.8].

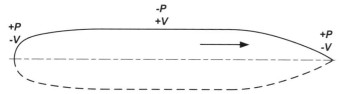

Figure 3.12. Pressure variations around a body.

3.1.5 Transverse Wave Interference

3.1.5.1 Waves
When a submerged body travels through a fluid, pressure variations are created around the body, Figure 3.12.

Near a free surface, the pressure variations manifest themselves by changes in the fluid level, creating waves, Figure 3.13. With a body moving through a stationary fluid, the waves travel at the same speed as the body.

3.1.5.2 Kelvin Wave Pattern
The Kelvin wave is a mathematical form of the wave system created by a travelling pressure point source at the free surface, see Chapter 7. The wave system formed is made up of transverse waves and divergent waves, Figure 3.14. The heights of the divergent cusps diminish at a slower rate than transverse waves, and the divergent waves are more predominant towards the rear. The wave system travels at a speed according to the gravity–wave speed relationship, see Appendix A1.8,

$$V = \sqrt{\frac{g\lambda}{2\pi}}, \qquad (3.23)$$

and with wavelength of the transverse waves,

$$\lambda = \frac{2\pi V^2}{g}. \qquad (3.24)$$

The ship wave system is determined mainly by the peaks of high and low pressure, or pressure points, that occur in the pressure distribution around the hull. The

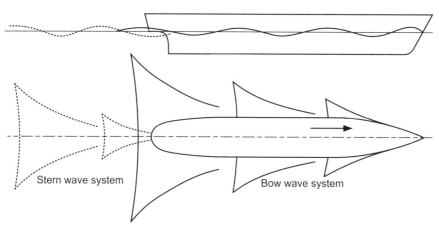

Figure 3.13. Ship waves.

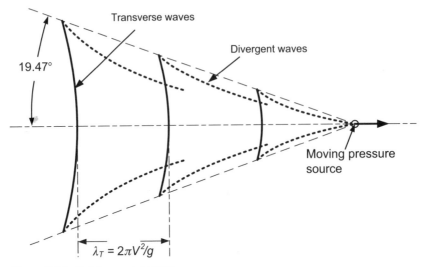

Figure 3.14. Kelvin wave pattern.

overall ship wave system may be considered as being created by a number of travelling pressure points, with the Kelvin wave pattern being a reasonable representation
of the actual ship wave system. The ship waves will not follow the Kelvin pattern
exactly due to the non-linearities in the waves and viscous effects (e.g. stern wave
damping) not allowed for in the Kelvin wave theory. This is discussed further in
Chapter 9. It should also be noted that the foregoing analysis is strictly for *deep*
water. Shallow water has significant effects on the wave pattern and these are discussed in Chapter 6.

3.1.5.3 Wave System Interference
The pressure peaks around the hull create a wave system with a crest (trough for
negative pressure peak) situated some distance behind the point of high pressure.
The wave system for a ship is built up of four components, Figure 3.15.

(1) Bow wave system, with a high pressure at entry, starting with a crest.
(2) Forward shoulder system, due to low pressure between forward and amidships.

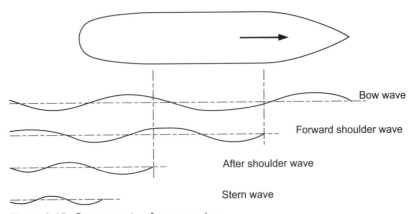

Figure 3.15. Components of wave system.

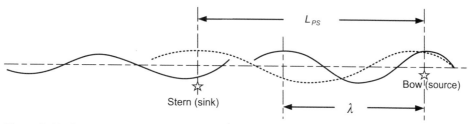

Figure 3.16. Bow and stern wave systems.

(3) After shoulder system, due to low pressure at amidships and higher pressure at stern.

(4) Stern wave system, due to rising pressure gradient and decreasing velocity, starting with a trough.

Components (1) and (4) will occur at 'fixed' places, whilst the magnitude and position of (2) and (3) will depend on the form and the position of the shoulders. Components (2) and (3) are usually associated with fuller forms with hard shoulders.

The amplitudes of the newly formed stern system are superimposed on those of the bow system. Waves move forward with the speed of the ship, and phasing will alter with speed, Figure 3.16. For example, as speed is increased, the length of the bow wave will increase (Equation (3.24)) until it coincides and interferes with the stern wave.

As a result of these wave interference effects, ship resistance curves exhibit an oscillatory nature with change in speed, that is, they exhibit humps and hollows, Figure 3.17. Humps will occur when the crests (or troughs) coincide (reinforcement), and hollows occur when a trough coincides with a crest (cancellation).

3.1.5.4 Speeds for Humps and Hollows

The analysis is based on the bow and stern transverse wave systems, Figure 3.16, where L_{PS} is the pressure source separation. Now $L_{PS} = k \times L_{BP}$, where k is typically 0.80–0.95, depending on the fullness of the vessel. The lengths λ of the individual waves are given by Equation (3.24).

It is clear from Figure 3.16 that when the bow wavelength λ is equal to L_{PS}, the crest from the bow wave is on the stern trough and wave cancellation will occur. At lower speeds there will be several wavelengths within L_{PS} and the following analysis indicates the speeds when favourable interference (hollows) or unfavourable interference (humps) are likely to occur.

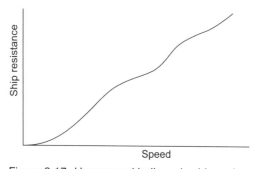

Figure 3.17. Humps and hollows in ship resistance curve.

Table 3.1. *Speeds for humps and hollows*

		Humps		
λ / L_{PS}	2	2/3	2/5	2/7
Fr	0.535	0.309	0.239	0.202
		Hollows		
λ / L_{PS}	1	1/2	1/3	1/4
Fr	0.378	0.268	0.219	0.189

Hollows: Bow crests coincide with stern troughs approximately at

$$\frac{\lambda}{L_{PS}} = 1, \frac{1}{2}, \frac{1}{3}, \frac{1}{4} \cdots\cdots$$

Humps: Bow crests coincide with stern crests approximately at:

$$\frac{\lambda}{L_{PS}} = 2, \frac{2}{3}, \frac{2}{5}, \frac{2}{7} \cdots\cdots$$

The speeds for humps and hollows may be related to the ship speed as follows:
Ship Froude number

$$Fr = \frac{V}{\sqrt{g \cdot L_{BP}}}$$

and

$$V^2 = Fr^2 g L_{BP} = Fr^2 g \frac{L_{PS}}{k}$$

and, from Equation (3.24), the wavelength $\lambda = \frac{2\pi}{g} V^2 = \frac{2\pi}{g} Fr^2 g \frac{L_{PS}}{k}$ and

$$Fr = \sqrt{\frac{k}{2\pi}} \cdot \sqrt{\frac{\lambda}{L_{PS}}}. \tag{3.25}$$

If say $k = 0.90$, then speeds for humps and hollows would occur as shown in Table 3.1. The values of speed in Table 3.1 are based only on the transverse waves. The diverging waves can change these speeds a little, leading to higher values for humps and hollows.

At the design stage it is desirable to choose the design ship length and/or speed to avoid resistance curve humps. This would, in particular, include avoiding the humps at or around Froude numbers of 0.30, the so-called prismatic hump and 0.50, the so-called main hump. Alternatively, a redistribution of displacement volume may be possible. For example, with a decrease in midship area (C_M) and an increase in prismatic coefficient (C_P), displacement is moved nearer the ends, with changes in the position of the pressure sources and, hence, wave interference. Furthermore, a decrease in bow wavemaking may be achieved by careful design of the entrance and sectional area curve (SAC) slope, or by the use of a bulbous bow for wave cancellation. These aspects are discussed further under hull form design in Chapter 14.

The wave characteristics described in this section are strictly for *deep* water. The effects of shallow water are discussed in Chapter 6.

3.1.6 Dimensional Analysis and Scaling

3.1.6.1 Dimensional Analysis

In addition to the physical approach to the separation of total resistance into identifiable components, the methods of dimensional analysis may also be applied [3.31], [3.32]. Physical variables and their dimensions are as follows:

Hull resistance:	R_T	$\dfrac{ML}{T^2}$	(force),
Hull speed:	V	$\dfrac{L}{T}$	(velocity),
Hull size:	L		
Fluid density:	ρ	$\dfrac{M}{L^3}$	(mass/unit volume),
Fluid viscosity:	μ	$\dfrac{M}{LT}$	(stress/rate of strain),
Acceleration due to gravity:	g	$\dfrac{M}{T^2}$	(acceleration).

According to the methods of dimensional analysis, the relationship between the quantities may be expressed as

$$f(R_T, V, L, \rho, \mu, g, \alpha_i) = 0 \text{ or}$$
$$R_T = f(V, L, \rho, \mu, g, \alpha_i), \tag{3.26}$$

where α_i are non-dimensional parameters of hull shape.

If geometrically similar models are considered, the method of dimensional analysis yields

$$\frac{R_T}{L^2 V^2 \rho} = k\left[\left(\frac{v}{VL}\right)^x \cdot \left(\frac{gL}{V^2}\right)^y\right], \tag{3.27}$$

where $v = \frac{\mu}{\rho}$, and

$$\frac{R_T}{L^2 V^2 \rho} = k\left[Re^x Fr^y\right], \tag{3.28}$$

or say $C_T = f(Re, Fr)$, where

$$C_T = \frac{R_T}{0.5 S V^2}, \tag{3.29}$$

where S = hull wetted surface area and $S \propto L^2$ for geometrically similar hulls (geosims), and 0.5 in the denominator does not disturb the non-dimensionality, where $Re = \frac{VL}{v}$ is the Reynolds number, and $Fr = \frac{V}{\sqrt{gL}}$ is the Froude number.

Hence, for complete dynamic similarity between two hull sizes (e.g. model and ship), three conditions must apply.

1. Shape parameters α_i must be the same. This implies that geometric similarity is required

2. Reynolds numbers must be the same
3. Froude numbers must be the same

The last two conditions imply:

$$\frac{V_1 L_1}{v_1} = \frac{V_2 L_2}{v_2} \quad \text{and} \quad \frac{V_1}{\sqrt{g_1 L_1}} = \frac{V_2}{\sqrt{g_2 L_2}}$$

i.e.

$$\frac{L_1 \sqrt{g_1 L_1}}{v_1} = \frac{L_2 \sqrt{g_2 L_2}}{v_2} \quad \text{or}$$

$$\frac{g_1 L_1^3}{v_1} = \frac{g_2 L_2^3}{v_2} \quad \text{or}$$

$$\frac{v_1}{v_2} = \left[\frac{L_1}{L_2}\right]^{3/2} \tag{3.30}$$

for constant g. Hence, complete similarity cannot be obtained without a large change in g or fluid viscosity v, and the ship *scaling problem* basically arises because complete similarity is not possible between model and ship.

The scaling problem can be simplified to some extent by keeping Re or Fr constant. For example, with constant Re,

$$\frac{V_1 L_1}{v_1} = \frac{V_2 L_2}{v_2},$$

i.e.

$$\frac{V_1}{V_2} = \frac{L_2 v_1}{L_1 v_2} = \frac{L_2}{L_1}$$

if the fluid medium is water in both cases. If the model scale is say $\frac{L_2}{L_1} = 25$, then $\frac{V_1}{V_2} = 25$, i.e. the model speed is 25 times faster than the ship, which is impractical.

With constant Fr,

$$\frac{V_1}{\sqrt{g_1 L_1}} = \frac{V_2}{\sqrt{g_2 L_2}},$$

i.e.

$$\frac{V_1}{V_2} = \sqrt{\frac{L_1}{L_2}} = \frac{1}{5}$$

for 1/25th scale, then the model speed is 1/5th that of the ship, which is a practical solution.

Hence, *in practice*, scaling between model and ship is carried out at constant Fr. At constant Fr,

$$\frac{Re_1}{Re_2} = \frac{V_1 L_1}{v_1} \cdot \frac{v_2}{V_2 L_2} = \frac{v_2}{v_1} \cdot \frac{L_1^{3/2}}{L_2^{3/2}}.$$

Hence with a 1/25th scale, $\frac{Re_1}{Re_2} = \frac{1}{125}$ and Re for the ship is much larger than Re for the model.

The scaling equation, Equation (3.28): $C_T = k[Re^x \cdot Fr^y]$ can be expanded and rewritten as

$$C_T = f_1(Re) + f_2(Fr) + f_3(Re \cdot Fr), \tag{3.31}$$

that is, parts dependent on Reynolds number (broadly speaking, identifiable with viscous resistance), Froude number (broadly identifiable with wave resistance) and a remainder dependent on both Re and Fr.

In practice, a physical breakdown of resistance into components is not available, and such components have to be identified from the character of the *total resistance* of the model. The methods of doing this assume that $f_3(Re \cdot Fr)$ is negligibly small, i.e. it is assumed that

$$C_T = f_1(Re) + f_2(Fr). \tag{3.32}$$

It is noted that if gravitational effects are neglected, dimensional analysis yields $C_T = f(Re)$ and if viscous effects are neglected, $C_T = f(Fr)$; hence, the breakdown as shown is not unreasonable.

However, Re and Fr can be broadly identified with viscous resistance and wavemaking resistance, but since the stern wave is suppressed by boundary layer growth, the wave resistance is not independent of Re. Also, the viscous resistance depends on the pressure distribution around the hull, which is itself dependent on wavemaking. Hence, viscous resistance is not independent of Fr, and the dimensional breakdown of resistance generally assumed for practical scaling is not identical to the actual physical breakdown.

3.1.6.2 Froude's Approach
The foregoing breakdown of resistance is basically that suggested by Froude working in the 1860s [3.19] and [3.33], although he was unaware of the dimensional methods discussed. He assumed that

total resistance = skin friction + {wavemaking and pressure form}

or skin friction + 'the rest', which he termed *residuary*,

i.e.

$$C_T = C_F + C_R. \tag{3.33}$$

The skin friction, C_F, is estimated from data for a flat plate of the same length, wetted surface and velocity of model or ship.

The difference between the skin friction resistance and the total resistance gives the residuary resistance, C_R. Hence, the part dependent on Reynolds number, Re, is separately determined and the model test is carried out at the corresponding velocity which gives equality of Froude number, Fr, for ship and model; hence, *dynamic similarity* for the wavemaking (or residuary) resistance is obtained. Hence, if the *residuary* resistance is considered:

$$R_R = \rho V^2 L^2 f_2\left[\frac{V}{\sqrt{gL}}\right].$$

For the model,

$$R_{Rm} = \rho V_m^2 L_m^2 f_2\left[\frac{V_m}{\sqrt{gL_m}}\right].$$

For the ship,

$$R_{Rs} = \rho V_s^2 L_s^2 f_2 \left[\frac{V_s}{\sqrt{g L_s}} \right],$$

where ρ is common, g is constant and f_2 is the same for the ship and model. It follows that if

$$\frac{V_m}{\sqrt{L_m}} = \frac{V_s}{\sqrt{L_s}}, \quad \text{or} \quad \frac{V_m^2}{V_s^2} = \frac{L_m}{L_s},$$

then

$$\frac{R_{Rm}}{R_{Rs}} = \frac{V_m^2}{V_s^2} \cdot \frac{L_m^2}{L_s^2} = \frac{L_m^3}{L_s^3} = \frac{\Delta_m}{\Delta_s}$$

which is Froude's law; that is, when the speeds of the ship and model are in the ratio of the square root of their lengths, then the resistance due to wavemaking varies as their displacements.

The speed in the ratio of the square root of lengths is termed the *corresponding speed*. In coefficient form,

$$C_T = \frac{R_T}{0.5 S V^2} \quad C_F = \frac{R_F}{0.5 S V^2} \quad C_R = \frac{R_R}{0.5 S V^2}.$$

$S \propto L^2$, $V^2 \propto L$; hence, $0.5 S V^2 \propto L^3 \propto \Delta$.

$$\frac{C_{Rm}}{C_{Rs}} = \frac{R_{Rm}}{\Delta_m} \times \frac{\Delta_s}{R_{Rs}} = \frac{\Delta_m}{\Delta_s} \times \frac{\Delta_s}{\Delta_m} = 1$$

at constant $\frac{V}{\sqrt{gL}}$. Hence, at constant $\frac{V}{\sqrt{gL}}$, C_R is the same for model and ship and $C_{Rm} = C_{Rs}$.

Now, $C_{Tm} = C_{Fm} + C_{Rm}$, $C_{Ts} = C_{Fs} + C_{Rs}$, and $C_{Rm} = C_{Rs}$.
Hence,

$$C_{Tm} - C_{Ts} = C_{Fm} - C_{Fs}$$

or

$$C_{Ts} = C_{Tm} - (C_{Fm} - C_{Fs}), \tag{3.34}$$

that is, change in total resistance coefficient is the change in friction coefficient, and the change in the total resistance depends on the Reynolds number, Re. In practical terms, C_{Tm} is derived from a model test with a measurement of total resistance, see Section 3.1.4, C_{Fm} and C_{Fs} are derived from published skin friction data, see Section 4.3, and estimates of C_{Ts} and ship resistance R_{Ts} can then be made.

Practical applications of this methodology are described in Chapter 4, model-ship extrapolation, and a worked example is given in Chapter 17.

3.2 Other Drag Components

3.2.1 Appendage Drag

3.2.1.1 Background
Typical appendages found on ships include rudders, stabilisers, bossings, shaft brackets, bilge keels and water inlet scoops and all these items give rise to additional

Table 3.2. *Resistance of appendages, as a percentage of hull naked resistance*

Item	% of naked resistance
Bilge keels	2–3
Rudder	up to about 5 (e.g. about 2 for a cargo vessel) but may be included in hull resistance tests
Stabiliser fins	3
Shafting and brackets, or bossings	6–7
Condenser scoops	1

resistance. The main appendages on a single-screw ship are the rudder and bilge keels, with a total appendage drag of about 2%–5%. On twin-screw vessels, the main appendages are the twin rudders, twin shafting and shaft brackets, or bossings, and bilge keels. These may amount to as much as 8%–25% depending on ship size. The resistance of appendages can be significant and some typical values, as a percentage of calm-water test resistance, are shown in Table 3.2. Typical total resistance of appendages, as a percentage of hull naked resistance, are shown in Table 3.3.

3.2.1.2 Factors Affecting Appendage Drag

With careful alignment, the resistance of appendages will result mainly from skin friction, based on the wetted area of the appendage. With poor alignment and/or badly designed bluff items, separated flow may occur leading to an increase in resistance. An appendage relatively near the surface may create wave resistance. Careful alignment needs a knowledge of the local flow direction. Model tests using paint streaks, tufts, flags or particle image velocimetry (PIV) and computational fluid dynamics (CFD) may be used to determine the flow direction and characteristics. This will normally be for the one design speed condition, whereas at other speeds and trim conditions cross flow may occur with a consequent increase in drag.

In addition to the correct alignment to flow, other features that affect the appendage drag include the thickness of the boundary layer in which the appendage is working and the local flow velocity past the appendage; for example, an increase of up to about 10% may occur over the bilges amidships, a decrease of up to 10% near the bow and an even bigger decrease at the stern.

Further features that affect the measurement and assessment of appendage drag at model and full scale are the type of flow over the appendage, separated flow on the appendage and velocity gradients in the flow.

Table 3.3. *Total resistance of appendages as a percentage of hull naked resistance*

Vessel type	% of naked resistance
Single screw	2–5
Large fast twin screw	8–14
Small fast twin screw	up to 25

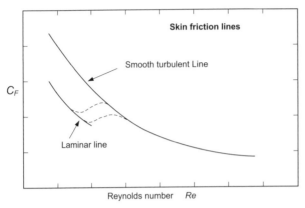

Figure 3.18. Skin friction lines.

3.2.1.3 Skin Friction Resistance

During a model test, the appendages are running at a much smaller Reynolds number than full scale, see Section 3.1.6. As a consequence, a model appendage may be operating in laminar flow, whilst the full-scale appendage is likely to be operating in turbulent flow. The skin friction resistance is lower in laminar flow than in turbulent flow, Figure 3.18, and that has to be taken into account when scaling model appendage resistance to full size.

Boundary layer velocity profiles for laminar and turbulent flows are shown in Figure 3.19. The surface shear stress τ_w is defined as

$$\tau_w = \mu \left[\frac{\partial u}{\partial y} \right]_{y=0}, \tag{3.35}$$

where μ is the fluid dynamic viscosity and $[\frac{\partial u}{\partial y}]$ is the velocity gradient at the surface.

The local skin friction coefficient C_F is defined as

$$C_F = \frac{\tau_w}{0.5\rho U^2}. \tag{3.36}$$

3.2.1.4 Separation Resistance

Resistance in separated flow is higher in laminar flow than in turbulent flow, Figure 3.20. This adds further problems to the scaling of appendage resistance.

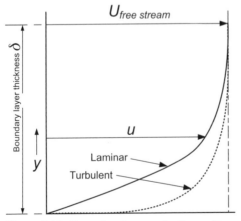

Figure 3.19. Boundary layer velocity profiles.

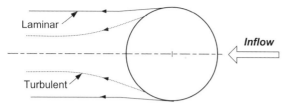

Figure 3.20. Separated flow.

3.2.1.5 Velocity Gradient Effects

It should be borne in mind that full-scale boundary layers are, when allowing for scale, about half as thick as model ones. Hence, the velocity gradient effects are higher on the model than on the full-scale ship.

Appendages which are wholly inside the model boundary layer may project through the ship boundary layer, Figure 3.21, and, hence, laminar conditions can exist full scale outside the boundary layer which do not exist on the model. This may increase or decrease the drag depending on whether the flow over the append- age separates. This discrepancy in the boundary layer thickness can be illustrated by the following approximate calculations for a 100 m ship travelling at 15 knots and a 1/20th scale 5 m geometrically similar tank test model.

For turbulent flow, using a 1/7th power law velocity distribution in Figure 3.19, an approximation to the boundary layer thickness δ on a flat plate is given as

$$\frac{\delta}{x} = 0.370 \ Re^{-1/5}, \tag{3.37}$$

where x is the distance from the leading edge.

For a 100 m ship, $Re = VL/v = 15 \times 0.5144 \times 100 \ / \ 1.19 \times 10^{-6} = 6.48 \times 10^{8}$.

The approximate boundary layer thickness at the aft end of the ship is $\delta = x \times 0.370 \ Re^{-1/5} = 100 \times 0.370 \times (6.48 \times 10^{8})^{-1/5} = 640$ mm.

For the 5 m model, corresponding speed $= 15 \times 0.5144 \times 1/\sqrt{20} = 1.73$ m/s.

The model $Re = VL/v = 1.73 \times 5 \ / \ 1.14 \times 10^{-6} = 7.59 \times 10^{6}$. The approximate boundary layer thickness for the model is $\delta = 5 \times 0.370 \ (7.59 \times 10^{6})^{-1/5} = 78$ mm.

Scaling geometrically (scale 1/20) from ship to model (as would the propeller and appendages) gives a required model boundary layer thickness of only 32 mm.

Hence, the model boundary layer thickness at 78 mm is more than twice as thick as it should be, as indicated in Figure 3.21. It should be noted that Equation (3.37) is

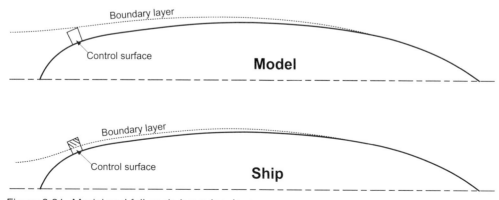

Figure 3.21. Model and full-scale boundary layers.

effectively for a flat plate, but may be considered adequate for ship shapes as a first approximation.

3.2.1.6 Estimating Appendage Drag
The primary methods are the following:

(i) Test the hull model with and without appendages. The difference in model C_T with and without appendages represents the appendage drag which is then scaled to full size.
(ii) Form factor approach. This is similar to (i), but a form factor is used, $C_{Ds} = (1 + k) C_{Dm}$, where the form factor $(1 + k)$ is derived from a geosim set of appended models of varying scales. The approach is expensive and time consuming, as noted for geosim tests, in general, in Section 4.2.
(iii) Test a larger separate model of the appendage at higher speeds. With large models of the appendages and high flow speeds, for example in a tank, circulating water channel or wind tunnel, higher Reynolds numbers, closer to full-scale values, can be achieved. This technique is typically used for ship rudders and control surfaces [3.34] and yacht keels [3.8].
(iv) Use of empirical data and equations derived from earlier model (and limited full-scale) tests.

3.2.1.7 Scaling Appendage Drag
When measuring the drag of appendages attached to the model hull, each appendage runs at its own Re and has a resistance which will, theoretically, scale differently to full size. These types of effects, together with flow type, separation and velocity gradient effects mentioned earlier, make appendage scaling uncertain. The International Towing Tank Conference (ITTC) has proposed the use of a scale effect factor β where, for appendages,

$$C_{D\text{ship}} = \beta C_{D\text{model}}. \tag{3.38}$$

The factor β varies typically from about 0.5 to 1.0 depending on the type of appendage.

There is a small amount of actual data on scaling. In the 1940s Allan at NPL tested various scales of model in a tank [3.35]. The British Ship Research Association (BSRA), in the 1950s, jet propelled the 58 m ship *Lucy Ashton*, fitted with various appendages. The jet engine was mounted to the hull via a load transducer, allowing direct measurements of thrust, hence resistance, to be made. The ship results were compared with six geosim models tested at NPL, as reported by Lackenby [3.36]. From these types of test results, the β factor is found to increase with larger-scale models, ultimately tending to 1.0 as the model length approaches the ship length. A summary of the β values for A-brackets and open shafts, derived from the *Lucy Ashton* tests, are given in Table 3.4. These results are relatively consistent, whereas some of the other test results did not compare well with earlier work. The resulting information from the *Lucy Ashton* and other such tests tends to be inconclusive and the data have been reanalysed many times over the years.

In summary, typical tank practice is to run the model naked, then with appendages and to apply an approximate scaling law to the difference. Typical practice

Table 3.4. *β values for A-brackets and open shafts from the Lucy Ashton tests*

Ship speed (knots)	Model					
	2.74 m	3.66 m	4.88 m	6.10 m	7.32 m	9.15 m
8	0.48	0.52	0.56	0.58	0.61	0.67
12	0.43	0.47	0.52	0.54	0.57	0.61
14.5	0.33	0.37	0.41	–	0.46	0.51

when using the ITTC $β$ factor is to take bilge keels and rudders at full model value, $β = 1$, and to halve the resistance of shafts and brackets, $β = 0.5$. For example, streamlined appendages placed favourably along streamlines may be expected to experience frictional resistance only. This implies (with more laminar flow over appendages likely with the ship) less frictional resistance for the ship, hence half of model values, $β = 0.5$, are used as an approximation.

3.2.1.8 Appendage Drag Data

LOCAL FLOW SPEED. When carrying out a detailed analysis of the appendage drag, local flow speed and boundary layer characteristics are required. Approximations to speed, such as ±10% around hull, and boundary layer thickness mentioned earlier, may be applied. For appendages in the vicinity of the propeller, wake speed may be appropriate, i.e. $Va = Vs (1 - w_T)$ (see Chapter 8). For rudders downstream of a propeller, Va is accelerated by 10%–20% due to the propeller and the use of Vs as a first approximation might be appropriate.

(a) Data

A good source of data is Hoerner [3.37] who provides drag information on a wide range of items such as:

- Bluff bodies such as sonar domes.
- Struts and bossings, including root interference drag.
- Shielding effects of several bodies in line.
- Local details: inlet heads, plate overlaps, gaps in flush plating.
- Scoops, inlets.
- Spray, ventilation, cavitation, normal and bluff bodies and hydrofoils.
- Separation control using vortex generator guide vanes.
- Rudders and control surfaces.

Mandel [3.38] discusses a number of the hydrodynamic aspects of appendage design.

The following provides a number of equations for estimating the drag of various appendages.

(b) Bilge keels

The sources of resistance are the following:

- Skin friction due to additional wetted surface.
- Interference drag at junction between bilge keel and hull.

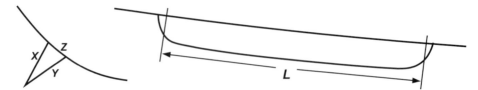

Figure 3.22. Geometry of bilge keel.

ITTC recommends that the total resistance be multiplied by the ratio $(S + S_{BK})/S$, where S is the wetted area of the hull and S_{BK} is the wetted area of the bilge keels.

Drag of bilge keel according to Peck [3.39], and referring to Figure 3.22, is

$$D_B = \frac{1}{2}\rho S V^2 C_F \left[2 - \frac{2Z}{X + Y} \right], \tag{3.39}$$

where S is the wetted surface of the bilge keel and L is the average length of the bilge keel to be used when calculating C_F. When Z is large, interference drag tends to zero; when Z tends to zero (a plate bilge keel), interference drag is assumed to equal skin friction drag.

(c) Rudders, shaft brackets and stabiliser fins

Sources of drag are the following:

(1) Control surface or strut drag, D_{CS}
(2) Spray drag if rudder or strut penetrates water surface, D_{SP}
(3) Drag of palm, D_P
(4) Interference drag of appendage with hull, D_{INT}

Total drag may be defined as

$$D_{AP} = D_{CS} + D_{SP} + D_P + D_{INT}. \tag{3.40}$$

Control surface drag, D_{CS}, as proposed by Peck [3.39], is

$$D_{CS} = \frac{1}{2}\rho S V^2 C_F \left[1.25\frac{Cm}{Cf} + \frac{S}{A} + 40 \left(\frac{t}{Ca} \right)^3 \right] \times 10^{-1}, \tag{3.41}$$

where Cm is the mean chord length which equals $(Cf + Ca)$, Figure 3.23, used for calculation of C_F, S is the wetted area, A is frontal area of maximum section, t is the maximum thickness and V is the ship speed.

Control surface drag as proposed by Hoerner [3.37], for 2D sections,

$$C_D = C_F \left[1 + 2 \left(\frac{t}{c} \right) + 60 \left(\frac{t}{c} \right)^4 \right], \tag{3.42}$$

where c is the chord length used for the calculation of C_F.

A number of alternative formulae are proposed by Kirkman and Kloetzli [3.40] when the appendages of the models are running in laminar or partly laminar flow.

Spray drag, D_{SP}, as proposed by Hoerner [3.37], is

$$D_{SP} = 0.24\tfrac{1}{2}\rho V^2 t_w^2, \tag{3.43}$$

where t_w is the maximum section thickness at the water surface.

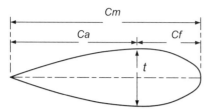

Figure 3.23. Geometry of strut or control surface.

Drag of palm, D_P, according to Hoerner [3.37], is

$$D_P = 0.75 C_{DPalm} \left(\frac{h_P}{\delta}\right)^{1/3} W h_P \frac{1}{2}\rho V^2, \qquad (3.44)$$

where h_P is the height of palm above surface, W is the frontal width of palm, δ is the boundary layer thickness, C_{DPalm} is 0.65 if the palm is rectangular with rounded edges and V is the ship speed.

Interference drag, D_{INT}, according to Hoerner [3.37], is

$$D_{INT} = \frac{1}{2}\rho V^2 t^2 \left[0.75\frac{t}{c} - \frac{0.0003}{(t/c)^2}\right], \qquad (3.45)$$

where t is the maximum thickness of appendage at the hull and c is the chord length of appendage at the hull.

Extensive drag data suitable for rudders, fin stabilisers and sections applicable to support struts may be found in Molland and Turnock [3.34]. A practical working value of rudder drag coefficient at zero incidence with section thickness ratio $t/c = 0.20$–0.25 is found to be $C_{D0} = 0.013$ [3.34], based on profile area (span × chord), *not* wetted area of both sides. This tends to give larger values of rudder resistance than Equations (3.41) and 3.48).

Further data are available for particular cases, such as base-ventilating sections, Tulin [3.41] and rudders with thick trailing edges, Rutgersson [3.42].

(d) Shafts and bossings

Propeller shafts are generally inclined at some angle to the flow, Figure 3.24, which leads to lift and drag forces on the shaft and shaft bracket. Careful alignment of the shaft bracket strut is necessary in order to avoid cross flow.

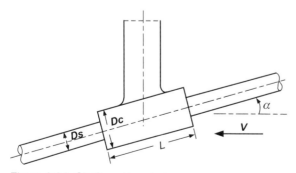

Figure 3.24. Shaft and bracket.

The sources of resistance are

(1) Drag of shaft, D_{SH}
(2) Pressure drag of cylindrical portion, C_{DP}
(3) Skin friction of cylindrical portion, C_F
(4) Drag of forward and after ends of the cylinder, C_{DE}

The drag of shaft, for $Re < 5 \times 10^5$ (based on diameter of shaft), according to Hoerner [3.37], is

$$D_{SH} = \tfrac{1}{2}\rho L_{SH} Ds\, V^2 (1.1 \sin^3\alpha + \pi C_F), \tag{3.46}$$

where L_{SH} is the total length of shaft and bossing, Ds is the diameter of shaft and bossing and α is the angle of flow in degrees relative to shaft axis, Figure 3.24.

For the cylindrical portions, Kirkman and Kloetzli [3.40] offer the following equations. The equations for pressure drag, C_{DP}, are as follows:

For $Re < 1 \times 10^5$, $C_{DP} = 1.1 \sin^3\alpha$
For $1 \times 10^5 < Re < 5 \times 10^5$, and $\alpha > \beta$, $C_{DP} = -0.7154 \log_{10} Re + 4.677$, and
For $\alpha < \beta$, $C_{DP} = (-0.754 \log_{10} Re + 4.677)[\sin^3(1.7883 \log_{10} Re - 7.9415)\,\alpha]$
For $Re > 5 \times 10^5$, and $0 < \alpha < 40°$, $C_{DP} = 0.60 \sin^3(2.25\,\alpha)$ and for $40° < \alpha < 90°$, $C_{DP} = 0.60$

where $Re = VDc/\upsilon$, $\beta = -71.54 \log_{10} Re + 447.7$ and the reference area is the cylinder projected area which is $(L \times Dc)$.

For friction drag, C_F, the equations are as follows:

$$
\begin{aligned}
&\text{For } Re < 5 \times 10^5, C_F = 1.327\, Re^{-0.5} \\
&\text{For } Re > 5 \times 10^5, C_F = \frac{1}{(3.461 \log_{10} Re - 5.6)^2} - \frac{1700}{Re}
\end{aligned}
\tag{3.47}
$$

where $Re = VLc/\upsilon$, $Lc = L/\tan\alpha$, $Lc > L$ and the reference area is the wetted surface area which equals $\pi \times$ length $\times$ diameter.

Equations for drag of ends, if applicable, C_{DE}, are the following:

For support cylinder with sharp edges, $C_{DE} = 0.90 \cos^3\alpha$
For support cylinder with faired edges, $C_{DE} = 0.01 \cos^3\alpha$.

Holtrop and Mennen [3.43] provide empirical equations for a wide range of appendages and these are summarised as follows:

$$R_{\text{APP}} = \tfrac{1}{2}\rho V_S^2 C_F (1+k_2)_E \sum S_{\text{APP}} + R_{BT}, \tag{3.48}$$

where V_S is ship speed, C_F is for the ship and is determined from the ITTC1957 line and S_{APP} is the wetted area of the appendage(s). The equivalent $(1+k_2)$ value for the appendages, $(1+k_2)_E$, is determined from

$$(1+k_2)_E = \frac{\sum (1+k_2)S_{\text{APP}}}{\sum S_{\text{APP}}}. \tag{3.49}$$

Table 3.5. *Appendage form factors* $(1 + k_2)$

Appendage type	$(1 + k_2)$
Rudder behind skeg	1.5–2.0
Rudder behind stern	1.3–1.5
Twin-screw balanced rudders	2.8
Shaft brackets	3.0
Skeg	1.5–2.0
Strut bossings	3.0
Hull bossings	2.0
Shafts	2.0–4.0
Stabiliser fins	2.8
Dome	2.7
Bilge keels	1.4

The appendage resistance factors $(1 + k_2)$ are defined by Holtrop as shown in Table 3.5. The term R_{BT} in Equation (3.48) takes account of bow thrusters, if fitted, and is defined as

$$R_{BT} = \pi \rho V_S^2 d_T C_{BTO}, \tag{3.50}$$

where d_T is the diameter of the thruster and the coefficient C_{BTO} lies in the range 0.003–0.012. When the thruster lies in the cylindrical part of the bulbous bow, $C_{BTO} \rightarrow 0.003$.

(e) Summary

In the absence of hull model tests (tested with and without appendages), detailed estimates of appendage drag may be carried out at the appropriate Reynolds number using the equations and various data described. Alternatively, for preliminary powering estimates, use may be made of the approximate data given in Tables 3.2 and 3.3. Examples of appendage drag estimates are given in Chapter 17.

3.2.2 Air Resistance of Hull and Superstructure

3.2.2.1 Background

A ship travelling in still air experiences air resistance on its above-water hull and superstructure. The level of air resistance will depend on the size and shape of the superstructure and on ship speed. Some typical values of air resistance for different ship types, as a percentage of calm water hull resistance, are given in Table 3.6.

The air drag of the above-water hull and superstructure is generally a relatively small proportion of the total resistance. However, for a large vessel consuming large quantities of fuel, any reductions in air drag are probably worth pursuing. The air drag values shown are for the ship travelling in still air. The proportion will of course rise significantly in any form of head wind.

The air drag on the superstructure and hull above the waterline may be treated as the drag on a bluff body. Typical values of C_D for bluff bodies for $Re > 10^3$ are given in Table 3.7.

Table 3.6. *Examples of approximate air resistance*

Type	L_{BP} (m)	C_B	Dw (tonnes)	Service speed (knots)	Service power (kW)	Fr	Air drag (%)
Tanker	330	0.84	250,000	15	24,000	0.136	2.0
Tanker	174	0.80	41,000	14.5	7300	0.181	3.0
Bulk carrier	290	0.83	170,000	15	15,800	0.145	2.5
Bulk carrier	180	0.80	45,000	14	7200	0.171	3.0
Container	334	0.64	100,000 10,000 TEU	26	62,000	0.234	4.5
Container	232	0.65	37,000 3500 TEU	23.5	29,000	0.253	4.0
Catamaran ferry	80	0.47	650 pass 150 cars	36	23,500	0.661	4.0
Passenger ship	265	0.66	2000 pass GRT90,000	22	32,000	0.222	6.0

When travelling into a wind, the ship and wind velocities and the relative velocity are defined as shown in Figure 3.25.

The resistance is

$$R_A = \tfrac{1}{2}\rho_A C_D A_P V_A^2, \tag{3.51}$$

where A_p is the projected area perpendicular to the relative velocity of the wind to the ship, V_A is the relative wind and, for air, $\rho_A = 1.23$ kg/m^3, see Table A1.1.

It is noted later that results of wind tunnel tests on models of superstructures are normally presented in terms of the drag force in the ship fore and aft direction (X-axis) and based on A_T, the transverse frontal area.

3.2.2.2 Shielding Effects
The wake behind one superstructure element can shield another element from the wind, Figure 3.26(a), or the wake from the sheerline can shield the superstructure, Figure 3.26(b).

3.2.2.3 Estimation of Air Drag
In general, the estimation of the wind resistance involves comparison with model data for a similar ship, or performing specific model tests in a wind tunnel. It can

Table 3.7. *Approximate values of drag coefficient for bluff bodies, based on frontal area*

Item	C_D
Square plates	1.1
Two-dimensional plate	1.9
Square box	0.9
Sphere	0.5
Ellipsoid, end on ($Re\ 2 \times 10^5$)	0.16

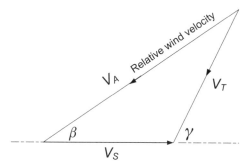

Figure 3.25. Vector diagram.

be noted that separation drag is not sensitive to Re, so scaling from model tests is generally acceptable, on the basis that $C_{Ds} = C_{Dm}$.

A typical air drag diagram for a ship model is broadly as shown in Figure 3.27. Actual wind tunnel results for different deckhouse configurations [3.44] are shown in Figure 3.28. In this particular case, C_X is a function of (A_T/L^2).

Wind drag data are usually referred to the frontal area of the hull plus superstructure, i.e. transverse area A_T. Because of shielding effects with the wind ahead, the drag coefficient may be lower at $0°$ wind angle than at $30°$ wind angle, where C_D is usually about maximum, Figure 3.27. The drag is the fore and aft drag on the ship centreline X-axis.

In the absence of other data, wind tunnel tests on ship models indicate values of about $C_D = 0.80$ for a reasonably streamlined superstructure, and about $C_D = 0.25$ for the main hull. ITTC recommends that, if no other data are available, air drag may be approximated from $C_{AA} = 0.001 \, A_T/S$, see Chapter 5, where A_T is the transverse projected area above the waterline and S is the ship hull wetted area. In this case, $D_{air} = C_{AA} \times \frac{1}{2} \rho_W S V^2$. Further typical air drag values for commercial

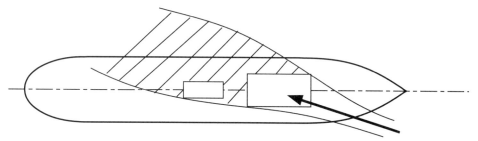

Figure 3.26. (a) Shielding effects of superstructure.

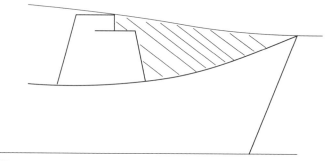

Figure 3.26. (b) Shielding effects of the sheerline.

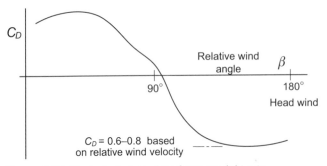

Figure 3.27. Typical air drag data from model tests.

ships can be found in Shearer and Lynn [3.45], White [3.46], Gould [3.47], Isherwood [3.48], van Berlekom [3.49], Blendermann [3.50] and Molland and Barbeau [3.51].

The regression equation for the Isherwood air drag data [3.48] in the longitudinal X-axis is

$$C_X = A_0 + A_1 \left(\frac{2A_L}{L^2}\right) + A_2 \left(\frac{2A_T}{B^2}\right) + A_3 \left(\frac{L}{B}\right)$$
$$+ A_4 \left(\frac{S_P}{L}\right) + A_5 \left(\frac{C}{L}\right) + A_6 (M), \tag{3.52}$$

where

$$C_X = \frac{F_X}{0.5\rho_A A_T V_R^2}, \tag{3.53}$$

and ρ_A is the density of air (Table A1.1 in Appendix A1), L is the length overall, B is the beam, A_L is the lateral projected area, A_T is the transverse projected area, S_P is the length of perimeter of lateral projection of model (ship) excluding waterline and slender bodies such as masts and ventilators, C is the distance from the bow

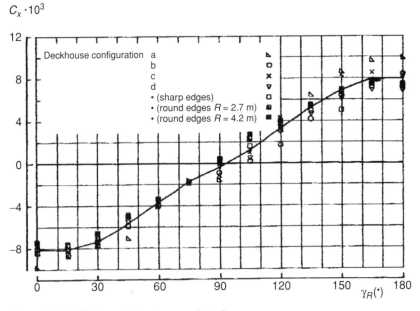

Figure 3.28. Wind coefficient curves [3.44].

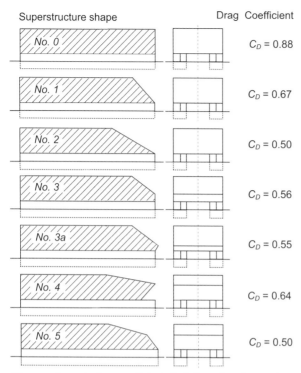

Superstructure shape   Drag Coefficient

No. 0 $C_D = 0.88$

No. 1 $C_D = 0.67$

No. 2 $C_D = 0.50$

No. 3 $C_D = 0.56$

No. 3a $C_D = 0.55$

No. 4 $C_D = 0.64$

No. 5 $C_D = 0.50$

The aerodynamic drag coefficient C_D is based on the total transverse frontal area of superstructure and hulls

Figure 3.29. Drag on the superstructures of fast ferries [3.51].

of the centroid of the lateral projected area, M is the number of distinct groups of masts or king posts seen in the lateral projection.

The coefficients A_0–A_6 are tabulated in Appendix A3, Table A3.1. Note, that according to the table, for 180° head wind, A_4 and A_6 are zero, and estimates of S_P and M are not required. For preliminary estimates, C/L can be taken as 0.5.

Examples of C_D from wind tunnel tests on representative superstructures of fast ferries [3.51] are shown in Figure 3.29. These coefficients are suitable also for monohull fast ferries.

3.2.2.4 CFD Applications

CFD has been used to investigate the flow over superstructures. Most studies have concentrated on the flow characteristics rather than on the forces acting. Such studies have investigated topics such as the flow around funnel uptakes, flow aft of the superstructures of warships for helicopter landing and over leisure areas on the top decks of passenger ships (Reddy *et al.* [3.52], Sezer-Uzol *et al.* [3.53], Wakefield *et al.* [3.54]). Moat *et al.* [3.55, 3.56] investigated, numerically and experimentally, the effects of flow distortion created by the hull and superstructure and the influences on actual onboard wind speed measurements. Few studies have investigated the actual air drag forces numerically. A full review of airwakes, including experimental and computational fluid dynamic approaches, is included in ITTC [3.57].

3.2.2.5 Reducing Air Drag

Improvements to the superstructure drag of commercial vessels with box-shaped superstructures may be made by rounding the corners, leading to reductions in drag.

It is found that the rounding of sharp corners can be beneficial, in particular, for box-shaped bluff bodies, Hoerner [3.37] and Hucho [3.58]. However, a rounding of at least $r/B_S = 0.05$ (where r is the rounding radius and B_S is the breadth of the superstructure) is necessary before there is a significant impact on the drag. At and above this rounding, decreases in drag of the order of 15%–20% can be achieved for rectangular box shapes, although it is unlikely such decreases can be achieved with shapes which are already fairly streamlined. It is noted that this procedure would conflict with design for production, and the use of 'box type' superstructure modules.

A detailed investigation into reducing the superstructure drag on large tankers is reported in [3.59].

Investigations by Molland and Barbeau [3.51] on the superstucture drag of large fast ferries indicated a reduction in drag coefficient (based on frontal area) from about 0.8 for a relatively bluff fore end down to 0.5 for a well-streamlined fore end, Figure 3.29.

3.2.2.6 Wind Gradient Effects

It is important to distinguish between *still air* resistance and resistance in a natural wind gradient. It is clear that, as air drag varies as the relative air speed squared, there will be significant increases in air drag when travelling into a wind. This is discussed further in Section 3.2.4. The relative air velocity of a ship travelling with speed Vs in still air is shown in Figure 3.30(a) and that of a ship travelling into a wind with speed Vw is shown in Figure 3.30(b).

Normally, relative wind measurements are made high up, for example, at mast head or bridge wings. Relative velocities near the water surface are much lower.

An approximation to the natural wind gradient is

$$\frac{V}{V_0} = \left(\frac{h}{h_0}\right)^n \tag{3.54}$$

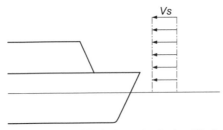

Figure 3.30. (a) Relative velocity in still air.

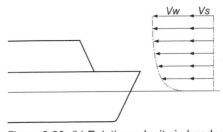

Figure 3.30. (b) Relative velocity in head wind.

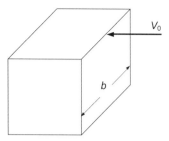

Figure 3.31. Illustration of wind gradient effect.

where n lies between 1/5 and 1/9. This applies over the sea; the index n varies with surface condition and temperature gradient.

3.2.2.7 Example of Gradient Effect

Consider the case of flow over a square box, Figure 3.31. V_0 is measured at the top of the box ($h = h_0$). Assume $V/V_0 = (h/h_0)^{1/7}$ and b and C_D are constant up the box.

Resistance in a wind gradient is

$$R = \frac{1}{2}\rho b C_D V_0^2 \int_0^{h_0} \left(\frac{h}{h_0}\right)^{2/7} dh$$

i.e.

$$R = \frac{1}{2}\rho b C_D \frac{V_0^2}{h_0^{2/7}} \left[h^{9/7} \cdot \frac{7}{9} \right]_0^{h_0}$$

and

$$R = \tfrac{1}{2}\rho b C_D V_0^2 \cdot \tfrac{7}{9} h_0 = \tfrac{7}{9} R_0 = 0.778 R_0. \qquad (3.55)$$

Comparative measurements on models indicate R/R_0 of this order. Air drag corrections as applied to ship trial results are discussed in Section 5.4.

3.2.2.8 Other Wind Effects

1. With the wind off the bow, forces and moments are produced which cause the hull to make leeway, leading to a slight increase in hydrodynamic resistance; rudder angle, hence, a drag force, is required to maintain course. These forces and moments may be defined as wind-induced forces and moments but will, in general, be very small relative to the direct wind force (van Berlekom [3.44], [3.49]). Manoeuvring may be adversely affected.
2. The wind generates a surface drift on the sea of the order of 2%–3% of wind velocity. This will reduce or increase the ship speed over the ground.

3.2.3 Roughness and Fouling

3.2.3.1 Background

Drag due to hull roughness is separation drag behind each individual item of roughness. Turbulent boundary layers have a thin laminar sublayer close to the surface and this layer can smooth out the surface by flowing round small roughness without

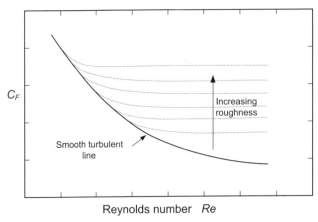

Figure 3.32. Effect of roughness on skin friction coefficient.

separating. Roughness only causes increasing drag if it is large enough to project through the sublayer. As Re increases (say for increasing V), the sublayer gets thinner and eventually a point is reached at which the drag coefficient ceases to follow the smooth turbulent line and becomes approximately constant, Figure 3.32. From the critical Re, Figure 3.33, increasing separation drag offsets falling C_F. It should be noted that surface undulations such as slight ripples in plating will not normally cause a resistance increase because no separation is caused.

3.2.3.2 Density of Roughness
As the density of the roughness increases over the surface, the additional resistance caused rises until a point is reached at which shielding of one 'grain' by another takes place. Further increase in roughness density can, in fact, then reduce resistance.

3.2.3.3 Location of Roughness
Boundary layers are thicker near the stern than at the bow and thinner at the bilge than at the waterline. Roughness has more effect where the boundary layer is thin. It also has the most effect where the local flow speed is high. For small yachts and models, roughness can cause early transition from laminar to turbulent flow, but

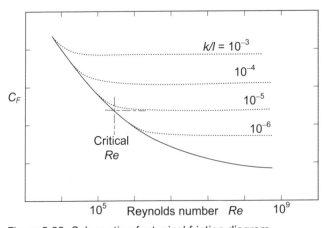

Figure 3.33. Schematic of a typical friction diagram.

Table 3.8. *Roughness of different materials*

Quality of surface	Grain size, μm ($= 10^{-6}$ m)
Plate glass	10^{-1}
Bare steel plate	50
Smooth marine paint	50*
Marine paint + antifouling etc.	100–150
Galvanised steel	150
Hot plastic coated	250
Bare wood	500
Concrete	1000
Barnacles	5000

* Possible with airless sprays and good conditions.

this is not significant for normal ship forms since transition may, in any case, occur as close as 1 m from the bow.

A typical friction diagram is shown in Figure 3.33. The roughness criterion, k/l, is defined as grain size (or equivalent sand roughness) / length of surface. The critical $Re = (90 \text{ to } 120)/(k_s/x)$, where x is distance from the leading edge. The numerator can be taken as 100 for approximate purposes. At Re above critical, C_F is constant and approximately equal to the smooth C_F at the critical Re. Some examples of roughness levels are shown in Table 3.8.

For example, consider a 200 m hull travelling at 23 knots, having a paint surface with $k_s = 100 \times 10^{-6}$ m. For this paint surface,

$$\text{Critical } Re = \frac{100}{100 \times 10^{-6}/200} = 2.0 \times 10^8. \tag{3.56}$$

The Re for the 200 m hull at 23 knots $= VL/v = 23 \times 0.5144 \times 200/1.19 \times 10^{-6} = 2.0 \times 10^9$.

Using the ITTC1957 friction formula, Equation (4.15) at $Re = 2.0 \times 10^8$, $C_F = 1.89 \times 10^{-3}$; at $Re = 2.0 \times 10^9$, $C_F = 1.41 \times 10^{-3}$ and the approximate increase due to roughness $\Delta C_F = 0.48 \times 10^{-3} \approx 34\%$. The traditional allowance for roughness for new ships, in particular, when based on the Schoenherr friction line (see Section 4.3), has been 0.40×10^{-3}. This example must be considered only as an illustration of the phenomenon. The results are very high compared with available ship results.

Some ship results are described by the Bowden–Davison equation, Equation (3.57), which was derived from correlation with ship thrust measurements and which gives lower values.

$$\Delta C_F = \left[105 \left(\frac{k_S}{L} \right)^{1/3} - 0.64 \right] \times 10^{-3}. \tag{3.57}$$

This formula was originally recommended by the ITTC for use in the 1978 Performance Prediction Method, see Chapter 5. If roughness measurements are not available, a value of $k_S = 150 \times 10^{-6}$ m is recommended, which is assumed to be the approximate roughness level for a newly built ship. The Bowden–Davison equation, Equation (3.57), was intended to be used as a correlation allowance including roughness, rather than just a roughness allowance, and should therefore *not* be used to predict the resistance increase due to change in hull roughness.

k_S is the mean apparent amplitude (MAA) as measured over 50 mm. A similar criterion is average hull roughness (AHR), which attempts to combine the individual MAA values into a single parameter defining the hull condition. It should be noted that Grigson [3.60] considers it necessary to take account of the 'texture', that is the *form* of the roughness, as well as k_S. Candries and Atlar [3.61] discuss this aspect in respect to self-polishing and silicone-based foul release coatings, where the self-polishing paint is described as having a more 'closed' spiky texture, whereas the foul release surface may be said to have a 'wavy' open texture.

It has also been determined that ΔC_F due to roughness is not independent of *Re* since ships do not necessarily operate in the 'fully rough' region; they will be forward, but not necessarily aft. The following equation, incorporating the effect of *Re*, has been proposed by Townsin [3.62]:

$$\Delta C_F = \left\{ 44\left[\left(\frac{k_S}{L}\right)^{1/3} - 10\,Re^{-1/3}\right] + 0.125 \right\} \times 10^{-3}. \tag{3.58}$$

More recently, it has been recommended that, if roughness measurements are available, this equation should be used in the ITTC Performance Prediction Method (ITTC [3.63]), together with the original Bowden–Davison equation (3.57), in order to estimate ΔC_F due only to roughness, see Chapter 5.

3.2.3.4 Service Conditions

In service, metal hulls deteriorate and corrosion and flaking paint increase roughness. Something towards the original surface quality can be recovered by shot blasting the hull back to bare metal. Typical values of roughness for actual ships, from Townsin *et al.* [3.64], for initial (new) and in-service increases are as follows:

Initial roughness, 80~120 μm
Annual increase, 10 μm for high-performance coating and cathodic protection,
  75~150 μm with resinous coatings and no cathodic protection and up to -3 μm
  for self-polishing.

The approximate equivalent power increases are 1% per 10 μm increase in roughness (based on a relatively smooth hull, 80~100 μm) or about 0.5% per 10 μm starting from a relatively rough hull (say, 200~300 μm).

3.2.3.5 Hull Fouling

Additional 'roughness' is caused by fouling, such as the growth of weeds and barnacles. The total increase in 'roughness' (including fouling) leads typically to increases in C_F of about 2%–4% C_F/month, e.g. see Aertssen [3.65–3.69]. If $C_F \approx$ 60% C_T, increase in $C_T \approx$ 1%~2%/month, i.e. 10%~30%/year (approximately half roughness, half fouling).

The initial rate of increase is often higher than this, but later growth is slower. Fouling growth rates depend on the ports being used and the season of the year. Since growth occurs mainly in fresh and coastal waters, trade patterns and turn-around times are also important. The typical influence of the growth of roughness and fouling on total resistance is shown in Figure 3.34.

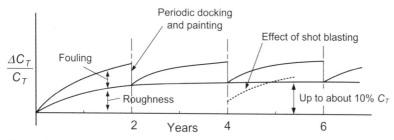

Figure 3.34. Growth of roughness and fouling.

The period of docking and shot blasting, whilst following statutory and classific-ation requirements for frequency, will also depend on the economics of hull surface finish versus fuel saved [3.64], [3.70].

It is seen that minimising roughness and fouling is important. With relatively high fuel costs, large sums can be saved by good surface finishes when new, and careful bottom maintenance in service. Surface finish and maintenance of the pro-peller is also important, Carlton [3.71]. Consequently, much attention has been paid to paint and antifouling technology such as the development of constant emission toxic coatings, self-polishing paints and methods of applying the paint [3.72], [3.73] and [3.74].

Antifouling paints commonly used since the 1960s have been self-polishing and have contained the organotin compound tributyltin (TBT). Such paints have been effective. However, TBT has since been proven to be harmful to marine life. The International Maritime Organisation (IMO) has consequently introduced regula-tions banning the use of TBT. The International Convention for the Control of Antifouling Systems on Ships (AFS) came into effect in September 2008. Under the Convention, ships are not allowed to use organotin compounds in their antifouling systems.

Since the ban on the tin-based, self-polishing antifouling systems, new altern-atives have been investigated and developed. These include tin-free self-polishing coatings and silicone-based foul release coatings which discourage marine growth from occurring, Candries and Atlar [3.61]. It is shown that a reduction in skin friction resistance of 2%–5% can be achieved with foul release coatings compared with self-polishing. It is difficult to measure actual roughness of the 'soft surface' silicone-based foul release coatings and, hence, difficult to match friction reductions against roughness levels. In addition, traditional rough-brush cleaning can damage the silicone-based soft surface, and brushless systems are being developed for this purpose.

Technology is arriving at the possibility of preventing most fouling, although the elimination of slime is not always achievable, Candries and Atlar [3.61] and Okuno *et al.* [3.74]. Slime can have a significant effect on resistance. For example, a ΔC_F of up to 80% over two years due to slime was measured by Lewthwaite *et al.* [3.75].

3.2.3.6 Quantifying Power/Resistance Increases
Due to Roughness and Fouling

GLOBAL INFORMATION. The methods used for global increases in resistance entail the use of voyage analysis techniques, that is, the analysis of ship voyage power data over a period of time, corrected for weather. Rates of increase and actual increases

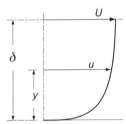

Figure 3.35. Boundary layer velocity profile.

in power (hence, resistance) can be monitored. The work of Aertssen [3.66] and Aertssen and Van Sluys [3.68], discussed earlier, uses such techniques. The results of such analyses can be used to estimate the most beneficial frequency of docking and to estimate suitable power margins.

DETAILED INFORMATION. A detailed knowledge of the changes in the *local* skin friction coefficient C_f, due say to roughness and fouling and hence, increase in resistance, can be gained from a knowledge of the local boundary layer profile, as described by Lewthwaite *et al.* [3.75]. Such a technique might be used to investigate the properties of particular antifouling systems.

The boundary layer velocity profile, Figure 3.35, can be measured by a Pitôt static tube projecting through the hull of the ship, or a laser doppler anemometer (LDA) projected through glass panels in the ship's hull.

The inner 10% of boundary layer is known as the *inner region* and a logarithmic relationship for the velocity distribution is satisfactory.

$$\frac{u}{u_0} = \frac{1}{k} \log_e \left(\frac{y u_0}{v} \right) + Br, \tag{3.59}$$

where $u_0 = \sqrt{\frac{\tau_0}{\rho}}$ is the wall friction velocity, k is the Von Karman constant, Br is a roughness function and

$$\tau_0 = \tfrac{1}{2} \rho C_f U^2. \tag{3.60}$$

From a plot of $\frac{u}{u_0}$ against $\log_e(yU/v)$, the slope of the line can be obtained, and it can be shown that $C_f = 2 \times (\text{slope} \times k)^2$, whence the local skin friction coefficient, C_f, can be derived.

Boundary layer profiles can be measured on a ship over a period of time and, hence, the influence of roughness and fouling on local C_f monitored. Other examples of the use of such a technique include Cutland [3.76], Okuno *et al.* [3.74] and Aertssen [3.66], who did not analyse the boundary layer results.

3.2.3.7 Summary

Equations (3.57) and (3.58) provide approximate values for ΔC_F, which may be applied to new ships, see model-ship correlation, Chapter 5.

Due to the continuing developments of new coatings, estimates of in-service roughness and fouling and power increases can only be approximate. For the purposes of estimating power margins, average annual increases in power due to roughness and fouling may be assumed. In-service monitoring of power and speed may be used to determine the frequency of underwater cleaning and/or docking, see

Chapter 13. Further extensive reviews of the effects of roughness and fouling may be found in Carlton [3.71] and ITTC2008 [3.63].

3.2.4 Wind and Waves

3.2.4.1 Background

Power requirements increase severely in rough weather, in part, because of wave action and, in part, because of wind resistance. Ultimately, ships slow down voluntarily to avoid slamming damage or excessive accelerations.

Ships on scheduled services tend to operate at constant speed and need a sufficient power margin to maintain speed in reasonable service weather. Other ships usually operate at maximum continuous rated power and their nominal service speed needs to be high enough to offset their average speed losses in rough weather.

Whatever the mode of operation, it is necessary, at the design stage, to be able to estimate the power increases due to wind and waves at a particular speed. This information will be used to estimate a suitable power margin for the main propulsion machinery. It also enables climatic design to be carried out, Satchwell [3.77], and forms a component of weather routeing. Climatic design entails designing the ship for the wind and wave conditions measured over a previous number of years for the relevant sea area(s). Weather routeing entails using forecasts of the likely wind and waves in a sea area the ship is about to enter. Both scenarios have the common need to be able to predict the likely ship speed loss or power increase for given weather conditions.

The influence of wind and waves on ship speed and power can be estimated by experimental and theoretical methods. The wind component will normally be estimated using the results of wind tunnel tests for a particular ship type, for example, van Berlekom *et al.* [3.44] (see also, Section 3.2.2). The wave component can be estimated as a result of tank tests and/or theoretical calculations, Townsin and Kwon [3.78], Townsin *et al.* [3.79], and ITTC2008 [3.80]. A common alternative approach is to analyse ship voyage data, for example, Aertssen [3.65], [3.66], and Aertssen and Van Sluys [3.68]. Voyage analysis is discussed further in Chapter 13. Whatever approach is used, the ultimate aim is to be able to predict the increase in power to maintain a particular speed, or the speed loss for a given power.

For ship trials, research and seakeeping investigations, the sea conditions such as wind speed, wave height, period and direction will be measured with a wave buoy. For practical purposes, the sea condition is normally defined by the Beaufort number, *BN*. The Beaufort scale of wind speeds, together with approximate wave heights, is shown in Table 3.9.

Typical speed loss curves, to a base of *BN*, are shown schematically in Figure 3.36. There tends to be little speed loss in following seas. Table 3.10 gives an example of head sea data for a cargo ship, extracted from Aertssen [3.65].

Considering head seas, the proportions of wind and wave action change with increasing *BN*, Figure 3.37, derived using experimental and theoretical estimates extracted from [3.66] and [3.67], and discussion of [3.49] indicates that, at $BN = 4$, about 10%–20% of the power increase at constant speed (depending on hull fullness and ship type) is due to wave action, whilst at $BN = 7$, about 80% is due to wave action. The balance is due to wind resistance. A detailed investigation of wave action and wave–wind proportions was carried out by Townsin *et al.* [3.79].

Table 3.9. *Beaufort scale*

Beaufort number BN	Description	Limits of speed		Approximate wave height (m)
		knots	m/s	
0	Calm	1	0.3	–
1	Light air	1–3	0.3–1.5	–
2	Light breeze	4–6	1.6–3.3	0.7
3	Gentle breeze	7–10	3.4–5.4	1.2
4	Moderate breeze	11–16	5.5–7.9	2.0
5	Fresh breeze	17–21	8.0–10.7	3.1
6	Strong breeze	22–27	10.8–13.8	4.0
7	Near gale	28–33	13.9–17.1	5.5
8	Gale	34–40	17.2–20.7	7.1
9	Strong gale	41–47	20.8–24.4	9.1
10	Storm	48–55	24.5–28.4	11.3
11	Violent storm	56–63	28.5–32.6	13.2
12	Hurricane	64 and over	32.7 and over	–

3.2.4.2 Practical Data

The following formulae are suitable for estimating the speed loss in particular sea conditions.

Aertssen formula [3.78], [3.81]:

$$\frac{\Delta V}{V} \times 100\% = \frac{m}{L_{BP}} + n, \tag{3.61}$$

where m and n vary with Beaufort number but do not account for ship type, condition or fullness. The values of m and n are given in Table 3.11.

Townsin and Kwon formulae [3.78], [3.82] and updated by Kwon in [3.83]: The percentage speed loss is given by

$$\alpha \cdot \mu \frac{\Delta V}{V} 100\%, \tag{3.62}$$

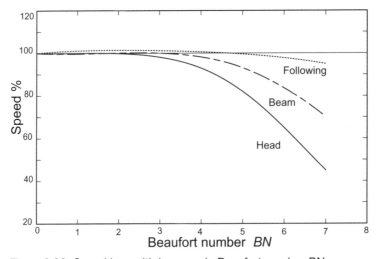

Figure 3.36. Speed loss with increase in Beaufort number BN.

Table 3.10. *Typical speed loss data for a cargo vessel,*
Aerrtssen [3.65]

Beaufort number BN	ΔP (%)	ΔV (%)	Approximate wave height (m)
0	0	–	–
1	1	–	–
2	2	–	0.2
3	5	1	0.6
4	15	3	1.5
5	32	6	2.3
6	85	17	4.2
7	200	40	8.2

where $\Delta V / V$ is the speed loss in head weather given by Equations (3.63, 3.64, 3.65), α is a correction factor for block coefficient (C_B) and Froude number (Fr) given in Table 3.12 and μ is a weather reduction factor given by Equations (3.66).

For all ships (with the exception of containerships) laden condition, $C_B = 0.75$, 0.80 and 0.85, the percentage speed loss is

$$\frac{\Delta V}{V} 100\% = 0.5\,BN + \frac{BN^{6.5}}{2.7\nabla^{2/3}}. \tag{3.63}$$

For all ships (with the exception of containerships) ballast condition, $C_B = 0.75, 0.80$ and 0.85, the percentage speed loss is

$$\frac{\Delta V}{V} 100\% = 0.7\,BN + \frac{BN^{6.5}}{2.7\nabla^{2/3}}. \tag{3.64}$$

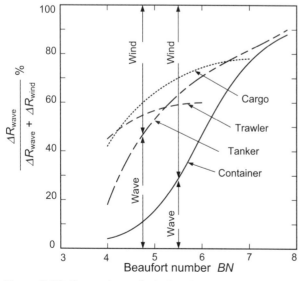

Figure 3.37. Proportions of wind and wave action.

Table 3.11. *Aertssen values for m and n*

	Head sea		Bow sea		Beam sea		Following sea	
BN	m	*n*	m	*n*	m	*n*	m	*n*
5	900	2	700	2	350	1	100	0
6	1300	6	1000	5	500	3	200	1
7	2100	11	1400	8	700	5	400	2
8	3600	18	2300	12	1000	7	700	3

where Head sea = up to 30° off bow; Bow sea = 30°–60° off bow; Beam sea = 60^0–150° off bow; Following sea = 150°–180° off bow.

For containerships, normal condition, $C_B = 0.55, 0.60, 0.65$ and 0.70, the percentage speed loss is

$$\frac{\Delta V}{V} 100\% = 0.7 BN + \frac{BN^{6.5}}{22\nabla^{2/3}}, \tag{3.65}$$

where BN is the Beaufort number and ∇ is the volume of displacement in m³.

The weather reduction factors are

$$2\mu_{\text{bow}} = 1.7 - 0.03(BN-4)^2 \qquad 30°-60° \tag{3.66a}$$

$$2\mu_{\text{beam}} = 0.9 - 0.06(BN-6)^2 \qquad 60°-150° \tag{3.66b}$$

$$2\mu_{\text{following}} = 0.4 - 0.03(BN-8)^2 \qquad 150°-180°. \tag{3.66c}$$

There is reasonable agreement between the Aertssen and Townsin-Kwon formulae as shown in Table 3.13, where Equations (3.61) and (3.65) have been compared for a container ship with a length of 220 m, $C_B = 0.600$, $\nabla = 36,500$ m³ and $Fr = 0.233$.

3.2.4.3 Derivation of Power Increase and Speed Loss
If increases in hull resistance have been calculated or measured in certain conditions and if it is assumed that, for small changes, resistance R varies as V^2, then

$$\frac{\Delta V}{V} = \left[1 + \frac{\Delta R}{R}\right]^{1/2} - 1, \tag{3.67}$$

where V is the calm water speed and R is the calm water resistance.

Table 3.12. *Values of correction factor α*

C_B	Condition	Correction factor α
0.55	Normal	$1.7 - 1.4Fr - 7.4(Fr)^2$
0.60	Normal	$2.2 - 2.5Fr - 9.7(Fr)^2$
0.65	Normal	$2.6 - 3.7Fr - 11.6(Fr)^2$
0.70	Normal	$3.1 - 5.3Fr - 12.4(Fr)^2$
0.75	Laden or normal	$2.4 - 10.6Fr - 9.5(Fr)^2$
0.80	Laden or normal	$2.6 - 13.1Fr - 15.1(Fr)^2$
0.85	Laden or normal	$3.1 - 18.7Fr + 28.0(Fr)^2$
0.75	Ballast	$2.6 - 12.5Fr - 13.5(Fr)^2$
0.80	Ballast	$3.0 - 16.3Fr - 21.6(Fr)^2$
0.85	Ballast	$3.4 - 20.9Fr + 31.8.4(Fr)^2$

Table 3.13. *Comparison of Aertssen and Townsin–Kwon formulae*

Beaufort number BN	Aertssen $\Delta V/V$ (%)	Townsin–Kwon $\Delta V/V$ (%)
5	6.1	5.4
6	11.9	9.7
7	20.5	19.4
8	34.4	39.5

For small changes, power and thrust remain reasonably constant. Such an equation has typically been used to develop approximate formulae such as Equations (3.63, 3.64, 3.65). For larger resistance increases, and for a more correct interpretation, changes in propeller efficiency should also be taken into account. With such increases in resistance, hence a required increase in thrust at a particular speed, the propeller is clearly working off-design. Off-design propeller operation is discussed in Chapter 13. Taking the changes in propeller efficiency into account leads to the following relationship, van Berlekom [3.49], Townsin *et al.* [3.79]:

$$\frac{\Delta P}{P} = \frac{\Delta R/R}{1 + \Delta \eta_0/\eta_0} - 1, \quad (3.68)$$

where $\Delta \eta_0$ is the change in propeller efficiency η_0 due to change in propeller loading.

3.2.4.4 Conversion from Speed Loss to Power Increase

An approximate conversion from speed loss ΔV at a constant power to a power increase ΔP at a constant speed may be made as follows, using the assumption that power P varies as V^3, Figure 3.38:

$$V_1 = V_S\left(1 - \frac{\Delta V}{V_S}\right)$$

$$\frac{\text{New } P}{\text{Old } P} = \frac{V_S^3}{V_1^3} = \frac{V_S^3}{V_S^3\left(1 - \frac{\Delta V}{V_S}\right)^3} = \frac{1}{\left(1 - \frac{\Delta V}{V_S}\right)^3},$$

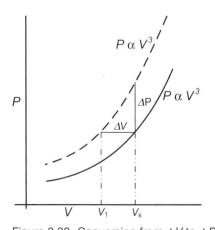

Figure 3.38. Conversion from ΔV to ΔP.

then

$$\frac{\Delta P}{P} = \frac{1}{\left(1 - \dfrac{\Delta V}{V_S}\right)^3} - 1 \qquad (3.69)$$

also

$$\frac{\Delta V}{V} = 1 - \sqrt[3]{\frac{1}{1 + \dfrac{\Delta P}{P}}}. \qquad (3.70)$$

Townsin and Kwan [3.78] derive the following approximate conversion:

$$\frac{\Delta P}{P} = (n + 1)\frac{\Delta V}{V}, \qquad (3.71)$$

where $\Delta P / P$ has been derived from Equation (3.68) and n has typical values, as follows:

VLCC laden $n = 1.91$
VLCC ballast $n = 2.40$
Container $n = 2.16$

Comparison of Equations (3.69) and (3.71) with actual data would suggest that Equation (3.71) underestimates the power increase at higher BN. For practical purposes, either Equation (3.69) or (3.71) may be applied.

3.2.4.5 Weighted Assessment of Average Increase in Power

It should be noted that in order to assess correctly the influences of weather on a certain route, power increases and/or speed losses should be judged in relation to the frequency with which the wave conditions occur (i.e. the occurrence of Beaufort number, BN, or significant wave height, $H_{1/3}$), Figure 3.39. Such weather conditions for different parts of the world may be obtained from [3.84], or the updated version [3.85]. Then, the weighted average power increase = Σ (power increase × frequency) = Σ ($\Delta P \times \sigma$).

Satchwell [3.77] applies this approach to climatic design and weather routeing. This approach should also be used when assessing the necessary power margin for a ship operating on a particular route.

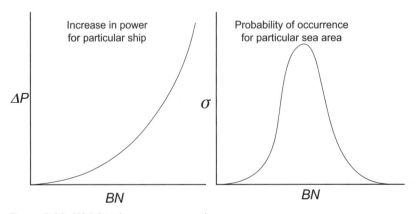

Figure 3.39. Weighted average power increase.

Table 3.14. *Weighted average power*

Beaufort number BN	Increase in power ΔP	Wave occurrence σ	$\Delta P \times \sigma$
0	0	0	0
1	1	0.075	0.075
2	2	0.165	0.330
3	5	0.235	1.175
4	15	0.260	3.900
5	32	0.205	6.560
6	85	0.060	5.100
7	200	0	0
		$\Sigma \, \Delta P \times \sigma$	17.14

A numerical example of such an approach is given in Table 3.14. Here, the power increase is taken from Table 3.10, but could have been derived using Equations (3.61) to (3.65) for a particular ship. The sea conditions for a particular sea area may typically be derived from [3.85]. It is noted that the average, or mean, power increase in this particular example is 17.14% which would therefore be a suitable power margin for a ship operating solely in this sea area. Margins are discussed further in Section 3.2.5.

3.2.5 Service Power Margins

3.2.5.1 Background

The basic ship power estimate will entail the estimation of the power to drive the ship at the required speed with a clean hull and propeller in calm water. In service, the hull will roughen and foul, leading to an increase in resistance and power, and the ship will encounter wind and waves, also leading to an increase in resistance. Some increase in resistance will occur from steering and coursekeeping, but this is likely to be relatively small. An increase in resistance will occur if the vessel has to operate in a restricted water depth; this would need to be taken into account if the vessel has to operate regularly in such conditions. The effects of operating in shallow water are discussed in Chapter 6. Thus, in order to maintain speed in service, a margin must be added to the basic clean hull calm water power, allowing the total installed propulsive power to be estimated. Margins and their estimation have been reviewed by ITTC [3.86], [3.63].

3.2.5.2 Design Data

ROUGHNESS AND FOULING. As discussed in Section 3.2.3, because of changes and ongoing developments in antifouling coatings, it is difficult to place a precise figure on the resistance increases due to roughness and fouling. Rate of fouling will also depend very much on factors such as the area of operation of the ship, time in port, local sea temperatures and pollution. Based on the voyage data described in Section 3.2.3, it might be acceptable to assume an annual increase in frictional resistance of say 10%, or an increase in total resistance and power of about 5%. If the hull were to be cleaned say every two years, then an assumed margin for roughness and fouling would be 10%.

WIND AND WAVES. The increase in power due to wind and waves will vary widely, depending on the sea area of operation. For example, the weather margin for a ship operating solely in the Mediterranean might be 10%, whilst to maintain a speed trans-Atlantic westbound, a ship might need a weather margin of 30% or higher. As illustrated in Section 3.2.4, a rigorous weighted approach is to apply the likely power increase for a particular ship to the wave conditions in the anticipated sea area of operation. This will typically lead to a power increase of 10%–30%. Methods of estimating added resistance in waves are reviewed in ITTC [3.80].

TOTAL. Based on the foregoing discussions, an approximate overall total margin will be the sum of the roughness-fouling and wind–wave components, typically say 10% plus 15%, leading to a total margin of 25%. This is applicable to approximate preliminary estimates. It is clear that this figure might be significantly larger or smaller depending on the frequency of underwater hull and propeller cleaning, the sea areas in which the ship will actually operate and the weather conditions it is likely to encounter.

3.2.5.3 Engine Operation Margin

The engine operation margin describes the mechanical and thermodynamic reserve of power for the economical operation of the main propulsion engine(s) with respect to reasonably low fuel and maintenance costs. Thus, an operator may run the engine(s) up to the continuous service rating (CSR), which is say 10% below the maximum continuous rating (MCR). Even bigger margins may be employed by the operator, see Woodyard [3.87] or Molland [3.88]. Some margin on revolutions will also be made to allow for changes in the power–rpm relationship in service, see propeller-engine matching, Chapter 13.

REFERENCES (CHAPTER 3)

3.1 Couser, P.R., Wellicome, J.F. and Molland, A.F. Experimental measurement of sideforce and induced drag on catamaran demihulls. *International Shipbuilding Progress*, Vol. 45(443), 1998, pp. 225–235.

3.2 Lackenby, H. An investigation into the nature and interdependence of the components of ship resistance. *Transactions of the Royal Institution of Naval Architects*, Vol. 107, 1965, pp. 474–501.

3.3 Insel, M. and Molland, A.F. An investigation into the resistance components of high speed displacement catamarans. *Transactions of the Royal Institution of Naval Architects*, Vol. 134, 1992, pp. 1–20.

3.4 Molland, A.F. and Utama, I.K.A.P. Experimental and numerical investigations into the drag characteristics of a pair of ellipsoids in close proximity. *Proceedings of the Institution of Mechanical Engineers. Journal of Engineering for the Maritime Environment*, Vol. 216, Part M, 2002, pp. 107–115.

3.5 Savitsky, D., DeLorme, M.F. and Datla, R. Inclusion of whisker spray drag in performance prediction method for high-speed planing hulls. *Marine Technology*, SNAME, Vol. 44, No. 1, 2007, pp. 35–36.

3.6 Clayton, B.R. and Bishop, R.E.D. *Mechanics of Marine Vehicles*. E. and F.N. Spon, London, 1982.

3.7 Faltinsen, O.M. *Hydrodynamics of High-Speed Marine Vehicles*. Cambridge University Press, Cambridge, UK, 2005.

3.8 Claughton, A.R., Wellicome, J.R. and Shenoi, R.A. (eds). *Sailing Yacht Design*. Vol. 1 *Theory*, Vol. 2 *Practice*. The University of Southampton, Southampton, UK, 2006.

3.9 Larsson, L. and Eliasson, R. *Principles of Yacht Design*. 3rd Edition. Adlard Coles Nautical, London, 2007.

3.10 Crewe, P.R. and Eggington, W.J. The hovercraft – a new concept in marine transport. *Transactions of the Royal Institution of Naval Architects*, Vol. 102, 1960, pp. 315–365.

3.11 Crewe, P.R. The hydrofoil boat: its history and future prospects. *Transactions of the Royal Institution of Naval Architects*, Vol. 100, 1958, pp. 329–373.

3.12 Yun, L. and Bliault, A. *Theory and Design of Air Cushion Craft*. Elsevier Butterworth-Heinemann, Oxford, UK, 2005.

3.13 Froude, R.E. On the 'constant' system of notation of results of experiments on models used at the Admiralty Experiment Works, *Transactions of the Royal Institution of Naval Architects*, Vol. 29, 1888, pp. 304–318.

3.14 Taylor, D.W. *The Speed and Power of Ships*. U.S. Government Printing Office, Washington, DC, 1943.

3.15 Lackenby, H. On the presentation of ship resistance data. *Transactions of the Royal Institution of Naval Architects*, Vol. 96, 1954, pp. 471–497.

3.16 Telfer, E.V. The design presentation of ship model resistance data. *Transactions of the North East Coast Institution of Engineers and Shipbuilders*, Vol. 79, 1962–1963, pp. 357–390.

3.17 Doust, D.J. Optimised trawler forms. *Transactions of the North East Coast Institution of Engineers and Shipbuilders*, Vol. 79, 1962–1963, pp. 95–136.

3.18 Sabit, A.S. Regression analysis of the resistance results of the BSRA Series. *International Shipbuilding Progress*, Vol. 18, No. 197, January 1971, pp. 3–17.

3.19 Froude, W. Experiments on the surface-friction experienced by a plane moving through water. 42nd Report of the British Association for the Advancement of Science, Brighton, 1872.

3.20 ITTC. Recommended procedure for the resistance test. Procedure 7.5–0.2-0.2-0.1 Revision 01, 2002.

3.21 ITTC. Recommended procedure for ship models. Procedure 7.5-01-01-01, Revision 01, 2002.

3.22 Hughes, G. and Allan, J.F. Turbulence stimulation on ship models. *Transactions of the Society of Naval Architects and Marine Engineers*, Vol. 59, 1951, pp. 281–314.

3.23 Barnaby, K.C. *Basic Naval Architecture*. Hutchinson, London, 1963.

3.24 Molland, A.F., Wellicome, J.F. and Couser, P.R. Resistance experiments on a systematic series of high speed displacement catamaran forms: Variation of length-displacement ratio and breadth-draught ratio. University of Southampton, Ship Science Report No. 71, 1994.

3.25 Hughes, G. Tank boundary effects on model resistance. *Transactions of the Royal Institution of Naval Architects*, Vol. 103, 1961, pp. 421–440.

3.26 Scott, J.R. A blockage corrector. *Transactions of the Royal Institution of Naval Architects*, Vol. 108, 1966, pp. 153–163.

3.27 Scott, J.R. Blockage correction at sub-critical speeds. *Transactions of the Royal Institution of Naval Architects*, Vol. 118, 1976, pp. 169–179.

3.28 Coleman, H.W. and Steele, W.G. *Experimentation and Uncertainty Analysis for Engineers*. 2nd Edition. Wiley, New York, 1999.

3.29 ITTC. Final report of The Specialist Committee on Model Tests of High Speed Marine Vehicles. *Proceedings of the 22nd International Towing Tank Conference*, Seoul and Shanghai, 1999.

3.30 Renilson, M. Predicting the hydrodynamic performance of very high-speed craft: A note on some of the problems. *Transactions of the Royal Institution of Naval Architects*, Vol. 149, 2007, pp. 15–22.

3.31 Duncan, W.J., Thom, A.S. and Young, A.D. *Mechanics of Fluids*. Edward Arnold, Port Melbourne, Australia, 1974.

3.32 Massey, B.S. and Ward-Smith J. *Mechanics of Fluids*. 8th Edition. Taylor and Francis, London, 2006.

3.33 Froude, W. Report to the Lords Commissioners of the Admiralty on experiments for the determination of the frictional resistance of water on a surface, under various conditions, performed at Chelston Cross, under the Authority of their Lordships. 44th Report by the British Association for the Advancement of Science, Belfast, 1874.

3.34 Molland, A.F. and Turnock, S.R. *Marine Rudders and Control Surfaces*. Butterworth-Heinemann, Oxford, UK, 2007.

3.35 Allan, J.F. Some results of scale effect experiments on a twin-screw hull series. *Transactions of the Institute of Engineers and Shipbuilders in Scotland*, Vol. 93, 1949/50, pp. 353–381.

3.36 Lackenby, H. BSRA resistance experiments on the Lucy Ashton. Part III. The ship model correlation for the shaft appendage conditions. *Transactions of the Royal Institution of Naval Architects*, Vol. 97, 1955, pp. 109–166.

3.37 Hoerner, S.F. *Fluid-Dynamic Drag*. Published by the Author. Washington, DC, 1965.

3.38 Mandel, P. Some hydrodynamic aspects of appendage design. *Transactions of the Society of Naval Architects and Marine Engineers*, Vol. 61, 1953, pp. 464–515.

3.39 Peck, R.W. The determination of appendage resistance of surface ships. AEW Technical Memorandum, 76020, 1976.

3.40 Kirkman, K.L. and Kloetzli, J.N. Scaling problems of model appendages, *Proceedings of the 19th ATTC*, Ann Arbor, Michigan, 1981.

3.41 Tulin, M.P. Supercavitating flow past foil and struts. *Proceedings of Symposium on Cavitation in Hydrodynamics*. NPL, 1955.

3.42 Rutgersson, O. Propeller-hull interaction problems on high-speed ships: on the influence of cavitation. *Symposium on Small Fast Warships and Security Vessels*. The Royal Institution of Naval Architects, London, 1982.

3.43 Holtrop, J. and Mennen, G.G.J. An approximate power prediction method. *International Shipbuilding Progress*, Vol. 29, No. 335, July 1982, pp. 166–170.

3.44 van Berlekom, W.B., Trägårdh, P. and Dellhag, A. Large tankers – wind coefficients and speed loss due to wind and waves. *Transactions of the Royal Institution of Naval Architects*, Vol. 117, 1975, pp. 41–58.

3.45 Shearer, K.D.A. and Lynn, W.M. Wind tunnel tests on models of merchant ships. *Transactions of the North East Coast Institution of Engineers and Shipbuilders*, Vol. 76, 1960, pp. 229–266.

3.46 White, G.P. Wind resistance – suggested procedure for correction of ship trial results. NPL TM116, 1966.

3.47 Gould, R.W.F. The Estimation of wind loadings on ship superstructures. *RINA Marine Technology Monograph* 8, 1982.

3.48 Isherwood, R.M. Wind resistance of merchant ships. *Transactions of the Royal Institution of Naval Architects*, Vol. 115, 1973, pp. 327–338.

3.49 van Berlekom, W.B. Wind forces on modern ship forms–effects on performance. *Transactions of the North East Coast Institute of Engineers and Shipbuilders*, Vol. 97, No. 4, 1981, pp. 123–134.

3.50 Blendermann, W. Estimation of wind loads on ships in wind with a strong gradient. *Proceedings of the 14th International Conference on Offshore Mechanics and Artic Engineering (OMAE)*, New York, ASME, Vol. 1-A, 1995, pp. 271–277.

3.51 Molland, A.F. and Barbeau, T.-E. An investigation into the aerodynamic drag on the superstructures of fast catamarans. *Transactions of the Royal Institution of Naval Architects*, Vol. 145, 2003, pp. 31–43.

3.52 Reddy, K.R., Tooletto, R. and Jones, K.R.W. Numerical simulation of ship airwake. *Computers and Fluids*, Vol. 29, 2000.

3.53 Sezer-Uzol, N., Sharma, A. and Long, L.N. Computational fluid dynamics simulations of ship airwake. *Proceedings of the Institution of Mechanical Engineers. Journal of Aerospace Engineering*, Vol. 219, Part G, 2005, pp. 369–392.

3.54 Wakefield, N.H., Newman, S.J. and Wilson, P.A. Helicopter flight around a ship's superstructure. *Proceedings of the Institution of Mechanical Engineers, Journal of Aerospace Engineering*, Vol. 216, Part G, 2002, pp. 13–28.

3.55 Moat, I, Yelland, M., Pascal, R.W. and Molland, A.F. Quantifying the airflow distortion over merchant ships. Part I. Validation of a CFD Model. *Journal of Atmospheric and Oceanic Technology*, Vol. 23, 2006, pp. 341–350.

3.56 Moat, I, Yelland, M., Pascal, R.W. and Molland, A.F. Quantifying the airflow distortion over merchant ships. Part II. Application of the model results. *Journal of Atmospheric and Oceanic Technology*, Vol. 23, 2006, pp. 351–360.

3.57 ITTC. Report of Resistance Committee. 25th ITTC, Fukuoka, Japan, 2008.

3.58 Hucho, W.H. (ed.) *Aerodynamics of Road Vehicles*. 4th Edition. Society of Automotive Engineers, Inc., Warrendale, PA, USA, 1998.

3.59 Anonymous. Emissions are a drag. *The Naval Architect*, RINA, London, January 2009, pp. 28–31.

3.60 Grigson, C.W.B. The drag coefficients of a range of ship surfaces. *Transactions of the Royal Institution of Naval Architects*, Vol. 123, 1981, pp. 195–208.

3.61 Candries, M. and Atlar, M. On the drag and roughness characteristics of antifoulings. *Transactions of the Royal Institution of Naval Architects*, Vol. 145, 2003, pp. 107–132.

3.62 Townsin, R.L. The ITTC line–its genesis and correlation allowance. *The Naval Architect*, RINA, London, September 1985, pp. E359–E362.

3.63 ITTC. Report of the Specialist Committee on Powering Performance Prediction, *Proceedings of the 25th International Towing Tank Conference*, Vol. 2, Fukuoka, Japan, 2008.

3.64 Townsin, R.L., Byrne, D., Milne, A. and Svensen, T. Speed, power and roughness – the economics of outer bottom maintenance. *Transactions of the Royal Institution of Naval Architects*, Vol. 122, 1980, pp. 459–483.

3.65 Aertssen, G. Service-performance and seakeeping trials on MV Lukuga. *Transactions of the Royal Institution of Naval Architects*, Vol. 105, 1963, pp. 293–335.

3.66 Aertssen, G. Service-performance and seakeeping trials on MV Jordaens. *Transactions of the Royal Institution of Naval Architects*, Vol. 108, 1966, pp. 305–343.

3.67 Aertssen, G., Ferdinande, V. and de Lembre, R. Service-performance and seakeeping trials on a stern trawler. *Transactions of the North East Coast Institution of Engineers and Shipbuilders*, Vol. 83, 1966–1967, pp. 13–27.

3.68 Aertssen, G. and Van Sluys, M.F. Service-performance and seakeeping trials on a large containership. *Transactions of the Royal Institution of Naval Architects*, Vol. 114, 1972, pp. 429–447.

3.69 Aertssen, G. Service performance and trials at sea. App.V Performance Committee, 12th ITTC, 1969.

3.70 Townsin, R.L., Moss, B, Wynne, J.B. and Whyte, I.M. Monitoring the speed performance of ships. *Transactions of the North East Coast Institution of Engineers and Shipbuilders*, Vol. 91, 1975, pp. 159–178.

3.71 Carlton, J.S. *Marine Propellers and Propulsion*. 2nd Edition. Butterworth-Heinemann, Oxford, UK, 2007.

3.72 Atlar, M., Glover, E.J., Candries, M., Mutton, R. and Anderson, C.D. The effect of foul release coating on propeller performance. *Proceedings of 2nd International Conference on Marine Research for Environmental Sustainability ENSUS 2002*, University of Newcastle upon Tyne, UK, 2002.

3.73 Atlar, M., Glover, E.J., Mutton, R. and Anderson, C.D. Calculation of the effects of new generation coatings on high speed propeller performance. *Proceedings of 2nd International Warship Cathodic Protection Symposium and Equipment Exhibition*. Cranfield University, Shrivenham, UK, 2003.

3.74 Okuno, T., Lewkowicz, A.K. and Nicholson, K. Roughness effects on a slender ship hull. *Second International Symposium on Ship Viscous Resistance*, SSPA, Gothenburg, 1985, pp. 20.1–20.27.

3.75 Lewthwaite, J.C., Molland, A.F. and Thomas, K.W. An investigation into the variation of ship skin frictional resistance with fouling. *Transactions of the Royal Institution of Naval Architects*, Vol. 127, 1985, pp. 269–284.

3.76 Cutland, R.S. Velocity measurements in close proximity to a ship's hull. *Transactions of the North East Coast Institution of Engineers and Shipbuilders*. Vol. 74, 1958, pp. 341–356.

3.77 Satchwell, C.J. Windship technology and its application to motor ships. *Transactions of the Royal Institution of Naval Architects*, Vol. 131, 1989, pp. 105–120.

3.78 Townsin, R.L. and Kwon, Y.J. Approximate formulae for the speed loss due to added resistance in wind and waves. *Transactions of the Royal Institution of Naval Architects*, Vol. 125, 1983, pp. 199–207.

3.79 Townsin, R.L., Kwon, Y.J., Baree, M.S. and Kim, D.Y. Estimating the influence of weather on ship performance. *Transactions of the Royal Institution of Naval Architects*, Vol. 135, 1993, pp. 191–209.

3.80 ITTC. Report of the Seakeeping Committee, *Proceedings of the 25th International Towing Tank Conference*, Vol. 1, Fukuoka, Japan, 2008.

3.81 Aertssen, G. The effect of weather on two classes of container ship in the North Atlantic. *The Naval Architect*, RINA, London, January 1975, p. 11.

3.82 Kwon, Y.J. Estimating the effect of wind and waves on ship speed and performance. *The Naval Architect*, RINA, London, September 2000.

3.83 Kwon, Y.J. Speed loss due to added resistance in wind and waves. *The Naval Architect*, RINA, London, March 2008, pp. 14–16.

3.84 Hogben, N. and Lumb, F.E. *Ocean Wave Statistics*. Her Majesty's Stationery Office, London, 1967.

3.85 Hogben, N., Dacunha, N.M.C. and Oliver, G.F. Global Wave Statistics. Compiled and edited by British Maritime Technology, Unwin Brothers, Old Woking, UK, 1985.

3.86 ITTC. Report of The Specialist Committee on Powering Performance and Prediction. *Proceedings of the 24th International Towing Tank Conference*, Vol. 2, Edinburgh, UK, 2005.

3.87 Woodyard, D.F. *Pounder's Marine Diesel Engines and Gas Turbines*. 8th Edition. Butterworth-Heinemann, Oxford, UK, 2004.

3.88 Molland, A.F. (ed.) *The Maritime Engineering Reference Book*. Butterworth-Heinemann, Oxford, UK, 2008.

4 Model-Ship Extrapolation

4.1 Practical Scaling Methods

When predicting ship power by the use of model tests, the resistance test results have to be scaled, or extrapolated, from model to ship. There are two main methods of extrapolation, one due to Froude which was introduced in the 1870s [4.1–4.3] and the other due to Hughes introduced in the 1950s [4.4] and later adopted by the International Towing Tank Conference (ITTC).

4.1.1 Traditional Approach: Froude

The basis of this approach is described in Section 3.1.6 and is summarised as follows:

$$C_T = C_F + C_R \qquad (4.1)$$

where C_F is for an equivalent flat plate of a length equal to the model or ship, C_R is the residuary resistance and is derived from the model test as:

$$C_R = C_{Tm} - C_{Fm}.$$

For the same Froude number, and following Froude's law,

$$C_{Rs} = C_{Rm}$$

and

$$C_{Ts} = C_{Fs} + C_{Rs} = C_{Fs} + C_{Rm} = C_{Fs} + [C_{Tm} - C_{Fm}]$$

i.e.

$$C_{Ts} = C_{Tm} - (C_{Fm} - C_{Fs}). \qquad (4.2)$$

This traditional approach is shown schematically in Figure 4.1, with $C_{Rs} = C_{Rm}$ at the ship Re corresponding to the same (model = ship) Froude number. $(C_{Fm} - C_{Fs})$ is a skin friction correction to C_{Tm} for the model. The model is not run at the correct Reynolds number (i.e. the model Re is much smaller than that for the ship, Figure 4.1) and consequently the skin friction coefficient for the model is higher

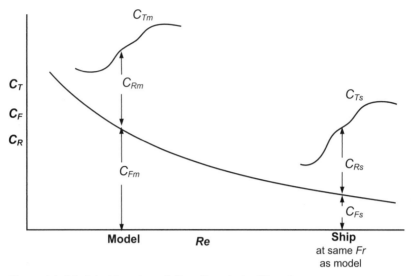

Figure 4.1. Model-ship extrapolation: Froude traditional.

than that for the ship. The method is still used by some naval architects, but it tends to overestimate the power for very large ships.

It should be noted that the C_F values developed by Froude were not explicitly defined in terms of Re, as suggested in Figure 4.1. This is discussed further in Section 4.3.

4.1.2 Form Factor Approach: Hughes

Hughes proposed taking form effect into account in the extrapolation process. The basis of the approach is summarised as follows:

$$C_T = (1 + k)C_F + C_W \qquad (4.3)$$

or

$$C_T = C_V + C_W, \qquad (4.4)$$

where

$$C_V = (1 + k)C_F,$$

and $(1 + k)$ is a form factor which depends on hull form, C_F is the skin friction coefficient based on flat plate results, C_V is a viscous coefficient taking account of both skin friction and viscous pressure resistance and C_W is the wave resistance coefficient. The method is shown schematically in Figure 4.2.

On the basis of Froude's law,

$$C_{Ws} = C_{Wm}$$

and

$$C_{Ts} = C_{Tm} - (1 + k)(C_{Fm} - C_{Fs}). \qquad (4.5)$$

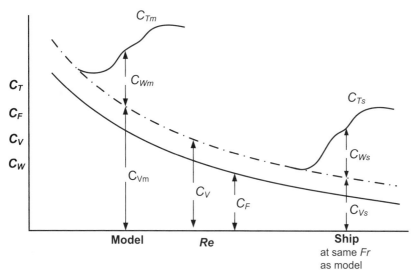

Figure 4.2. Model-ship extrapolation: form factor approach.

This method is recommended by ITTC and is the one adopted by most naval architects. A form factor approach may not be applied for some high-speed craft and for yachts.

The form factor $(1 + k)$ depends on the hull form and may be derived from low-speed tests when, at low Fr, wave resistance C_W tends to zero and $(1 + k) = C_{Tm}/C_{Fm}$. This and other methods of obtaining the form factor are described in Section 4.4.

It is worth emphasising the fundamental difference between the two scaling methods described. Froude assumes that all resistance in excess of C_F (the residuary resistance C_R) scales according to Froude's law, that is, as displacement at the same Froude number. This is not physically correct because the viscous pressure (form) drag included within C_R should scale according to Reynolds' law. Hughes assumes that the total viscous resistance (friction and form) scales according to Reynolds' law. This also is not entirely correct as the viscous resistance interferes with the wave resistance which is Froude number dependent. The form factor method (Hughes) is, however, much closer to the actual physical breakdown of components than Froude's approach and is the method now generally adopted.

It is important to note that both the Froude and form factor methods rely very heavily on the *level* and *slope* of the chosen skin friction, C_F, line. Alternative skin friction lines are discussed in Section 4.3.

4.2 Geosim Series

In order to identify f_1 and f_2 in Equation (3.26), $C_T = f_1(Re) + f_2(Fr)$, from measurements of *total resistance only*, several experimenters have run resistance tests for a range of differently sized models of the same geometric form. Telfer [4.5–4.7] coined the term 'Geosim Series', or 'Geosims' for such a series of models. The results of such a series of tests, plotted on a Reynolds number base, would appear as shown schematically in Figure 4.3.

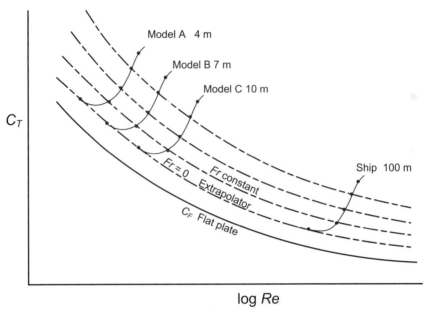

Figure 4.3. Schematic layout of Geosim test results.

The successive sets of C_T measurements at increasing Reynolds numbers show successively lower C_T values at corresponding Froude numbers, the individual resistance curves being approximately the same amount above the resistance curve for a flat plate of the same wetted area. In other words, the ship resistance is estimated directly from models, without separation into frictional and residuary resistance. Lines drawn through the same Froude numbers should be parallel with the friction line. The slope of the extrapolator can be determined experimentally from the models, as shown schematically in Figure 4.3. However, whilst Geosim tests are valuable for research work, such as validating single model tests, they tend not to be cost-effective for routine commercial testing. Examples of actual Geosim tests for the *Simon Bolivar* and *Lucy Ashton* families of models are shown in Figures 4.4 and 4.5 [4.8, 4.9, 4.10].

4.3 Flat Plate Friction Formulae

The level and slope of the skin friction line is fundamental to the extrapolation of resistance data from model to ship, as discussed in Section 4.1. The following sections outline the principal skin friction lines employed in ship resistance work.

4.3.1 Froude Experiments

The first systematic experiments to determine frictional resistance in water of thin flat planks were carried out in the late 1860s by W. Froude. He used planks 19 in deep, 3/16 in thick and lengths of 2 to 50 ft, coated in different ways [4.1, 4.2] A mechanical dynamometer was used to measure the total model resistance, Barnaby [4.11], using speeds from 0 to 800 ft/min (4 m/s).

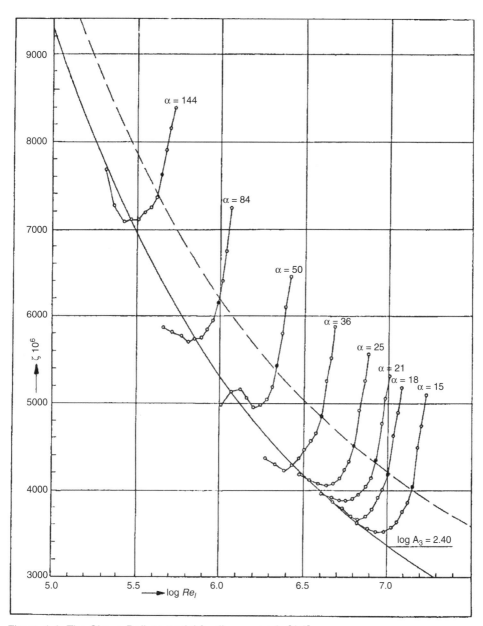

Figure 4.4. The *Simon Bolivar* model family, α = scale [4.8].

Froude found that he could express the results in the empirical formula

$$R = f \cdot S \cdot V^{n}. \tag{4.6}$$

The coefficient f and index n were found to vary for both type and length of surface. The original findings are summarised as follows:

1. The coefficient f decreased with increasing plank length, with the exception of very short lengths.

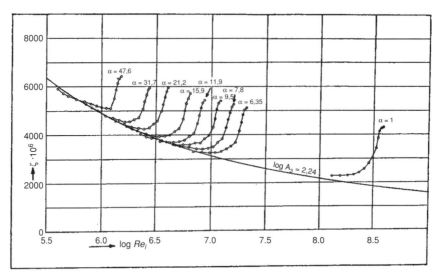

Figure 4.5. The *Lucy Ashton* model family, α = scale [4.8].

2. The index n is appreciably less than 2 with the exception of rough surfaces when it approaches 2 (is >2 for a very short/very smooth surface).
3. The degree of roughness of the surface has a marked influence on the magnitude of f.

Froude summarised his values for f and n for varnish, paraffin wax, fine sand and coarse sand for plank lengths up to 50 ft (for >50 ft Froude suggested using f for 49–50 ft).

R. E. Froude (son of W. Froude) re-examined the results obtained by his father and, together with data from other experiments, considered that the results of planks having surfaces corresponding to those of clean ship hulls or to paraffin wax models could be expressed as the following:

$$R_F = f \cdot S \cdot V^{1.825},\qquad(4.7)$$

with associated table of f values, see Table 4.1.

If Froude's data are plotted on a Reynolds Number base, then the results appear as follows:

$$R = f \cdot S \cdot V^{1.825}.$$
$$C_F = R/\tfrac{1}{2}\rho S V^2 = 2 \cdot f \cdot V^{-0.175}/\rho$$
$$= 2 \cdot f \cdot V^{-0.175} \cdot Re^{-0.175}/\rho L^{-0.175},$$

then

$$C_F = f' \cdot Re^{-0.175},$$

Table 4.1. *R.E. Froude's skin friction f values*

Length (m)	f	Length (m)	f	Length (m)	f
2.0	1.966	11	1.589	40	1.464
2.5	1.913	12	1.577	45	1.459
3.0	1.867	13	1.566	50	1.454
3.5	1.826	14	1.556	60	1.447
4.0	1.791	15	1.547	70	1.441
4.5	1.761	16	1.539	80	1.437
5.0	1.736	17	1.532	90	1.432
5.5	1.715	18	1.526	100	1.428
6.0	1.696	19	1.520	120	1.421
6.5	1.681	20	1.515	140	1.415
7.0	1.667	22	1.506	160	1.410
7.5	1.654	24	1.499	180	1.404
8.0	1.643	26	1.492	200	1.399
8.5	1.632	28	1.487	250	1.389
9.0	1.622	30	1.482	300	1.380
9.5	1.613	35	1.472	350	1.373
10.0	1.604				

where f' depends on length. According to the data, f' increases with length as seen in Figure 4.6 [4.12].

On dimensional grounds this is not admissible since C_F should be a function of Re only. It should be noted that Froude was unaware of dimensional analysis, or of the work of Reynolds [4.13].

Although it was not recognised at the time, the Froude data exhibited three boundary layer characteristics. Referring to the classical work of Nikuradse,

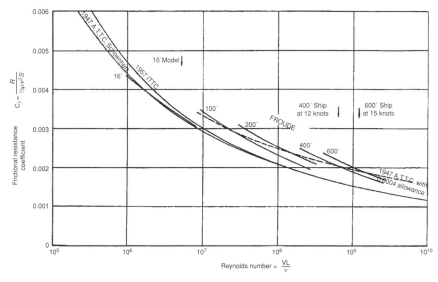

Figure 4.6. Comparison of different friction formulae.

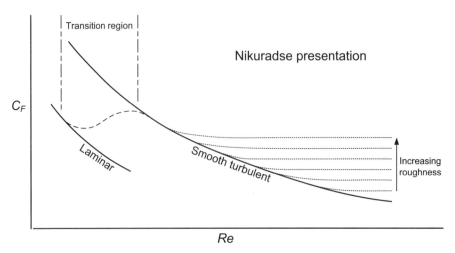

Figure 4.7. Effects of laminar flow and roughness on C_F.

Figure 4.7, and examining the Froude results in terms of Re, the following characteristics are evident.

(i) The Froude results for lengths < 20 ft are influenced by laminar or transitional flow; Froude had recorded anomalies.
(ii) At high Re, C_F for rough planks becomes constant independent of Re at a level that depends on roughness, Figure 4.7; C_F constant implies $R \propto V^2$ as Froude observed.
(iii) Along sharp edges of the plank, the boundary layer is thinner; hence C_F is higher. Hence, for the constant plank depth used by Froude, the edge effect is more marked with an increase in plank length; hence f' increases with length.

The Froude values of f, hence C_F, for higher ship length (high Re), lie well above the smooth turbulent line, Figure 4.6. The Froude data are satisfactory up to about 500 ft (152 m) ship length, and are still in use, but are obviously in error for large ships when the power by Froude is overestimated (by up to 15%). Froude f values are listed in Table 4.1, where L is waterline length (m) and units in Equation (4.7) are as follows: V is speed (m/s), S is wetted area (m^2) and R_F is frictional resistance (N).

A reasonable approximation (within 1.5%) to the table of f values is

$$f = 1.38 + 9.4/[8.8 + (L \times 3.28)] \ (L \text{ in metres}). \tag{4.8}$$

R. E. Froude also established the circular non-dimensional notation, [4.14], together with the use of 'O' values for the skin friction correction, see Sections 3.1 and 10.3.

4.3.2 Schoenherr Formula

In the early 1920s Von Karman deduced a friction law for flat plates based on a two-dimensional analysis of turbulent boundary layers. He produced a theoretical

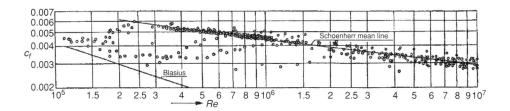

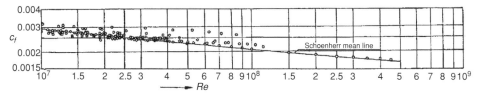

Figure 4.8. The Schoenherr mean line for C_F.

'smooth turbulent' friction law of the following form:

$$1/\sqrt{C_F} = A + B \, \mathrm{Log}\,(Re \cdot C_F), \tag{4.9}$$

where A and B were two undetermined constants. Following the publication of this formula, Schoenherr replotted all the available experimental data from plank experiments both in air and water and attempted to determine the constants A and B to suit the available data, [4.15]. He determined the following formula:

$$1/\sqrt{C_F} = 4.13 \log_{10}\,(Re \cdot C_F) \tag{4.10}$$

The use of this formula provides a better basis for extrapolating beyond the range of the experimental data than does the Froude method simply because of the theoretical basis behind the formula. The data published by Schoenherr as a basis for his line are shown in Figure 4.8 from [4.16]. The data show a fair amount of scatter and clearly include both transition and edge effects, and the mean line shown must be judged in this light. The Schoenherr line was adopted by the American Towing Tank Conference (ATTC) in 1947. When using the Schoenherr line for model-ship extrapolation, it has been common practice to add a roughness allowance $\Delta C_F = 0.0004$ to the ship value, see Figure 4.6.

The Schoenherr formula is not very convenient to use since C_F is not explicitly defined for a given Re. In order to determine C_F for a given Re, it is necessary to assume a range of C_F, calculate the corresponding Re and then interpolate. Such iterations are, however, simple to carry out using a computer or spreadsheet. A reasonable fit to the Schoenherr line (within 1%) for preliminary power estimates is given in [4.17]

$$C_F = \frac{1}{(3.5 \log_{10} Re - 5.96)^2}. \tag{4.11}$$

Table 4.2. *Variation in* C_F *with Re*

Re	C_F	$\log_{10} Re$	$\log_{10} C_F$
10^5	8.3×10^{-3}	5	-2.06
10^9	1.53×10^{-3}	9	-2.83

4.3.3 The ITTC Formula

Several proposals for a more direct formula which approximates the Schoenherr values have been made. The Schoenherr formula (Equation (4.10)) can be expanded as follows:

$$1/\sqrt{C_F} = 4.13 \log_{10}(Re.C_F) = 4.13(\log_{10} Re + \log_{10} C_F). \qquad (4.12)$$

C_F and log C_F vary comparatively slowly with Re as shown in Table 4.2. Thus, a formula of the form

$$1/\sqrt{C_F} = A(\log_{10} Re - B)$$

may not be an unreasonable approximation with B assumed as 2. The formula can then be rewritten as:

$$C_F = \frac{A'}{(\log_{10} Re - 2)^2}. \qquad (4.13)$$

There are several variations of this formula type which are, essentially, approximations of the Schoenherr formula.

In 1957 the ITTC adopted one such formula for use as a 'correlation line' in powering calculations. It is termed the 'ITTC1957 model-ship correlation line'. This formula was based on a proposal by Hughes [4.4] for a two-dimensional line of the following form:

$$C_F = \frac{0.066}{(\log_{10} Re - 2.03)^2}. \qquad (4.14)$$

The ITTC1957 formula incorporates some three-dimensional friction effects and is defined as:

$$C_F = \frac{0.075}{(\log_{10} Re - 2)^2}. \qquad (4.15)$$

It is, in effect, the Hughes formula (Equation (4.14)) with a 12% form effect built in.

A comparison of the ITTC correlation line and the Schoenherr formula, Figure 4.6, indicates that the ITTC line agrees with the Schoenherr formula at ship Re values, but is above the Schoenherr formula at small Re values. This was deliberately built into the ITTC formula because experience with using the Schoenherr formula indicated that the smaller models were overestimating ship powers in comparison with identical tests with larger models.

At this point it should be emphasised that there is no pretence that these various formulae represent the drag of flat plates (bearing in mind the effects of roughness and edge conditions) and certainly not to claim that they represent the skin friction (tangential shear stress) resistance of an actual ship form, although they may be a tolerable approximation to the latter for most forms. These lines are used simply as *correlation lines* from which to judge the scaling allowance to be made between model and ship and between ships of different size.

The ITTC1978 powering prediction procedure (see Chapter 5) recommends the use of Equation (4.15), together with a form factor. The derivation of the form factor $(1 + k)$ is discussed in Section 4.4.

4.3.4 Other Proposals for Friction Lines

4.3.4.1 Grigson Formula

The most serious alternative to the Schoenherr and ITTC formulae is a proposal by Grigson [4.18], who argues the case for small corrections to the ITTC formula, Equation (4.15), at low and high Reynolds numbers. Grigson's proposal is as follows:

$$C_F = \left[0.93 + 0.1377(\log Re - 6.3)^2 - 0.06334(\log Re - 6.3)^4\right]$$
$$\times \frac{0.075}{(\log_{10} Re - 2)^2}, \tag{4.16}$$

for $1.5 \times 10^6 < Re < 2 \times 10^7$.

$$C_F = \left[1.032 + 0.02816(\log Re - 8) - 0.006273(\log Re - 8)^2\right]$$
$$\times \frac{0.075}{(\log_{10} Re - 2)^2}, \tag{4.17}$$

for $10^8 < Re < 4 \times 10^9$.

It seems to be agreed, in general, that the Grigson approach is physically more correct than the existing methods. However, the differences and improvements between it and the existing methods tend to be small enough for the test tank community not to adopt it for model-ship extrapolation purposes, ITTC [4.19, 4.20]. Grigson suggested further refinements to his approach in [4.21].

4.3.4.2 CFD Methods

Computational methods have been used to simulate a friction line. An example of such an approach is provided by Date and Turnock [4.22] who used a Reynolds averaged Navier–Stokes (RANS) solver to derive friction values over a plate for a range of speeds, and to develop a resistance correlation line. The formula produced was very close to the Schoenherr line. This work demonstrated the ability of computational fluid dynamics (CFD) to predict skin friction reasonably well, with the potential also to predict total viscous drag and form factors. This is discussed further in Chapter 9.

4.4 Derivation of Form Factor (1 + *k*)

It is clear from Equation (4.5) that the size of the form factor has a direct influence on the model to ship extrapolation process and the size of the ship resistance estimate. These changes occur because of the change in the proportion of viscous to wave resistance components, i.e. Re and Fr dependency. For example, when extrapolating model resistance to full scale, an increase in derived (or assumed) model $(1 + k)$ will result in a decrease in C_W and a decrease in the estimated full-scale resistance. Methods of estimating $(1 + k)$ include experimental, numerical and empirical.

4.4.1 Model Experiments

There are a number of model experiments that allow the form factor to be derived directly or indirectly. These are summarised as follows:

1. The model is tested at very low Fr until C_T runs parallel with C_F, Figure 4.9. In this case, C_W tends to zero and $(1 + k) = C_T/C_F$.
2. C_W is extrapolated back at low speeds. The procedure assumes that:

$$R_W \propto V^6 \quad \text{or} \quad C_W \propto R_W/V^2 \propto V^4$$

that is

$$C_W \propto Fr^4, \quad \text{or} \quad C_W = A Fr^4,$$

where A is a constant. Hence, from two measurements of C_T at *relatively* low speeds, and using $C_T = (1 + k) C_F + A Fr^4$, $(1 + k)$ can be found. Speeds as low as $Fr = 0.1 \sim 0.2$ are necessary for this method and a problem exists in that it is generally difficult to achieve accurate resistance measurements at such low speeds.

The methods described are attributable to Hughes. Prohaska [4.23] uses a similar technique but applies more data points to the equation as follows:

$$C_T/C_F = (1 + k) + A Fr^4/C_F, \tag{4.18}$$

where the intercept is $(1 + k)$, and the slope is A, Figure 4.10.

For full form vessels the points may not plot on a straight line and a power of Fr between 4 and 6 may be more appropriate.

A later ITTC recommendation as a modification to Prohaska is

$$C_T/C_F = (1 + k) + A Fr^n/C_F, \tag{4.19}$$

where n, A and k are derived from a least-squares approximation.

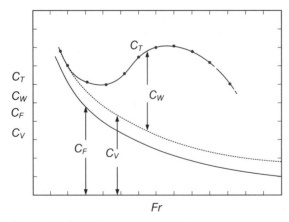

Figure 4.9. Resistance components.

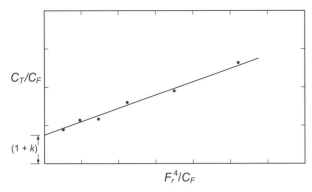

Figure 4.10. Prohaska plot.

3. $(1 + k)$ from direct physical measurement of resistance components:

$$C_T = (1 + k)C_F + C_W$$
$$= C_V + C_W.$$

(a) Measurement of total viscous drag, C_V (e.g. from a wake traverse; see Chapter 7):

$$C_V = (1 + k)\,C_F, \text{ and } (1 + k) = C_V/C_F.$$

(b) Measurement of wave pattern drag, C_W (e.g. using wave probes, see Chapter 7):

$$(1 + k)\,C_F = C_T - C_W, \text{ and } (1 + k) = (C_T - C_W)\,/C_F.$$

Methods 3(a) and 3(b) are generally used for research purposes, rather than for routine testing, although measurement of wave pattern drag on a routine basis is a practical option. It should be noted that methods 3(a) and 3(b) allow the derivation of $(1 + k)$ over the whole speed range and should indicate any likely changes in $(1 + k)$ with speed.

4.4.2 CFD Methods

CFD may be employed to derive viscous drag and form factors. The derivation of a friction line using a RANS solver is discussed in Section 4.3.4.2. Form factors for ellipsoids, both in monohull and catamaran configurations, were estimated by Molland and Utama [4.24] using a RANS solver and wind tunnel tests. The use of CFD for the derivation of viscous drag and skin friction drag is discussed further in Chapter 9.

4.4.3 Empirical Methods

Several investigators have developed empirical formulae for $(1 + k)$ based on model test results. The following are some examples which may be used for practical powering purposes.

Table 4.3. C_{stern} *parameter*

Afterbody form	C_{stern}
Pram with gondola	−25
V-shaped sections	−10
Normal section shape	0
U-shaped sections with Hogner stern	10

Watanabe:

$$k = -0.095 + 25.6 \frac{C_B}{\left[\frac{L}{B}\right]^2 \sqrt{\frac{B}{T}}}. \tag{4.20}$$

Conn and Ferguson [4.9]:

$$k = 18.7 \left[C_B \frac{B}{L} \right]^2. \tag{4.21}$$

Grigson [4.21], based on a slightly modified ITTC line:

$$k = 0.028 + 3.30 \left[\frac{S}{L^2} \sqrt{C_B \frac{B}{L}} \right]. \tag{4.22}$$

Holtrop regression [4.25]:

$$(1 + k) = 0.93 + 0.487118(1 + 0.011 C_{\text{stern}}) \times (B/L)^{1.06806} (T/L)^{0.46106}$$

$$\times (L_{WL}/L_R)^{0.121563} (L_{WL}^3/\nabla)^{0.36486} \times (1 - C_P)^{-0.604247}. \tag{4.23}$$

If the length of run L_R is not known, it may be estimated using the following formula:

$$L_R = L_{WL} \left[1 - C_P + \frac{0.06 C_P LCB}{(4C_P - 1)} \right], \tag{4.24}$$

where LCB is a percentage of L_{WL} forward of $0.5L_{WL}$. The stern shape parameter C_{stern} for different hull forms is shown in Table 4.3.

Wright [4.26]:

$$(1 + k) = 2.480 \ C_B^{0.1526} (B/T)^{0.0533} (B/L_{BP})^{0.3856}. \tag{4.25}$$

Couser *et al.* [4.27], suitable for round bilge monohulls and catamarans:

$$\text{Monohulls: } (1 + k) = 2.76 (L/\nabla^{1/3})^{-0.4}. \tag{4.26}$$

$$\text{Catamarans: } (1 + \beta k) = 3.03 (L/\nabla^{1/3})^{-0.40}. \tag{4.27}$$

For practical purposes, the form factor is assumed to remain constant over the speed range and between model and ship.

4.4.4 Effects of Shallow Water

Millward [4.28] investigated the effects of shallow water on form factor. As a result of shallow water tank tests, he deduced that the form factor increases as water depth decreases and that the increase in form factor could be approximated by the relationship:

$$\Delta k = 0.644(T/h)^{1.72}, \tag{4.28}$$

where T is the ship draught (m) and h the water depth (m).

REFERENCES (CHAPTER 4)

4.1 Froude, W. Experiments on the surface-friction experienced by a plane moving through water, *42nd Report of the British Association for the Advancement of Science*, Brighton, 1872.

4.2 Froude, W. Report to the Lords Commissioners of the Admiralty on experiments for the determination of the frictional resistance of water on a surface, under various conditions, performed at Chelston Cross, under the Authority of their Lordships, *44th Report of the British Association for the Advancement of Science*, Belfast, 1874.

4.3 Froude, W. *The Papers of William Froude.* The Royal Institution of Naval Architects, 1955.

4.4 Hughes, G. Friction and form resistance in turbulent flow and a proposed formulation for use in model and ship correlation. *Transactions of the Royal Institution of Naval Architects*, Vol. 96, 1954, pp. 314–376.

4.5 Telfer, E.V. Ship resistance similarity. *Transactions of the Royal Institution of Naval Architects*, Vol. 69, 1927, pp. 174–190.

4.6 Telfer, E.V. Frictional resistance and ship resistance similarity. *Transactions of the North East Coast Institution of Engineers and Shipbuilders*, 1928/29.

4.7 Telfer, E.V. Further ship resistance similarity. *Transactions of the Royal Institution of Naval Architects*, Vol. 93, 1951, pp. 205–234.

4.8 Lap, A.J.W. Frictional drag of smooth and rough ship forms. *Transactions of the Royal Institution of Naval Architects*, Vol. 98, 1956, pp. 137–172.

4.9 Conn, J.F.C. and Ferguson, A.M. Results obtained with a series of geometrically similar models. *Transactions of the Royal Institution of Naval Architects*, Vol. 110, 1968, pp. 255–300.

4.10 Conn, J.F.C., Lackenby, H. and Walker, W.P. BSRA Resistance experiments on the Lucy Ashton. *Transactions of the Royal Institution of Naval Architects*, Vol. 95, 1953, pp. 350–436.

4.11 Barnaby, K.C. *Basic Naval Architecture.* Hutchinson, London, 1963.

4.12 Clements, R.E. An analysis of ship-model correlation using the 1957 ITTC line. *Transactions of the Royal Institution of Naval Architects*, Vol. 101, 1959, pp. 373–402.

4.13 Reynolds, O. An experimental investigation of the circumstances which determine whether the motion of water shall be direct or sinuous, and the law of resistance in parallel channels. *Philosophical Transactions of the Royal Society*, Vol. 174, 1883, pp. 935–982.

4.14 Froude, R.E. On the 'constant' system of notation of results of experiments on models used at the Admiralty Experiment Works, *Transactions of the Royal Institution of Naval Architects*, Vol. 29, 1888, pp. 304–318.

4.15 Schoenherr, K.E. Resistance of flat surfaces moving through a fluid. *Transactions of the Society of Naval Architects and Marine Engineers*. Vol. 40, 1932.

4.16 Lap, A.J.W. Fundamentals of ship resistance and propulsion. Part A Resistance. Publication No. 129a of the Netherlands Ship Model Basin, Wageningen. Reprinted in *International Shipbuilding Progress*.

4.17 Zborowski, A. Approximate method for estimating resistance and power of twin-screw merchant ships. *International Shipbuilding Progress*, Vol. 20, No. 221, January 1973, pp. 3–11.

4.18 Grigson, C.W.B. An accurate smooth friction line for use in performance prediction. *Transactions of the Royal Institution of Naval Architects*, Vol. 135, 1993, pp. 149–162.

4.19 ITTC. Report of Resistance Committee, p. 64, *23rd International Towing Tank Conference*, Venice, 2002.

4.20 ITTC. Report of Resistance Committee, p. 38, *25th International Towing Tank Conference*, Fukuoka, 2008.

4.21 Grigson, C.W.B. A planar friction algorithm and its use in analysing hull resistance. *Transactions of the Royal Institution of Naval Architects*, Vol. 142, 2000, pp. 76–115.

4.22 Date, J.C. and Turnock, S.R. Computational fluid dynamics estimation of skin friction experienced by a plane moving through water. *Transactions of the Royal Institution of Naval Architects*, Vol. 142, 2000, pp. 116–135.

4.23 ITTC Recommended Procedure, Resistance Test 7.5-02-02-01, 2008.

4.24 Molland, A.F. and Utama, I.K.A.P. Experimental and numerical investigations into the drag characteristics of a pair of ellipsoids in close proximity. *Proceedings of the Institution of Mechanical Engineers*, Vol. 216, Part M. *Journal of Engineering for the Maritime Environment*, 2002.

4.25 Holtrop, J. A statistical re-analysis of resistance and propulsion data. *International Shipbuilding Progress*, Vol. 31, November 1984, pp. 272–276.

4.26 Wright, B.D.W. Apparent viscous levels of resistance of a series of model geosims. BSRA Report WG/H99, 1984.

4.27 Couser, P.R., Molland, A.F., Armstrong, N.A. and Utama, I.K.A.P. Calm water powering prediction for high speed catamarans. *Proceedings of 4th International Conference on Fast Sea Transportation, FAST'97*, Sydney, 1997.

4.28 Millward, A. The effects of water depth on hull form factor. *International Shipbuilding Progress*, Vol. 36, No. 407, October 1989.

5 Model-Ship Correlation

5.1 Purpose

When making conventional power predictions, no account is usually taken of scale effects on:

(1) Hull form effect,
(2) Wake and thrust deduction factors,
(3) Scale effect on propeller efficiency,
(4) Uncertainty of scaling laws for appendage drag.

Experience shows that power predictions can be in error and corrections need to be applied to obtain a realistic trials power estimate. Suitable correction (or correlation) factors have been found using voyage analysis techniques applied to trials data. The errors in predictions are most significant with large, slow-speed, high C_B vessels.

Model-ship correlation should not be confused with model-ship extrapolation. The extrapolation process entails extrapolating the model results to full scale to create the ship power prediction. The correlation process compares the full-scale ship power prediction with measured or expected full-scale ship results.

5.2 Procedures

5.2.1 Original Procedure

5.2.1.1 Method
Predictions of power and propeller revolutions per minute (rpm) are corrected to give the best estimates of trial-delivered power P_D and revs N, i.e.

$$P_{Ds} = (1 + x)P_D \qquad (5.1)$$

and

$$N_S = (1 + k_2)N \qquad (5.2)$$

where P_D and N are tank predictions, P_{Ds} and N_S are expected ship values, $(1 + x)$ is the power correlation allowance (or ship correlation factor, SCF) and $(1 + k_2)$ is the rpm correction factor.

Factors used by the British Ship Research Association (BSRA) and the UK towing tanks for single-screw ships [5.1, 5.2, 5.3] have been derived from an analysis of more than 100 ships (mainly tankers) in the range 20 000–100 000 TDW, together with a smaller amount of data from trawlers and smaller cargo vessels. This correlation exercise involved model tests, after the trials, conducted in exactly the condition (draught and trim) of the corresponding ship trial. Regression analysis methods were used to correct the trial results for depth of water, sea condition, wind, time out of dock and measured hull roughness. The analysis showed a scatter of about 5% of power about the mean trend as given by the regression equation. This finding is mostly a reflection of measurement accuracies and represents the basic level of uncertainty in any power prediction.

5.2.1.2 Values of $(1 + x)$ (SCF) and $(1 + k_2)$

VALUES OF $(1 + x)$ These values vary greatly with ship size and the basic C_F formula used. Although they are primarily functions of ship length, other parameters, such as draught and C_B can have significant influences.

Typical values for these overall correction (correlation) factors are contained in [5.1], and some values for $(1 + x)$ for 'average hull/best trial' are summarised in Table 5.1.

A suitable approximation to the Froude friction line SCF data is:

$$\text{SCF} = 1.2 - \frac{\sqrt{L_{BP}}}{48}; \tag{5.3}$$

hence, estimated ship-delivered power

$$P_{Ds} = (P_E/\eta_D) \times (1 + x). \tag{5.4}$$

VALUES OF $(1 + k_2)$. These values vary slightly depending on ship size (primarily length) and the method of analysis (torque or thrust identity) but, in general, they are of the order of 1.02; hence, estimated ship rpm

$$N_s = N_{\text{model}} \times (1 + k_2) \tag{5.5}$$

In 1972–1973 the UK tanks published further refinements to the factors [5.2, 5.3]. The predictions were based on $(1 + x)_{\text{ITTC}}$ of unity, with corrections for roughness and draught different from assumed standard values. The value of k_2 is based on length, plus corrections for roughness and draught.

Table 5.1. *Typical values for ship correlation factor SCF $(1 + x)$*

L_{BP} (m)	122	150	180	240	300
Froude friction line	0.97	0.93	0.90	0.86	0.85
ITTC friction line	1.17	1.12	1.08	1.04	1.02

Note: for L < 122 m, SCF = 1.0 assumed.

Scott [5.4, 5.5] carried out multiple regression analyses on available data and determined $(1 + x)$ and k_2 in terms of length, hull roughness, Fr, C_B and so on. Some small improvements were claimed for each of the above methods.

5.2.2 ITTC1978 Performance Prediction Method

Recommendations of the International Towing Tank Conference (ITTC) through the 1970s led to a proposed new 'unified' method for power prediction. This method attempts to separate out and correct the various elements of the prediction process, rather than using one overall correlation factor such as $(1 + x)$. This was generally accepted by most test tanks across the world in 1978 and the procedure is known as 'The 1978 ITTC Performance Prediction Method for Single Screw Ships' [5.6].

The process comprises three basic steps:

(1) Total resistance coefficient for ship, C_{TS}

$$C_{TS} = (1 + k)C_{FS} + C_R + \Delta C_F + C_{AA}, \tag{5.6}$$

(Note, ITTC chose to use C_R rather than C_W)
where the form factor $(1 + k)$ is based on the ITTC line,

$$C_F = \frac{0.075}{[\log_{10} Re - 2]^2}. \tag{5.7}$$

The residual coefficient C_R is the same for the model and ship and is derived as:

$$C_R = C_{TM} - (1 + k)C_{FM} \tag{5.8}$$

The roughness allowance ΔC_F is:

$$\Delta C_F = \left[105 \left(\frac{k_S}{L} \right)^{1/3} - 0.64 \right] \times 10^{-3}. \tag{5.9}$$

If roughness measurements are lacking, $k_S = 150 \times 10^{-6}$ m is recommended.

The following equation, incorporating the effect of Re, has been proposed by Townsin [5.7]:

$$\Delta C_F = \left\{ 44 \left[\left(\frac{k_S}{L} \right)^{1/3} - 10 \, Re^{-1/3} \right] + 0.125 \right\} \times 10^{-3} \tag{5.10}$$

It was a recommendation of the 19th ITTC (1990), and discussed in [5.8, 5.9], that if roughness measurements are available, then the Bowden – Davison formula, Equation (5.9), should be replaced by Townsin's formula, Equation (5.10). It should be recognised that Equation (5.9) was recommended as a correlation allowance, including effects of roughness, rather than solely as a roughness allowance. Thus, the difference between Equations (5.9) and (5.10) may be seen as a component that is not accounted for elsewhere. This component amounts to:

$$[\Delta C_F]_{\text{Bowden}} - [\Delta C_F]_{\text{Townsin}} = [5.68 - 0.6 \log_{10} Re] \times 10^{-3}. \tag{5.11}$$

Air resistance C_{AA} is approximated from Equation (5.12), when better information is not available, as follows.

$$C_{AA} = 0.001 \frac{A_T}{S}, \tag{5.12}$$

where A_T is the transverse projected area above the waterline and S is the ship wetted area. See also Chapter 3 for methods of estimating air resistance.

If the ship is fitted with bilge keels, the total resistance is increased by the ratio:

$$\frac{S + S_{BK}}{S}$$

where S is the wetted area of the naked hull and S_{BK} is the wetted area of the bilge keels.

(2) Propeller characteristics

The values of K_T, K_Q and η_0 determined in open water tests are corrected for the differences in drag coefficient C_D between the model and full-scale ship.

$C_{DM} > C_{DS}$; hence, for a given J, K_Q full scale is lower and K_T higher than in the model case and η_0 is larger full scale.

The full-scale characteristics are calculated from the model characteristics as follows:

$$K_{TS} = K_{TM} + \Delta K_T, \tag{5.13}$$

and

$$K_{QS} = K_{QM} - \Delta K_Q, \tag{5.14}$$

where

$$\Delta K_T = \Delta C_D \cdot 0.3 \frac{P}{D} \frac{c \cdot Z}{D}, \tag{5.15}$$

$$\Delta K_Q = \Delta C_D \cdot 0.25 \frac{c \cdot Z}{D} \tag{5.16}$$

The difference in drag coefficient is

$$\Delta C_D = C_{DM} - C_{DS} \tag{5.17}$$

where

$$C_{DM} = 2 \left(1 + 2\frac{t}{c}\right) \left[\frac{0.04}{(Re_{co})^{1/6}} - \frac{5}{(Re_{co})^{2/3}}\right], \tag{5.18}$$

and

$$C_{DS} = 2 \left(1 + 2\frac{t}{c}\right) \left[1.89 + 1.62 \log_{10} \frac{c}{k_p}\right]^{-2.5}. \tag{5.19}$$

In the above equations, Z is the number of blades, P/D is the pitch ratio, c is the chord length, t is the maximum thickness and Re_{co} is the local Reynolds number at a non-dimensional radius $x = 0.75$. The blade roughness is set at $k_p = 30 \times 10^{-6}$ m. Re_{co} must not be lower than 2×10^5 at the open-water test.

When estimating Re_{co} $(=V_R \cdot c/\nu)$, an approximation to the chord ratio at $x = 0.75$ $(= 0.75\text{R})$, based on the Wageningen series of propellers (Figure 16.2) is:

$$\left(\frac{c}{D}\right)_{0.75R} = X_1 \times BAR, \tag{5.20}$$

where $X_1 = 0.732$ for three blades, 0.510 for four blades and 0.413 for five blades.

An approximate estimate of the thickness t may be obtained from Table 12.4, and V_R is estimated as

$$V_R = \sqrt{Va^2 + (0.75\,\pi nD)^2}. \tag{5.21}$$

It can also be noted that later regressions of the Wageningen propeller series data include corrections for Re, see Chapter 16.

(3) Propulsive coefficients $\eta_H = (1 - t)/(1 - w_T)$ and η_R

Propulsive coefficients η_H and η_R determined from the self-propulsion (SP) test are corrected as follows. t and η_R are to be assumed the same for the ship and the model. The full-scale wake fraction w_T is calculated from the model wake fraction and thrust deduction factor as follows:

$$w_{TS} = (t + 0.04) + (w_{TM} - t - 0.04)\frac{(1+k)\,C_{FS} + \Delta C_F}{(1+k)\,C_{FM}}, \tag{5.22}$$

where 0.04 takes into account the rudder effect and ΔC_F is the roughness allowance as given by Equation (5.9).

The foregoing gives an outline of the 'ITTC Performance Prediction Method' for P_D and N. The final trial prediction is obtained by multiplying P_D and N by trial prediction coefficients C_P and C_N (or by introducing individual ΔC_F and Δw_T corrections). C_P and C_N are introduced to account for any remaining differences between the predicted and the trial (in effect, C_P replaces $(1 + x)$ and C_N replaces $(1 + k_2)$). The magnitude of these corrections depends on the model and trial test procedures used as well as the choice of prediction margin.

A full account of the ITTC1978 Procedure is given in [5.6]. Further reviews, discussions and updates are provided by the ITTC Powering Performance Committee [5.8, 5.9].

5.2.2.1 Advantages of the Method
A review by SSPA [5.10] indicates that the advantages of the ITTC1978 method are as follows:

No length correction is necessary for C_P and C_N.
The same correction is satisfactory for load and ballast.
The standard deviation is better than the original method, although the scatter in C_P and C_N is still relatively large (within 6% and 2% of mean).

5.2.2.2 Shortcomings of the Method
The methods of estimating form factor $(1 + k)$ (e.g. low-speed tests or assuming that $C_W \propto Fr^4$) lead to errors and it may also not be correct to assume that $(1 + k)$ is independent of Fr.

ΔC_F is empirical and approximate.

ΔC_D correction to propeller is approximate, and a C_L correction is probably required because there is some change with Re.

η_R has a scale effect which may be similar to measurement errors.

w_T correction is empirical and approximate. However, CFD is being used to predict model and full-scale wake distributions (see Chapters 8 and 9) and full-scale LDV measurements are being carried out which should contribute to improving the model–full-scale correlation.

It should be noted that a number of tanks and institutions have chosen to use C_A as an overall 'correlation allowance' rather than to use ΔC_F. In effect, this is defining Equation (5.9) as C_A. Some tanks choose to include air drag in C_A. Regression analysis of test tank model resistance data, such as those attributable to Holtrop [5.11], tend to combine ΔC_F and C_{AA} into C_A as an overall model-ship correlation allowance (see Equations (10.24) and (10.34)).

The ITTC1978 method has in general been adopted by test tanks, with some local interpretations and with updates of individual component corrections being applied as more data are acquired. Bose [5.12] gives a detailed review of variations from the ITTC method used in practice.

5.2.3 Summary

The use of the ITTC1978 method is preferred as it attempts to scale the individual components of the power estimate. It also allows updates to be made to the individual components as new data become available.

The original method, using an overall correlation factor such as that shown in Table 5.1, is still appropriate for use with results scaled using the Froude friction line(s), such as the BSRA series and other data of that era.

5.3 Ship Speed Trials and Analysis

5.3.1 Purpose

The principal purposes of ship speed trials may be summarised as follows:

(1) to fulfil contractual obligations for speed, power and fuel consumption.
(2) to obtain performance and propulsive characteristics of the ship:
   • speed through the water under trials conditions
   • power against speed
   • power against rpm
   • speed against rpm for in-service use.
(3) to obtain full-scale hull–propeller interaction/wake data.
(4) to obtain model-ship correlation data.

Detailed recommendations for the conduct of speed/power trials and the analysis of trials data are given in ITTC [5.8, 5.13, 5.14] and [5.15].

5.3.2 Trials Conditions

The preferred conditions may be summarised as:

- zero wind
- calm water
- deep water
- minimal current and tidal influence.

5.3.3 Ship Condition

This will normally be a newly completed ship with a clean hull and propeller. It is preferable to take hull roughness measurements prior to the trials, typically leading to AHR values of 80–150 μm. The ITTC recommends an AHR not greater than 250 μm.

5.3.4 Trials Procedures and Measurements

Measurements should include:

(1) Water depth
(2) Seawater SG and temperature
(3) Wind speed and direction and estimated wave height
(4) Ship draughts (foreward, aft and amidships for large ships); hence, trim and displacement (should be before and after trials, and an average is usually adequate)
(5) Propeller rpm (N)
(6) Power (P): possibly via BMEP, preferably via torque
(7) Torque (Q): preferably via torsionmeter (attached to shaft) or strain gauge rosette on shaft and power $P = 2\pi NQ$
(8) Thrust measurement (possibly from main shaft thrust bearing/load cells): direct strain gauge measurements are generally for research rather than for routine commercial trials
(9) Speed: speed measurements normally at *fixed/constant rpm*

Speed is derived from recorded time over a fixed distance (mile). Typically, time measurements are taken for four runs over a measured mile at a fixed heading (e.g. E → W → E) in order to cancel any effects of current, Figure 5.1. The mile is measured from posts on land or GPS. 1 Nm = 6080 ft = 1853.7 m; 1 mile = 5280 ft.

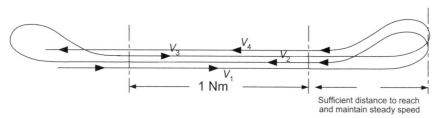

Figure 5.1. Typical runs on a measured mile.

Table 5.2. *Analysis of speed, including change in current*

No. of run	Speed over ground	V_1	V_2	V_3	V_4	Final	Current
1E	6.50						+1.01
2W	8.52	7.51					−1.01
3E	6.66	7.59	7.55				+0.85
4W	8.09	7.38	7.48	7.52			−0.58
5E	7.28	7.69	7.53	7.51	7.51	7.51	0.23
6W	7.43	7.36	7.52	7.53	7.52		−0.08

An analysis of speed, from distance/time and using a 'mean of means', is as follows:

$$
\left.\begin{array}{l} V_1 \\[2em] V_2 \\[2em] V_3 \\[2em] V_4 \end{array}\right\}
\left.\begin{array}{l} \dfrac{V_1 + V_2}{2} \\[1.5em] \dfrac{V_2 + V_3}{2} \\[1.5em] \dfrac{V_3 + V_4}{2} \end{array}\right\}
\left.\begin{array}{l} \sum V/2 \\[2em] \sum V/2 \end{array}\right\}
\Big\} \sum V/2 \,,
\tag{5.23}
$$

i.e. mean speed $V_m = \{V_1 + 3V_2 + 3V_3 + V_4\} \times \frac{1}{8}$. In principle, this eliminates the effect of current, see Table 5.2.

The process is repeated at different rpm, hence speed, to develop $P - V$, $P - N$ and $V - N$ relationships.

(10) Record the use of the rudder during measured speed runs (typically varies up to 5 deg for coursekeeping)

5.3.5 Corrections

5.3.5.1 Current

This is carried out noting that current can change with time, Figure 5.2.

- testing when low current changes, or assuming linear change over the period of trial
- running with/against current and using 'mean of means' speed effectively minimises/eliminates the problem, Table 5.2.

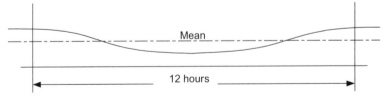

Figure 5.2. Change in current with time.

5.3.5.2 Water Depth

Potential shallow water effects are considered relative to water depth h and depth Froude number

$$Fr_h = \frac{V}{\sqrt{gh}},$$

where $Fr_h < 1.0$ is subcritical and $Fr_h > 1.0$ is supercritical. Once operating near or approaching $Fr_h = 1$, corrections will be required, usually based on water depth and ship speed.

The recommended limit on trial water depths, according to SNAME 73/21st ITTC code for sea trials, is a water depth $(h) \geq 10\ T\ V/\sqrt{L}$. According to the 12th/22nd ITTC, the recommended limit is the greater of $h \geq 3\ (B \times T)^{0.5}$ and $h \geq 2.75\ V^2/g$. According to the ITTC procedure, it is the greater of $h \geq 6.0\ A_M^{0.5}$ and $h \geq 0.5\ V^2$.

At lower depths of water, shallow water corrections should be applied, such as that attributable to Lackenby [5.16]:

$$\frac{\Delta V}{V} = 0.1242 \left[\frac{A_M}{h^2} - 0.05 \right] + 1 - \left[\tanh \left(\frac{gh}{V^2} \right) \right]^{0.5}, \tag{5.24}$$

where h is the depth of water, A_M is the midship area under water and ΔV is the speed loss due to the shallow water effect.

A more detailed account of shallow water effects is given in Chapter 6.

5.3.5.3 Wind and Weather

It is preferable that ship trials *not* be carried out in a sea state > Beaufort No. 3 and/or wind speed > 20 knots. For waves up to 2.0 m ITTC [5.14] recommends a resistance increase corrector, according to Kreitner, as

$$\Delta R_T = 0.64 \xi_W^2 B^2 C_B \rho\ 1/L, \tag{5.25}$$

where ξ_W is the wave height. Power would then be corrected using Equations (3.67), (3.68), and (3.69).

BSRA WIND CORRECTION. The BSRA recommends that, for ship trials, the results should be corrected to still air conditions, including a velocity gradient allowance.

The trials correction procedure entails deducting the wind resistance (hence, power) due to relative wind velocity (taking account of the velocity gradient and using the C_D from model tests for a similar vessel) to derive the corresponding power in a vacuum. To this vacuum condition is added the power due to basic air resistance caused by the uniform wind generated by the ship forward motion.

If ship speed $= V$, head wind $= U$ and natural wind gradient, Figure 5.3, is say

$$\frac{u}{U} = \left(\frac{h}{H} \right)^{1/5}$$

i.e.

$$u = U \left(\frac{h}{H} \right)^{1/5},$$

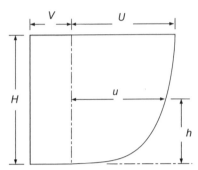

Figure 5.3. Wind velocity gradient.

$$\text{the correction} = \left\{ -\int_0^H (V+u)^2 \, dh + V^2 H \right\} \times \frac{1}{2}\rho \frac{A_T}{H} C_D,$$

where $\frac{A_T}{H} = B$. Note that the first term within the brackets corrects to a vacuum, and the second term corrects back to still air.

$$= \left\{ -\int_0^H \left(V+U\left[\frac{h}{H}\right]^{1/5}\right)^2 \, dh + V^2 H \right\} \times \dotsb$$

$$= \left\{ -\int_0^H V^2 + 2\frac{VU}{H^{1/5}}h^{1/5} + \frac{U^2}{H^{2/5}}h^{2/5} dh + V^2 H \right\} \times \dotsb$$

$$= \left\{ -\left[V^2 h + 2\frac{VU}{H^{1/5}}h^{6/5} \cdot \frac{5}{6} + \frac{U^2}{H^{2/5}}h^{7/5} \cdot \frac{5}{7} \right]_0^H + V^2 H \right\} \times \dotsb \qquad (5.26)$$

i.e. the correction to trials resistance to give 'still air' resistance.

$$\text{Correction} = \left\{ -\frac{5}{7}U^2 - \frac{5}{3}VU \right\} \times \frac{1}{2}\rho A_T C_D$$

$$= \left\{ -\left[V^2 + 2VU \cdot \frac{5}{6} + U^2 \cdot \frac{5}{7} \right] H + V^2 H \right\} \times \dotsb \qquad (5.27)$$

Breaks in area can be accounted for by integrating vertically in increments, e.g. 0 to H_1, H_1 to H_2 etc. A worked example application in Chapter 17 illustrates the use of the wind correction.

5.3.5.4 Rudder
Calculate and subtract the added resistance due to the use of the rudder(s). This is likely to be small, in particular, in calm conditions.

5.3.6 Analysis of Correlation Factors and Wake Fraction

5.3.6.1 Correlation Factor
The measured ship power for a given speed may be compared with the model prediction. The process may need a displacement ($\Delta^{2/3}$) correction to full-scale resistance (power) if the ship Δ is not the same as the model, or the model may be retested at trials Δ if time and costs allow.

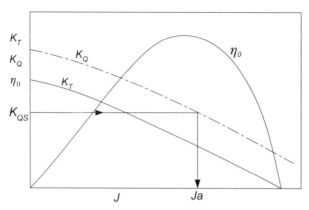

Figure 5.4. *Ja* from torque identity.

5.3.6.2 Wake Fraction

Assuming that thrust measurements have not been made on trial, which is usual for most commercial tests, the wake fraction will be derived using a *torque* identity method, i.e. using measured ship torque Q_S at revs n_S,

$$K_{QS} = Q_S/\rho n_S^2 D^5.$$

The propeller open water chart is entered, at the correct P/D for this propeller, with ship K_{QS} to derive the ship value of *Ja*, Figure 5.4.

The full-scale ship wake fraction is then derived as follows:

$$Ja = \frac{Va}{nD} = \frac{Vs\,(1 - w_T)}{nD}$$

and

$$Js = \frac{Vs}{nD},$$

hence,

$$(1 - w_T) = \frac{Ja\,nD}{Vs} = \frac{Ja}{Js}$$

and

$$w_T = 1 - \frac{Ja}{Js} = 1 - \frac{Va}{Vs}. \tag{5.28}$$

A worked example application in Chapter 17 illustrates the derivation of a full-scale wake fraction.

It can be noted that most test establishments use a thrust identity in the analysis of model self-propulsion tests, as discussed in Chapter 8.

5.3.7 Summary

The gathering of full-scale data under controlled conditions is very important for the development of correct scaling procedures. There is still a lack of good quality full-scale data, which tends to inhibit improvements in scaling methods.

REFERENCES (CHAPTER 5)

5.1 NPL. BTTP 1965 standard procedure for the prediction of Ship performance from model experiments, *NPL Ship TM 82*. March 1965.

5.2 NPL. Prediction of the performance of SS ships on measured mile trials, *NPL Ship Report 165*, March 1972.

5.3 NPL. Performance prediction factors for T.S. ships, *NPL Ship Report 172*, March 1973.

5.4 Scott, J.R. A method of predicting trial performance of single screw merchant ships. *Transactions of the Royal Institution of Naval Architects*. Vol. 115, 1973, pp. 149–171.

5.5 Scott, J.R. A method of predicting trial performance of twin screw merchant ships. *Transactions of the Royal Institution of Naval Architects*, Vol. 116, 1974, pp. 175–186.

5.6 ITTC Recommended Procedure. 1978 Performance Prediction Method, Procedure Number 7.5-02-03-01.4, 2002.

5.7 Townsin, R.L. The ITTC line – its genesis and correlation allowance. *The Naval Architect*. RINA, London, September 1985.

5.8 ITTC Report of Specialist Committee on Powering Performance and Prediction, *24th International Towing Tank Conference*, Edinburgh, 2005.

5.9 ITTC Report of Specialist Committee on Powering Performance Prediction, *25th International Towing Tank Conference*, Fukuoka, 2008.

5.10 Lindgren, H. and Dyne, G. Ship performance prediction, SSPA Report No. 85, 1980.

5.11 Holtrop, J. A statistical re-analysis of resistance and propulsion data. *International Shipbuilding Progress*, Vol. 31, 1984, pp. 272–276.

5.12 Bose, N. *Marine Powering Predictions and Propulsors*. The Society of Naval Architects and Marine Engineers, New York, 2008.

5.13 ITTC Recommended Procedure. Full scale measurements. Speed and power trials, Preparation and conduct of speed/power trials. Procedure Number 7.5-04-01-01.1, 2005.

5.14 ITTC Recommended Procedure. Full scale measurements. Speed and power trials. Analysis of speed/power trial data, Procedure Number 7.5-04-01-01.2, 2005.

5.15 ITTC, Report of Specialist Committee on Speed and Powering Trials, *23rd International Towing Tank Conference*, Venice, 2002.

5.16 Lackenby, H. Note on the effect of shallow water on ship resistance, BSRA *Report* No. 377, 1963.

6 Restricted Water Depth and Breadth

6.1 Shallow Water Effects

When a ship enters water of restricted depth, termed shallow water, a number of changes occur due to the interaction between the ship and the seabed. There is an effective increase in velocity, backflow, decrease in pressure under the hull and significant changes in sinkage and trim. This leads to increases in potential and skin friction drag, together with an increase in wave resistance. These effects can be considered in terms of the water depth, ship speed and wave speed. Using wave theory [6.1], and outlined in Appendix A1.8, wave velocity c can be developed in terms of h and λ, where h is the water depth from the still water level and λ is the wave length, crest to crest.

6.1.1 Deep Water

When h/λ is large,

$$c = \sqrt{\frac{g\lambda}{2\pi}}.$$ (6.1)

This deep water relationship is suitable for approximately $h/\lambda \geq 1/2$.

6.1.2 Shallow Water

When h/λ is small,

$$c = \sqrt{gh}.$$ (6.2)

The velocity now depends only on the water depth and waves of different wavelength propagate at the same speed. This shallow water relationship is suitable for approximately $h/\lambda \leq 1/20$ and $c = \sqrt{gh}$ is known as the critical speed.

It is useful to discuss the speed ranges in terms of the depth Froude number, noting that the waves travel at the same velocity, c, as the ship speed V. The depth Froude number is defined as:

$$Fr_h = \frac{V}{\sqrt{gh}}.$$ (6.3)

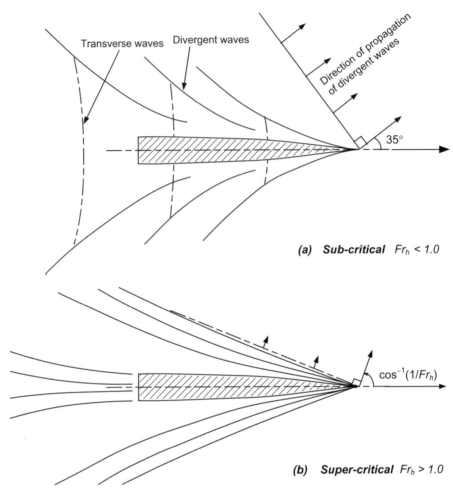

Figure 6.1. Sub-critical and super-critical wave patterns.

At the critical speed, or critical Fr_h, $Fr_h = 1.0$.

Speeds $< Fr_h = 1.0$ are known as subcritical speeds;
Speeds $> Fr_h = 1.0$ are known as supercritical speeds.

Around the critical speed the motion is unsteady and, particularly in the case of a model in a test tank with finite width, solitary waves (solitons) may be generated that move ahead of the model, [6.2]. For these sorts of reasons, some authorities define a region with speeds in the approximate range $0.90 < Fr_h < 1.1$ as the transcritical region.

At speeds well below $Fr_h = 1.0$, the wave system is as shown in Figure 6.1(a), with a transverse wave system and a divergent wave system propagating away from the ship at an angle of about 35°. See also the Kelvin wave pattern, Figure 3.14. As the ship speed approaches the critical speed, $Fr_h = 1.0$, the wave angle approaches 0°, or perpendicular to the track of the ship. At speeds greater than the critical speed, the diverging wave system returns to a wave propagation angle of about $\cos^{-1}(1/Fr_h)$, Figure 6.1(b). It can be noted that there are now no transverse waves.

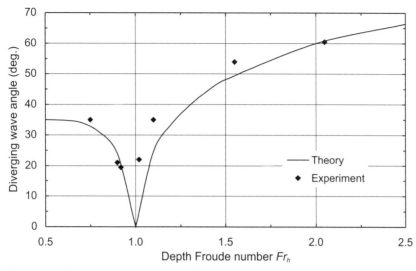

Figure 6.2. Change in wave angle with speed.

Because a gravity wave cannot travel at $c > \sqrt{gh}$ the transverse wave system is left behind and now only divergent waves are present. The changes in divergent wave angle with speed are shown in Figure 6.2. Experimental values [6.2] show reasonable agreement with the theoretical predictions.

As the speed approaches the critical speed, $Fr_h = 1.0$, a significant amplification of wave resistance occurs. Figure 6.3 shows the typical influence of shallow water on the resistance curve, to a base of length Froude number, and Figure 6.4 shows the ratio of shallow to deep water wave resistance to a base of depth Froude number. At speeds greater than critical, the resistance reduces again and can even fall to a little less than the deep water value. In practice, the maximum interference occurs at a Fr_h a little less than $Fr_h = 1.0$, in general in the range 0.96–0.98. At speeds around critical, the increase in resistance, hence required propeller thrust, leads also to a decrease in propeller efficiency as the propeller is now working well off design.

The influence of shallow water on the resistance of high-speed displacement monohull and catamaran forms is described and discussed by Molland *et al.* [6.2] and test results are presented for a series of models. The influence of a solid boundary

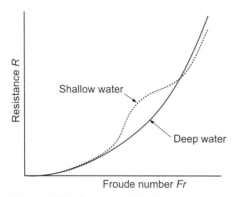

Figure 6.3. Influence of shallow water on the resistance curve.

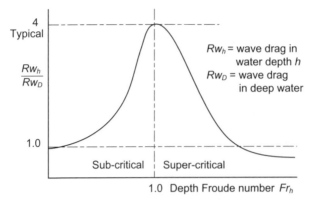

Figure 6.4. Amplification of wave drag at $Fr_h = 1.0$.

on the behaviour of high-speed ship forms was investigated by Millward and Bevan [6.3].

In order to describe fully the effects of shallow water, it is necessary to use a parameter such as T/h or L/h as well as depth Fr_h. The results of resistance experiments, to a base of length Fr, for changes in L/h are shown in Figure 6.5 [6.2]. The increases in resistance around $Fr_h = 1.0$, when $Fr = 1/\sqrt{L/h}$, can be clearly seen.

6.2 Bank Effects

The effects of a bank, or restricted breadth, on the ship are similar to those experienced in shallow water, and exaggerate the effects of restricted depth.

Corrections for bank effects may be incorporated with those for restricted depth, such as those described in Section 6.3.

6.3 Blockage Speed Corrections

Corrections for the effect of shallow water are generally suitable for speeds up to about $Fr_h = 0.7$. They are directed at the influences of potential and skin friction drag, rather than at wave drag whose influence is weak below about $Fr_h = 0.7$.

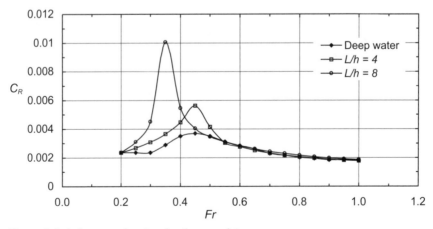

Figure 6.5. Influence of water depth on resistance.

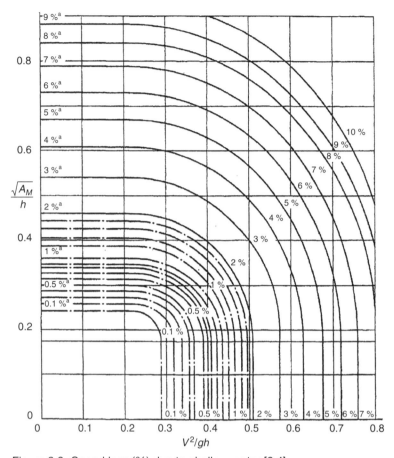

Figure 6.6. Speed loss (%) due to shallow water [6.4].

A commonly used correction is that due to Lackenby [6.4], shown in Figure 6.6. This amounts to a correction formula, attributable to Lackenby [6.5] of the following form:

$$\frac{\Delta V}{V} = 0.1242 \left[\frac{A_M}{h^2} - 0.05 \right] + 1 - \left[\tanh \left(\frac{gh}{V^2} \right) \right]^{0.5}, \qquad (6.4)$$

which is recommended by the International Towing Tank Conference (ITTC) as a correction for the trials procedure (Section 5.3).

For higher speeds, a simple shallow water correction is not practicable due to changes in sinkage and trim, wave breaking and other non-linearities. Experimental and theoretical data, such as those found in [6.2, 6.3, 6.6 and 6.7] provide some guidance, for higher-speed ship types, on likely increases in resistance and speed loss in more severe shallow water conditions.

Figure 6.6 and Equation (6.4) apply effectively to water of infinite breadth. A limited amount of data is available for the influence of finite breadth. Landweber [6.8] carried out experiments and developed corrections for the effects of different sized rectangular channels. These data are presented in [6.9]. An approximate curve fit to the data is

$$\frac{V_h}{V_\infty} = 1 - 0.09 \left[\frac{\sqrt{A_M}}{R_H} \right]^{1.5}, \qquad (6.5)$$

where V_∞ is the speed in deep water, V_h is the speed in shallow water of depth h and R_H is the hydraulic radius, defined as the area of cross section of a channel divided by its wetted perimeter, that is:

$$R_H = bh/(b + 2h)$$

It is seen that as the breadth of the channel b becomes large, R_H tends to h. When a ship or model is in a rectangular channel, then

$$R_H = (bh - A_M)/(b + 2h + p),$$

where A_M is the maximum cross-sectional area of the hull and p is the wetted girth of the hull at this section. It is found that if R_H is set equal to h (effectively infinite breadth), then Equation (6.5) is in satisfactory agreement with Figure 6.6 and Equation (6.4) up to about $\sqrt{A_M}/h = 0.70$ and $V^2/gh = 0.36$, or $Fr_h = 0.60$, up to which the corrections tend to be independent of speed.

Example: A cargo vessel has $L = 135$ m, $B = 22$ m and $T = 9.5$ m. For a given power, the vessel travels at 13 knots in deep water. Determine the speed loss, (a) when travelling at the same power in water of infinite breadth and with depth of water $h = 14$ m and, (b) in a river with a breadth of 200 m and depth of water $h = 14$ m when travelling at the same power as in deep water at 8 knots. Neglect any changes in propulsive efficiency.

$$Fr = V/\sqrt{gL} = 13 \times 0.5144/\sqrt{9.81 \times 135} = 0.184.$$
$$Fr_h = V/\sqrt{gh} = 13 \times 0.5144/\sqrt{9.81 \times 14} = 0.571.$$
$$A_M = 22 \times 9.5 = 209 \text{ m}^2; \sqrt{A_M}/h = \sqrt{209}/14 = 1.033.$$
$$V^2/gh = (13 \times 0.5144)^2/(9.81 \times 14) = 0.325; \text{and } gh/V^2 = 3.07.$$

For water with infinite breadth:

Using Equation (6.4), speed loss $\Delta V/V = 0.126 = 12.6\%$ and speed $= 11.4$ knots.

Using Equation (6.5) and $R_H = h = 14$ m, $V_h/V_\infty = 0.906$, or $\Delta V/V = 9.4\%$ and speed $= 11.8$ knots.

For water with finite breadth 200 m:

Wetted girth $p = (B + 2T) = 22 + (2 \times 9.5) = 41$ m.
$R_H = (200 \times 14 - 209)/(200 + 2 \times 14 + 41) = 9.63$ m.
$\sqrt{A_M}/R_H = \sqrt{209}/9.63 = 1.501.$

Using Equation (6.5), $V_h/V_\infty = 0.834$, or $\Delta V/V = 16.6\%$ and speed decreases from 8 to 6.7 knots.

A blockage corrector for canals was developed by Dand [6.10]. The analysis and tank tests include sloping banks and flooded banks. The corrections entail some complex reductions, but allow changes in width, changes in depth or combinations of the two to be investigated.

Hoffman and Kozarski [6.11] applied the theoretical work of Strettensky [6.12] to develop shallow water resistance charts including the critical speed region. Their results were found to show satisfactory agreement with published model data.

Blockage correctors developed primarily for the correction of model resistance tests are discussed in Section 3.1.4.

6.4 Squat

When a ship proceeds through shallow water there is an effective increase in flow speed, backflow, under the vessel and a drop in pressure. This drop in pressure leads to squat which is made up of vertical sinkage together with trim by the bow or stern. If a vessel is travelling too fast in shallow water, squat will lead to a loss of underkeel clearance and possible grounding. Various investigations into squat have been carried out, such as [6.13 and 6.14]. The following simple formula has been proposed by Barrass and Derrett [6.15] for estimating maximum squat $\delta_{\max}$ in a confined channel such as a river:

$$\delta_{\max} = \frac{C_B \times S_B^{0.81} \times V_S^{2.08}}{20} \text{ metres,} \qquad (6.6)$$

where C_B is the block coefficient, S_B is a blockage factor, being the ratio of the ship's cross section to the cross section of the channel, and V_S is the ship speed in knots.

Maximum squat will be at the bow if $C_B > 0.700$ and at the stern if $C_B < 0.700$.

Equation (6.6) may be used for estimating preliminary values of squat and indicating whether more detailed investigations are necessary.

Example: Consider a bulk carrier with breadth 40 m, draught 11 m and $C_B = 0.80$, proceeding at 5 knots along a river with breadth 200 m and depth of water 14 m.

$S_B = (B \times T)/(B_{RIV} \times h) = (40 \times 11)/(200 \times 14) = 0.157.$
$\delta_{\max} = 0.80 \times 0.157^{0.81} \times 5^{2.08}/20 = 0.25$ m.

The squat of 0.25 m will be at the bow, since $C_B > 0.700$.

Barrass [6.16] reports on an investigation into the squat for a large passenger cruise liner, both for open water and for a confined channel. Barrass points out that squat in confined channels can be over twice that measured in open water.

6.5 Wave Wash

The waves generated by a ship propagate away from the ship and to the shore. In doing so they can have a significant impact on the safety of smaller craft and on the local environment. This is particularly important for vessels operating anywhere near the critical depth Froude number, $Fr_h = 1.0$, when very large waves are generated, Figure 6.4. Operation at or near the critical Froude number may arise from high speed and/or operation in shallow water. Passenger-car ferries are examples of vessels that often have to combine high speed with operation in relatively shallow water. A full review of wave wash is carried out in ITTC [6.17] with further discussions in ITTC [6.18, 6.19].

In assessing wave wash, it is necessary to

- estimate the wave height at or near the ship,
- estimate its direction of propagation and,
- estimate the rate of decay in the height of the wave between the ship and shore, or area of interest.

The wave height in the near field, say within 0.5 to 1.0 ship lengths of the ship's track, may be derived by experimental or theoretical methods [6.20, 6.21, 6.22, 6.23, 6.24]. In this way the effect of changes in hull shape, speed, trim and operational conditions can be assessed. An approximation for the direction of propagation may be obtained from data such as those presented in Figure 6.2. Regarding wave decay, in *deep* water the rate of decay can be adequately described by Havelock's theoretical prediction of decay [6.25], that is:

$$H \propto \gamma y^{-n}, \qquad (6.7)$$

where H is wave height (m), n is 0.5 for transverse wave components and n is 0.33 for divergent waves. The value of γ can be determined experimentally based on a wave height at an initial value (offset) of y (m) from the ship and as a function of the speed of the vessel. Thus, once the maximum wave height is measured close to the ship's track, it can be calculated at any required distance from the ship.

It is noted that the transverse waves decay at a greater rate than the divergent waves. At a greater distance from the vessel, the divergent waves will therefore become more prominent to an observer than the transverse waves. As a result, it has been suggested that the divergent waves are more likely to cause problems in the far field [6.26]. It is generally found that in *deep* water the divergent waves for real ships behave fairly closely to the theoretical predictions.

In *shallow* water, further complications arise and deep water decay rates are no longer valid. Smaller values of n in Equation (6.7) between 0.2 and 0.4 may be applicable, depending on the wave period and the water depth/ship length ratio. The decay rates for shallow water waves (supercritical with $Fr_h > 1.0$) are less than for deep water and, consequently, the wave height at a given distance from the ship is greater than that of the equivalent height of a wave in deep water (sub-critical, $Fr_h < 1.0$), Doyle *et al.* [6.27]. Robbins *et al.* [6.28] carried out experiments to determine rates of decay at different depth Froude numbers.

River, port, harbour and coastal authorities are increasingly specifying maximum levels of acceptable wave wash. This in turn allows such authorities to take suitable actions where necessary to regulate the speed and routes of ships. It is therefore necessary to apply suitable criteria to describe the wave system on which the wave or wave system may then be judged. The most commonly used criterion is maximum wave height H_M. This is a simple criterion, is easy to measure and understand and can be used to compare one ship with another. In [6.29] it is argued that the criterion should be based on the wave height immediately before breaking.

The energy (E) in the wave front may also be used as a criterion. It can be seen as a better representation of the potential damaging effects of the waves since it

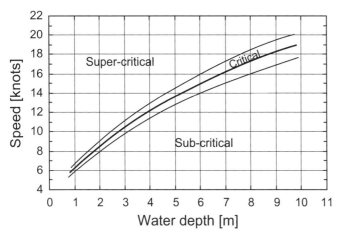

Figure 6.7. Sub-critical and super-critical operating regions.

combines the effects of wave height and speed. For deep water, the energy is

$$E = \rho g^2 H^2 T^2 / 16\pi, \tag{6.8}$$

where H is the wave height (m) and T is the wave period (s). This approach takes account, for example, of those waves with long periods which, as they approach more shallow water, may be more damaging to the environment.

In the case of shallow water,

$$E = \rho g H^2 \lambda / 8, \tag{6.9}$$

where $\lambda = (gT^2/2\pi)\tanh(2\pi h/L_W)$, h is the water depth and L_W is the wave length.

In shallow water, most of the wave energy is contained in a single long-period wave with a relatively small decay of wave energy and wave height with distance from the ship. In [6.27] it is pointed out that if energy alone is used, the individual components of wave height and period are lost, and it is recommended that the description of wash waves in shallow water should include both maximum wave height and maximum wave energy.

Absolute values need to be applied to the criteria if they are to be employed by port, harbour or coastal authorities to regulate the speeds and courses of ships in order to control the impact of wave wash. A typical case may require a maximum wave height of say 280 mm at a particular location, [6.30], or 350 mm for 3 m water depth and wave period 9 s [6.29].

From the ship operational viewpoint, it is recommended that ships likely to operate frequently in shallow water should carry a graph such as that shown in Figure 6.7. This indicates how the ship should operate well below or well above the critical speed for a particular water depth. Phillips and Hook [6.31] address the problems of operational risks and give an outline of the development of risk assessment passage plans for fast commercial ships.

REFERENCES (CHAPTER 6)

6.1 Lamb, H. *Hydrodynamics*. Cambridge University Press, Cambridge, 1962.
6.2 Molland, A.F., Wilson, P.A., Taunton, D.J., Chandraprabha, S. and Ghani, P.A. Resistance and wash measurements on a series of high speed

displacement monohull and catamaran forms in shallow water. *Transactions of the Royal Institution of Naval Architects*, Vol. 146, 2004, pp. 97–116.

6.3 Millward, A. and Bevan, M.G. The behaviour of high speed ship forms when operating in water restricted by a solid boundary. *Transactions of the Royal Institution of Naval Architects*, Vol. 128, 1986, pp. 189–204.

6.4 Lackenby, H. The effect of shallow water on ship speed. *Shipbuilder and Marine Engine Builder*. Vol. 70, 1963.

6.5 Lackenby, H. Note on the effect of shallow water on ship resistance. BSRA Report No. 377, 1963.

6.6 Millward, A. The effect of shallow water on the resistance of a ship at high sub-critical and super-critical speeds. *Transactions of the Royal Institution of Naval Architects*, Vol. 124, 1982, pp. 175–181.

6.7 Millward, A. Shallow water and channel effects on ship wave resistance at high sub-critical and super-critical speeds. *Transactions of the Royal Institution of Naval Architects*, Vol. 125, 1983, pp. 163–170.

6.8 Landweber, L. Tests on a model in restricted channels. EMB Report 460, 1939.

6.9 Comstock, J.P. (Ed.) *Principles of Naval Architecture*, SNAME, New York, 1967.

6.10 Dand, I.W. On ship-bank interaction. *Transactions of the Royal Institution of Naval Architects*, Vol. 124, 1982, pp. 25–40.

6.11 Hofman, M. and Kozarski, V. Shallow water resistance charts for preliminary vessel design. *International Shipbuilding Progress*, Vol. 47, No. 449, 2000, pp. 61–76.

6.12 Srettensky, L.N. Theoretical investigations of wave-making resistance. (in Russian). Central Aero-Hydrodynamics Institute Report 319, 1937.

6.13 Dand, I.W. and Ferguson, A.M. The squat of full ships in shallow water. *Transactions of the Royal Institution of Naval Architects*, Vol. 115, 1973, pp. 237–255.

6.14 Gourlay, T.P. Ship squat in water of varying depth. *Transactions of the Royal Institution of Naval Architects*, Vol. 145, 2003, pp. 1–14.

6.15 Barrass, C.B. and Derrett, D.R. *Ship Stability for Masters and Mates*. 6th Edition. Butterworth-Heinemann, Oxford, UK, 2006.

6.16 Barrass, C.B. Maximum squats for Victoria. *The Naval Architect*, RINA, London, February 2009, pp. 25–34.

6.17 ITTC Report of Resistance Committee. *23rd International Towing Tank Conference*, Venice, 2002.

6.18 ITTC Report of Resistance Committee. *24th International Towing Tank Conference*, Edinburgh, 2005.

6.19 ITTC Report of Resistance Committee. *25th International Towing Tank Conference*, Fukuoka, 2008.

6.20 Molland, A.F., Wilson, P.A., Turnock, S.R., Taunton, D.J. and Chandraprabha, S. The prediction of the characterstics of ship generated near-field wash waves. *Proceedings of Sixth International Conference on Fast Sea Transportation, FAST'2001*, Southampton, September 2001, pp. 149–164.

6.21 Day, A.H. and Doctors, L.J. Rapid estimation of near- and far-field wave wake from ships and application to hull form design and optimisation. *Journal of Ship Research*, Vol. 45, No. 1, March 2001, pp. 73–84.

6.22 Day, A.H. and Doctors, L.J. Wave-wake criteria and low-wash hullform design. *Transactions of the Royal Institution of Naval Architects*, Vol. 143, 2001, pp. 253–265.

6.23 Macfarlane, G.J. Correlation of prototype and model wave wake characteristics at low Froude numbers. *Transactions of the Royal Institution of Naval Architects*, Vol. 148, 2006, pp. 41–56.

6.24 Raven, H.C. Numerical wash prediction using a free-surface panel code. *International Conference on the Hydrodynamics of High Speed Craft, Wake Wash and Motion Control*, RINA, London, 2000.

6.25 Havelock, T.H. The propagation of groups of waves in dispersive media, with application to waves produced by a travelling disturbance. *Proceedings of the Royal Society, London, Series A*, 1908, pp. 398–430.

6.26 Macfarlane, G.J. and Renilson, M.R. Wave wake – a rational method for assessment. *Proceedings of International Conference on Coastal Ships and Inland Waterways* RINA, London, February 1999, Paper 7, pp. 1–10.

6.27 Doyle, R., Whittaker, T.J.T. and Elsasser, B. A study of fast ferry wash in shallow water. *Proceedings of Sixth International Conference on Fast Sea Transportation, FAST'2001*, Southampton, September 2001, Vol. 1, pp. 107–120.

6.28 Robbins, A., Thomas, G., Macfarlane, G.J., Renilson, M.R. and Dand, I.W. The decay of catamaran wave wake in deep and shallow water. *Proceedings of Ninth International Conference on Fast Sea Transportation, FAST'2007*, Shanghai, 2007, pp. 184–191.

6.29 Kofoed-Hansen, H. and Mikkelsen, A.C. Wake wash from fast ferries in Denmark. *Fourth International Conference on Fast Sea Transportation, FAST'97*, Sydney, 1997.

6.30 Stumbo, S., Fox, K. and Elliott, L. Hull form considerations in the design of low wash catamarans. *Proceedings of Fifth International Conference on Fast Sea Transportation, FAST'99*, Seattle, 1999, pp. 83–90.

6.31 Phillips, S. and Hook, D. Wash from ships as they approach the coast. *International Conference on Coastal Ships and Inland Waterways*. RINA, London, 2006.

7 Measurement of Resistance Components

7.1 Background

The accurate experimental measurement of ship model resistance components relies on access to high-quality facilities. Typically these include towing tanks, cavitation tunnels, circulating water channels and wind tunnels. Detailed description of appropriate experimental methodology and uncertainty analysis are contained within the procedures and guidance of the International Towing Tank Conference (ITTC) [7.1]. There are two approaches to understanding the resistance of a ship form. The first examines the direct body forces acting on the surface of the hull and the second examines the induced changes to pressure and velocity acting at a distance away from the ship. It is possible to use measurements at model scale to obtain global forces and moments with the use of either approach. This chapter considers experimental methods that can be applied, typically at model scale, to measure pressure, velocity and shear stress. When applied, such measurements should be made in a systematic manner that allows quantification of uncertainty in all stages of the analysis process. Guidance on best practice can be found in the excellent text of Coleman and Steele [7.2], the processes recommended by the International Standards Organisation (ISO) [7.3] or in specific procedures of the ITTC, the main ones of which are identified in Table 7.1.

In general, if the model is made larger (smaller scale factor), the flow will be steadier, and if the experimental facility is made larger, there will be less uncertainty in the experimental measurements. Facilities such as cavitation tunnels, circulating water channels and wind tunnels provide a steady flow regime more suited to measurements at many spatially distributed locations around and on ship hulls. Alternatively, the towing tank provides a straightforward means of obtaining global forces and moments as well as capturing the unsteady interaction of a ship with a head or following sea.

7.2 Need for Physical Measurements

Much effort has been devoted to the direct experimental determination of the various components of ship resistance. This is for three basic reasons:

(1) To obtain a better understanding of the physical mechanism
(2) To formulate more accurate scaling procedures

Table 7.1. *ITTC procedures of interest to ship resistance measurement*

Section number	Topic of recommended procedure
7.5-01-01-01	Ship models
7.5-02-02-01	Resistance tests
7.5-02-05-01	Resistance tests: high-speed marine vehicles
7.5-02-01-02	Uncertainty analysis in EFD: guideline for resistance towing tank test

(3) To support theoretical methods which may, for example, be used to minimise certain resistance components and derive more efficient hull forms.

The experimental methods used are:

(a) Measurement of total head loss across the wake of the hull to determine the total 'viscous' resistance.
(b) Measurements of velocity profile through the boundary layer.
(c) Measurements of wall shear stress using the Preston tube technique to measure 'skin friction' resistance.
(d) Measurement of surface pressure distribution to determine the 'pressure' resistance.
(e) Measurement of the wave pattern created by the hull to determine the 'wave pattern' resistance (as distinct from total 'wave' resistance, which may include wave breaking).
(f) Flow visualisation observations to determine the basic character of the flow past the model using wool tufts, neutral buoyancy particles, dye streaks and paint streaks etc. Particular interest is centred on observing separation effects.

The total resistance can be broken down into a number of physically identifiable components related to one of three basic causes:

(1) Boundary layer growth,
(2) Wave making,
(3) Induced drag due to the trailing vortex system.

As discussed in Chapter 3, Section 3.1.1, when considering the basic components of hull resistance, it is apparent that the total resistance of the ship may be determined from the resolution of the forces acting at each point on the hull, i.e. tangential and normal forces (summation of fore and aft components of tangential forces = frictional resistance, whilst a similar summation of resolved normal forces gives the pressure resistance) *or* by measuring energy dissipation (in the waves and in the wake).

Hence, these experimentally determined components may be summarised as:

1. Shear stress (friction) drag
   +   } forces acting
2. Pressure drag

3. Viscous wake (total viscous resistance)
   +   } energy dissipation
4. Wave pattern resistance

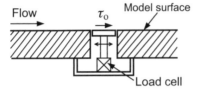

Figure 7.1. Schematic layout of transducer for direct measurement of skin friction.

7.3 Physical Measurements of Resistance Components

7.3.1 Skin Friction Resistance

The shear of flow across a hull surface develops a force typically aligned with the flow direction at the edge of the boundary layer and proportional to the viscosity of the water and the velocity gradient normal to the surface, see Appendix A1.2. Measurement of this force requires devices that are sufficiently small to resolve the force without causing significant disturbance to the fluid flow.

7.3.1.1 Direct Method

This method uses a transducer, as schematically illustrated in Figure 7.1, which has a movable part flush with the local surface. A small displacement of this surface is a measure of the tangential force. Various techniques can be used to measure the calibrated displacement. With the advent of microelectromechanical systems (MEMS) technology [7.4] and the possibility of wireless data transmission, such measurements will prove more attractive. This method is the most efficient as it makes no assumptions about the off-surface behaviour of the boundary layer. As there can be high curvature in a ships hull, a flat transducer surface may cause a discontinuity. Likewise, if there is a high longitudinal pressure gradient, the pressures in the gaps each side of the element are different, leading to possible errors in the transducer measurements.

In wind tunnel applications it is possible to apply a thin oil film and use optical interference techniques to measure the thinning of the film as the shear stress varies [7.5].

7.3.1.2 Indirect Methods

A number of techniques make use of the known behaviour of boundary layer flow characteristics in order to infer wall shear stress and, hence, skin friction [7.4, 7.5]. All these devices require suitable calibration in boundary layers of known velocity profile and sufficiently similar to that experienced on the hull.

(1) HOT-FILM PROBE. The probe measures the electrical current required to maintain a platinum film at a constant temperature when placed on surface of a body, Figure 7.2. Such a device has a suitably sensitive time response so that it is also used for measuring turbulence levels. The method is sensitive to temperature variations in the water and to surface bubbles. Such probes are usually insensitive to direction and measure total friction at a point. Further experiments are required to determine flow direction from which the fore and aft force components are then derived. Calibration is difficult; a rectangular duct is used in which the pressure drop along

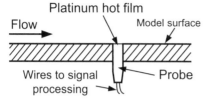

Platinum hot film
Flow
Model surface
Wires to signal processing — Probe

Figure 7.2. Hot-film probe.

a fixed length is accurately measured, then equated to the friction force. Such a calibration is described for the Preston tube. The method is relatively insensitive to pressure gradient and is therefore good for ship models, where adverse pressure gradients aft plus separation are possible. The probes are small and can be mounted flush with the hull, Figure 7.2. Hot films may also be surface mounted, being similar in appearance to a strain gauge used for measuring surface strain in a material. An example application of surface-mounted hot films is shown in Figure 7.3. In this case the hot films were used to detect the transition from laminar to turbulent flow on a rowing scull.

(2) STANTON TUBE. The Stanton tube is a knife-edged Pitôt tube, lying within the laminar sublayer, see Appendix A1.6. The height is adjusted to give a convenient reading at maximum velocity. The height above the surface is measured with a feeler gauge. Clearances are generally too small for ship model work, taking into account surface undulations and dirt in the water. The method is more suitable for wind tunnel work.

(3) PRESTON TUBE. The layout of the Preston tube is shown in Figure 7.4. A Pitôt tube measures velocity by recording the difference between static press P_O and total pressure P_T. If the tube is in contact with a hull surface then, since the velocity is zero at the wall, any such 'velocity' measurement will relate to velocity gradient at the surface and, hence, surface shear stress. This principle was first used by Professor Preston in the early 1950s.

For the inner region of the boundary layer,

$$\frac{u}{u_\tau} = f\left[\frac{yu_\tau}{\nu}\right]. \tag{7.1}$$

This is termed the inner velocity law or, more often, 'law of the wall', where the friction velocity $u_\tau = (\frac{\tau_0}{\rho})^{1/2}$ and the shear stress at the wall is $\tau_0 = \mu\frac{du}{dy}$ and noting

$$\frac{u}{u_\tau} = A\log\left[\frac{yu_\tau}{\nu}\right] + B. \tag{7.2}$$

Since Equation (7.1) holds, it must be in a region in which quantities depend only on ρ, ν, τ_0 and a suitable length. Thus, if a Pitôt tube of circular section and outside diameter d is placed in contact with the surface and wholly immersed in the 'inner' region, the difference between the total Pitôt press P_T and static press P_0 must depend only on ρ, ν, τ_0 and d where

τ_0 = wall shear stress
ρ = fluid density
ν = kinematic viscosity
d = external diameter of Preston tube

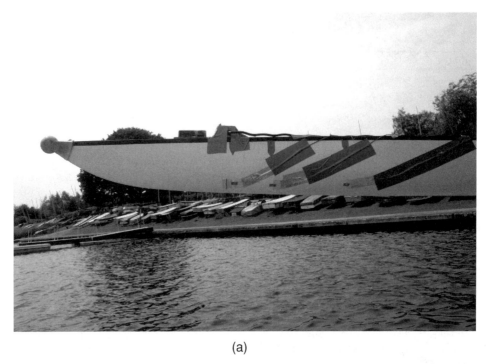

(a)

(b)

Figure 7.3. Hot-film surface-mounted application to determine the location of laminar-turbulent transition on a rowing scull. Photographs courtesy of WUMTIA.

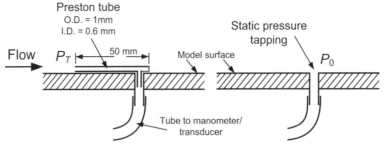

Figure 7.4. Layout of the Preston tube.

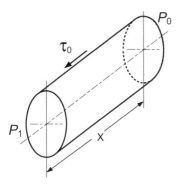

Figure 7.5. Calibration pipe with a known static pressure drop between the two measurement locations.

It should be noted that the diameter of the tube (d) must be small enough to be within the inner region of the boundary layer (about 10% of boundary layer thickness or less).

It can be shown by using dimensional methods that

$$\frac{(p_T - p_0)\, d^2}{\rho v^2} \quad \text{and} \quad \frac{\tau_0 d^2}{\rho v^2}$$

are dimensionless and, hence, the calibration of the Preston tube is of the following form:

$$\frac{\tau_0 d^2}{\rho v^2} = F\left[\frac{(p_T - p_0)\, d^2}{\rho v^2}\right].$$

The calibration of the Preston tube is usually carried out inside a pipe with a fully turbulent flow through it. The shear stress at the wall can be calculated from the *static* pressure gradient along the pipe as shown in Figure 7.5.

$$(p_1 - p_0)\, \frac{\pi D^2}{4} = \tau_0 \pi D x,$$

where D is the pipe diameter and

$$\tau_0 = \frac{(P_1 - P_0)}{x} \cdot \frac{D}{4} = \frac{D}{4} \cdot \frac{dp}{dx}. \tag{7.3}$$

Preston's original calibration was as follows:

Within pipes

$$\log_{10} \frac{\tau_0 d^2}{4\rho v^2} = -1.396 + 0.875 \log_{10}\left[\frac{(P_T - P_0)\, d^2}{4\rho v^2}\right] \tag{7.4}$$

Flat plates

$$= -1.366 + 0.877 \log_{10}\left[\frac{(P_T - P_0)\, d^2}{4\rho v^2}\right] \tag{7.5}$$

hence, if P_T and P_0 deduced at a point then τ_0 can be calculated and

$$C_F = \frac{\tau_0}{\frac{1}{2}\rho U^2}$$

where U is model speed (not local). For further information on the calibration of Preston tubes see Patel [7.6].

Measurements at the National Physical Laboratory (NPL) of skin friction on ship models using Preston tubes are described by Steele and Pearce [7.7] and Shearer and Steele [7.8]. Some observations on the use of Preston tubes based on [7.7] and [7.8] are as follows:

(a) The experiments were used to determine trends rather than an absolute measure of friction. Experimental accuracy was within about $\pm 5\%$.
(b) The method is sensitive to pressure gradients as there are possible deviations from the 'Law of the wall' in favourable pressure gradients.
(c) Calibration is valid only in turbulent flow; hence, the distance of total turbulence from the bow is important, to ensure transition has occurred.
(d) Ideally, the Preston tube total and static pressures should be measured simultaneously. For practical reasons, this is not convenient; hence, care must be taken to repeat identical conditions.
(e) There are difficulties in measuring the small differences in water pressure experienced by this type of experiment.
(f) Flow direction experiments with wool tufts or surface ink streaks are required to precede friction measurements. The Preston tubes are aligned to the direction of flow at each position in order to measure the maximum skin friction.
(g) Findings at NPL for *water* in pipes indicated a calibration close to that of Preston.

Results for a tanker form [7.8] are shown in Figure 7.6. In Figure 7.6, the local C_f has been based on *model* speed not on *local* flow speed. It was observed that the waviness of the measured C_f closely corresponds to the hull wave profile, but is inverted, that is, a high C_f in a trough and low C_f at a crest. The C_f variations are therefore primarily a local speed change effect due to the waves. At the deeper measurements, where the wave orbital velocities are less effective, it is seen that the undulations in measured C_f are small.

The Hughes local C_f is based on the differentiation of the ITTC formula for a flat plate. The trend clearly matches that of the mean line through the measured C_f. It is also noted that there is a difference in the distribution of skin friction between the raked (normal) bow and the bulbous bow.

(4) MEASUREMENTS OF BOUNDARY LAYER PROFILE. These are difficult to make at model scale. They have, however, been carried out at ship scale in order to derive local C_f, as discussed in Chapter 3, Section 3.2.3.6.

(5) LIQUID CRYSTALS. These can be designed to respond to changes in surface temperature. A flow over a surface controls the heat transfer rate and, hence, local surface temperature. Hence, the colour of a surface can be correlated with the local shear rate. This is related to the temperature on the surface, see Ireland and Jones [7.9]. To date, in general, practical applications are only in air.

7.3.1.3 Summary
Measurements of surface shear are difficult and, hence, expensive to make and are generally impractical for use as a basis for global integration of surface shear. However, they can provide insight into specific aspects of flow within a local area

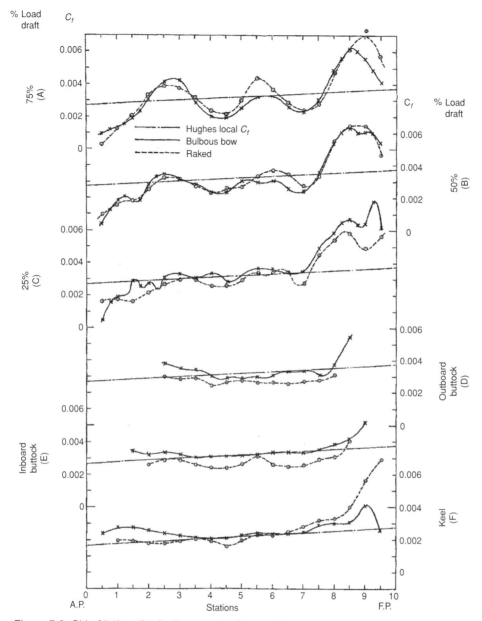

Figure 7.6. Skin friction distribution on a tanker model.

as, for example, in identifying areas of higher shear stress and in determining the location of transition from laminar to turbulent flow. Such measurements require a steady flow best achieved either in a circulating water channel or a wind tunnel.

7.3.2 Pressure Resistance

The normal force imposed on the hull by the flow around it can be measured through the use of static pressure tappings and transducers. Typically, 300–400 static pressure points are required to be distributed over the hull along waterlines in order that

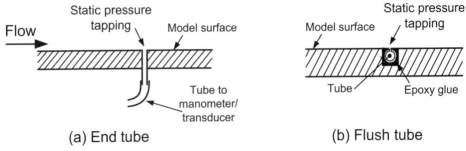

Figure 7.7. Alternative arrangements for surface pressure measurements.

sufficient resolution of the fore-aft force components can be made. As shown in Figure 7.7(a) each tapping comprises a tube of internal diameter of about 1–1.5 mm mounted through the hull surface. This is often manufactured from brass, glued in place and then sanded flush with the model surface. A larger diameter PVC tube is then sealed on the hull inside and run to a suitable manometer bank or multiport scanning pressure transducer. An alternative that requires fewer internal pressure tubes but more test runs uses a tube mounted in a waterline groove machined in the surface, see Figure 7.7(b), and backfilled with a suitable epoxy. A series of holes are drilled along the tube. For a given test, one of the holes is left exposed with the remainder taped.

It is worth noting that pressure measurements that rely on a water-filled tube are notorious for difficulties in ensuring that there are no air bubbles within the tube. Typically, a suitable pump system is required to flush the tubes once they are immersed in the water and/or a suitable time is required to allow the air to enter solution. If air remains, then its compressibility prevents accurate transmission of the surface pressure to the measurement device.

Typical references describing such measurements include Shearer and Cross [7.10], Townsin [7.11] and Molland and Turnock [7.12, 7.13] for models in a wind tunnel. In water, there can be problems with the waves generated when in motion, for example, leaving pressure tappings exposed in a trough. Hence, pressures at the upper part of the hull may have to be measured by diaphragm/electric pressure gauges, compared with a water manometry system. Such electrical pressure sensors [7.4] need to be suitably water proof and are often sensitive to rapid changes in temperature [7.14].

There are some basic experimental difficulties which concern the need for:

(a) Very accurate measurement of hull trim β for resolving forces and
(b) The measurement of wave surface elevation, for pressure integration.

In Figure 7.8, the longitudinal force (drag) is $Pds \cdot \sin\theta = Pds'$. Similarly, the vertical force is Pdv'. The horizontal force is

$$R_p = \int Pds' \cos\beta + \int Pdv' \sin\beta. \qquad (7.6)$$

The vertical force

$$R_v = \int Pds' \sin\beta - \int Pdv' \cos\beta = W\,(=-B). \qquad (7.7)$$

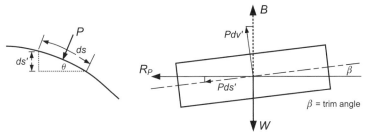

Figure 7.8. Measurement of surface pressure.

From Equation (7.7)

$$\int Pdv' = \int Pds' \tan \beta - W/\cos \beta,$$

and substituting in Equation (7.6), hence

$$\text{Total horizontal force} = \int Pds' (\cos \beta + \sin \beta \tan \beta) - W \tan \beta. \qquad (7.8)$$

W is large, hence the *accurate* measurement of trim β is important, possibly requiring the use, for example, of linear displacement voltmeters at each end of the model.

In the analysis of the data, the normal pressure (P) is projected onto the midship section for pressure tappings along a particular waterline, Figure 7.9. The Pds' values are then integrated to give the total pressure drag.

The local pressure measurements for a tanker form, [7.8], were integrated to give the total pressure resistance, as have the local C_f values to give the total skin friction. Wave pattern measurements and total resistance measurements were also made, see Section 7.3.4. The resistance breakdown for this tanker form is shown in Figure 7.10.

The following comments can be made on the resistance breakdown in Figure 7.10:

(1) There is satisfactory agreement between measured pressure + skin friction resistance and total resistance.
(2) Measured wave pattern resistance for the tanker is small (6% of total).
(3) Measured total C_F is closely comparable with the ITTC estimate, but it shows a slight dependence on Froude number, that is, the curve undulates.

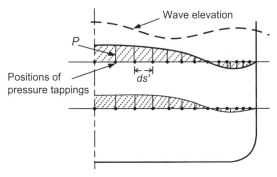

Figure 7.9. Projection of pressures on midship section.

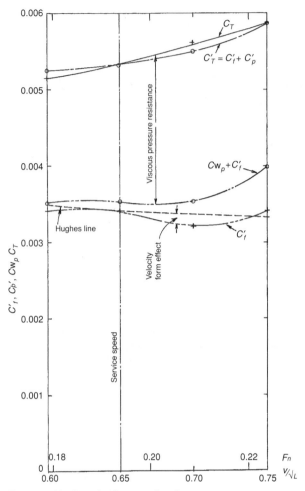

Figure 7.10. Results for a tanker form.

(4) There was a slight influence of hull form on measured C_F. Changes in C_F of about 5% can occur due to changes in form. Changes occurred mainly at the fore end.

In summary, the measurement and then integration of surface pressure is not a procedure to be used from day to day. It is expensive to acquire sufficient points to give an accurate value of resistance, especially as it involves the subtraction of two quantities of similar magnitude.

7.3.3 Viscous Resistance

The use of a control volume approach to identify the effective change in fluid momentum and, hence, the resistance of the hull has many advantages in comparison with the direct evaluation of shear stress and surface normal pressure. It is widely applied in the wind tunnel measurement of aircraft drag. More recently, it has also been found to exhibit less susceptibility to issues of surface mesh definition in computational fluid dynamics (CFD) [7.15], see Chapter 9. Giles and Cummings [7.16] give the full derivation of all the relevant terms in the control volume. In the case of CFD evaluations, the effective momentum exchange associated with the turbulent wake Reynolds averaged stress terms should also be included. In more practical experimentation, these terms are usually neglected, although the optical laser-based flow field measurement techniques (Section 7.4) do allow their measurement.

In Chapter 3, Equation (3.14) gives the total viscous drag for the control volume as the following:

$$R_V = \iint\limits_{\text{wake}} \left\{ \Delta p + \frac{1}{2}\rho(u'^2 - u^2) \right\} dz\, dy, \tag{7.9}$$

where Δp and u' are found from the following equations:

$$\frac{p_B}{\rho} + \frac{1}{2}[(U + u')^2 + v^2 + w^2] + gz_B + \frac{\Delta p}{\rho} = \frac{1}{2}U^2 \tag{7.10}$$

and

$$\tfrac{1}{2}\rho\,(U + u')^2 = \tfrac{1}{2}\rho\,(U + u)^2 + \Delta p, \tag{7.11}$$

remembering u' is the equivalent velocity that includes the pressure loss along the streamline. This formula is the same as the Betz formula for viscous drag, but it is generally less convenient to use for the purpose of experimental analysis.

An alternative formula is that originally developed by Melville Jones when measuring the viscous drag of an aircraft wing section:

$$R_V = \rho U^2 \sqrt{g - p}\, dy\, dz, \tag{7.12}$$

where two experimentally measured non-dimensional quantities

$$p = \frac{p_B - p_0}{\frac{1}{2}\rho U^2} \quad \text{and} \quad g = p + \left(\frac{U + u}{U}\right)^2$$

can be found through measurement of total head and static pressure loss downstream of the hull using a rake or traverse of Pitôt and static probes.

7.3.3.1 Derivation of Melville Jones Formula

In this derivation, viscous stresses are neglected as are wave-induced velocity components as these are assumed to be negligible at the plane of interest. Likewise, the influence of vorticity is assumed to be small. Figure 7.11 illustrates the two downstream measurement planes 1 and 2 in a ship fixed system with the upstream plane 0, the undisturbed hydrostatic pressure field P_0 and ship speed u. Far downstream at plane 2, any wave motion is negligible and $P_2 = P_0$.

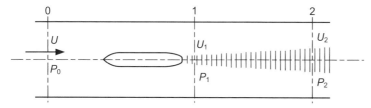

Figure 7.11. Plan view of ship hull with two wake planes identified.

The assumption is that, along a streamtube between 1 and 2, no total head loss occurs (e.g. viscous mixing is minimal) so that the total head $H_2 = H_1$. The assumption of no total head loss is not strictly true, since the streamlines/tubes will not be strictly ordered, and there will be some frictional losses. The total viscous drag R_V will then be the rate of change of momentum between stations 0 and 2 (no net pressure loss), i.e.

$$R_V = \rho \iint u_2(u - u_2)\, dS_2$$

(where dS_2 is the area of the streamtube). For mass continuity along streamtube,

$$u_1 dS_1 = u_2 dS_2$$

$$\therefore R_V = \rho \iint u_1(u - u_2)\, dS_1 \text{ over plane 1}$$

as

$$H_0 = \tfrac{1}{2}\rho u^2 + P_0$$

$$H_2 = \tfrac{1}{2}\rho u_2^2 + P_2 = \frac{1}{2}\rho u_2^2 + P_0 = \tfrac{1}{2}\rho u_1^2 + P_1$$

$$\frac{H_0 - H_2}{\tfrac{1}{2}\rho u^2} = 1 - \left(\frac{u_2}{u}\right)^2 \quad \text{or}$$

$$\frac{u_2}{u} = \sqrt{g},$$

where

$$g = 1 - \frac{H_0 - H_2}{\tfrac{1}{2}\rho u^2} \quad \text{also}$$

$$\left(\frac{u_1}{u}\right)^2 = \left(\frac{u_2}{u}\right)^2 - \frac{(P_1 - P_0)}{\tfrac{1}{2}\rho u^2}$$

$$= g - p$$

where

$$p = \frac{P_1 - P_0}{\tfrac{1}{2}\rho u^2},$$

$$\frac{u_1}{u} = \sqrt{g - p} \quad \text{and} \quad g = p + \left(\frac{u_1}{u}\right)^2$$

Substitute for u_1 and u_2 for R_V to get the following:

$$R_V = \rho u^2 \iint_{\text{wake}} (1 - \sqrt{g})(\sqrt{g - p})\, dy\, dz \text{ over plane at 1.}$$

At the edge of the wake u_2 or $u_1 = u$ and $g = 1$ and the integrand goes to zero; measurements must extend to edge of the wake to obtain this condition.

The Melville Jones formula, as derived, does not include a free surface, but experimental evidence and comparison with the Betz formula indicates satisfactory use, Townsin [7.17]. For example, experimental evidence indicates that the Melville

Jones and Betz formulae agree at a distance above about 5% of body length down-stream of the model aft end (a typical measurement position is 25% downstream).

7.3.3.2 Experimental Measurements

Examples of this analysis applied to ship models can be found in Shearer and Cross [7.10], Townsin [7.17, 7.18] and, more recently, Insel and Molland [7.19] who applied the methods to monohulls and catamarans. In these examples, for convenience, pressures are measured relative to a still-water datum, whence

$$p = \frac{P_1}{\frac{1}{2}\rho u^2} \quad \text{and} \quad g = \frac{P_1 + \frac{1}{2}\rho u_1^2}{\frac{1}{2}\rho u^2},$$

where P_1 is the local static head (above P_0), $[P_1 + \frac{1}{2}\rho u_1^2]$ is the local total head and u is the free-stream velocity.

Hence, for a complete wake integration behind the model, total and static heads are required over that part of the plane within which total head differs from that in the free stream. Figure 7.12 illustrates a pressure rake. This could combine a series of total and static head probes across the wake. Alternatively, the use of static caps fitted over the total head tubes can be used, and the data taken from a pair of matched runs at a given depth. The spacing of the probes and vertical increments should be chosen to capture the wake with sufficient resolution.

If used in a towing tank, pressures should be measured only during the steady phase of the run by using pressure transducers with suitably filtered averaging applied. Measurement of the local transverse wave elevation is required to obtain the local static pressure deficit $p = 2g\zeta/u^2$, where ζ is the wave elevation above the still water level at the rake position.

A typical analysis might be as follows (shown graphically in Figure 7.13). For a particular speed, $2\sqrt{g - p}(1 - \sqrt{g})$ is computed for each point in the field and plotted to a base of y for each depth of immersion. Integration of these curves yields the viscous resistance R_V.

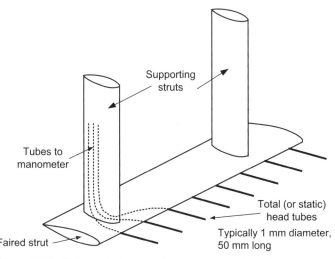

Figure 7.12. Schematic layout of a pressure probe rake.

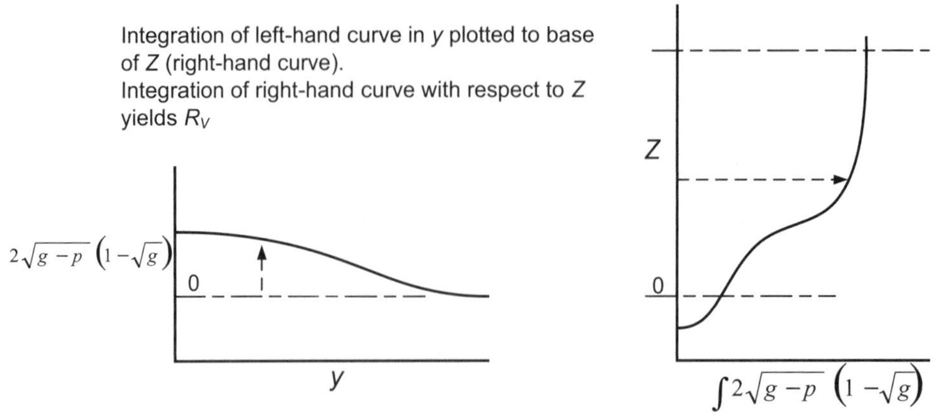

Integration of left-hand curve in *y* plotted to base of *Z* (right-hand curve).
Integration of right-hand curve with respect to *Z* yields R_V

$2\sqrt{g-p}\left(1-\sqrt{g}\right)$

Z

0

0

y

$\int 2\sqrt{g-p}\left(1-\sqrt{g}\right)$

Figure 7.13. Schematic sketches of wake integration process.

7.3.3.3 Typical Wake Distribution for a High C_B Form

The typical wake distribution for a high C_B form is shown in Figure 7.14. This shows that, besides the main hull boundary layer wake deficit, characteristic side lobes may also be displayed. These result from turbulent 'debris' due to a breaking bow wave. They may contain as much as 5% of total resistance (and may be comparable with wave pattern resistance for high C_B forms).

Similar characteristics may be exhibited by high-speed multihulls [7.19], both outboard of the hulls and due to interaction and the breaking of waves between the hulls, Figure 7.15.

7.3.3.4 Examples of Results of Wake Traverse and Surface Pressure Measurements

Figure 7.16 shows the results of wake traverse and surface pressure measurements on a model of the *Lucy Ashton* (Townsin [7.17]). Note that the frictional resistance is obtained from the total resistance minus the pressure resistance. The results show the same general trends as other measurements of resistance components, such as Preston tube C_F measurements, where C_F is seen to be comparable to normal flat plate estimates but slightly Froude number dependent (roughly reciprocal

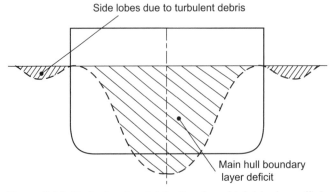

Side lobes due to turbulent debris

Main hull boundary layer deficit

Figure 7.14. Typical wake distribution for a high block coefficient form.

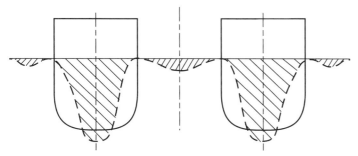

Figure 7.15. Typical wake distribution for a catamaran form.

with humps and hollows in total drag). The form drag correction is approximately of the same order as the Hughes/ITTC-type correction $[(1 + k)\,C_F]$.

7.3.4 Wave Resistance

From a control-volume examination of momentum exchange the ship creates a propagating wave field that in steady motion remains in a fixed position relative to the ship. Measurement of the energy associated with this wave pattern allows the wave resistance to be evaluated. This section explains the necessary analysis of

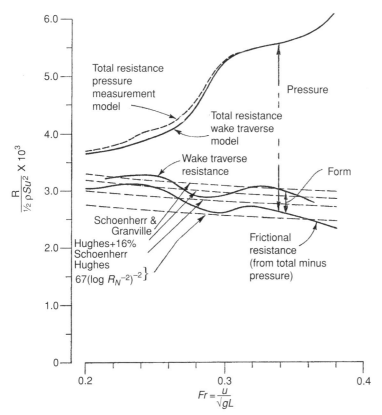

Figure 7.16. A comparison of undulations in the wake traverse resistance and the frictional resistance (Townsin [7.17]).

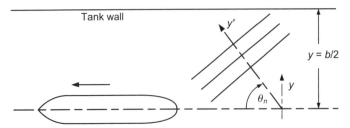

Figure 7.17. Schematic view of a of ship model moving with a wave system.

the wave pattern specifically tailored to measurements made in a channel of finite depth and width. A more detailed explanation of the underpinning analysis is given, for example, by Newman [7.20].

7.3.4.1 Assumed Character of Wave Pattern

Figure 7.17 shows the case of a model travelling at uniform speed down a rectangular channel. The resultant wave pattern can be considered as being composed of a set of plane gravity waves travelling at various angles θ_n to the model path. The ship fixed system is chosen such that:

(a) The wave pattern is symmetrical and stationary,
(b) The wave pattern moves with the model (wave speed condition),
(c) The wave pattern reflects so there is no flow through the tank walls.

The waves are generated at the origin $x = 0$, $y = 0$. The wave components propagate at an angle θ_n and hold a fixed orientation relative to each other and to the ship when viewed in a ship fixed axis system.

(A) WAVE PATTERN. Each wave of angle θ_n can be expressed as a sinusoidally vary-ing surface elevation ζ_n which is a function of distance y' along its direction of propagation. $\zeta_n = A_n \cos(\gamma_n y' + \varepsilon_n)$ say, where A_n and ε_n are the associated amplitude and phase shift, and γ_n is the wave number. The distance along the wave can be expressed as a surface elevation ζ_n which is a function of y'. Now, $y' = y \sin \theta_n - x \cos \theta_n$. Expressing this in terms of the lateral distance y gives the following:

$$\zeta_n = A_n \cos(y\gamma_n \sin \theta - x\gamma_n \cos \theta_n + \varepsilon_n)$$
$$= A_n [\cos(x\gamma_n \cos \theta_n - \varepsilon_n) \cos(y\gamma_n \sin \theta_n) + \sin(x\gamma_n \cos \theta_n - \varepsilon_n) \sin(y\gamma_n \sin \theta_n)].$$

In order for the wave system to be symmetric, every component of wave angle θ_n is matched by a component of angle $-\theta_n$; for which

$$\zeta_n' = A_n [\cos(\ )\cos(\ ) - \sin(\ )\sin(\ )].$$

Hence, adding the two components, a symmetric wave system consists of the terms:

$$\zeta_n = 2A_n \cos(x\gamma_n \cos \theta_n - \varepsilon_n) \cos(y\gamma_n \sin \theta_n)$$
$$= [\xi_n \cos(x\gamma_n \cos \theta_n) + \eta_n \sin(x\gamma_n \cos \theta_n)] \cos(y\gamma_n \sin \theta_n), \qquad (7.13)$$

where ξ_n, η_n are modified wave amplitude coefficients and

$$\xi_n = 2A_n \cos\varepsilon_n \quad \eta_n = 2A_n \sin\varepsilon_n.$$

The complete wave system is considered to be composed of a sum of a number of waves of the above form, known as the Eggers Series, with a total elevation as follows:

$$\zeta = \sum_{n=0}^{\infty} [\xi_n \cos(x\gamma_n \cos\theta_n) + \eta_n \sin(x\gamma_n \cos\theta_n)] \cos(y\gamma_n \sin\theta_n). \quad (7.14)$$

(B) WAVE SPEED CONDITION. For water of finite depth h, a gravity wave will move with a speed c_n of $c_n^2 = \frac{g}{\gamma_n}\tanh(\gamma_n h)$, see Appendix A1.8. The wave system travels with the model, and $c_n = c\cos\theta_n$, where c is the model speed. Hence,

$$\gamma_n \cos^2\theta_n = \frac{g}{c^2}\tanh(\gamma_n h). \quad (7.15)$$

(C) WALL REFLECTION. At the walls $y = \pm b/2$, the transverse components of velocities are zero, and $\frac{d\zeta_n}{dy} = 0$. Hence from Equation (7.13) $\sin\left(\frac{b}{2}\gamma_n \sin\theta_n\right) = 0$, i.e.

$$\frac{b}{2}\gamma_n \sin\theta_n = 0, \pi, 2\pi, 3\pi, \cdots$$

from which

$$\gamma_n \sin\theta_n = \frac{2\pi m}{b}, \quad (7.16)$$

where $m = 0, 1, 2, 3 \cdots$ From Equations (7.15) and (7.16), noting $\cos^2\theta + \sin^2\theta = 1$ and eliminating θ_n, the wave number needs to satisfy

$$\gamma_n^2 = \frac{g}{c^2}\gamma_n \tanh(\gamma_n h) + \left(\frac{2m\pi}{b}\right)^2. \quad (7.17)$$

For infinitely deep water, $\tanh(\gamma_n h) \to 1$ and Equation (7.17) becomes a quadratic equation.

It should be noted that there are a number of discrete sets of values of γ_n and θ_n for a channel of finite width, where γ_n can be found from the roots of Equation (7.17) and θ_n can be found by substituting in Equation (7.16). As the channel breadth increases, the wave angles become more numerous and ultimately the distribution becomes a continuous spectrum.

It is worth examining a typical set of values for θ_n and γ_n which are shown in Table 7.2. These assume that $g/c^2 = 2$, $b = 10$, deep water, $h = \infty$.

Note the way that $(\theta_n - \theta_{n-1})$ becomes much smaller as n becomes larger. It will be shown in Section 7.3.4.3 that the transverse part of the Kelvin wave system corresponds to $\theta_n < 35°$. The above example is typical of a ship model in a (large) towing tank, and it is to be noted how few components there are in this range of angles for a model experiment.

Table 7.2. *Typical sets of allowable wave components for a finite-width tank of infinite depth*

n	γ_n	θ_n	n	γ_n	θ_n
0	2	0°	10	7.35	59.0°
1	2.18	16.8°	15	10.5	63.5°
2	2.60	28.9°	20	13.6	67.1°
3	3.13	37.1°	25	16.7	69.6°
4	3.76	42.0°	30	19.8	71.5°
5	4.3	47.0°			

7.3.4.2 Restriction on Wave Angles in Shallow Water

It can be shown that for small $\gamma_n h$, $\tanh(\gamma_n h) < \gamma_n h$ and so, from Equation (7.15),

$$\gamma_n \cos^2 \theta_n = \frac{g}{c^2} \tanh(\gamma_n h) < \frac{g}{c^2} \gamma_n h$$

$$\therefore \; \cos \theta_n < \frac{\sqrt{gh}}{c}. \tag{7.18}$$

If $c < \sqrt{gh}$ this creates no restriction (since $\cos \theta \le 1.0$ for $0-90°$). Above $c = \sqrt{gh}$, θ_n must be restricted to lie in the range as follows:

$$\theta_n > \cos^{-1}\left(\frac{\sqrt{gh}}{c}\right).$$

Speeds of $c < \sqrt{gh}$ are called sub-critical speeds and of $c > \sqrt{gh}$ are called super-critical speeds.

At super-critical speeds part of the transverse wave system must vanish, as a gravity wave cannot travel at speeds greater than $\sqrt{gh}$.

As an example for shallow water assume that $h = 1$, $g = 9.81$ $\sqrt{gh} = 3.13$ and $c = 4$.

Take

$$\frac{\sqrt{gh}}{c} = \frac{3.13}{4} = 0.783$$

i.e.

$$\cos \theta < 0.783 \quad \text{or} \quad \theta > 38°$$

If c is reduced to 3.13, $\cos \theta \le 1$, $\theta > 0$ and all angles are now included.

7.3.4.3 Kelvin Wave System

It can be shown theoretically that the wave system generated by a point source is such that, for all components, $\eta_n = 0$, so all the wave components will have a crest at the point $x = y = 0$ above the source position. This fact can be used to construct the Kelvin wave pattern from a system of plane waves.

If a diagram is drawn for the wave system, Figure 7.18, it is found that the crest lines of the wave components cross over each other and there is one region where many wave crests (or troughs) come together to produce a large crest (or trough) in the overall system.

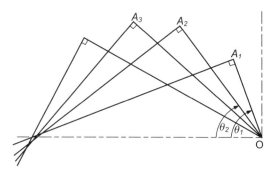

Figure 7.18. Graphical representation of wave components showing relative change in wave-length and intersection of crests.

Figure 7.18 defines the location of the wave crests. Let OA_1 be a given multiple of one wave length for a wave angle θ_1, and OA_2 be the same multiple of wave length for wave angle θ_2 etc. The corresponding wave crest lines overlay to produce the envelope shown.

If A-A is a crest line in waves from 0, in order to define the wave envelope in deep water, the equation of any given crest line A-A associated with a wave angle θ is required, where A-A is a crest line m waves from 0, Figure 7.19.

For a stationary wave pattern, the wave speed is $C_n(\theta) = c \cos \theta$ and $\lambda = 2\pi c^2/g$. In wave pattern, m lengths at θ_n along OP

$$= \frac{2\pi c_n^2 m}{g} = \frac{2\pi c^2 m \cos^2 \theta}{g} = m\lambda \cos^2 \theta.$$

Hence, the distance of A-A from source origin 0 is $\lambda \cos^2 \theta$, for $m = 1$, (since source waves all have crest lines through 0, and $m = 1, 2, 3 \ldots$).The co-ordinates of P are $(-\lambda \cos^3 \theta \quad \lambda \cos^2 \theta \sin \theta)$ and the slope of A-A is $\tan(\pi/2 - \theta) = \cot \theta$. Hence, the equation for A-A is $y - y_p = (x - x_p)\cot \theta$. Substitution for x_p, y_p gives the following:

$$y = \frac{\lambda \cos^2 \theta}{\sin \theta} + x \cot \theta. \tag{7.19a}$$

In order to find the equation of the wave envelope it is required to determine the point where this line meets a neighbouring line at wave angle $\theta + \delta\theta$. The equation

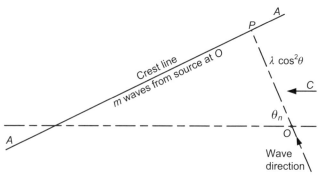

Figure 7.19. Geometrical representation of a wave crest relative to the origin.

of this neighbouring crest line is

$$y' = y + \frac{dy}{d\theta}\delta\theta.$$

Now, as

$$y = (x + \lambda\cos\theta)\cot\theta$$

$$\frac{dy}{d\theta} = -(x + \lambda\cos\theta)\operatorname{cosec}^2\theta - \lambda\sin\theta\cot\theta$$

$$= -x\operatorname{cosec}^2\theta - \lambda\cos\theta\operatorname{cosec}^2\theta - \lambda\sin\theta\cot\theta$$

$$= -x\operatorname{cosec}^2\theta - \lambda\cos\theta(\operatorname{cosec}^2\theta + 1)$$

hence,

$$y' = y + [-x\operatorname{cosec}^2\theta - \lambda\cos\theta(\operatorname{cosec}^2\theta + 1)]\delta\theta$$

$$= y + \frac{1}{\sin^2\theta}[-x - \lambda\cos\theta(1 + \sin^2\theta)]\delta\theta$$

In order for the wave crests for wave angles θ and $\theta + \delta\theta$ to intersect, $y' = y$ and, hence, $[-x - \lambda\cos\theta(1 + \sin^2\theta)] = 0$. Thus, the intersection is at the point:

$$x = -\lambda\cos\theta(1 + \sin^2\theta)$$

and

$$y = -\lambda\cos^2\theta\sin\theta. \tag{7.19b}$$

These parametric Equations (7.19b) represent the envelope of the wave crest lines, shown schematically in Figure 7.20.

On differentiating with respect to θ,

$$\frac{\mathrm{d}x}{\mathrm{d}\theta} = \lambda\sin\theta(1 + \sin^2\theta) - \lambda\cos\theta 2\sin\theta\cos\theta$$

$$= -\lambda\sin\theta(1 - 3\sin^2\theta).$$

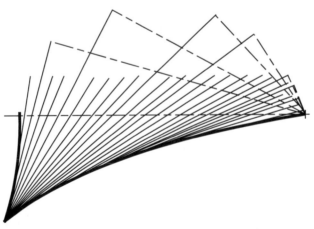

Figure 7.20. Overlay of crest lines and wave envelope.

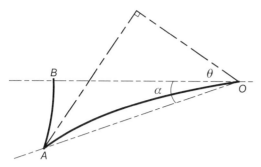

Figure 7.21. Deep water wave envelope with cusp located at A.

Similarly

$$\frac{dy}{d\theta} = -\lambda 2 \cos\theta \sin\theta \sin\theta - \cos\theta\lambda \cos^2\theta$$
$$= -\lambda(-2\cos\theta \sin^2\theta + \cos^3\theta)$$
$$= -\lambda \cos\theta(-2\sin^2\theta + \cos^2\theta)$$
$$= -\lambda \cos\theta(1 - 3\sin^2\theta)$$

and

$$\frac{dx}{d\theta} = \frac{dy}{d\theta} = 0 \quad \text{at} \quad \theta = \sin^{-1}(1/\sqrt{3}) = 35.3°.$$

The point A corresponding to $\theta = \sin^{-1}(1/\sqrt{3})$ is a cusp on the curve as shown in Figure 7.21. $\theta = 0$ corresponds to $x = -\lambda$, $y = 0$ (point B on Figure 7.21) and $\theta = \pi/2$ corresponds to $x = y = 0$ the origin. Hence, the envelope has the appearance as shown.

By substituting the co-ordinates of A,

$$x = \frac{-4\sqrt{2}}{3\sqrt{3}}\lambda$$
$$y = \frac{-2}{3\sqrt{3}}\lambda.$$

The slope of line OA is such that

$$\alpha = \tan^{-1}\left(\frac{y}{x}\right) = \tan^{-1}\left(\frac{1}{2\sqrt{2}}\right) = 19°47' \text{ or}$$
$$\alpha = \sin^{-1}\left(\frac{1}{3}\right)$$

Figure 7.22 summarises the preceding description with a graphical representation of a deep water Kelvin wave. As previously noted in Chapter 3, Section 3.1.5, a ship hull can be considered as a number of wave sources acting along its length, typically dominated by the bow and stern systems.

Figure 7.22 shows the construction of the complete wave system for a Kelvin wave source. Varying values of $m\lambda$ correspond to successive crest lines and a whole series of geometrically similar crest lines are formed to give the complete Kelvin pattern, Figure 7.23.

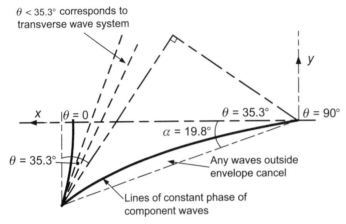

Figure 7.22. Kelvin wave system development.

7.3.4.4 Eggers Formula for Wave Resistance (Summary)

The following is a summary of the wave resistance analysis given in more detail in Appendix A2. The analysis is also explained in some detail in the publications of Hogben [7.21, 7.22 and 7.23], together with the use of wave probes to measure wave resistance.

From the momentum analysis of the flow around a hull (see Chapter 3, Equation (3.10)), it can be deduced that

$$R = \left\{ \frac{1}{2}\rho g \int\limits_{-b/2}^{b/2} \zeta_B^2 dy + \frac{1}{2}\rho \int\limits_{-b/2}^{b/2} \int\limits_{-h}^{\zeta_B} (v^2 + w^2 - u^2)dz\,dy \right\}$$
$$+ \int\limits_{-b/2}^{b/2} \int\limits_{-h}^{\zeta_B} \Delta p\, dz\, dy, \qquad (7.20)$$

where the first two terms are broadly associated with wave pattern drag, although perturbation velocities v, w, u are due partly to viscous shear in the boundary layer.

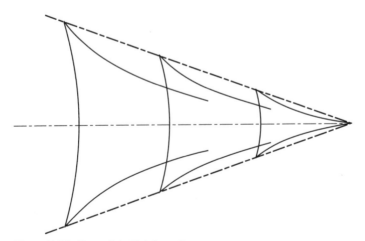

Figure 7.23. Complete Kelvin pattern.

Thus, from measurements of wave elevation ζ and perturbation velocity components u, v, w over the downstream plane, the wave resistance could be determined. However, measurements of subsurface velocities u, v, w would be difficult to make, so linearised potential theory is used, in effect, to deduce these velocities from the more conveniently measured wave pattern (height ζ).

It has been shown that the wave elevation may be expressed as the Eggers series, as follows:

$$\zeta = \sum_{n=0}^{\infty} [\xi_n \cos(x\gamma_n \cos\theta_n) + \eta_n \sin(x\gamma_n \cos\theta_n)] \cos\left(\frac{2\pi ny}{b}\right), \qquad (7.21)$$

as

$$\gamma_n \sin\theta_n = \frac{2\pi n}{b}.$$

Linearising the free-surface pressure condition for small waves yields the following:

$$c\frac{\partial\theta}{\partial x} + g\zeta = 0 \quad \text{or} \quad \zeta = -\frac{c}{g}\frac{\partial\theta}{\partial x}\Big|_{z=0}.$$

by using this result, the velocity potential for the wave pattern can be deduced as the following:

$$\phi = \frac{g}{c}\sum_{n=0}^{\infty} \frac{\cosh\gamma_n(z+h)}{\lambda_n\cosh(\gamma_n h)} [\eta_n \cos\lambda_n x - \xi_n \sin\lambda_n x] \cos\frac{2\pi ny}{b}, \qquad (7.22)$$

where

$$\lambda_n = \gamma_n \cos\theta_n.$$

From the momentum analysis, Equation (7.20), wave resistance will be found from

$$R_w = \left\{ \frac{1}{2}\rho g \int_{-b/2}^{b/2} \zeta_B^2 dy + \int_{-b/2}^{b/2}\int_{-h}^{\zeta_B} (v^2 + w^2 - u^2)dz\,dy \right\}, \qquad (7.23)$$

now

$$u = \frac{\partial\phi}{\partial x} \qquad v = \frac{\partial\phi}{\partial y} \qquad w = \frac{\partial\phi}{\partial z},$$

which can be derived from Equation (7.22).

Hence, substituting these values of u, v, w into Equation (7.23) yields the Eggers formula for wave resistance R_w in terms of ξ_n and η_n, i.e. for the deep water case:

$$R_w = \frac{1}{4}\rho g b \left\{ (\xi_0^2 + \eta_0^2) + \sum_{n=1}^{\infty} (\xi_n^2 + \eta_n^2)\left(1 - \frac{1}{2}\cos^2\theta_n\right) \right\}. \qquad (7.24)$$

If the coefficients γ_n and θ_n have been determined from Equations (7.17) and (7.16), the wave resistance may readily be found from (7.24) once the coefficients ξ_n and η_n have been determined. The coefficients ξ_n and η_n can be found by measuring the wave pattern elevation. They can also be obtained theoretically, as described in Chapter 9.

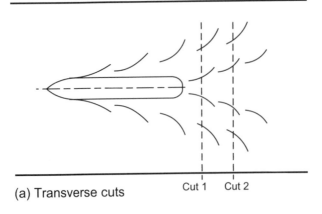

(a) Transverse cuts   Cut 1   Cut 2

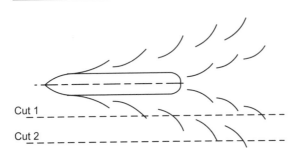

Cut 1

Cut 2

(b) Longitudinal cuts

Figure 7.24. Possible wave cuts to determine wave resistance.

7.3.4.5 Methods of Wave Height Measurement and Analysis

Figure 7.24 shows schematically two possible methods of measuring wave elevation (transverse and longitudinal cuts) that can be applied to determine wave resistance.

(A) TRANSVERSE CUT. In this approach the wave elevation is measured for at least two positions behind the model, Figure 7.24(a). Each cut will be a Fourier series in y.

$$\zeta = \sum_{n=0}^{\infty} [\xi_n \cos(x\gamma_n \cos\theta_n) + \eta_n \sin(x\gamma_n \cos\theta_n)] \cos\left(\frac{2\pi ny}{b}\right).$$

For a fixed position x

$$\zeta = \sum A_n \cos\frac{2\pi ny}{b},$$

and for cut 1 $A_{n1} = \xi_n \cos(x_1\gamma_n \cos\theta_n) + \eta_n \sin(x_1\gamma_n \cos\theta_n)$.

For cut 2 $A_{n2} = \xi_n \cos(x_2\gamma_n \cos\theta_n) + \eta_n \sin(x_2\gamma_n \cos\theta_n)$.

Values of A_{n1}, A_{n2} are obtained from a Fourier analysis of

$$\zeta = \sum A_n \cos\frac{2\pi ny}{b}$$

hence, two equations from which ξ_n and η_n can be found for various known values of θ_n.

This is generally not considered a practical method. Stationary probes fixed to the towing tank are not efficient as a gap must be left for the model to pass through. Probes moving with the carriage cause problems such as non-linear velocity effects for typical resistance or capacitance two-wire wave probes. Carriage-borne mechanical pointers can be used, but the method is very time consuming. The method is theoretically the most efficient because it correctly takes account of the tank walls.

(B) LONGITUDINAL CUT. The cuts are made parallel to the centreline, Figure 7.24(b). The model is driven past a single wave probe and measurements are made at equally spaced intervals of time to give the spatial variation.

Only one cut is required. In practice, up to four cuts are used to eliminate the possible case of the term $\cos 2\pi ny/b$ tending to zero for that n, i.e. is a function of ny, hence a different cut (y value) may be required to get a reasonable value of ζ.

$$\zeta = \sum_{n=0}^{\infty} [\xi_n \cos (x\gamma_n \cos \theta_n) + \eta_n \sin (x\gamma_n \cos \theta_n)] \cos \left(\frac{2\pi ny}{b}\right). \qquad (7.25)$$

In theory, for a particular value of y, the ζ values can be measured for different values of x and simultaneous equations for ξ_n and η_n solved. In practice, this approach tends to be inaccurate, and more rigorous analysis methods are usually used to overcome this deficiency [7.21–7.23].

Current practice is to use (multiple) longitudinal cuts to derive ξ_n and η_n and, hence, find R_w from the Eggers resistance formula. Analysis techniques differ, and multiple longitudinal cuts are sometimes referred to as 'matrix' methods [7.23].

7.3.4.6 Typical Results from Wave Pattern Analysis
Figures 7.25–7.27 give typical wave resistance contributions for given wave components. Summation of the resistance components gives the total wave resistance for a finite width tank. Work on the performance of the technique for application to catamaran resistance is described in the doctoral theses of Insel [7.24], Couser [7.25] and Taunton [7.26].

Figure 7.25 shows that the wave energy is a series of humps dying out at about 75°. The largest hump extends to higher wave angles as speed increases. Energy due to the transverse wave system lies between $\theta = 0°$ and approximately 35°. At low speeds, the large hump lies within the transverse part of the wave pattern and, there, transverse waves predominate, but at higher speeds the diverging waves become more significant. Figure 7.26 shows low wave resistance associated with transverse wave interference.

In shallow water, $\cos \theta_n < \sqrt{gh}/c$ and above $c = \sqrt{gh}$, θ is restricted to lie in the range $\theta_n > \cos^{-1} \sqrt{gh}/c$. Above $c = \sqrt{gh}$ (super-critical), part of the transverse wave system must vanish, Figure 7.27, since a gravity wave cannot travel at speeds $> \sqrt{gh}$. At these speeds, only diverging waves are present, Figure 7.27.

7.3.4.7 Example Results of Wake Traverse and Wave Pattern Measurements
Insel and Molland [7.19] carried out a detailed study of the resistance components of semi-displacement catamarans using wake traverse and wave resistance measurements. Figure 7.28 demonstrates the relative importance of each resistance

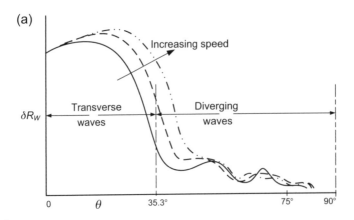

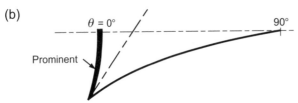

Figure 7.25. Typical wave energy distribution and prominent part of wave pattern.

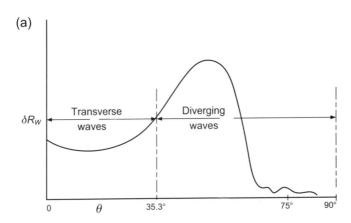

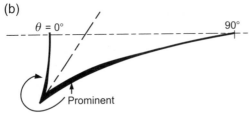

Figure 7.26. Wave energy distribution: effect of transverse wave interference and prominent parts of wave pattern.

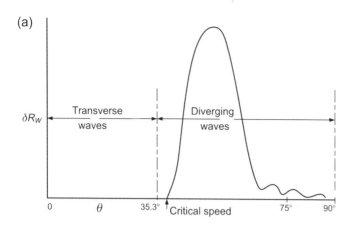

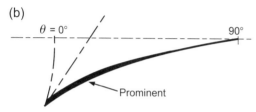

Figure 7.27. Wave energy distribution: influence of shallow water and prominent part of wave pattern.

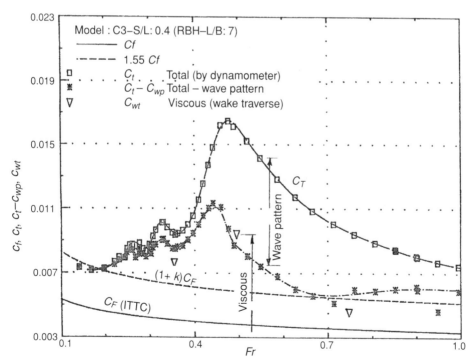

Figure 7.28. Resistance components of C3 catamaran with hull separation ratio of 0.4 [7.19].

component. It is noted that broad agreement is achieved between the total measured drag (by dynamometer) and the sum of the viscous and wave pattern drags. In this particular research programme one of the objectives was to deduce the form factors of catamaran models, both by measuring the total viscous drag (by wake traverse) C_V, whence $C_V = (1 + k)C_F$, hence $(1 + k)$, and by measuring wave pattern drag C_{WP}, whence $C_V = (C_T - C_{WP}) = (1 + k)C_F$, hence, $(1 + k)$. See also Chapter 4, Section 4.4 for a discussion of the derivation of form factors.

7.4 Flow Field Measurement Techniques

The advent of significant computational power and development of coherent (laser) light sources has made possible non-invasive measurements of the flow field surrounding a ship hull. Although these techniques are usually too expensive to be applied to measure the resistance components directly, they are invaluable in providing data for validation of CFD-based analysis. As an example of the development of such datasets, Kim *et al.* [7.27] report on the use of a five-hole Pitôt traverse applied to the towing tank tests of two crude carriers and a container ship hull forms. Wave pattern and global force measurements were also applied. Associated tests were also carried out in a wind tunnel using laser Doppler velocimetry (LDV) for the same hull forms. This dataset formed part of the validation dataset for the ITTC related international workshops on CFD held in Gothenburg (2000) and Tokyo (2005).

The following sections give a short overview of available techniques, including both the traditional and the newer non-invasive methods.

7.4.1 Hot-Wire Anemometry

The hot wire is used in wind tunnel tests and works in the same manner as the hot-film shear stress gauge, that is, the passage of air over a fine wire through which an electric current flows, which responds to the rapid changes in heat transfer associated fluctuations in velocity. Measurement of the current fluctuations and suitable calibration allows high-frequency velocity field measurements to be made [7.28]. A single wire allows measurement of the mean flow U and fluctuating component u'. The application of two or three wires at different orientations allows the full mean and Reynolds stress components to be found. The sensitivity of the wire is related to its length and diameter and, as a result, tends to be vulnerable to damage. The wires would normally be moved using an automated traverse.

7.4.2 Five-Hole Pitôt Probe

A more robust device for obtaining three mean velocity components is a five-hole Pitôt probe. As the name suggests, these consist of five Pitôt probes bound closely together. Figure 7.29 illustrates the method of construction and a photograph of an example used to measure the flow components in a wind tunnel model of a waterjet inlet, Turnock *et al.* [7.29].

There are two methods of using these probes. In the first, two orthogonal servo drives are moved to ensure that there is no pressure difference between the

(a)

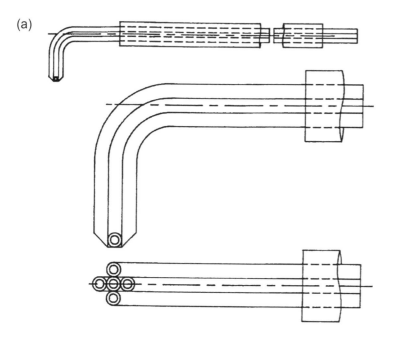

(b)

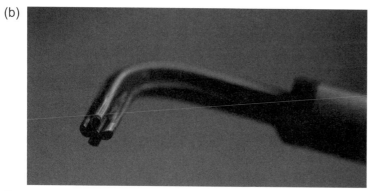

Figure 7.29. Five-hole Pitôt.

vertical and transverse pressure pairs. Measuring the dynamic pressure and the two resultant orientations of the whole probe allows the three velocity components to be found. In the normal approach, an appropriate calibration map of pressure differences between the side pairs of probes allows the flow direction and magnitude to be found.

Total pressure measurements are only effective if the onset flow is towards the Pitôt probe. Caution has to be taken to ensure that the probe is not being used in a region of separated flow. The earlier comments about the measurement of pressure in water similarly apply to use of a five-hole probe. The pressure measurement is most responsive for larger diameter and small runs of pressure tube.

7.4.3 Photogrammetry

The recent advances in the frame rate and pixel resolution of digital cameras, both still and moving, offer new opportunities for capturing free-surface wave elevations.

Capture rates of greater than 5000 frames per second are now possible, with typical colour image sizes of 5–10 Mbit. Lewis *et al.* [7.14] used such a camera to capture the free-surface elevation as a free-falling two-dimensional wedge impacted still water. Glass microparticles were used to enhance the contrast of the free surface. Good quality images rely on application of suitable strength light sources. Again, recent improvements in light-emitting diodes (LEDs) allow much more intense light to be created without the usual problems with halogen bulbs of high power, and the need to dissipate heat which is difficult underwater.

Alternative application of the technology can be applied to capture the free-surface elevation of a wider area or along a hull surface. One of the difficulties is the transparency of water. Methods to overcome this problem include the methods used by competition divers where a light water mist is applied to the free surface to improve contrast for determining height, or a digital data projector is used to project a suitable pattern onto the water surface. Both of these allow image recognition software to infer surface elevation. Such methods are still the subject of considerable development.

7.4.4 Laser-Based Techniques

The first applications of the newly developed single frequency, coherent (laser) light sources to measurements in towing tanks took place in the early 1970s (see for example Halliwell [7.30] who used single component laser Doppler velocimetry in the Lamont towing tank at the University of Southampton). In the past decade there has been a rapid growth in their area of application and in the types of technique available. They can be broadly classed into two different types of system as follows.

7.4.4.1 Laser Doppler Velocimetry

In this technique the light beam from a single laser source is split. The two separate beams are focussed to intersect in a small volume in space. As small particles pass through this volume, they cause a Doppler shift in the interference pattern between the beams. Measurement of this frequency shift allows the instantaneous velocity of the particle to be inferred. If sufficient particles pass through the volume, the frequency content of the velocity component can be determined. The use of three separate frequency beams, all at difficult angles to the measurement volume, allows three components of velocity to be measured. If enough passages of a single particle can be captured simultaneously on all three detectors, then the correlated mean and all six Reynolds stress components can be determined. This requires a high density of seeded particles. A further enhancement for rotating propellers is to record the relative location of the propeller and to phase sort the data into groups of measurements made with the propeller at the same relative orientation over many revolutions. Such measurements, for instance, can give significant insight into the flow field interaction between a hull, propeller and rudder. Laser systems can also be applied on full-scale ships with suitable boroscope or measurement windows placed at appropriate locations on a ship hull, for instance, at or near the propeller plane. In this case, the system usually relies on there being sufficient existing particulates within the water.

The particles chosen have to be sufficiently small in size and mass that they can be assumed to be moving with the underlying flow. One of the main difficulties is in ensuring that sufficient particles are 'seeded' within the area of interest. A variety of particles are available and the technique can be used in air or water. Within water, a good response has been found with silver halide-based particles. These can, however, be expensive to seed at a high enough density throughout a large towing tank as well as imposing environmental constraints on the eventual disposal of water from the tank. In wind tunnel applications, smoke generators, as originally used in theatres, or vapourised vegetable oil particles can be applied.

Overall LDV measurement can provide considerable physical insight into the time-varying flow field at a point in space. Transverse spatial distributions can only be obtained by traversing the whole optical beam head/detector system so that it is focussed on another small volume. These measurement volumes are of the order of a cubic millimetre. Often, movement requires slight re-alignment of multiple beams which can often be time consuming. Guidance as to the uncertainty associated with such measurements in water based facilities, and general advice with regard to test processes, can be found in ITTC report 7.5-01-03-02 [7.31].

7.4.4.2 Particle Image Velocimetry

A technology being more rapidly adopted is that of particle image velocimetry (PIV) and its many variants. Raffel *et al.* [7.32] give a thorough overview of all the possible techniques and designations and Gui *et al.* [7.33] give a description of its application in a towing tank environment. The basic approach again relies on the presence of suitable seeded particles within the flow. A pulsed beam of laser light is passed through a lens that produces a sheet of light. A digital camera is placed whose axis of view is perpendicular to the plane of the sheet, Figure 7.30. The lens of the camera is chosen such that the focal plane lies at the sheet and that the capture area maps across the whole field of view of the camera. Two images are captured in short succession. Particles which are travelling across the laser sheet will produce a bright flash at two different locations. An area based statistical correlation technique is usually used to infer the likely transverse velocity components for each interrogation area. As a result, the derivation of statistically satisfactory results requires the results from many pairs of images to be averaged.

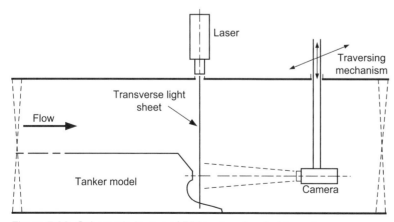

Figure 7.30. Schematic layout of PIV system in wind tunnel.

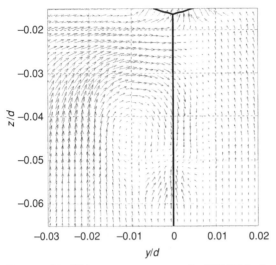

Figure 7.31. PIV measurements on the KVLCC hull.

The main advantage of this method is that the average velocity field can be found across an area of a flow. The resolution of these pairs of transverse velocity components is related to the field of view and the pixel size of the charge-coupled device (CCD) camera. Larger areas can be constructed using a mosaic of overlapping sub-areas. Again, the physical insight gained can be of great importance. For instance, the location of an off-body flow feature, such as a bilge vortex, can be readily identified and its strength assessed.

A restriction on earlier systems was the laser pulse recharge rate so that obtaining sufficient images of approximately 500 could require a long time of continuous operation of the experimental facility. The newer laser systems allow many more dynamic measurements to be made and, with the application of multiple cameras and intersecting laser sheets, all velocity components can be found at a limited number of locations.

Figure 7.30 shows the schematic layout of the application of a PIV system to the measurement of the flow field at the propeller plane of a wind tunnel model (1 m long) in the 0.9 m × 0.6 m open wind tunnel at the University of Southampton [7.34].

Figure 7.31 gives an example of PIV measurements on the KVLCC hull at a small yaw angle, clearly showing the presence of a bilge vortex on the port side [7.34], [7.35].

7.4.5 Summary

The ability of the experimenter to resolve the minutiae of the flow field around as well as on a hull model surface allows a much greater depth of understanding of the fluid dynamic mechanisms of resistance and propulsion. The drawback of such detail is the concomitant cost in terms of facility hire, model construction and experimenter expertise. Such quality of measurement is essential if the most is to be made of the CFD-based analysis tools described in Chapters 9 and 15. It is vital that the uncertainty associated with the test environment, equipment, measurements and subsequent analysis are known.

In summarising, it is worth noting that measurements of total viscous and wave pattern resistance yield only overall effects. These indicate how energy dissipation is modified by hull form variation, although they do not indicate the local origins of the effects. However, local surface measurements of pressure and frictional resistance allow an examination of the distribution of forces to be made and, hence, an indication of the effect due to specific hull and appendage modifications. Although pressure measurements are reasonably straightforward, friction measurements are extremely difficult and only a few tests on this component have been carried out.

Particular problems associated with the measurement of the individual components of resistance include the following:

(a) When measuring pressure resistance it is very important to measure model trim and to take this into account in estimating local static pressures.
(b) Wave breaking regions can be easily overlooked in making wake traverse experiments.
(c) Measurements of wave patterns can (incorrectly) be made in the local hull disturbance region and longitudinal cuts made for too short a spatial distance.

As a general comment it is suggested that wave resistance measurements should be made as a matter of course during the assessment of total resistance and in self-propulsion tests. This incurs little additional cost and yet provides considerable insight into any possible Froude number dependence of form factor and flow regimes where significant additional viscous or induced drag components exist.

REFERENCES (CHAPTER 7)

7.1 ITTC *International Towing Tank Conference*. Register of Recommended Procedures. Accessed via www.sname.org. Last accessed January 2011.
7.2 Coleman, H.W. and Steele, W.G. *Experimentation and Uncertainty Analysis for Engineers*. 2nd Edition. Wiley, New York, 1999.
7.3 ISO (1995) Guide to the expression of uncertainty in measurement. International Organisation for Standardisation, Genève, Switzerland. ISO ISBN 92-67-10188-9, 1995.
7.4 Lofdahl, L. and Gad-el-Hak, M., MEMS-based pressure and shear stress sensors for turbulent flows. *Measurement Science Technology*, Vol. 10, 1999, pp. 665–686.
7.5 Fernholz, H.H., Janke, G., Schober, M., Wagner, P.M. and Warnack, D., New developments and applications of skin-friction measuring techniques. *Measurement Science Technology*, Vol. 7, 1996, pp. 1396–1409.
7.6 Patel, V.C. Calibration of the Preston tube and limitations on its use in pressure gradients. *Journal of Fluid Mechanics*, Vol. 23, 1965, pp. 185–208.
7.7 Steele, B.N. and Pearce, G.B. Experimental determination of the distribution of skin friction on a model of a high speed liner. *Transactions of the Royal Institution of Naval Architects*, Vol. 110, 1968, pp. 79–100.
7.8 Shearer, J.R. and Steele, B.N. Some aspects of the resistance of full form ships. *Transactions of the Royal Institution of Naval Architects*, Vol. 112, 1970, pp. 465–486.

7.9 Ireland, P.T. and Jones, T.V. Liquid crystal measurements of heat transfer and surface shear stress. *Measurement Science Technology*, Vol. 11, 2000, pp. 969–986.

7.10 Shearer, J.R. and Cross, J.J. The experimental determination of the components of ship resistance for a mathematical model. *Transactions of the Royal Institution of Naval Architects*, Vol. 107, 1965, pp. 459–473.

7.11 Townsin, R.L. The frictional and pressure resistance of two 'Lucy Ashton' geosims. *Transactions of the Royal Institution of Naval Architects*, Vol. 109, 1967, pp. 249–281.

7.12 Molland A.F. and Turnock, S.R. Wind tunnel investigation of the influence of propeller loading on ship rudder performance. *University of Southampton, Ship Science Report No. 46*, 1991.

7.13 Molland A.F., Turnock, S.R. and Smithwick, J.E.T. Wind tunnel tests on the influence of propeller loading and the effect of a ship hull on skeg-rudder performance, *University of Southampton, Ship Science Report No. 90*, 1995.

7.14 Lewis, S.G., Hudson, D.A., Turnock, S.R. and Taunton, D.J. Impact of a freefalling wedge with water: synchronised visualisation, pressure and acceleration measurements, *Fluid Dynamics Research*, Vol. 42, No. 3, 2010.

7.15 van Dam, C.P. Recent experience with different methods of drag prediction. *Progress in Aerospace Sciences*, Vol. 35, 1999, pp. 751–798.

7.16 Giles, M.B. and Cummings, R.M. Wake integration for three-dimensional flowfield computations: theoretical development, *Journal of Aircraft*, Vol. 36, No. 2, 1999, pp. 357–365.

7.17 Townsin, R.L. Viscous drag from a wake survey. Measurements in the wake of a 'Lucy Ashton' model. *Transactions of the Royal Institution of Naval Architects*, Vol. 110, 1968, pp. 301–326.

7.18 Townsin, R.L. The viscous drag of a 'Victory' model. Results from wake and wave pattern measurements. *Transactions of the Royal Institution of Naval Architects*, Vol. 113, 1971, pp. 307–321.

7.19 Insel, M. and Molland, A.F. An investigation into the resistance components of high speed displacement catamarans. *Transactions of the Royal Institution of Naval Architects*, Vol. 134, 1992, pp. 1–20.

7.20 Newman, J. *Marine Hydrodynamics*. MIT Press, Cambridge, MA, 1977.

7.21 Gadd, G.E. and Hogben, N. The determination of wave resistance from measurements of the wave profile. *NPL Ship Division Report No. 70*, November 1965.

7.22 Hogben, N. and Standing, B.A. Wave pattern resistance from routine model tests. *Transactions of the Royal Institution of Naval Architects*, Vol. 117, 1975, pp. 279–299.

7.23 Hogben, N. Automated recording and analysis of wave patterns behind towed models. *Transactions of the Royal Institution of Naval Architects*, Vol. 114, 1972, pp. 127–153.

7.24 Insel, M. An investigation into the resistance components of high speed displacement catamarans. Ph.D. thesis, University of Southampton, 1990.

7.25 Couser, P. An investigation into the performance of high-speed catamarans in calm water and waves. Ph.D. thesis, University of Southampton, 1996.

7.26 Taunton, D.J. Methods for assessing the seakeeping performance of high speed displacement monohulls and catamarans. Ph.D. thesis, University of Southampton, 2001.

7.27 Kim, W.J., Van, S.H., Kim, D.H., Measurement of flows around modern commercial ship models, *Experiments in Fluids*, Vol. 31, 2001, pp. 567–578.

7.28 Bruun, H.H. *Hot-Wire Anemometry: Principles and Signal Processing*. Oxford University Press, Oxford, UK, 1995.

7.29 Turnock, S.R., Hughes, A.W., Moss, R. and Molland, A.F. *Investigation of hull-waterjet flow interaction. Proceedings of Fourth International Conference on Fast Sea Transportation, FAST'97*, Sydney, Australia. Vol. 1, Baird Publications, South Yarra, Australia, July 1997, pp. 51–58.

7.30 Halliwell, N.A. A laser anemometer for use in ship research. *University of Southampton, Ship Science Report No. 1*, 1975.

7.31 International Towing Tank Conference. Uncertainty analysis: laser Doppler velocimetry calibration. Document 7.5-01-03-02, 2008, 14 pp.

7.32 Raffel, M., Willert, C., Werely, S. and Kompenhans, J. *Particle Image Velocimetry: A Practicle Guide*. 2nd Edition. Springer, New York, 2007.

7.33 Gui, L., Longo, J. and Stern, F, Towing tank PIV measurement system, data and uncertainty assessment for DTMB Model 5512. *Experiments in Fluids*, Vol. 31, 2001, pp. 336–346.

7.34 Pattenden, R.J.,Turnock, S.R., Bissuel, M. and Pashias, C. Experiments and numerical modelling of the flow around the KVLCC2 hullform at an angle of yaw. *Proceedings of 5th Osaka Colloquium on Advanced Research on Ship Viscous Flow and Hull Form Design*, 2005, pp. 163–170.

7.35 Bissuel, M. Experimental investigation of the flow around a KVLCC2 hull for CFD validation. M.Sc. thesis, University of Southampton, 2004.

8 Wake and Thrust Deduction

8.1 Introduction

An interaction occurs between the hull and the propulsion device which affects the propulsive efficiency and influences the design of the propulsion device. The components of this interaction are wake, thrust deduction and relative rotative efficiency.

Direct detailed measurements of wake velocity at the position of the propeller plane can be carried out in the absence of the propeller. These provide a detailed knowledge of the wake field for detailed aspects of propeller design such as radial pitch variation to suit a particular wake, termed wake adaption, or prediction of the variation in load for propeller strength and/or vibration purposes.

Average wake values can be obtained indirectly by means of model open water and self-propulsion tests. In this case, an integrated average value over the propeller disc is obtained, known as the effective wake. It is normally this average effective wake, derived from self-propulsion tests or data from earlier tests, which is used for basic propeller design purposes.

8.1.1 Wake Fraction

A propeller is situated close to the hull in such a position that the flow into the propeller is affected by the presence of the hull. Thus, the average speed of flow into the propeller (Va) is different from (usually less than) the speed of advance of the hull (Vs), Figure 8.1. It is usual to refer this change in speed to the ship speed, termed the Taylor wake fraction w_T, where w_T is defined as

$$w_T = \frac{(Vs - Va)}{Vs} \qquad (8.1)$$

and

$$Va = Vs(1 - w_T). \qquad (8.2)$$

Figure 8.1. Wake speed *Va*.

8.1.2 Thrust Deduction

The propulsion device (e.g. propeller) accelerates the flow ahead of itself, thereby (a) increasing the rate of shear in the boundary layer and, hence, increasing the frictional resistance of the hull and (b) reducing pressure (Bernouli) over the rear of the hull, and hence, increasing the pressure resistance. In addition, if separation occurs in the afterbody of the hull when towed without a propeller, the action of the propeller may suppress the separation by reducing the unfavourable pressure gradient over the afterbody. Hence, the action of the propeller is to alter the resistance of the hull (usually to increase it) by an amount that is approximately proportional to thrust. This means that the thrust will exceed the naked resistance of the hull. Physically, this is best understood as a resistance augment. In practice, it is taken as a thrust deduction, where the thrust deduction factor t is defined as

$$t = \frac{(T - R)}{T} \qquad (8.3)$$

and

$$T = \frac{R}{(1 - t)}. \qquad (8.4)$$

8.1.3 Relative Rotative Efficiency η_R

The efficiency of a propeller in the wake behind the ship is not the same as the efficiency of the same propeller under the conditions of the open water test. There are two reasons for this. (a) The level of turbulence in the flow is low in an open water test in a towing tank, whereas it is very high in the wake behind a hull and (b) the flow behind a hull is non-uniform so that flow conditions at each radius are different from the open water test.

The higher turbulence levels tend to reduce propeller efficiency, whilst a propeller deliberately designed for a radial variation in wake can gain considerably when operating in the wake field for which it was designed.

The derivation of relative rotative efficiency in the self-propulsion test is described in Section 8.7 and empirical values are given with the propeller design data in Chapter 16.

8.2 Origins of Wake

The wake originates from three sources: potential flow effects, the effects of friction on the flow around the hull and the influence of wave subsurface velocities.

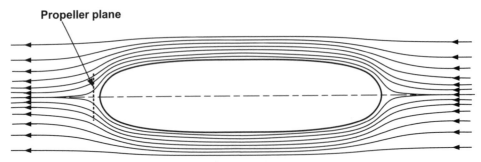

Figure 8.2. Potential wake.

8.2.1 Potential Wake: w_P

This arises in a frictionless or near-frictionless fluid. As the streamlines close in aft there is a rise in pressure and decrease in velocity in the position of the propeller plane, Figure 8.2.

8.2.2 Frictional Wake: w_F

This arises due to the hull surface skin friction effects and the slow-moving layer of fluid (boundary layer) that develops on the hull and increases in thickness as it moves aft. Frictional wake is usually the largest component of total wake. The frictional wake augments the potential wake. Harvald [8.1] discusses the estimation of potential and frictional wake.

8.2.3 Wave Wake: w_W

This arises due to the influence of the subsurface orbital motions of the waves, see Appendix A1.8. In single-screw vessels, this component is likely to be small. It can be significant in twin-screw vessels where the propeller may be effectively closer to the free surface. The direction of the wave component will depend on whether the propeller is located under a wave crest or a wave trough, which in turn will change with speed, see Section 3.1.5 and Appendix A1.8.

8.2.4 Summary

Typical values for the three components of wake fraction, from [8.1], are

    Potential wake: 0.08–0.12
    Frictional wake: 0.09–0.23
    Wave wake: 0.03–0.05
    With total wake fraction being 0.20–0.40.

A more detailed account of the components of wake is given in Harvald [8.2].

8.3 Nominal and Effective Wake

The *nominal* wake is that measured in the vicinity of the propeller plane, but *without* the propeller present.

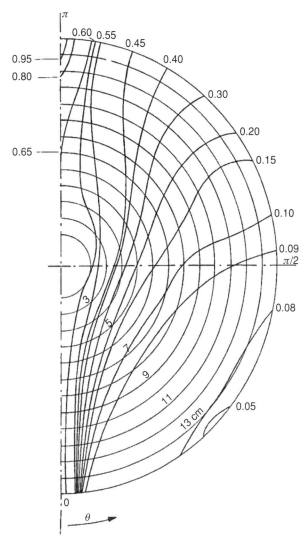

Figure 8.3. Wake distribution: single-screw vessel.

The *effective* wake is that measured in the propeller plane, with the propeller present, in the course of the self-propulsion experiment (see Section 8.7).

Because the propeller influences the boundary layer properties and possible separation effects, the nominal wake will normally be larger than the effective wake.

8.4 Wake Distribution

8.4.1 General Distribution

Due to the hull shape at the aft end and boundary layer development effects, the wake distribution is non-uniform in the general vicinity of the propeller. An example of the wake distribution (contours of constant wake fraction w_T) for a single-screw vessel is shown in Figure 8.3 [8.3].

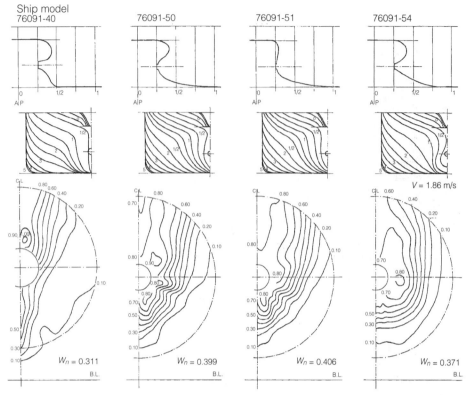

Figure 8.4. Influence of afterbody shape on wake distribution.

Different hull aft end shapes lead to different wake distributions and this is illustrated in Figure 8.4 [8.4]. It can be seen that as the stern becomes more 'bulbous', moving from left to right across the diagram, the contours of constant wake become more 'circular' and concentric. This approach may be adopted to provide each radial element of the propeller blade with a relatively uniform circumferential inflow velocity, reducing the levels of blade load fluctuations. These matters, including the influences of such hull shape changes on both propulsion *and* hull resistance, are discussed in Chapter 14.

A typical wake distribution for a twin-screw vessel is shown in Figure 8.5 [8.5], showing the effects of the boundary layer and local changes around the shafting and bossings. The average wake fraction for twin-screw vessels is normally less than for single-screw vessels.

8.4.2 Circumferential Distribution of Wake

The circumferential wake fraction, w_T'', for a single-screw vessel, at a particular propeller blade radius is shown schematically in Figure 8.6.

It is seen that there are high wake values at top dead centre (TDC) and bottom dead centre (BDC) as the propeller blade passes through the slow-moving water near the centreline of the ship. The value is lower at about 90° where the propeller blade passes closer to the edge of the boundary layer, and this effect is more apparent towards the blade tip.

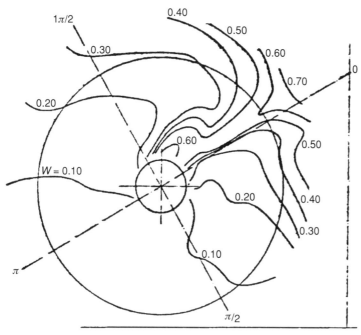

Figure 8.5. Wake distribution: twin-screw vessel.

8.4.3 Radial Distribution of Wake

Typical mean values of wake fraction w_T' for a single-screw vessel, when plotted radially, are shown in Figure 8.7. Twin-screw vessels tend to have less variation and lower average wake values. Integration of the average value at each radius yields the overall average wake fraction, w_T, in way of the propeller disc.

8.4.4 Analysis of Detailed Wake Measurements

Detailed measurements of wake are described in Section 8.5. These detailed measurements can be used to obtain the circumferential and radial wake distributions, using a volumetric approach, as follows:

Assume the local wake fraction, Figures 8.3 and 8.6, derived from the detailed measurements, to be denoted w_T'', the radial wake fraction, Figure 8.7, to be

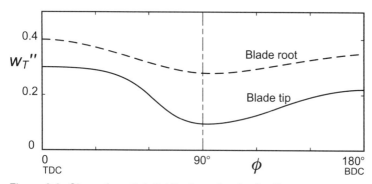

Figure 8.6. Circumferential distribution of wake fraction.

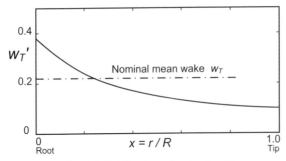

Figure 8.7. Radial distribution of wake fraction.

denoted w_T' and the overall average or nominal mean wake to be w_T. The volumetric mean wake fraction w_T' at radius r is

$$w_T' = \frac{\int_0^{2\pi} w_T'' \cdot r \cdot d\theta}{\int_0^{2\pi} r \cdot d\theta} \tag{8.5}$$

or

$$w_T' = \frac{1}{2\pi} \int_0^{2\pi} w_T'' \cdot d\theta, \tag{8.6}$$

where w_T' is the circumferential mean at each radius, giving the radial wake distribution, Figure 8.7.

The radial wake can be integrated to obtain the nominal mean wake w_T, as follows:

$$w_T = \frac{\int_{r_B}^{R} w_T' \cdot 2\pi r \cdot dr}{\int_{r_B}^{R} 2\pi r \cdot dr} = \frac{\int_{r_B}^{R} w_T' \cdot r \cdot dr}{\frac{1}{2}(R^2 - r_B^2)}, \tag{8.7}$$

where R is the propeller radius and r_B is the boss radius.

If a radial distribution of screw loading is adopted, and if the effective mean wake w_{Te} is known, for example from a self-propulsion test, then a suitable variation in radial wake would be

$$(1 - w_T') \times \frac{(1 - w_{Te})}{(1 - w_T)}. \tag{8.8}$$

Such a radial distribution of wake would be used in the calculations for a wake-adapted propeller, as described in Section 15.6.

8.5 Detailed Physical Measurements of Wake

8.5.1 Circumferential Average Wake

The two techniques that have been used to measure circumferential average wake (w_T' in Figure 8.7) are as follows:

(a) Blade wheels: The model is towed with a series of light blade wheels freely rotating behind the model. Four to five small blades (typically 1-cm square) are set at an angle to the spokes, with the wheel diameter depending on the model size. The rate of rotation of the blade wheel is measured and compared with an open water calibration, allowing the estimation of the mean wake over a range of diameters.

(b) Ring meters: The model is towed with various sizes of ring (resembling the duct of a ducted propeller) mounted at the position of the propeller disc and the drag of the ring is measured. By comparison with an open water drag calibration of drag against speed, a mean wake can be determined. It is generally considered that the ring meter wake value (compared with the blade wheel) is nearer to that integrated by the propeller.

8.5.2 Detailed Measurements

Detailed measurements of wake may be carried out in the vicinity of the propeller plane. The techniques used are the same as, or similar to, those used to measure the flow field around the hull, Chapter 7, Section 7.4.

(a) Pitôt static tubes: These may be used to scan a grid of points at the propeller plane. An alternative is to use a rake of Pitôt tubes mounted on the propeller shaft which can be rotated through 360°. The measurements provide results such as those shown in Figures 8.3 to 8.6.
(b) Five-holed Pitôt: This may be used over a grid in the propeller plane to provide measurements of flow direction as well as velocity. Such devices will determine the tangential flow across the propeller plane. A five-holed Pitôt is described in Chapter 7.
(c) LDV: Laser Doppler velocimetry (or LDA, laser Doppler anemometry) may be used to determine the local velocity at a point in the propeller plane. Application of LDV is discussed further in Chapter 7.
(d) PIV: Particle image velocimetry can be used to determine the distribution of velocity over a plane, providing a more detailed image of the overall flow. This is discussed further in Chapter 7.

Experimental methods of determining the wake field are fully reviewed in ITTC2008 [8.6]. Examples of typical experimental investigations into wake distribution include [8.7] and [8.8].

8.6 Computational Fluid Dynamics Predictions of Wake

The techniques used are similar to those used to predict the flow around the hull, Chapter 9. Cuts can be made in the propeller plane to provide a prediction of the detailed distribution of wake. Typical numerical investigations into model and full-scale wake include [8.9] and [8.10].

8.7 Model Self-propulsion Experiments

8.7.1 Introduction

The components of propulsive efficiency (wake, thrust deduction and relative rotative efficiency) can be determined from a set of propulsion experiments with models. A partial analysis can also be made from an analysis of ship trial performance, provided the trial takes place in good weather on a deep course and the ship is adequately instrumented.

A complete set of performance experiments would comprise the following:

(i) A set of model resistance experiments: to determine C_{TM} as a function of Fr, from which ship C_{TS} can be found by applying appropriate scaling methods, see Chapter 4.

(ii) A propeller open water test: to determine the performance of the model propeller. This may possibly be backed by tests in a cavitation tunnel, see Chapter 12.

(iii) A self-propulsion test with the model, or a trial result corrected for tide, wind, weather, shallow water etc.

The ITTC recommended procedure for the standard propulsion test is described in ITTC2002 [8.11]

8.7.2 Resistance Tests

The model total resistance is measured at various speeds, as described in Section 3.1.4.

8.7.3 Propeller Open Water Tests

Open water tests may be made either in a towing tank under cavitation conditions appropriate to the model, or in a cavitation tunnel at cavitation conditions appropriate to the ship. These are described in Chapter 12. Thrust and torque are measured at various J values, usually at constant speed of advance, unless bollard conditions ($J = 0$) are required, such as for a tug.

8.7.4 Model Self-propulsion Tests

The model is towed at various speeds and *at each speed* a number of tests are made at differing propeller revolutions, spanning the self-propulsion condition *for the ship*.

For each test, propeller revolutions, thrust and torque are measured, together with resistance dynamometer balance load and model speed. The measurements made, as described in [8.11], are summarised in Figure 8.8.

8.7.4.1 Analysis of Self-propulsion Tests
In theory, the case is required when thrust = resistance, $R = T$ or $R - T = 0$.

In practice, it is difficult to obtain this condition in one run. Common practice is to carry out a series of runs at constant speed with different revolutions, hence, different values of $R - T$ passing through zero. In the case of the model, the model self-propulsion point is as shown in Figure 8.9.

8.7.4.2 Analysis for Ship
If the *total* resistance obeyed Froude's law, then the ship self-propulsion point would be the same as that for the model. However, $C_{TM} > C_{TS}$, Figure 8.10, where C_{TS} is

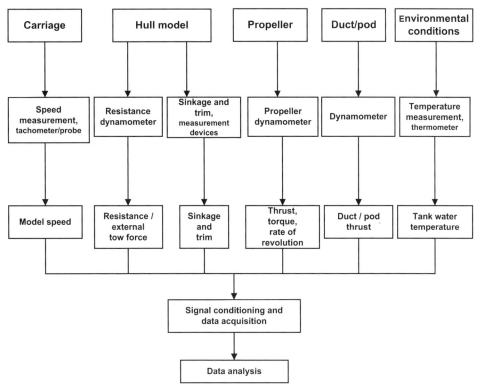

Figure 8.8. Propulsion test measurements.

the ship prediction (which may include allowances for C_V scaling of hull, append-ages, hull roughness and fouling, temperature and blockage correction to tank resist-ance, shallow water effects and weather allowance full scale). This difference ($C_{TM} - C_{TS}$) has to be offset on the resistance dynamometer balance load, or on the dia-gram, Figure 8.9, in order to determine the ship self-propulsion point. If R_{Tm} is the

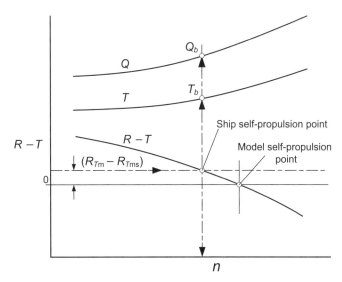

Figure 8.9. Model and ship self-propulsion points.

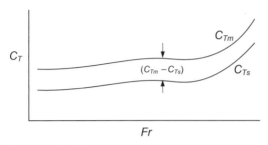

Figure 8.10. Model and ship C_T values.

model resistance corresponding to C_{TM} and R_{Tms} is the model resistance corresponding to C_{TS}, then the ship self-propulsion point is at $(R - T) = (R_{Tm} - R_{Tms})$. This then allows the revolutions n, behind thrust T_b and behind torque Q_b to be obtained for the ship self-propulsion point.

In order to determine the wake fraction and thrust deduction factor an equivalent propeller open water condition must be assumed. The equivalent condition is usually taken to be that at which the screw produces either

 (i) the same thrust as at the self-propulsion test revolutions per minute (rpm), known as *thrust identity* or
(ii) the same torque as at the self-propulsion test rpm, known as *torque identity*.

The difference between the analysed wake and thrust deduction values from these two analyses is usually quite small. The difference depends on the relative rotative efficiency η_R and disappears for $\eta_R = 1.0$.

8.7.4.3 Procedure: Thrust Identity

(a) From the resistance curves, Figure 8.10, $(C_{Tm} - C_{Ts})$ can be calculated to allow for differences between the model- and ship-predicted C_T values. (Various loadings can be investigated to allow for the effects of fouling and weather etc.)
(b) n and J_b $(= \frac{V_s}{n.D})$ can be determined for the self-propulsion point and T_b and Q_b; hence, K_{Tb} and K_{Qb} can be obtained from the self-propulsion data, Figure 8.9.
(c) Thrust identity analysis assumes $K_{To} = K_{Tb}$. The open water curve, Figure 8.11(a), is entered with K_{Tb} to determine the corresponding J_o, K_{Qo} and η_o.

The suffix 'b' indicates values behind the model and suffix 'o' values in the open water test.

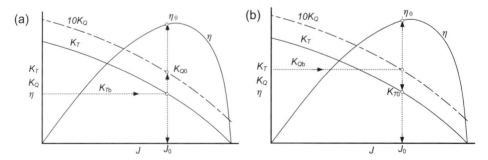

Figure 8.11. Open water curve, (a) showing thrust identity, (b) showing torque identity.

The wake fraction is given by the following:

$$w_T = \frac{(V_S - V_O)}{V_S} = 1 - \frac{V_O}{V_S} = 1 - \frac{nDJ_o}{nDJ_b} = 1 - \frac{J_o}{J_b}, \qquad (8.9)$$

where $J_b = \frac{V_S}{nD}$ and V_S is ship speed.

The thrust deduction is given by the following:

$$t = \frac{(T_b - R)}{T_b} = 1 - \frac{R}{T_b} = 1 - \frac{0.5\rho SV^2 C_{TS}}{\rho n^2 D^4 K_{Tb}} = 1 - \frac{0.5 J_b^2 S C_{TS}}{D^2 K_{Tb}}. \qquad (8.10)$$

The relative rotative efficiency is given by the following:

$$\eta_R = \frac{\eta_b}{\eta_o} = \frac{J_o K_{Tb}}{2\pi K_{Qb}} \cdot \frac{2\pi K_{Qo}}{J_o K_{To}} = \frac{K_{Tb}}{K_{To}} \cdot \frac{K_{Qo}}{K_{Qb}}. \qquad (8.11)$$

For thrust identity,

$$K_{To} = K_{Tb} \quad \text{and} \quad \eta_R = \frac{K_{Qo}}{K_{Qb}}. \qquad (8.12)$$

For torque identity, $J_o = J$ for which $K_{Qo} = K_{Qb}$ and $\eta_R = K_{Tb}/K_{To}$. Most commercial test tanks employ the thrust identity method.

Finally, all the components of the quasi-propulsive coefficient (QPC) η_D are now known and η_D can be assembled as

$$\eta_D = \eta_o \eta_H \eta_R = \eta_o \frac{(1 - t)}{(1 - w_T)} \eta_R. \qquad (8.13)$$

8.7.5 Trials Analysis

Ship trials and trials analysis are discussed in Chapter 5. Usually only torque, revolutions and speed are available on trials so that ship analysis wake fraction w_T values are obtained on a torque identity basis. Thrust deduction t can only be estimated on the basis of scaled model information (e.g. [8.12]), and η_R can only be obtained using estimated thrust from effective power (P_E), see worked example application 3, Chapter 17.

8.7.6 Wake Scale Effects

The model boundary layer when scaled (see Figure 3.21) is thicker than the ship boundary layer. Hence, the wake fraction w_T tends to be smaller for the ship, although extra ship roughness compensates to a certain extent. Equation (5.22) was adopted by the ITTC in its 1978 Performance Prediction Method to allow for wake scale effect. Detailed full-scale measurements of wake are relatively sparse. Work, such as by Lübke [8.13], is helping to shed some light on scale effects, as are the increasing abilities of CFD analyses to predict aft end flows at higher Reynolds numbers [8.9, 8.10]. Lübke describes an investigation into the estimation of wake at model and full scale. At model scale the agreement between computational fluid dynamics (CFD) and experiment was good. The comparisons of CFD with experiment at full scale indicated that further validation was required.

8.8 Empirical Data for Wake Fraction and Thrust Deduction Factor

8.8.1 Introduction

Empirical data for wake fraction w_T and thrust deduction factor t suitable for preliminary design purposes are summarised. It should be noted that the following data are mainly for models. The data are generally nominal values and are not strictly correct due to scale effects and dependence on detail not included in the formulae.

Empirical data for η_R, the third component of hull–propeller interaction, are included in Chapter 16.

8.8.2 Single Screw

8.8.2.1 Wake Fraction w_T
Wake fraction data attributable to Harvald for single-screw vessels, reproduced in [8.3], are shown in Figure 8.12. This illustrates the dependence of w_T on C_B, L/B, hull shape and propeller diameter.

A satisfactory fit to the Harvald single-screw data is the following:

$$w_T = \left[1.095 - 3.4C_B + 3.3C_B^2\right] + \left[\frac{0.5C_B^2(6.5 - L/B)}{L/B}\right], \qquad (8.14)$$

suitable for C_B range 0.525–0.75 and L/B range 5.0–8.0.

An earlier formula, attributable to Taylor [8.14], is the following:

$$w_T = 0.50C_B - 0.05, \qquad (8.15)$$

noting that Equation (8.15) tends to give low values of w_T at high C_B.

The British Ship Research Association (BSRA) wake data regression [8.15, 8.16] gives the following:

$$w_T = -0.0458 + 0.3745C_B^2 + 0.1590D_W - 0.8635Fr + 1.4773Fr^2, \qquad (8.16)$$

where D_W is the wake fraction parameter defined as

$$D_W = \frac{B}{\nabla^{1/3}} \cdot \sqrt{\frac{\nabla^{1/3}}{D}},$$

suitable for C_B range 0.55–0.85 and Fr range 0.12–0.36.

Holtrop wake data regression [8.17] gives the following:

$$w_T = c_9 c_{20} C_V \frac{L}{T_A}\left(0.050776 + 0.93405c_{11}\frac{C_V}{(1 - C_{P1})}\right)$$
$$+ 0.27915c_{20}\sqrt{\frac{B}{L(1 - C_{P1})}} + c_{19}c_{20}. \qquad (8.17)$$

The coefficient c_9 depends on the coefficient c_8, defined as:

$c_8 = B\,S/(L\,D\,T_A)$ when $B/T_A \le 5$
  $= S(7\,B/T_A - 25)\,/\,(LD(B/T_A - 3))$ when $B/T_A > 5$
$c_9 = c_8$ when $c_8 \le 28$
  $= 32 - 16/(c_8 - 24)$ when $c_8 > 28$

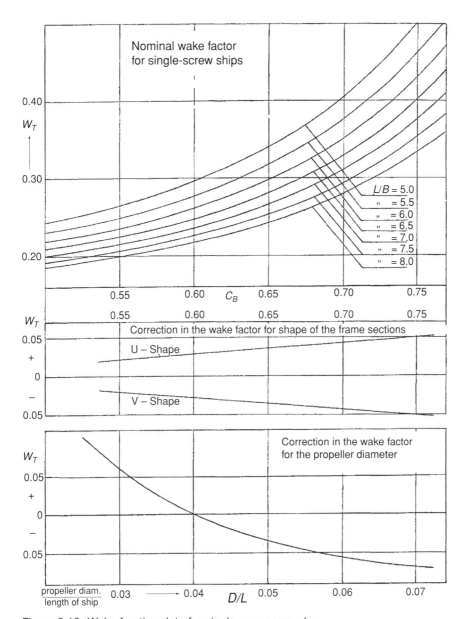

Figure 8.12. Wake fraction data for single-screw vessels.

$c_{11} = T_A/D$ when $T_A/D \le 2$
$\quad = 0.0833333(T_A/D)^3 + 1.33333$ when $T_A/D > 2$
$c_{19} = 0.12997/(0.95 - C_B) - 0.11056/(0.95 - C_P)$ when $C_P \le 0.7$
$\quad = 0.18567/(1.3571 - C_M) - 0.71276 + 0.38648C_P$ when $C_P > 0.7$
$c_{20} = 1 + 0.015 C_{\text{stern}}$
$C_{P1} = 1.45C_P - 0.315 - 0.0225 \, LCB$ (where LCB is LCB forward of 0.5L as a
    percentage of L)
C_V is the viscous resistance coefficient with $C_V = (1 + k)C_F + C_A$ and C_A is the
    correlation allowance coefficient, discussed in Chapters 5 and 10, Equation
    (10.34). S is wetted area, D is propeller diameter and T_A is draught aft.

Table 8.1. C_{stern} parameter

Afterbody form	C_{stern}
Pram with gondola	−25
V-shaped sections	−10
Normal section shape	0
U-shaped sections with Hogner stern	10

8.8.2.2 Thrust Deduction (t)

For single-screw vessels a good first approximation is the following:

$$t = k_R \cdot w_T, \tag{8.18}$$

where k_R varies between 0.5 for thin rudders and 0.7 for thick rudders [8.3].

BSRA thrust deduction regression [8.16] gives Equation (8.19a) which is the preferred expression and Equation (8.19b) which is an alternative if the pitch ratio (P/D) is not available.

$$t = -0.2064 + 0.3246C_B^2 - 2.1504C_B(LCB/L_{BP})$$
$$+ 0.1705(B/\nabla^{1/3}) + 0.1504(P/D). \tag{8.19a}$$

$$t = -0.5352 - 1.6837C_B + 1.4935C_B^2$$
$$- 1.6625(LCB/L_{BP}) + 0.6688D_t, \tag{8.19b}$$

where D_t is the thrust deduction parameter defined as $\frac{B}{\nabla^{1/3}} \cdot \frac{D}{\nabla^{1/3}}$, B is the breadth (m) and ∇ is the displaced volume in m^3. Equations are suitable for C_B range 0.55–0.85 and P/D range 0.60–1.10.

Holtrop thrust deduction regression [8.17] gives the following:

$$t = 0.25014(B/L)^{0.28956} \left(\sqrt{BT/D}\right)^{0.2624} / (1 - C_P + 0.0225LCB)^{0.01762}$$
$$+ 0.0015C_{stern} \tag{8.20}$$

where C_{stern} is given in Table 8.1.

8.8.2.3 Tug Data

Typical approximate mean values of w_T and t from Parker – Dawson [8.18] and Moor [8.19] are given in Table 8.2. Further data for changes in propeller diameter and hull form are given in [8.18 and 8.19].

8.8.2.4 Trawler Data

See BSRA [8.20]–[8.22].

Typical approximate values of w_T and t for trawler forms, from BSRA [8.21], are given in Table 8.3. Speed range $Fr = 0.29$–0.33. The influence of $L/\nabla^{1/3}$, B/T, LCB and hull shape variations are given in BSRA [8.20 and 8.22].

Table 8.2. *Wake fraction and thrust deduction for tugs*

Source	Fr	w_T	t	Case
[8.18] $C_B = 0.503$	0.34	0.21	0.23	Free running
	0.21	–	0.12	Towing
	0.15	–	0.10	Towing
	0.09	–	0.07	Towing
	0 (bollard)	–	0.02	Bollard
[8.19] $C_B = 0.575$	0.36	0.20	0.25	Free running
	0.21	0.20	0.15	Towing
	0.12	0.25	0.12	Towing
	0 (bollard)	–	0.07	Bollard

8.8.3 Twin Screw

8.8.3.1 Wake Fraction (w_T)

Wake fraction data attributable to Harvald for twin-screw vessels, reproduced in [8.3], are shown in Figure 8.13. A satisfactory fit to the Harvald twin-screw data is the following:

$$w_T = [0.71 - 2.39C_B + 2.33C_B^2] + [0.12C_B^4(6.5 - L/B)], \tag{8.21}$$

suitable for C_B range 0.525–0.675 and L/B range 6.0–7.0.

An earlier formula, attributable to Taylor [8.14] is the following:

$$w_T = 0.55C_B - 0.20, \tag{8.22}$$

noting that Equation (8.22) tends to give high values at high C_B.

The Holtrop [8.17] wake data regression analysis for twin-screw ships gives the following:

$$w_T = 0.3095C_B + 10C_VC_B - \frac{D}{\sqrt{BT}}, \tag{8.23}$$

where C_V is the viscous resistance coefficient with $C_V = (1 + k)C_F + C_A$, $(1 + k)$ is the form factor, Chapter 4, and C_A is the correlation allowance coefficient (discussed in Chapters 5 and 10, Equation (10.34)). Equation (8.23) is suitable for C_B range 0.55–0.80, see Table 10.2.

The Flikkema *et al.* [8.23] wake data regression analysis for podded units gives the following:

$$w_{Tp} = -0.21035 + 0.18053C_B + 56.724C_VC_B$$

Table 8.3. *Wake fraction and thrust deduction for trawlers*

C_B	w_T	t
0.53	0.153	0.195
0.57	0.178	0.200
0.60	0.200	0.230

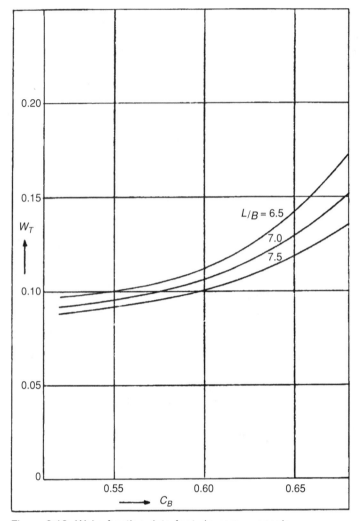

Figure 8.13. Wake fraction data for twin-screw vessels.

$$+ 0.18566\frac{D}{\sqrt{BT}} + 0.090198\frac{C_{\text{Tip}}}{D}, \tag{8.24}$$

where C_V is as defined for Equation (8.23) and C_{Tip} is the tip clearance which was introduced to account for the degree in which the pod is embedded in the hull boundary layer; a typical value for C_{Tip}/D is 0.35 (Z/D in Table 16.4).

8.8.3.2 Thrust Deduction (t)
For twin screws, a suitable first approximation is as follows:

$$t = w_T. \tag{8.25}$$

The Holtrop [8.17] thrust deduction regression analysis for twin-screw ships gives the following:

$$t = 0.325C_B - 0.1885\frac{D}{\sqrt{BT}}. \tag{8.26}$$

Table 8.4. *Wake fraction and thrust deduction for round bilge semi-displacement forms*

C_B range	Fr_∇	w_T	t
$C_B \leq 0.45$	0.6	0	0.12
	1.4	−0.04	0.07
	2.6	0	0.08
$C_B > 0.45$	0.6	0.08	0.15
	1.4	−0.02	0.07
	2.2	0.04	0.06

The Flikkema *et al.* [8.23] thrust deduction regression analysis for podded units gives the following:

$$t_P = 0.21593 + 0.099768 C_B - 0.56056 \frac{D}{\sqrt{BT}}. \tag{8.27}$$

8.8.3.3 Round Bilge Semi-displacement Craft

NPL ROUND BILGE SERIES (BAILEY [8.24]). Typical approximate mean values of w_T and t for round bilge forms are given in Table 8.4. The data are generally applicable to round bilge forms in association with twin screws. Speed range $Fr_\nabla = 0.58-2.76$, where $Fr_\nabla = 0.165 \, V/\Delta^{1/6}$ (*V* in knots, Δ in tonnes) and C_B range is 0.37–0.52. Regression equations are derived for w_T and t in [8.24].

ROUND BILGE SKLAD SERIES (GAMULIN [8.25]).
For $C_B \leq 0.45$,

Speed range $Fr_\nabla = 0.60-1.45$
$$w_T = 0.056-0.066 \, Fr_\nabla, \tag{8.28}$$
Speed range $Fr_\nabla = 1.45-3.00$
$$w_T = 0.04 \, Fr_\nabla - 0.10. \tag{8.29}$$

For $C_B \leq 0.45$,

Speed range $Fr_\nabla = 0.60-1.45$
$$t = 0.15-0.08 \, Fr_\nabla, \tag{8.30}$$
Speed range $Fr_\nabla = 1.45-3.00$
$$t = 0.02 \, Fr_\nabla. \tag{8.31}$$

8.8.4 Effects of Speed and Ballast Condition

8.8.4.1 Speed
Wake fraction tends to decrease a little with increasing speed, but is usually assumed constant for preliminary calculations. Equation (8.16) provides an indication of the influence of speed (*Fr*) for single-screw vessels.

8.8.4.2 Ballast (or a Part Load) Condition

The wake fraction in the ballast, or a part load, condition tends to be 5–15% larger than the wake fraction in the loaded condition.

Moor and O'Connor [8.26] provide equations which predict the change in wake fraction and thrust deduction with draught ratio $(T)_R$, as follows:

$$(1 - w_T)_R = 1 + [(T)_R - 1](0.2882 + 0.1054\theta), \tag{8.32}$$

where θ is the trim angle expressed as $\theta = (100 \times \text{trim by bow})/L_{BP}$.

$$(1 - t)_R = 1 + [(T)_R - 1](0.4322 - 0.4880\,C_B), \tag{8.33}$$

where

$$(1 - w_T)_R = \frac{(1 - w_T)_{\text{Ballast}}}{(1 - w_T)_{\text{Load}}} \quad (1 - t)_R = \frac{(1 - t)_{\text{Ballast}}}{(1 - t)_{\text{Load}}} \text{ and } (T)_R = \left(\frac{T_{\text{Ballast}}}{T_{\text{Load}}}\right).$$

Example: Consider a ship with $L_{BP} = 150$ m, $C_B = 0.750$ and w_T and t values in the loaded condition of $w_T = 0.320$ and $t = 0.180$.

In a ballast condition, the draught ratio $(T)_R = 0.70$ and trim is 3.0 m by the stern.

Trim angle $\theta = 100 \times (-3/150) = -2.0$
Using Equation (8.32),
$(1 - w_T)_R = 1 + [0.70 - 1](0.2882 + 0.1054 \times (-2.0)) = 0.977$
$(1 - w_T)_{\text{Ballast}} = (1 - 0.320) \times 0.977 = 0.664$ and $w_{T\text{Ballast}} = 0.336$
Using Equation (8.33),
$(1 - t)_R = 1 + [0.70 - 1](0.4322 - 0.4880 \times 0.750) = 0.980$
$(1 - t)_{\text{Ballast}} = (1 - 0.18) \times 0.980 = 0.804$ and $t_{\text{Ballast}} = 0.196$

8.9 Tangential Wake

8.9.1 Origins of Tangential Wake

The preceding sections of this chapter have considered only the axial wake as this is the predominant component as far as basic propeller design is concerned. However, in most cases, there is also a tangential flow across the propeller plane. For example, in a single-screw vessel there is a general upflow at the aft end leading to an axial component plus an upward or tangential component V_T, Figure 8.14.

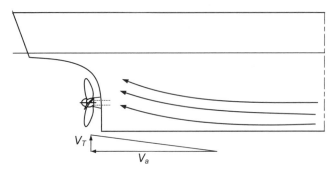

Figure 8.14. General upflow at aft end of single-screw vessel.

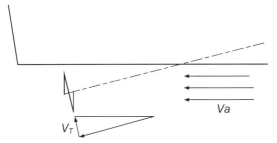

Figure 8.15. Tangential flow due to inclined shaft.

In the case of an inclined shaft, often employed in smaller higher-speed craft, the propeller encounters an axial flow together with a tangential component V_T, Figure 8.15. Cyclic load variations of the order of 100% can be caused by shaft inclinations.

8.9.2 Effects of Tangential Wake

The general upflow across the propeller plane, Figure 8.14, decreases blade angles of attack, hence forces, as the blade rises towards TDC and increases angles of attack as the blades descend away from TDC, Figure 8.16.

For a propeller rotating clockwise, viewed from aft, the load on the starboard side is higher than on the port side. The effect is to offset the centre of thrust to starboard, Figure 8.17. It may be offset by as much as 33% of propeller radius. The effect of the varying torque force is to introduce a vertical load on the shaft. The forces can be split into a steady-state load together with a time varying component.

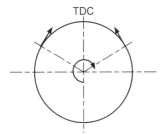

Figure 8.16. Effect of upflow at propeller plane.

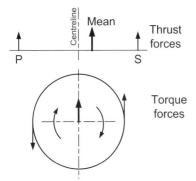

Figure 8.17. Thrust and torque forces due to tangential wake.

The forces resulting from an inclined shaft are shown in Figure 16.21 and the effects on blade loadings are discussed in Chapter 16, Section 16.2.8. A blade element diagram including tangential flow is described in Chapter 15.

REFERENCES (CHAPTER 8)

8.1   Harvald, S.A. Potential and frictional wake of ships. *Transactions of the Royal Institution of Naval Architects*, Vol. 115, 1973, pp. 315–325.

8.2   Harvald, S.A. *Resistance and Propulsion of Ships*. Wiley Interscience, New York, 1983.

8.3   Van Manen, J.D. Fundamentals of ship resistance and propulsion. Part B Propulsion. Publication No. 129a of NSMB, Wageningen. Reprinted in *International Shipbuilding Progress*.

8.4   Harvald, S.A. Wake distributions and wake measurements. *Transactions of the Royal Institution of Naval Architects*, Vol. 123, 1981, pp. 265–286.

8.5   Van Manen, J.D. and Kamps, J. The effect of shape of afterbody on propulsion. *Transactions of the Society of Naval Architects and Marine Engineers*, Vol. 67, 1959, pp. 253–289.

8.6   ITTC. Report of the specialist committee on wake fields. *Proceedings of 25th ITTC*, Vol. II, Fukuoka, 2008.

8.7   Di Felice, F., Di Florio, D., Felli, M. and Romano, G.P. Experimental investigation of the propeller wake at different loading conditions by particle image velocimetry. *Journal of Ship Research*, Vol. 48, No. 2, 2004, pp. 168–190.

8.8   Felli, M. and Di Fellice, F. Propeller wake analysis in non uniform flow by LDV phase sampling techniques. *Journal of Marine Science and Technology*, Vol. 10, 2005.

8.9   Visonneau, M., Deng, D.B. and Queutey, P. Computation of model and full scale flows around fully-appended ships with an unstructured RANSE solver. *26th Symposium on Naval Hydrodynamics*, Rome, 2005.

8.10 Starke, B., Windt, J. and Raven, H. Validation of viscous flow and wake field predictions for ships at full scale. *26th Symposium on Naval Hydrodynamics*, Rome, 2005.

8.11 ITTC. Recommended procedure for the propulsion test. Procedure 7.5-02-03-01.1. Revision 01, 2002.

8.12 Dyne, G. On the scale effect of thrust deduction. *Transactions of the Royal Institution of Naval Architects*, Vol. 115, 1973, pp. 187–199.

8.13 Lübke, L. Calculation of the wake field in model and full scale. *Proceedings of International Conference on Ship and Shipping Research, NAV'2003*, Palermo, Italy, June 2003.

8.14 Taylor, D.W. *The Speed and Power of Ships*. Government Printing Office, Washington, DC, 1943.

8.15 Lackenby, H. and Parker, M.N. The BSRA methodical series – An overall presentation: variation of resistance with breadth-draught ratio and length-displacement ratio. *Transactions of the Royal Institution of Naval Architects*, Vol. 108, 1966, pp. 363–388.

8.16 Pattullo, R.N.M. and Wright, B.D.W. Methodical series experiments on single-screw ocean-going merchant ship forms. Extended and revised overall analysis. BSRA Report NS333, 1971.

8.17 Holtrop J. A statistical re-analysis of resistance and propulsion data. *International Shipbuilding Progress*, Vol. 31, 1984, pp. 272–276.

8.18 Parker, M.N. and Dawson, J. Tug propulsion investigation. The effect of a buttock flow stern on bollard pull, towing and free-running performance. *Transactions of the Royal Institution of Naval Architects*, Vol. 104, 1962, pp. 237–279.

8.19 Moor, D.I. An investigation of tug propulsion. *Transactions of the Royal Insti-
      tution of Naval Architects*, Vol. 105, 1963, pp. 107–152.
8.20 Pattulo, R.N.M. and Thomson, G.R. The BSRA Trawler Series (Part I). Beam-
      draught and length-displacement ratio series, resistance and propulsion tests.
      Transactions of the Royal Institution of Naval Architects, Vol. 107, 1965, pp.
      215–241.
8.21 Pattulo, R.N.M. The BSRA Trawler Series (Part II). Block coefficient and lon-
      gitudinal centre of buoyancy series, resistance and propulsion tests. *Transac-
      tions of the Royal Institution of Naval Architects*, Vol. 110, 1968, pp. 151–183.
8.22 Thomson, G.R. and Pattulo, R.N.M. The BSRA Trawler Series (Part III). Res-
      istance and propulsion tests with bow and stern variations. *Transactions of the
      Royal Institution of Naval Architects*, Vol. 111, 1969, pp. 317–342.
8.23 Flikkema, M.B., Holtrop, J. and Van Terwisga, T.J.C. A parametric power pre-
      diction model for tractor pods. *Proceedings of Second International Conference
      on Advances in Podded Propulsion, T-POD*. University of Brest, France, Octo-
      ber 2006.
8.24 Bailey, D. A statistical analysis of propulsion data obtained from models of
      high speed round bilge hulls. *Symposium on Small Fast Warships and Security
      Vessels*. RINA, London, 1982.
8.25 Gamulin, A. A displacement series of ships. *International Shipbuilding Pro-
      gress*, Vol. 43, No. 434, 1996, pp. 93–107.
8.26 Moor, D.I. and O'Connor, F.R.C. Resistance and propulsion factors of some
      single-screw ships at fractional draught. *Transactions of the North East Coast
      Institution of Engineers and Shipbuilders*. Vol. 80, 1963–1964, pp. 185–202.

9 Numerical Estimation of Ship Resistance

9.1 Introduction

The appeal of a numerical method for estimating ship hull resistance is in the ability to seek the 'best' solution from many variations in shape. Such a hull design optimisation process has the potential to find better solutions more rapidly than a conventional design cycle using scale models and associated towing tank tests.

Historically, the capability of the numerical methods has expanded as computers have become more powerful and faster. At present, there still appears to be no diminution in the rate of increase in computational power and, as a result, numerical methods will play an ever increasing role. It is worth noting that the correct application of such techniques has many similarities to that of high-quality experimentation. Great care has to be taken to ensure that the correct values are determined and that there is a clear understanding of the level of uncertainty associated with the results.

One aspect with which even the simplest methods have an advantage over traditional towing tank tests is in the level of flow field detail that is available. If correctly interpreted, this brings a greatly enhanced level of understanding to the designer of the physical behaviour of the hull on the flow around it. This chapter is intended to act as a guide, rather than a technical manual, regarding exactly how specific numerical techniques can be applied. Several useful techniques are described that allow numerical tools to be used most effectively.

The ability to extract flow field information, either as values of static pressure and shear stress on the wetted hull surface, or on the bounding surface of a control volume, allows force components or an energy breakdown to be used to evaluate a theoretical estimate of numerical resistance using the techniques discussed in Chapter 7. It should always be remembered that the uncertainty associated with experimental measurement is now replaced by the uncertainties associated with the use of numerical techniques. These always contain inherent levels of abstraction away from physical reality and are associated with the mathematical representation applied and the use of numerical solutions to these mathematical models.

This chapter is not intended to give the details of the theoretical background of all the available computational fluid dynamic (CFD) analysis techniques, but rather an overview that provides an appreciation of the inherent strengths and weaknesses

Table 9.1. *Evolution of CFD capabilities for evaluating ship powering*

Location	Year	ITTC Proc. Resistance Committee[1]	Methods used	Test cases	Ref.
Gothenburg	1980	16th (1981)	16 boundary layer-based methods (difference and integral) and 1 RANS	HSVA tankers	[9.4]
Gothenburg	1990	20th (1993)	All methods RANS except 1 LES and 1 boundary layer	HSVA tankers	[9.5]
Tokyo	1994	21st (1996)	Viscous and inviscid free-surface methods, and viscous at zero *Fr* (double hull)	Series 60 ($C_B = 0.6$) HSVA tanker	[9.6, 9.7]
Gothenburg	2000	23rd (2002)	All RANS but with/without free-surface both commercial and in-house codes	KVLCC2, KCS, DTMB5415	[9.8, 9.9]
Tokyo	2005	25th (2008)	RANS with self-propelled, at drift and in head seas	KVLCC2, KCS, DTMB5415	[9.10]
Gothenburg	2010	26th (2011)	RANS with a variety of turbulence models, free surface, dynamic heave and trim, propeller, waves and some LES	KVLCC2, KCS, DTMB5415	

Note: (1) Full text available via ITTC website, http://ittc.sname.org

and their associated costs. A number of publications give a good overview of CFD, for example, Ferziger and Peric [9.1], and how best to apply CFD to maritime problems [9.2, 9.3].

9.2 Historical Development

The development of numerical methods and their success or otherwise is well documented in the series of proceedings of the International Towing Tank Conference (ITTC). In particular, the Resistance Committee has consistently reported on the capabilities of the various CFD techniques and their developments. Associated with the ITTC have been a number of international workshops which aimed to benchmark the capability of the prediction methods against high-quality experimental test cases. The proceedings of these workshops and the associated experimental test cases still provide a suitable starting point for those wishing to develop such capabilities.

Table 9.1 identifies the main workshops and the state-of-the-art capability at the time. The theoretical foundations of most techniques have been reasonably well understood, but the crucial component of their subsequent development has been the rapid reduction in computational cost. This is the cost associated with both the processing power in terms of floating point operations per second of a given processor and the cost of the necessary memory storage associated with each processor. A good measure of what is a practical (industrial) timescale for a large-scale single CFD calculation has been that of the overnight run, e.g. the size of problem that can be set going when engineers go home in the evening and the answer is ready when they arrive for work the next morning. To a large extent this dictates the computational mesh size that can be applied to a given problem and the likely numerical accuracy.

9.3 Available Techniques

The flow around a ship hull, as previously described in Chapters 3 and 4, is primarily incompressible and inviscid. Viscous effects are confined to a thin boundary layer close to the hull and a resultant turbulent wake. What makes the flow particularly interesting and such a challenge to the designer of a ship hull is the interaction between the development of the hull boundary layer and the generation of free-surface gravity waves due to the shape of the hull. The challenge of capturing the complexity of this flow regime is described by Landweber and Patel [9.11]. In particular, they focus on the interaction at the bow between the presence, or otherwise, of a stagnation point and the creation of the bow wave which may or may not break and, even if it does not break, may still induce flow separation at the stem and create significant vorticity. It is only recently that computations of such complex flow interactions have become possible [9.12].

9.3.1 Navier–Stokes Equations

Ship flows are governed by the general conservation laws for mass, momentum and energy, collectively referred to as the Navier–Stokes equations.

- The continuity equation states that the rate of change of mass in an infinitesimally small control volume equals the rate of mass flux through its bounding surface.

$$\frac{\partial \rho}{\partial t} + \nabla \cdot (\rho V) = 0, \tag{9.1}$$

where ∇ is the differential operator $(\partial/\partial x, \partial/\partial y, \partial/\partial z)$.

- The momentum equation states that the rate of change of momentum for the infinitesimally small control volume is equal to the rate at which momentum is entering or leaving through the surface of the control volume, plus the sum of the forces acting on the volume itself.

$$\frac{\partial (\rho u)}{\partial t} + \nabla \cdot (\rho u \mathbf{V}) = -\frac{\partial p}{\partial x} + \frac{\partial \tau_{xx}}{\partial x} + \frac{\partial \tau_{yx}}{\partial y} + \frac{\partial \tau_{zx}}{\partial z} + \rho f_x$$

$$\frac{\partial (\rho v)}{\partial t} + \nabla \cdot (\rho v \mathbf{V}) = -\frac{\partial p}{\partial y} + \frac{\partial \tau_{xy}}{\partial x} + \frac{\partial \tau_{yy}}{\partial y} + \frac{\partial \tau_{zy}}{\partial z} + \rho f_y$$

$$\frac{\partial (\rho w)}{\partial t} + \nabla \cdot (\rho w \mathbf{V}) = -\frac{\partial p}{\partial z} + \frac{\partial \tau_{xz}}{\partial x} + \frac{\partial \tau_{yz}}{\partial y} + \frac{\partial \tau_{zz}}{\partial z} + \rho f_z, \tag{9.2}$$

where $\mathbf{V} = (u, v, w)$.

- The energy equation states that the rate of change in internal energy in the control volume is equal to the rate at which enthalpy is entering, plus work done on the control volume by the viscous stresses.

$$\frac{\partial}{\partial t}\left[\rho\left(e + \frac{V^2}{2}\right)\right] + \nabla \cdot \left[\rho\left(e + \frac{V^2}{2}\right)\mathbf{V}\right] = \rho\dot{q} + \frac{\partial}{\partial x}\left(k\frac{\partial T}{\partial x}\right)$$

$$+ \frac{\partial}{\partial y}\left(k\frac{\partial T}{\partial y}\right) + \frac{\partial}{\partial z}\left(k\frac{\partial T}{\partial z}\right) - \frac{\partial(up)}{\partial x} - \frac{\partial(vp)}{\partial y} - \frac{\partial(wp)}{\partial z} + \frac{\partial(u\tau_{xx})}{\partial x}$$

$$+ \frac{\partial(u\tau_{yx})}{\partial y} + \frac{\partial(u\tau_{zx})}{\partial z} + \frac{\partial(v\tau_{xy})}{\partial x} + \frac{\partial(v\tau_{yy})}{\partial y} + \frac{\partial(v\tau_{zy})}{\partial z} + \frac{\partial(w\tau_{xz})}{\partial x}$$

$$+ \frac{\partial(w\tau_{yz})}{\partial y} + \frac{\partial(w\tau_{zz})}{\partial z} + \rho f \cdot V. \tag{9.3}$$

and $V^2 = \mathbf{V} \cdot \mathbf{V}$.

The Navier–Stokes equations can only be solved analytically for just a few cases, see for example Batchelor [9.13] and, as a result, a numerical solution has to be sought.

In practice, it is possible to make a number of simplifying assumptions that can either allow an analytical solution to be obtained or to significantly reduce the computational effort required to solve the full Navier–Stokes equations. These can be broadly grouped into the three sets of techniques described in the following sections.

9.3.2 Incompressible Reynolds Averaged Navier–Stokes equations (RANS)

In these methods the flow is considered as incompressible, which simplifies Equations (9.1) and (9.2) and removes the need to solve Equation (9.3). The Reynolds averaging process assumes that the three velocity components can be represented as a rapidly fluctuating turbulent velocity around a slowly varying mean velocity. This averaging process introduces six new terms, known as Reynolds stresses. These represent the increase in effective fluid velocity due to the presence of turbulent eddies within the flow.

$$\frac{\partial U}{\partial t} + U\left(\frac{\partial U}{\partial x} + \frac{\partial V}{\partial x} + \frac{\partial W}{\partial x}\right)$$

$$= \frac{-1}{\rho}\frac{\partial P}{\partial x} + v\left(\frac{\partial^2 U}{\partial x^2} + \frac{\partial^2 U}{\partial y^2} + \frac{\partial^2 U}{\partial z^2}\right) - \left(\frac{\partial \overline{u'^2}}{\partial x^2} + \frac{\partial \overline{u'v'}}{\partial x\partial y} + \frac{\partial \overline{u'w'}}{\partial x\partial z}\right)$$

$$\frac{\partial V}{\partial t} + V\left(\frac{\partial U}{\partial y} + \frac{\partial V}{\partial y} + \frac{\partial W}{\partial y}\right)$$

$$= \frac{-1}{\rho}\frac{\partial P}{\partial y} + v\left(\frac{\partial^2 V}{\partial x^2} + \frac{\partial^2 V}{\partial y^2} + \frac{\partial^2 V}{\partial z^2}\right) - \left(\frac{\partial \overline{u'v'}}{\partial x\partial y} + \frac{\partial \overline{v'^2}}{\partial y^2} + \frac{\partial \overline{v'w'}}{\partial y\partial z}\right)$$

$$\frac{\partial W}{\partial t} + W\left(\frac{\partial U}{\partial z} + \frac{\partial V}{\partial z} + \frac{\partial W}{\partial z}\right)$$

$$= \frac{-1}{\rho}\frac{\partial P}{\partial z} + v\left(\frac{\partial^2 W}{\partial x^2} + \frac{\partial^2 W}{\partial y^2} + \frac{\partial^2 W}{\partial z^2}\right) - \left(\frac{\partial \overline{u'w'}}{\partial x\partial z} + \frac{\partial \overline{v'w'}}{\partial y\partial z} + \frac{\partial \overline{w'^2}}{\partial z^2}\right). \tag{9.4}$$

where $u = U + u'$, $v = V + v'$, $w = W + w'$.

In order to close this system of equations a turbulence model has to be introduced that can be used to represent the interaction between these Reynolds stresses and the underlying mean flow. It is in the appropriate choice of the model used to achieve turbulence closure that many of the uncertainties arise. Wilcox [9.14] discusses the possible approaches that range from a simple empirical relationship which introduces no additional unknowns to those which require six or more additional unknowns and appropriate auxiliary equations. Alternative approaches include:

(1) Large eddy simulation (LES), which uses the unsteady Navier–Stokes momentum equations and only models turbulence effects at length scales comparable with the local mesh size; or
(2) Direct numerical simulation (DNS) which attempts to resolve all flow features across all length and time scales.

LES requires a large number of time steps to derive a statistically valid solution and a very fine mesh for the boundary layer, whereas DNS introduces an extremely large increase in mesh resolution and a very small time step. In practice, zonal approaches, such as detached eddy simulation (DES), provide a reasonable compromise through use of a suitable wall boundary layer turbulence closure and application of an LES model through use of a suitable switch in separated flow regions [9.15].

In all of the above methods the flow is solved in a volume of space surrounding the hull. The space is divided up into contiguous finite volumes (FV) or finite elements (FE) within which the mass and momentum conservation properties, alongside the turbulence closure conditions, are satisfied. Key decisions are associated with how many such FV or FE are required, and their size and location within the domain.

9.3.3 Potential Flow

In addition to treating the flow as incompressible, if the influence of viscosity is ignored, then Equation (9.2) can be reduced to Laplace's equation, as follows:

$$\nabla^2\phi = 0, \tag{9.5}$$

where ϕ is the velocity potential. In this case, the flow is representative of that at an infinite Reynolds number. As the length based Reynolds number of a typical ship can easily be 10^9, this provides a reasonable representation of the flow. The advantage of this approach is that, through the use of an appropriate Green's function, the problem can be reduced to a solution of equations just on the wetted ship hull. Such boundary element (or surface panel) methods are widely applied [9.16, 9.17]. The selection of a Green's function that incorporates the free-surface boundary condition will give detailed knowledge of the wave pattern and associated drag. Section 9.5 gives a particular example of such an approach using thin ship theory.

The removal of viscous effects requires the use of an appropriate empiricism to estimate the full-scale resistance of a ship, for example, using the ITTC 1957 correlation line, Chapter 4. However, as in the main, the ship boundary layer is thin, it is possible to apply a zonal approach. In this zonal approach, the inner boundary layer is solved using a viscous method. This could include, at its simplest, an integral boundary layer method ranging to solution of the Navier–Stokes equations with thin boundary layer assumptions [9.4, 9.11]. The solution of the boundary layer requires a detailed knowledge of the surface pressure distribution and the ship hull geometry. This can be obtained directly using an appropriate surface panel method. These methods are best applied in an iterative manner with application of a suitable matching condition between the inner (viscous) and outer (inviscid) zone. Considerable effort went into the development of these techniques through the 1980s.

Typically, potential methods only require a definition of the hull surface in terms of panels mapped across its wetted surface. As a result, for a given resolution of force detail on the hull surface, the number of panels scale as N^2 compared with N^3 for a steady RANS calculation.

9.3.4 Free Surface

The inability of the free-surface interface to withstand a significant pressure differential poses a challenge when determining the flow around a ship. Until the flow field around a hull is known it is not possible to define the location of the free surface which in turn will influence the flow around a hull. The boundary conditions are [9.1] as follows:

(1) Kinematic: the interface is sharp with a local normal velocity of the interface that is the same as that of the normal velocity of the air and water at the interface.
(2) Dynamic or force equilibrium: the pressure difference across the interface is associated with that sustained due to surface tension and interface curvature, and the shear stress is equal and of opposite direction either side.

There are two approaches to determining the location of the free surface for RANS methods, as illustrated schematically in Figure 9.1. The first approach attempts to track the interface location by moving a boundary so that it is located where the sharp free-surface interface lies. This requires the whole mesh and boundary location to move as the solution progresses. The second captures the location implicitly through determining where, within the computational domain, the boundary between air and water is located. Typically this is done by introducing an extra conservation variable as in the volume of fluid approach which determines the proportion of water in the particular mesh cell, a value of one being assigned for full and zero for empty, [9.18], or in the level set method [9.19] where an extra scalar is a distance to the interface location.

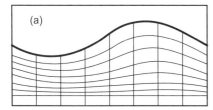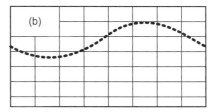

Figure 9.1. Location of free surface. (a) Tracking: mesh fitted to boundary. (b) Capture: boundary located across mesh elements.

For potential flow surface panel codes, which use a simple Rankine source/ dipole Green's function, the free surface is considered as a physical boundary with a distribution of panels [9.20] located on the free surface, or often in linearised methods on the static water level. It is possible to develop more advanced hull panel boundary Green's functions that satisfy the free-surface boundary condition automatically. Again, these can use a variety of linearisation assumptions: the boundary is on the static water level and the wave amplitudes are small. It is outside the scope of this book to describe the many variations and developments in this area and interested readers can consult Newman [9.21] for greater detail.

A difficulty for both potential and RANS approaches is for more dynamic flow regimes where the sinkage and trim of the hull become significant ($\sim Fn > 0.15$). An iterative approach has to be applied to obtain the dynamic balance of forces and moments on the hull for its resultant trim and heave.

In consideration of the RANS free-surface methods, there are a number of approaches to dealing with the flow conditions at the location of the air–water interface. These typically result in choices as to whether both air and water flow problems will be solved and as to whether the water and air are treated as incompressible or not. Godderidge *et al.* [9.22, 9.23] examined the various alternatives and suggest guidance as to which should be selected. The choice made reflects the level of fidelity required for the various resistance components and how much of the viscous free-surface interaction is to be captured.

9.4 Interpretation of Numerical Methods

9.4.1 Introduction

The art of effective CFD analysis is in being able to identify the inherent approximations and to have confidence that the level of approximation is acceptable. CFD tools should never replace the importance of sound engineering judgement in assessing the results of the analysis. Indeed, one of the inherent problems of the latest CFD methods is the wealth of data generated, and the ability to 'visualise' the implications of the results requires considerable skill. Due to this, interpretation is still seen as a largely subjective process based on personal experience of hydrodynamics. The subjective nature of the process can often be seen to imply an unknown level of risk. This is one of the reasons for the concern expressed by the maritime industry for the use of CFD as an integral part of the design process.

The ever reducing cost of computational resources has made available tools which can deliver results within a sufficiently short time span that they can be included within the design process. Uses of such tools are in concept design and parametric studies of main dimensions; optimization of hull form, appendages and propulsion systems; and detailed analysis of individual components and their interaction with the whole ship, for example, appendage alignment or sloshing of liquids in tanks. In addition to addressing these issues during the design process, CFD methods are often applied as a diagnostic technique for identifying the cause of a particular problem. Understanding the fluid dynamic cause of the problem also then allows possible remedies to be suggested.

The easy availability of results from complex computational analysis often fosters the belief that, when it comes to data, more detail implies more accuracy. Hence, greater reliance can be placed on the results. Automatic shape optimisation in particular exposes the ship design process to considerable risk. An optimum shape found using a particular computational implementation of a mathematical model will not necessarily be optimum when exposed to real conditions.

An oft assumed, but usually not stated, belief that small changes in input lead only to small changes in output [9.24] cannot be guaranteed for the complex, highly non-linear nature of the flow around vessels. Typical everyday examples of situations which violate this assumption include laminar-turbulent transition, flow separation, cavitation and breaking waves. It is the presence or otherwise of these features which can strongly influence the dependent parameters such as wave resistance, viscous resistance, wake fraction and so on, for which the shape is optimised. Not surprisingly, it is these features which are the least tractable for CFD analysis.

Correct dimensional analysis is essential to the proper understanding of the behaviour of a ship moving through water, see Chapter 3. Knowledge of the relative importance of the set of non-dimensional parameters, constructed from the independent variables, in controlling the behaviour of the non-dimensional dependent variables is the first step in reducing the ship design challenge to a manageable problem. It is the functional relationships of non-dimensional independent variables, based on the properties of the fluid, relative motions, shape parameters and relative size and position, which control ship performance.

The physical behaviour of moving fluids is well understood. However, understanding the complex interrelationship between a shape and how the fluid responds is central to ship design. The power required to propel a ship, the dynamic distortions of the structure of the ship and its response to imposed fluid motions are fundamental features of hydrodynamic design.

Engineers seek to analyse problems and then to use the information obtained to improve the design of artefacts and overall systems. Historically, two approaches have been possible in the analysis of fluid dynamic problems.

(1) Systematic experimentation can be used to vary design parameters and, hence, obtain an optimum design. However, the cost of such test programmes can be prohibitive. A more fundamental drawback is the necessity to carry out tests at model scale and extrapolate the results to full scale, see Chapter 4. These still cause a considerable level of uncertainty in the extrapolation of model results to that of full scale.

(2) An analytical approach is the second possibility. Closed form solutions exist for a few tightly specified flows. In addition, approximations can be made, for example, slender body theory, which at least gives reasonable predictions. In general, the more complex the flow the greater the level of mathematical detail required to specify and, if possible, to solve the problem. Errors and uncertainty arise from the assumptions made. Many 'difficult' integrals require asymptotic approximations to be made or equations are linearised based on the assumption that only small perturbations exist. These greatly restrict the range of applicability of the analytic solution and there is always the temptation to use results outside their range of validity.

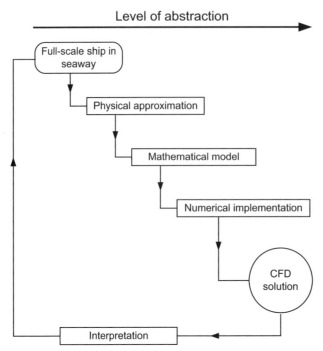

Figure 9.2. Levels of abstraction of CFD solution from physical reality.

The advent of powerful computers has, over the past five decades, allowed progressively more complex problems to be solved numerically. These computational techniques now offer the engineer a third, numerical alternative. In general, the continuous mathematical representation of a fluid is replaced by a discrete representation. This reduces the complexity of the mathematical formulation to such a level that it can be solved numerically through the repetitive application of a large number of mathematical operations. The result of the numeric analysis is a solution defined at discrete positions in time and/or space. The spatial and temporal resolution of the solution in some way is a measure of the usefulness and validity of the result. However, the cost of higher resolution is a greatly increased requirement for both data storage and computational power. The numerical approximation will also limit the maximum achievable resolution, as will the accuracy with which a computer can represent a real number.

The process of simplifying the complex unsteady flow regime around a full-scale ship can be considered to be one of progressive abstraction of simpler models from the complete problem, Figure 9.2. Each level of abstraction corresponds to the neglect of a particular non-dimensional parameter. Removal of these parameters can be considered to occur in three distinct phases: those which relate to physical parameters, those which relate to the assumptions made when deriving a continuous mathematical representation and, finally, those used in constructing a numerical (or discrete) representation of the mathematical model.

9.4.2 Validation of Applied CFD Methodology

In assessing ship resistance with the use of a numerical tool it is essential to be able to quantify the approximation in the different levels of interpretation applied. This

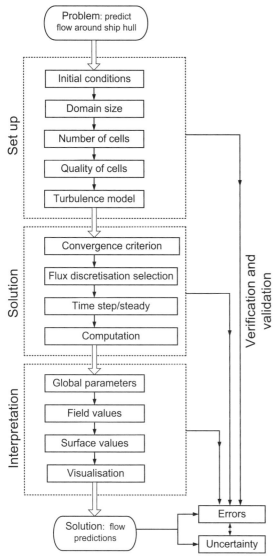

Figure 9.3. CFD process.

process of validation has been investigated in depth by the Resistance Committee of the ITTC and the workshops described in Table 9.1.

The process of validation can be seen as an attempt to eliminate or at least quantify these uncertainties. The process of code validation can be seen as a series of stages. Figure 9.3 illustrates the various stages required to solve the flow around a ship hull to obtain its resistance. Each of these stages requires use of an appropriate tool or analysis. Exactly how each stage is actually implemented depends on the numerical approach and the layout of the computational code.

Verification of the applied code implementation considers how well it represents the underlying mathematical formulation. This verification ensures that the code is free of error due to mistakes in expressing the mathematics in the particular computer language used. Ideally, the comparison should be made against an analytic solution, although often the comparison can only be made with other numerical codes.

CFD typically requires the user, or more often these days the code designer, to define a number of parameters for each stage of the process. Each of these parameters will introduce a solution dependence. Investigation of the sensitivity of the solution to all of these numerical parameters will require a significant investment of effort. It is in this area that an experienced user will be able to make rational and informed choices.

The most common form of dependence will be that due to the density and quality of the grid of points at which the governing equations are solved [9.25]. The process of grid, or often now mesh, generation [9.26] requires specialist software tools that ideally interface well to an underlying geometry definition. These tools typically will be from a general purpose computer–aided design (CAD) package, and they often struggle to work well when defining a ship hull and it is well to be conversant with methods for defining the complex curvature required in hull geometry definition [9.27].

The goal of effective mesh generation is to use *just* sufficient numbers of FV of the correct size, shape and orientation to resolve all the necessary flow features that control ship resistance. To date, it is rare that any practical computational problem can be said to have achieved this level of mesh resolution. However, with the reduction of computational cost, multimillion FV problems have been solved for steady flows and these appear to give largely mesh-independent solutions.

The final arbiter of performance will always be comparison with a physically measured quantity. It is in this comparison that the efforts of the maritime CFD communities, through the ongoing workshop series, Table 9.1, provides a valuable resource to the user of CFD for ship design. These publically available datasets provide a suitable series of test cases to develop confidence in the whole CFD process. As the majority of fluid dynamic codes are an approximation to the actual physics of the flow, differences will occur between the experimental and numerical results. Experimental data should always have a specified accuracy. This should then allow the difference between experiment and theory to be quantified. In many codes, however, some degree of empiricism is used to adjust the numerical model to fit specific experimental data. The extent to which such an empirically adjusted model can be said to be valid for cases run at different conditions requires careful consideration. A comparison will only be valid if both experiment and computation are at the same level of abstraction, i.e. all assumptions and values of non-dimensional parameters are the same.

9.4.3 Access to CFD

Users have four possible routes to using CFD.

(1) Development of their own bespoke computational code. This requires a significant investment of resources and time to achieve a level of performance comparable with those available through (2) and (4). It is unlikely that this route can still be recommended.
(2) Purchase of a commercial, usually general purpose, CFD flow solver. There are only a few commercial codes that can be applied to the problem of free-surface ship flows. Details of these vary and can typically be found via various

web-based CFD communities such as [9.28]. The likely commercial licence and training costs can be high. This still makes application of CFD techniques prohibitively expensive for small to medium scale enterprises unless they have employed individuals already conversant with use of CFD to a high standard and who can ensure a highly productive usage of the licence.

(3) Use of third party CFD consultants. As always with consultancy services, they cost a significant premium and there is often little knowledge transfer to the organisation. Such services, however, will provide detailed results that can be used as part of the design process and there is little wasted effort.

(4) Development of open-source CFD software. A number of these software products are now available. As in (1) and (2), they can require a significant training and organisational learning cost. The organisations that coordinate their development have an alternative business model which will still require investment. They do, however, offer a flexible route to bespoke computational analysis. This may have advantages because it allows a process tailored to the design task and one that can be readily adapted for use in automated design optimisation.

The remaining choice is then of the computational machine upon which the calculations are to be performed. As the price of computational resources is reduced, suitable machines are now affordable. Large scale computations can also be accessed via web-based computational resources at a reasonable cost.

9.5 Thin Ship Theory

9.5.1 Background

Potential flow theory provides a powerful approach for the calculation of wave resistance, as through the suitable choice of the Green's function in a boundary element method, the free-surface boundary condition can be automatically captured. Thin ship theory provides a direct method of determining the likely wave field around a hull form. The background and development of the theory is described in [9.29–9.31].

In the theory, it is assumed that the ship hull(s) will be slender, the fluid is inviscid, incompressible and homogeneous, the fluid motion is steady and irrotational, surface tension may be neglected and the wave height at the free surface is small compared with the wave length. For the theory in its basic form, ship shape bodies are represented by planar arrays of Kelvin sources on the local hull centrelines, together with the assumption of linearised free-surface conditions. The theory includes the effects of a channel of finite breadth and the effects of shallow water.

The strength of the source on each panel may be calculated from the local slope of the local waterline, Equation (9.6):

$$\sigma = \frac{-U}{2\pi}\frac{dy}{dx}dS, \tag{9.6}$$

where dy/dx is the slope of the waterline, σ is the source strength and S is the wetted surface area.

The hull waterline offsets can be obtained directly and rapidly as output from a commercial lines fairing package, such as ShipShape [9.32].

The wave system is described as a series using the Eggers coefficients as follows:

$$\zeta = \sum_{m=0}^{m} [\xi_m \cos(xk_m \cos\theta_m) + \eta_m \sin(xk_m \cos\theta_m)] \cos\frac{m\pi y}{W}. \qquad (9.7)$$

This is derived as Equation (7.21) in Chapter 7.

The wave coefficients ξ_m and η_m can be derived theoretically using Equation (9.8), noting that they can also be derived experimentally from physical measurements of ζ in Equation (9.7), as described in Chapter 7. This is an important property of the approach described.

$$\left| \begin{matrix} \xi_m \\ \eta_m \end{matrix} \right| = \frac{16\pi U}{Wg} \frac{k_0 + k_m \cos^2\theta_m}{1 + \sin^2\theta_m - k_0 h \sec h^2(k_m h)}$$

$$\times \sum_\sigma \left[\sigma_\sigma e^{-k_m h} \cosh[k_m(h+z_\sigma)] \left| \begin{matrix} \cos(k_m x_\sigma \cos\theta_m) \\ \sin(k_m x_\sigma \cos\theta_m) \end{matrix} \right| \begin{matrix} \cos\frac{m\pi y_\sigma}{W} \\ \sin\frac{m\pi y_\sigma}{W} \end{matrix} \right] \qquad (9.8)$$

The wave pattern resistance may be calculated from Equation (9.9) which describes the resistance in terms of the Eggers coefficients, as follows:

$$R_{WP} = \frac{\rho g W}{4} \left\{ (\xi_0^2 + \eta_0^2)\left(1 - \frac{2k_0 h}{\sinh(2k_0 h)}\right) \right.$$

$$\left. + \sum_{m=1}^{M} (\xi_m^2 + \eta_m^2)\left[1 - \frac{\cos^2\theta_m}{2}\left(1 + \frac{2k_m h}{\sinh(2k_m h)}\right)\right] \right\}. \qquad (9.9)$$

This is derived as Equation (7.24) in Chapter 7 and a full derivation is given in Appendix 2, Equation (A2.1). Note that the theory provides an estimate of the proportions of transverse and diverging content in the wave system, see Chapter 7, and that the theoretical predictions of the wave pattern and wave resistance can be compared directly with values derived from physical measurements of the wave elevation.

9.5.2 Distribution of Sources

The hull is represented by an array of sources on the hull centreline and the strength of each source is derived from the slope of the local waterline. It was found from earlier use of the theory, e.g. [9.31], that above about 18 waterlines and 30 sections the difference in the predicted results became very small as the number of panels was increased further. The main hull source distribution finally adopted for most of the calculations was derived from 20 waterlines and 50 sections. This number was also maintained for changes in trim and sinkage.

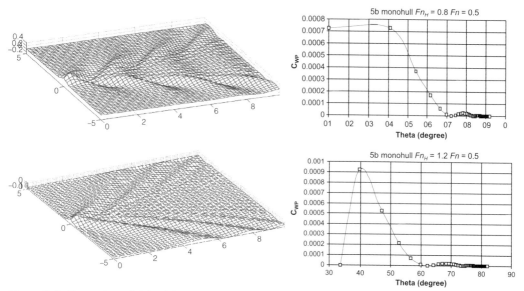

Figure 9.4. Examples of thin ship theory predictions of wave elevation and wave energy distribution.

9.5.3 Modifications to the Basic Theory

The basic theory was modified in order to facilitate the insertion of additional sources and sinks to simulate local pressure changes. These could be used, for example, to represent the transom stern, a bulbous bow and other discontinuities on the hull.

It had been noted from model tests and full-scale operation that trim and, hence, transom immersion can have a significant influence on the wave pattern and consequently on the wave resistance and wave wash. An important refinement to the basic theory, and a requirement of all wave resistance and wave wash theories, therefore does concern the need to model the transom stern in a satisfactory manner. A popular and reasonably satisfactory procedure had been to apply a hydrostatic ($\rho g H_T$) transom resistance correction [9.33]. Whilst this gives a reasonable correction to the resistance, it does not do so by correcting the wave system and is therefore not capable of predicting the wave pattern correctly. The creation of a virtual stern and associated source strengths [9.31] and [9.34] has been found to provide the best results in terms of wave pattern resistance and the prediction of wash waves.

9.5.4 Example Results

Examples of predicted wave patterns and distributions of wave energy using thin ship theory are shown in Figure 9.4 [9.35]. These clearly show the effects of shallow water on the wave system and on the distribution of wave energy, see Chapter 7, Section 7.3.4.6. These results were found to correlate well with measurements of wave height and wave resistance [9.35].

Table 9.2. *Computational parameters applied to the self-propulsion of the KVLCC2*

Parameter	Setting
Computing	64-bit desktop PC 4 GB of RAM
No. of elements	Approx. 2 million
Mesh type	Unstructured –hybrid (tetrahedra/prism)
Turbulence model	Shear stress transport
Advection scheme	CFX high resolution
Convergence control	RMS residual $<10^{-5}$
Pseudo time step	Automatic
Simulation time	Typically 5 hours
Wall modelling	CFX automatic wall modelling
y^+	~30

9.6 Estimation of Ship Self-propulsion Using RANS

9.6.1 Background

It is possible to model the performance of a ship propeller using a solution of the RANS equation, see Chapter 15. In practice this is a computationally expensive process, and it can often be more effective to represent the integrating effect of the propeller on the hull nominal wake. The process couples a RANS solution of the flow over the hull with a propeller analysis tool, see Chapter 15, that evaluates the axial and momentum changes for a series of annuli, typically 10–20. These momentum changes are then used as appropriate body force $\{f_x, f_y, f_z\}$ terms over the region of the propeller and the RANS equations resolved. If necessary, this process can be repeated until no significant changes in propeller thrust occur [9.36].

There are a number of alternative methods of evaluating the propeller momentum sources [9.37]. These range from a straightforward specified constant thrust, an empirically based thrust distribution through to distribution of axial and angular momentum derived from the methods described in Chapter 15. In the following example the fluid flow around the KVLCC2 hull form has been modelled using the commercial finite-volume code [9.38]. The motion of the fluid is modelled using the incompressible isothermal RANS equations (9.4) in order to determine the Cartesian flow (u, v, w) and pressure (p) field of the water around the KVLCC2 hull and rudder. Table 9.2 gives details of the computational model applied. Blade element-momentum theory (BEMT), as detailed in Section 15.5, is applied to evaluate the propeller performance.

9.6.2 Mesh Generation

A hybrid finite-volume unstructured mesh was built using tetrahedra in the far field and inflated prism elements around the hull with a first element thickness equating to a $y^+ = 30$, with 10–15 elements capturing the boundary layer of both hull and rudder. Separate meshes were produced for each rudder angle using a representation of the skeg (horn) rudder with gaps between the movable and fixed part of the rudder. Examples of various areas of the generated mesh are shown in Figure 9.5.

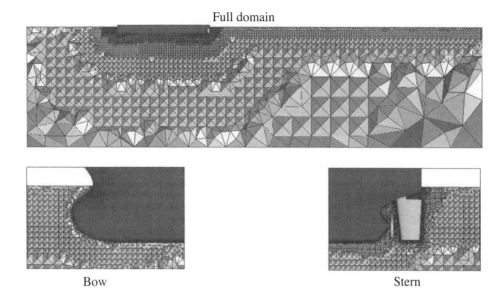

Figure 9.5. Mesh generated around KVLCC2.

9.6.3 Boundary Conditions

The solution of the RANS equations requires a series of appropriate boundary conditions to be defined. The hull is modelled using a no-slip wall condition. A Dirichlet inlet condition, one body length upstream of the hull, is defined where the inlet velocity and turbulence are prescribed explicitly. The model scale velocity is replicated in the CFD analyses and inlet turbulence intensity is set at 5%. A mass flow outlet is positioned $3\times L_{BP}$ downstream of the hull. The influence of the tank cross section (blockage effect) on the self-propulsion is automatically included through use of sidewall conditions with a free-slip wall condition placed at the locations of the floor and sides of the tank (16 m wide × 7 m deep) to enable direct comparison with the experimental results without having to account for blockage effects. The influence of a free surface is not included in these simulations due to the increase in computational cost, and the free surface is modelled with a symmetry plane. The Froude number is sufficiently low, $Fr = 0.14$, that this is a reasonable assumption.

Figure 9.6 shows an example including the free-surface flow in the stern region of a typical container ship (Korean container ship) with and without the application of a self-propulsion propeller model where free-surface effects are much more important. A volume of fluid approach is used to capture the free-surface location. The presence of the propeller influences the wave hump behind the stern and hence alters the pressure drag.

9.6.4 Methodology

In placing the propeller at the stern of the vessel the flow into the propeller is modified compared with the open water, see Chapter 8. The presence of the hull boundary layer results in the average velocity of the fluid entering the propeller disc (V_A) varying across the propeller disc. The propeller accelerates the flow ahead of itself,

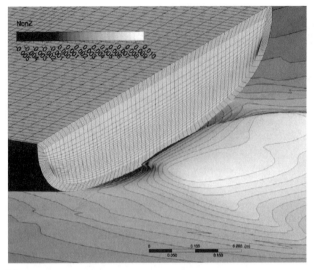

(a)

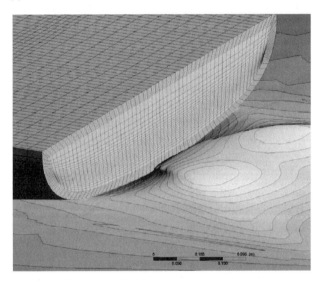

(b)

Figure 9.6. RANS CFD solution using ANSYS CFX v.12 [9.38] capturing the free-surface contours at the stern of the Korean container ship (KCS). (a) Free surface with propeller. (b) Free surface without propeller.

increasing the rate of shear in the boundary layer, leading to an increase in the skin friction resistance, and reducing the pressure over the rear of the hull, leading to an increase in pressure drag and a possible suppression of flow separation. Within the RANS mesh the propeller is represented as a cylindrical subdomain with a diameter equal to that of the propeller. The subdomain is divided into a series of ten annuli corresponding to ten radial slices (dr) along the blade. The appropriate momentum source terms from BEMT, a and a' in Section 15.5.5, are then applied over the subdomain in cylindrical co-ordinates to represent the axial and tangential influence of the propeller.

Table 9.3. *Force components for self-propulsion with rudder at 10°*

	Towing tank	Fine 2.1 M	Medium 1.5 M	Coarse 1.05 M
Longitudinal force, X (N)	−11.05	−11.74	−12.60	−13.82
Transverse force, Y (N)	6.79	7.6	7.51	7.33
Yaw moment, N (Nm)	−19.47	−18.75	−18.70	−18.35
Thrust, T (N)	10.46	12.53	12.37	12.08
Rudder X force, Rx (N)	−2.02	−1.83	−1.89	−1.94
Rudder Y force, Ry (N)	4.32	4.94	4.99	4.88

The following procedure is used to calculate the propeller performance and replicate it in the RANS simulations.

1. An initial converged stage of the RANS simulation (RMS residuals $< 1 \times 10^{-5}$) of flow past the hull is performed, without the propeller model. The local nominal wake fraction, w_T', is then determined for each annulus by calculating the average circumferential mean velocity at the corresponding annuli, as follows:

$$w_T' = \frac{1}{2\pi r} \int_0^{2\pi} \left(1 - \frac{U}{V_A}\right) r\, d\theta, \tag{9.10}$$

   where U is the axial velocity at a given r and θ.
2. A user specified Fortran module is used to export the set of local axial wake fractions to the BEMT code.
3. The BEMT code is used to calculate the thrust (dK_T) and torque (dK_Q) for the 10 radial slices based on ship speed, the local nominal wake fraction and the propeller rpm.
4. The local thrust and torque derived by the BEMT code are assumed to act uniformly over the annulus corresponding to each radial slice. The thrust is converted to axial momentum sources (momentum/time) distributed over the annuli by dividing the force by the volume of annuli. The torque is converted to tangential momentum sources by dividing the torque by the average radius of the annulus and the volume of the annulus.
5. These momentum sources are then returned to the RANS solver by a user Fortran Module which distributes them equally over the axial length of the propeller disc.
6. The RANS simulation is then restarted from the naked hull solution but now with the additional momentum sources. The final solution is assumed to have converged when the RMS residuals $< 1 \times 10^{-5}$.

Further refinements to the model add an iterative loop that uses the solution found in stage 6 by re-entering the wake fractions at stage 2 and, for manoeuvring use, a series of circumferential sectors to examine the influence of cross flow [9.36].

9.6.5 Results

As an example, the self-propulsion performance of the KVLCC2 hull is evaluated at model scale. The full-size ship design is 320 m and is modelled at 1:58 scale. A four

Table 9.4. *Influence of the rudder on propeller performance at the model self-propulsion point [9.36]*

	CFD – no rudder	CFD – rudder
Wake fraction, w_t	0.467	0.485
Thrust deduction factor, t	0.326	0.258
Rpm at model self-propulsion point	552	542
Advance coefficient, J	0.357	0.351
Thrust coefficient, K_T	0.226	0.233
Torque coefficient, K_Q	0.026	0.027
Efficiency, η	0.494	0.482

bladed fixed-pitch propeller with $P/D = 0.721$ and diameter of 9.86 m is used. The model propeller is operated at 515 rpm, the equivalent of full-scale self-propulsion. The advantage of the BEMT approach is that the influence of the rudder on propeller performance can be accurately captured [9.39]. Table 9.3 identifies the influence of mesh resolution on the evaluation of various force components. The finest mesh has 2.1M FV cells and a rudder angle of 10° is used. Convergent behaviour can be seen for all force components. Using the fine mesh, Table 9.4 illustrates the influence of the rudder on the self-propulsion point of the model. Figure 9.7 shows

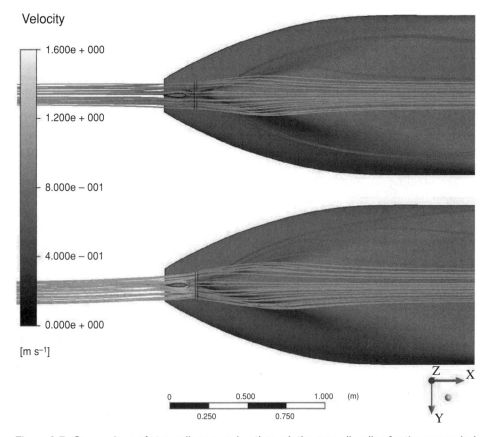

Figure 9.7. Comparison of streamlines passing through the propeller disc for the appended hull, no propeller model (top), and with propeller model on (bottom) [9.36].

the influence of the propeller model on streamlines passing through the propeller disk.

9.7 Summary

It is clear that numerical methods will provide an ever increasing role in the design of new ship hull forms. Their correct application will always rely on the correct interpretation of their result to the actual full-scale ship operating condition. It should also be recognised that a fully automated ship optimisation process will remain a computationally costly process. A range of computational tools ranging from simple thin ship theory and surface panel codes through to a self-propelled ship operating in a seaway solved using an unsteady RANS method will provide the designer with a hierarchical approach that will prove more time- and cost-effective.

REFERENCES (CHAPTER 9)

9.1 Ferziger, J.H. and Peric, M. *Computational Methods for Fluid Dynamics*. Axel-Springer Verlag, Berlin, 2002.

9.2 *Best Practice Guidelines for Marine Applications of Computational Fluid Dynamics*, W.S. Atkins Ltd., Epsom, UK, 2003.

9.3 Molland, A.F. and Turnock, S.R., *Marine Rudders and Control Surfaces*. Butterworth-Heinemann, Oxford, UK, 2007, Chapter 6. pp. 233–31.

9.4 Larsson, L. *SSPA-ITTC Workshop on Ship Boundary Layers*. SSPA Publication No. 90, Ship Boundary Layer Workshop Göteburg, 1980.

9.5 Larsson, L., Patel, V., Dyne, G. (eds.) *Proceedings of the 1990 SSPA-CTH-IIHR Workshop on Ship Viscous Flow*. Flowtech International Report No. 2, 1991.

9.6 Kodama, Y., Takeshi, H., Hinatsu, M., Hino, T., Uto, S., Hirata, N. and Murashige, S. (eds.) *Proceedings, CFD Workshop*. Ship Research Institute, Tokyo, Japan. 1994.

9.7 Tahara, Y. and Stern, F. A large-domain approach for calculating ship boundary layers and wakes and wave fields for nonzero Froude number. *Journal of Computational Physics*, Vol. 127, 1996, pp. 398–411.

9.8 Larsson, L., Stern F. and Bertram V. (eds.) *Gothenburg 2000 a Workshop on Numerical Ship Hydrodynamics*. Chalmers University of Technology, CHA/NAV/R-02/0073, 2000.

9.9 Larsson, L., Stern, F. and Bertram, V. Benchmarking of computational fluid dynamics for ship flows: the Gothenburg 2000 workshop. *Journal of Ship Research*, Vol. 47, No. 1, March 2003, pp. 63–81.

9.10 Hino, T. *Proceedings of the CFD Workshop Tokyo*. National Maritime Institute of Japan, 2005.

9.11 Landweber, L. and Patel V.C. Ship boundary layers. *Annual Review of Fluid Mechanics*, Vol. 11, 1979, pp. 173–205.

9.12 Wilson, R.V., Carrica, P.M. and Stern, F. Simulation of ship breaking bow waves and induced vortices and scars. *International Journal of Numerical Methods in Fluids*, Vol. 54, pp. 419–451.

9.13 Batchelor, G.K. *An Introduction to Fluid Dynamics*. Cambridge University Press, Cambride, UK, 1967.

9.14 Wilcox, D.C. *Turbulence Modeling for CFD*. 2nd Edition DCW Industries, La Canada, CA, 1998.

9.15 Pattenden, R.J., Bressloff, N.W., Turnock, S.R. and Zhang, X. Unsteady simulations of the flow around a short surface-mounted cylinder. *International Journal for Numerical Methods in Fluids*, Vol. 53, No. 6, 2007, pp. 895–914.

9.16 Hess, J.L. Panel methods in computational fluid dynamics. *Annual Review of Fluid Mechanics*, Vol. 22, 1990, pp. 255–274.

9.17 Katz, J. and Plotkin, A. *Low Speed Aerodynamics: From Wing Theory to Panel Methods*, Mcgraw-Hill, New York, 1991.

9.18 Hirt, C.W. and Nicolls, B.D. Volume of fluid (VOF) method for the dynamics of free boundaries. *Journal of Computational Physics*, Vol. 39, No. 1, January 1981, pp. 201–225.

9.19 Sethian, J.A. and Smereka, P. Level set method for fluid interfaces. *Annual Review of Fluid Mechanics*, Vol. 35, 2003, pp. 341–372.

9.20 Farmer, J., Martinelli, L. and Jameson, A. A fast multigrid method for solving incompressible hydrodynamic problems with free surfaces, *AIAA*-93-0767, pp. 15.

9.21 Newman, J. *Marine Hydrodynamics*. MIT Press, Cambridge, MA, 1977.

9.22 Godderidge, B., Turnock, S.R., Earl, C. and Tan, M. The effect of fluid compressibility on the simulation of sloshing impacts. *Ocean Engineering*, Vol. 36, No. 8, 2009, pp. 578–587.

9.23 Godderidge, B., Turnock, S.R., Tan, M. and Earl, C. An investigation of multiphase CFD modelling of a lateral sloshing tank. *Computers and Fluids*, Vol. 38, No. 2, 2009, pp.183–193.

9.24 Gleick, J. *Chaos, the Amazing Science of the Unpredictable*. William Heinemann, Portsmouth, NH, 1988.

9.25 Wright, A.M. Automated adaptation of spatial grids for flow solutions around marine bodies of complex geometry. Ph.D. thesis, University of Southampton, 2000.

9.26 Thompson, J.F., Soni, B.K. and Weatherill, N.P. (eds.) *Handbook of Grid Generation*, CRC Press, Boca Raton, FL, 1998.

9.27 Nowacki, H., Bloor, M.I.G. and Oleksiewicz, B. (eds.) Computational geometry for ships, *World Scientific Publishing*, London, 1995.

9.28 CFD Online. www.cfd-online.com Last accessed May 2010.

9.29 Insel, M. An investigation into the resistance components of high speed displacement catamarans. Ph.D. thesis, University of Southampton, 1990.

9.30 Couser, P.R. An investigation into the performance of high-speed catamarans in calm water and waves. Ph.D. thesis, University of Southampton, 1996.

9.31 Couser, P.R., Wellicome, J.F. and Molland, A.F. An improved method for the theoretical prediction of the wave resistance of transom stern hulls using a slender body approach. *International Shipbuilding Progress*, Vol. 45, No. 444, 1998, pp. 331–349.

9.32 *ShipShape User Manual*, Wolfson Unit MTIA, University of Southampton, 1990.

9.33 Insel, M., Molland, A.F. and Wellicome, J.F. Wave resistance prediction of a catamaran by linearised theory. *Proceedings of Fifth International Conference on Computer Aided Design, Manufacture and Operation, CADMO'94*. Computational Mechanics Publications, Southampton, UK, 1994.

9.34 Doctors, L.J. Resistance prediction for transom-stern vessels. *Proceedings of Fourth International Conference on Fast Sea Transportation, FAST'97*, Sydney, 1997.

9.35 Molland, A.F., Wilson, P.A. and Taunton, D.J. Theoretical prediction of the characteristics of ship generated near-field wash waves. Ship Science Report No. 125, University of Southampton, November 2002.

9.36 Phillips, A.B. Simulations of a self propelled autonomous underwater vehicle, Ph.D. thesis, University of Southampton, 2010.

9.37 Phillips, A.B., Turnock, S.R. and Furlong, M.E. Evaluation of manoeuvring coefficients of a self-propelled ship using a blade element momentum propeller model coupled to a Reynolds averaged Navier Stokes flow solver. *Ocean Engineering*, Vol. 36, 2009, pp. 1217–1225.

9.38 *Ansys CFX User Guide* v11, Ansys, Canonsburg, PA, 2007.
9.39 Phillips, A.B., Turnock, S.R. and Furlong, M.E. Accurate capture of rudder-propeller interaction using a coupled blade element momentum-RANS approach. *Ship Technology Research (Schiffstechnik)*, Vol. 57, No. 2, 2010, pp. 128–139.

10 Resistance Design Data

10.1 Introduction

Resistance data suitable for power estimates may be obtained from a number of sources. If model tests are not carried out, the most useful sources are standard series data, whilst regression analysis of model resistance test results provides a good basis for preliminary power estimates. Numerical methods can provide useful inputs for specific investigations of hull form changes and this is discussed in Chapter 9. Methods of presenting resistance data are described in Section 3.1.3. This chapter reviews sources of resistance data. Design charts or tabulations of data for a number of the standard series, together with coefficients of regression analyses, are included in Appendix A3.

10.2 Data Sources

10.2.1 Standard Series Data

Standard series data result from systematic resistance tests that have been carried out on particular series of hull forms. Such tests entail the systematic variation of the main hull form parameters such as C_B, $L/\nabla^{1/3}$, B/T and LCB. Standard series tests provide an invaluable source of resistance data for use in the power estimate, in particular, for use at the early design stage and/or when model tank tests have not been carried out. The data may typically be used for the following:

(1) Deriving power requirements for a given hull form,
(2) Selecting suitable hull forms for a particular task, including the investigation of the influence of changes in hull parameters such as C_B and B/T, and as
(3) A standard for judging the quality of a particular (non-series) hull form.

Standard series data are available for a large range of ship types. The following section summarises the principal series. Some sources are not strictly series data, but are included for completeness as they make specific contributions to the database. Design data, for direct use in making practical power predictions, have been extracted from those references marked with an asterisk *. These are described in Section 10.3.

10.2.1.1 Single-Screw Merchant Ship Forms

Series 60 [10.1], [10.2], [10.3]*.
British Ship Research Association (BSRA) Series [10.4], [10.5], [10.6]*.
Statens Skeppsprovingansalt (SSPA) series [10.7], [10.8].
Maritime Administration (US) MARAD Series [10.9].

10.2.1.2 Twin-Screw Merchant Ship Forms

Taylor–Gertler series [10.10]*.
Lindblad series [10.11], [10.12].
Zborowski Polish series [10.13]*.

10.2.1.3 Coasters

Dawson series [10.14], [10.15], [10.16], [10.17].

10.2.1.4 Trawlers

BSRA series [10.18], [10.19], [10.20], [10.21].
Ridgely–Nevitt series [10.22], [10.23], [10.24].

10.2.1.5 Tugs

Parker and Dawson tug investigations [10.25].
Moor tug investigations [10.26].

10.2.1.6 Semi-displacement Forms, Round Bilge

SSPA Nordström [10.27].
SSPA series, Lindgren and Williams [10.28].
Series 63, Beys [10.29].
Series 64, Yeh [10.30] *, [10.31], [10.32], [10.33].
National Physical Laboratory (NPL) series, Bailey [10.34]*.
Semi-planing series, Compton [10.35].
High-speed displacement hull forms, Robson [10.36].
Fast transom-stern hulls, Lahtiharju *et al.* [10.37].
SKLAD semi-displacement series, Gamulin [10.38], Radojcic *et al.* [10.39].

10.2.1.7 Semi-displacement Forms, Double Chine

National Technical University of Athens (NTUA) Series, Radojcic *et al.*
  [10.40] *, Grigoropoulos and Loukakis [10.41].

10.2.1.8 Planing Hulls

Series 62, Clement and Blount, [10.42]*, Keuning and Gerritsma [10.43].
United States Coast Guard (USCG) series, Kowalyshyn and Metcalf, [10.44].
Series 65, Hadler *et al.* [10.45].
Savitsky *et al.* [10.46], [10.47], [10.48]*, [10.49].

10.2.1.9 Multihulls

Southampton catamaran series. Insel and Molland *et al.* [10.50], [10.51]*,
  [10.52], [10.53].
Other multihull data, Steen, Cassella, Bruzzone *et al.* [10.54]–[10.58].
Versuchsanstalt für Wasserbau und Schiffbau Berlin (VWS) catamaran series,
  Müller-Graf [10.59], Zips, [10.60]*, Müller-Graf and Radojcic, [10.61].

10.2.1.10 Yachts

Delft series. Gerritsma and Keuning *et al.* [10.62] to [10.67]∗.

10.2.2 Other Resistance Data

Average ⓒ data, Moor and Small [10.70]∗.
Tanker and bulk carrier forms. Moor [10.71].
0.85 Block coefficient series, Clements and Thompson [10.72], [10.73].
Fractional draught data, Moor and O'Connor [10.74]∗.
Regressions:
Sabit regressions: BSRA series [10.75]∗, Series 60, [10.76]∗ and SSPA series
   [10.77].
Holtrop and Mennen [10.78], [10.79], [10.80], [10.81]∗.
Hollenbach [10.82]∗.
Radojcic [10.83], [10.84]∗.
Van Oortmerssen, small craft [10.85]∗.
Robinson, Wolfson Unit for Marine Technology and Industrial Aerodynamics
   (WUMTIA) small craft [10.86]∗.

10.2.3 Regression Analysis of Resistance Data

If sufficient data for a large number of independent designs exist in a standard
form (e.g. from tests on models of similar size in one towing tank), then statistical
treatment (regression analysis) gives an alternative to standard series which in principle allows the evaluation of optimum parameter combinations free from artificial
constraints.

Regression methods can only be applied in the long term to ships of closely similar type since upwards of 150 models may be required to provide an adequate analysis of non-linear combinations of parameters. Typical regressions of note include
those reported in [10.75–10.91] and the results of some of these are discussed in
Section 10.3.

A typical set of variables for ship resistance regression analysis might be as
follows:

$$C_T = f\left[C_B, L\nabla^{1/3}, B/T, LCB, \tfrac{1}{2}\alpha_E \text{ etc.}\right]$$

The references indicate the scope of published work on regression analysis. For
example, Sabit's regression of the BSRA series [10.75], uses:

$$CR_{400} = f[L/\nabla^{1/3}, B/T, C_B, LCB]$$

where

$$CR = \frac{R \cdot L}{\Delta \cdot V^2} \text{ and } CR = 2.4938 \, ⓒ \, L/\nabla^{1/3}$$

for each speed increment, and for three draught values (per series)

$$\text{and} \quad CR_{400} = a_1 + a_2 \, L/\nabla^{1/3} + a_3 \, B/T + \ldots..a_6 \, (L/\nabla^{1/3})^2 + a_7 \, (B/T)^2 + \ldots..$$

and coefficients a_n are published for each speed and draught.

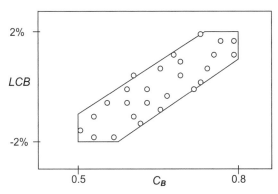

Figure 10.1. Typical limitations of database.

Holtrop [10.81] breaks down the resistance into viscous and wave, and includes speed (Fr) in the analysis. $C_F(1 + k)$ is derived using the International Towing Tank Conference (ITTC) C_F line and $(1 + k)$ by regression. C_W is based on Havelock's wavemaking theory:

$$R_w/\Delta = C_1 e^{-mFr**-2/9} + e^{-Fr**-2}\{C_2 + C_3 \cos(\lambda Fr^{-2})\}$$

C_1, C_2, C_3, λ and m are coefficients which depend on hull form and are derived by regression analysis.

Molland and Watson [10.90] use $\copyright = f[L/B, B/T, C_B, LCB, \tfrac{1}{2}\alpha_E]$ at each speed increment.

Lin *et al.* [10.91] include the slope properties of the sectional area curve in the C_W formulation.

The limitations of regression analysis are the following:

1. Analysis data should be for the correct ship type
2. Note the 'statistical quality' of the data, such as standard error
3. Great care must be taken that the prediction is confined to the limits of the database, in particular, where such a regression is used for hull form optimisation

Predictions should not be made for unrealistic combinations of hull parameters. For example, simply stating the limits as $0.5 < C_B < 0.8$ and $-2\% < LCB < +2\%$ may not be satisfactory, as the actual source data will probably be made up as shown in Figure 10.1. In other words, the regression should not be used, for example, to predict results for a hull form with a block coefficient C_B of 0.8 and an LCB of $-2\%L$ (2% aft).

10.2.4 Numerical Methods

Viscous resistance and wave resistance may be derived by numerical and theoretical methods. Such methods provide a powerful tool, allowing parametric changes in hull form to be investigated and the influence of Reynolds number to be explored. Raven *et al.* [10.92] describe a computational fluid dynamics (CFD)-based prediction of

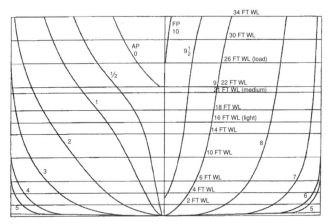

Figure 10.2. BSRA series body plan $C_B = 0.65$.

resistance, including an investigation of scale effects. CFD and numerical methods are outlined in Chapter 9.

10.3 Selected Design Data

10.3.1 Displacement Ships

10.3.1.1 BSRA Series

This series, suitable for single-screw merchant ships, was developed by the British Ship Research Association during the 1950s and 1960s.

The data for the BSRA series are summarised in [10.4, 10.5, 10.6]. These include full details of the body plans for the series, together with propulsion data. An example of a body plan for the series is shown in Figure 10.2 and the series covers the following range of speeds and hull parameters:

Speed: $V_k/\sqrt{L_f}$: 0.20–0.85 (V_k knots, L_f ft) [Fr: 0.06–0.25].
C_B: 0.60–0.85; B/T: 2–4; $L/\nabla^{1/3}$: 4.5–6.5; LCB: $-2\%L - +2\%L$.
Output: $\textcircled{C}_{400} = \frac{579.8\,P_E}{\Delta^{2/3}\,V^3}$.

$\textcircled{C}_{400}$ values are presented for a standard ship with dimensions (ft): $400 \times 55 \times 26$ (load draught) and standard $LCB = 20(C_B - 0.675)\ \%L$ forward of amidships, and at reduced draughts (ft) of 21, 16 and 16 trimmed.

The $\textcircled{C}_{400}$ data (for a 400 ft ship) are presented to a base of C_B for a range of speeds, as shown for the 26 ft load draught in Figure 10.3.

Charts are provided to correct changes from the standard ship dimensions to the actual ship for B/T, LCB and $L/\nabla^{1/3}$. Examples of these corrections are shown in Figure 10.4.

A skin friction correction $\delta\textcircled{C}$ must be applied to correct the 400 ft (122 m) ship value to the actual ship length value $\textcircled{C}_s$ as follows:

$$\textcircled{C}_s = \textcircled{C}_{400} \pm \delta\textcircled{C}.$$

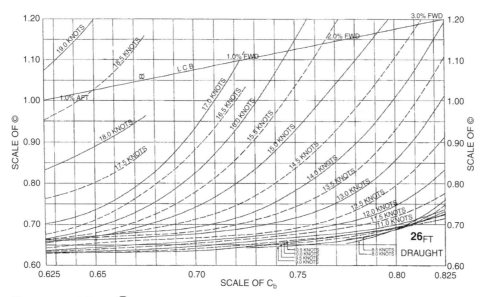

Figure 10.3. BSRA $\textcircled{C}_{400}$ values to a base of block coefficient.

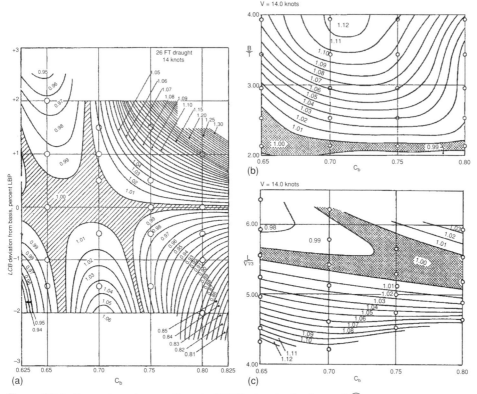

Figure 10.4. Examples of corrections to BSRA standard values of $\textcircled{C}_{400}$.

The correction is added for ships <122 m, and subtracted for ships >122 m. Using the Froude O_M and O_S values [10.93], the correction is

$$\delta \textcircled{C} = \textcircled{C}_{400} - \textcircled{C}_s = (O_{400} - O_S) \, \textcircled{S} - \textcircled{L}^{-0.175}$$

where $\textcircled{L} = 1.055 \, V_k / \sqrt{L_f}$.

A chart is provided in [10.4], reproduced in Figure 10.5, which is based on the Froude O_M and O_S values. This allows the $\delta \textcircled{C}$ correction to be derived for ship lengths other than 400 ft (122 m).

Approximations to $\delta \textcircled{C}$ based on Figure 10.5 and a mean $V_k / \sqrt{L_f} = 0.70$ are as follows:

For $L < 122$ m, $\delta \textcircled{C}$ is added to $\textcircled{C}_{400}$.

$$\delta \textcircled{C} = +0.54 \times (122 - L)^{1.63} \times 10^{-4}. \tag{10.1}$$

For $L \geq 122$ m, $\delta \textcircled{C}$ is subtracted from $\textcircled{C}_{400}$.

$$\delta \textcircled{C} = -\frac{0.10}{1 + \frac{188}{(L-122)}}. \tag{10.2}$$

When using Equations (10.1) and (10.2), for typical $\textcircled{C}$ values of 0.6–0.8, the error in $\textcircled{C}$ due to possible error in $\delta \textcircled{C}$ at a ship length of 250 m (820 ft) over the speed range $V_k / \sqrt{L_f} = 0.5$–0.9 is less than 0.5%. At 60 m (197 ft) the error is also less than 0.5%.

Finally, the effective power P_E may be derived using Equation (3.12) as

$$P_E = \textcircled{C}_s \times \Delta^{2/3} V^3 / 579.8, \tag{10.3}$$

where P_E is in kW, Δ is in tonne, V is in knots and using 1 knot $= 0.5144$ m/s.

A ship correlation (or load) factor (SCF), or $(1 + x)$, should be applied using Equation (5.3) as follows:

For $L \geq 122$ m,

$$(1 + x) = 1.2 - \frac{\sqrt{L}}{48}. \tag{10.4}$$

For $L < 122$ m, $(1 + x) = 1.0$ is recommended (Table 5.1).

$$P_{Eship} = P_{Emodel} \times (1 + x).$$

The basic $\textcircled{C}_{400}$ values and corrections for B/T, LCB and $L/\nabla^{1/3}$ are contained in [10.4–10.6].

BULBOUS BOW. Reference [10.6] contains charts which indicate whether or not the use of a bulbous bow would be beneficial. These are reproduced in Figure 10.6. The data cover a range of speeds and block coefficients and are based on results for the BSRA series. The data are for the loaded condition only and are likely to be suitable for many merchant ships, such as cargo and container ships, tankers and bulk carriers and the like. However, basing a decision on the loaded condition alone may not be suitable for tankers and bulk carriers, where the effect of a bulb in the ballast condition is likely to be advantageous. For this reason, most tankers and bulk carriers are fitted with a bulb. A more detailed discussion on the application of bulbous bows is included in Chapter 14.

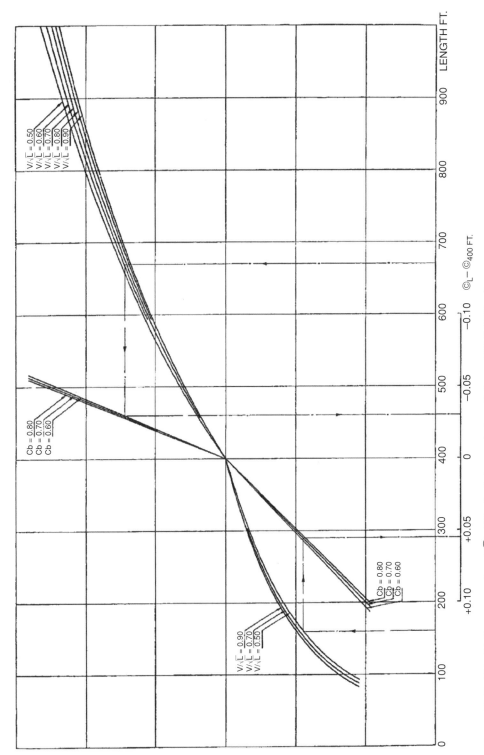

Figure 10.5. Skin friction correction $\delta\textcircled{C}$: Effect of change in length from 400 ft (122 m).

195

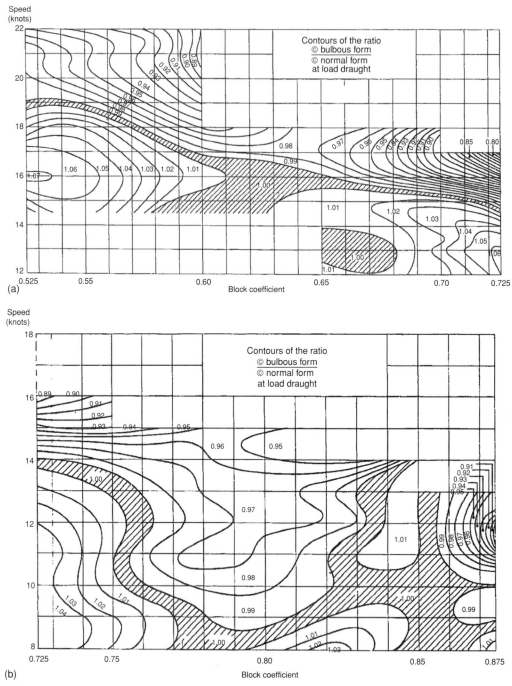

Figure 10.6. Effect of bulbous bow on resistance (speed in knots for 400 ft ship).

SABIT REGRESSION. A useful alternative that harnesses most of the BSRA series data is the regression analysis of the series carried out by Sabit [10.75]. These were carried out for the load, medium and light draught conditions.

The resistance data are presented in terms of

$$CR_{400} = f[L/\nabla^{1/3}, B/T, C_B, LCB], \qquad (10.5)$$

where

$$CR = \frac{R \cdot L}{\Delta \cdot V^2} \qquad (10.6)$$

and the suffix 400 denotes values for a 400 ft ship, and at speeds $V_k/\sqrt{L_f} = 0.50, 0.55,$ 0.60, 0.65, 0.70, 0.75, 0.80. The values of $L/\nabla^{1/3}$, C_B, B/T and LCB used in the analysis are for the load, medium and light draught conditions. Values of ∇ and LCB for the medium and light conditions, based on a draught ratio T_R (= intermediate draught/load draught) may be derived from the following equations. L_{BP} is assumed constant, B is assumed constant, $T_{medium} = 0.808\, T_{load}$ and $T_{light} = 0.616 T_{load}$. Displacement ratio Δ_R (= intermediate displacement/load displacement).
    For $C_B = 0.65$–0.725,

$$\Delta_R = 1.0 + (T_R - 1.0)\big[3.776 - 7.16C_B + 4.8C_B^2\big]. \qquad (10.7)$$

    For $C_B = 0.725$–0.80,

$$\Delta_R = 1.0 + (T_R - 1.0)\big[-1.1245 + 6.366C_B - 4.533C_B^2\big]. \qquad (10.8)$$

These two expressions can be used to derive the displacement and $L/\nabla^{1/3}$ at any intermediate draught. C_B is at load displacement.
    For $C_B = 0.65$–0.725,

$$LCB = LCB_{load} - (1 - T_R)\big[-124.335 + 328.98C_B - 218.93C_B^2$$
$$-10.553LCB_{load} + 27.42(LCB_{load} \times C_B) - 18.4\big(LCB_{load} \times C_B^2\big)\big]. \qquad (10.9)$$

    For $C_B = 0.725$–0.80,

$$LCB = LCB_{load} - (1 - T_R)\big[-169.975 + 449.74C_B - 298.667C_B^2$$
$$+ 3.855\, LCB_{load} - 12.56(LCB_{load} \times C_B)$$
$$+ 9.333\big(LCB_{load} \times C_B^2\big)\big]. \qquad (10.10)$$

These two expressions can be used to derive the LCB at any intermediate draught. C_B is at load displacement.
    The regression equations for the load, medium and light draughts take the following form.

LOAD DRAUGHT. Limits of parameters in the load condition are: $L/\nabla^{1/3}$ 4.2–6.4, B/T 2.2–4.0, C_B 0.65–0.80, LCB −2.0% − +3.5%. Extrapolation beyond these limits can result in relatively large errors.

$$Y_{400} = a1 \times X1 + a2 \times X2 + a3 \times X3 + a4 \times X4 + a5 \times X5 + a6 \times X6$$
$$+ a7 \times X7 + a8 \times X8 + a9 \times X9 + a10 \times X10 + a11 \times X11$$
$$+ a12 \times X12 + a13 \times X13 + a14 \times X14 + a15 \times X15$$
$$+ a16 \times X16, \qquad (10.11)$$

where:

$X1 = 1$	$X2 = (L/\nabla^{1/3} - 5.296)/1.064$
$X3 = 10(B/T - 3.025)/9.05$	$X4 = 1000(C_B - 0.725)/75$
$X5 = (LCB - 0.77)/2.77$	$X6 = X2^2$
$X7 = X3^2$	$X8 = X4^2$
$X9 = X5^2$	$X10 = X2 \times X3$
$X11 = X2 \times X4$	$X12 = X2 \times X5$
$X13 = X3 \times X4$	$X14 = X3 \times X5$
$X15 = X4 \times X5$	$X16 = X5 \times X4^2$

$$CR_{400} = (Y_{400} \times 5.1635) + 13.1035, \tag{10.12}$$

and from Equation (3.20), $\textcircled{C}_{400} = (CR_{400} \times \nabla^{1/3})/ (2.4938 \times L)$.

The coefficients a1 to a16 are given in Table A3.2, Appendix A3.

MEDIUM DRAUGHT. Limits of parameters in the medium condition are: $L/\nabla^{1/3}$ 4.6–6.9, B/T 2.6–4.9, C_B 0.62–0.78, LCB −1.6% – +3.9%. Extrapolation beyond these limits can result in relatively large errors.

$$
\begin{aligned}
Z_{400} = {} & b1 \times R1 + b2 \times R2 + b3 \times R3 + b4 \times R4 + b5 \times R5 + b6 \times R6 \\
& + b7 \times R7 + b8 \times R8 + b9 \times R9 + b10 \times R10 + b11 \times R11 \\
& + b12 \times R12 + b13 \times R13 + b14 \times R14 + b15 \times R15 \\
& + b16 \times R16,
\end{aligned}
\tag{10.13}
$$

where

$R1 = 1$	$R2 = (L/\nabla^{1/3} - 5.7605)/1.1665$
$R3 = (B/T - 3.745)/1.125$	$R4 = 100(C_B - 0.7035)/8.05$
$R5 = (LCB - 1.20)/2.76$	$R6 = R2^2$
$R7 = R3^2$	$R8 = R4^2$
$R9 = R5^2$	$R10 = R2 \times R3$
$R11 = R2 \times R4$	$R12 = R2 \times R5$
$R13 = R3 \times R4$	$R14 = R3 \times R5$
$R15 = R4 \times R5$	$R16 = R5 \times R4^2$

$$CR_{400} = (Z_{400} \times 6.449) + 15.010, \tag{10.14}$$

and from Equation (3.20), $\textcircled{C}_{400} = (CR_{400} \times \nabla^{1/3})/ (2.4938 \times L)$. The coefficients b1 to b16 are given in Table A3.3, Appendix A3. T, ∇, C_B and LCB are for the medium draught condition.

LIGHT DRAUGHT. Limits of parameters in the light condition are: $L/\nabla^{1/3}$ 5.1–7.7, B/T 3.4–6.4, C_B 0.59–0.77, LCB −1.1% – +4.3%. Extrapolation beyond these limits can result in relatively large errors.

$$
\begin{aligned}
S_{400} = {} & c1 \times T1 + c2 \times T2 + c3 \times T3 + c4 \times T4 + c5 \times T5 + c6 \times T6 \\
& + c7 \times T7 + c8 \times T8 + c9 \times T9 + c10 \times T10 + c11 \times T11 + c12 \times T12 \\
& + c13 \times T13 + c14 \times T14 + c15 \times T15 + c16 \times T16,
\end{aligned}
\tag{10.15}
$$

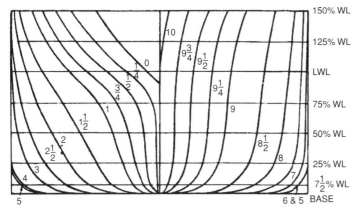

Figure 10.7. Series 60 body plan $C_B = 0.65$.

where

T1 = 1	T2 = $(L/\nabla^{1/3} - 6.4085)/1.3085$
T3 = $(B/T - 4.915)/1.475$	T4 = $100(C_B - 0.679)/8.70$
T5 = $LCB - 1.615)/2.735$	T6 = T2^2
T7 = T3^2	T8 = T4^2
T9 = T5^2	T10 = T2 × T3
T11 = T2 × T4	T12 = T2 × T5
T13 = T3 × T4	T14 = T3 × T5
T15 = T4 × T5	T16 = T5 × T4^2

$$CR_{400} = (S_{400} \times 7.826) + 17.417, \qquad (10.16)$$

and from Equation (3.20), $\copyright_{400} = (CR_{400} \times \nabla^{1/3})/(2.4938 \times L)$. The coefficients c1 to c16 are given in Table A3.4, Appendix A3. L, ∇, C_B and LCB are for the light draught condition.

The $\copyright_{400}$ will be corrected for skin friction and correlation, and P_E derived, in a manner similar to that described earlier for the BSRA series.

10.3.1.2 Series 60

The Series 60 was developed in the United States during the 1950s [10.1, 10.2]. A new presentation was proposed by Lackenby and Milton [10.3] and a regression analysis of the data was carried out by Sabit [10.76]. An example of a body plan for the series is shown in Figure 10.7 and the series covers the following range of speeds and hull parameters:

Speed: $V_k/\sqrt{L_f}$: 0.20–0.90 [Fr: 0.06–0.27].
C_B: 0.60–0.80; B/T: 2.5–3.5; L/B: 5.5–8.5; LCB: −2.5% − +3.5%.

Output, using Lackenby and Milton's presentation [10.3],

$$\copyright_{400} = \frac{579.8 P_E}{\Delta^{2/3} V^3}.$$

Lackenby and Milton [10.3] used both the Schoenherr and Froude friction lines.

SABIT REGRESSION. A useful tool is the regression analysis of the Series 60 carried out by Sabit [10.76]. The approach is similar to that used for the BSRA series, but using L/B rather than $L/\nabla^{1/3}$. The Froude C_F line was used to determine the 400 ft ship values.

The resistance data are presented in terms of

$$CR_{400} = f[L/B, B/T, C_B, LCB], \tag{10.17}$$

and at speeds $V_k/\sqrt{L_f} = 0.50, 0.55, 0.60, 0.65, 0.70, 0.75, 0.80, 0.85, 0.90$.

The regression equation for the load draught takes the following form:

$$\begin{aligned}
Y_{400} =\ & a1 \times X1 + a2 \times X2 + a3 \times X3 + a4 \times X4 + a5 \times X5 + a6 \times X6 \\
& + a7 \times X7 + a8 \times X8 + a9 \times X9 + a10 \times X10 + a11 \times X11 + a12 \times X12 \\
& + a13 \times X13 + a14 \times X14 + a15 \times X15 + a16 \times X16, \tag{10.18}
\end{aligned}$$

where

$X1 = 1$	$X2 = 2(L/B - 7.0)/3.0$
$X3 = 2(B/T - 3)$	$X4 = 10(C_B - 0.7)$
$X5 = (LCB - 0.515) / 2.995$	$X6 = X2^2$
$X7 = X3^2$	$X8 = X4^2$
$X9 = X5^2$	$X10 = X2 \times X3$
$X11 = X2 \times X4$	$X12 = X2 \times X5$
$X13 = X3 \times X4$	$X14 = X3 \times X5$
$X15 = X4 \times X5$	$X16 = X5 \times X4^2$

$$CR_{400} = (Y_{400} \times 8.3375) + 17.3505, \tag{10.19}$$

and from Equation (3.20), $\copyright_{400} = (CR_{400} \times \nabla^{1/3})/(2.4938 \times L)$. The coefficients a1 to a16 are given in Table A3.5, Appendix A3.

10.3.1.3 Average $\copyright$ Data

Moor and Small [10.70] gathered together many model resistance test data during the 1950s, including the results for the BSRA series. These data were cross faired and so-called average data were tabulated. These average values are given in Table A3.6 in Appendix A3. The data provide a good first estimate of resistance, but in many cases can be improved upon with small refinements to the hull parameters.

$\copyright_{400}$ values are presented for a standard ship with dimensions (ft): $400 \times 55 \times 26$ for a range of speed, C_B and LCB values.

Speed: $V_k/\sqrt{L_f}$: 0.50–0.90 [Fr: 0.15–0.27].
C_B: 0.625–0.825; LCB: $-2.0\% - +2.5\%$.

In order to correct for the dimensions of a proposed new ship, compared with the standard dimensions ($400 \times 55 \times 26$), Moor and Small propose the use of Mumford's indicies x and y. In this approach, it is assumed that $P_E \propto B^x \cdot T^y$ where the indicies x and y vary primarily with speed.

The correction becomes

$$P_{E2} = P_{E1} \times \left(\frac{B_2}{B_1}\right)^x \left(\frac{T_2}{T_1}\right)^y, \tag{10.20}$$

Table 10.1. *Mumford indicies*

$V/\sqrt{L}$	0.50	0.55	0.60	0.65	0.70	0.75	0.80	0.85	0.90
y	0.54	0.55	0.57	0.58	0.60	0.62	0.64	0.67	0.70
x	0.90	0.90	0.90	0.90	0.90	0.90	0.90	0.90	0.90

and, using the analysis of resistance data for many models, it is proposed that $x = 0.90$ and y has the Mumford values shown in Table 10.1.

If $\textcircled{C}$ is used, the correction becomes

$$\textcircled{C}_2 = \textcircled{C}_1 \times \left(\frac{B_2}{B_1}\right)^{x-\frac{2}{3}} \left(\frac{T_2}{T_1}\right)^{y-\frac{2}{3}}. \tag{10.21}$$

After correction for dimensions $[(B_2/B_1 \text{ and } (T_2/T_1)]$, the $\textcircled{C}_{400}$ will be corrected for skin friction and correlation, and P_E derived, in a manner similar to that described earlier for the BSRA series.

The $\textcircled{C}$ data specifically for full form vessels such as tankers and bulk carriers can be found in [10.71, 10.72, 10.73]. These $\textcircled{C}$ data can be derived and corrected in a manner similar to that described in this section.

10.3.1.4 Fractional Draught Data/Equations

Values of resistance at reduced draught (for example, at ballast draught), as fractions of the resistance at load draught for single-screw ships, have been published by Moor and O'Connor [10.74]. Equations were developed that predict the effective displacement and power ratios in terms of the draught ratio $(T)_R$, as follows:

$$\frac{\Delta_2}{\Delta_1} = (T)_R^{1.607-0.661C_B}, \tag{10.22}$$

where

$$(T)_R = \left(\frac{T_{\text{Ballast}}}{T_{\text{load}}}\right)$$

$$\frac{P_{E\,\text{ballast}}}{P_{E\,\text{load}}} = 1 + [(T)_R - 1]\left\{\left(0.789 - 0.270[(T)_R - 1] + 0.529\,C_B\left(\frac{L}{10T}\right)^{0.5}\right)\right.$$

$$+ V/\sqrt{L}\left(2.336 + 1.439[(T)_R - 1] - 4.605\,C_B\left(\frac{L}{10T}\right)^{0.5}\right)$$

$$\left. + (V/\sqrt{L})^2\left(-2.056 - 1.485[(T)_R - 1] + 3.798\,C_B\left(\frac{L}{10T}\right)^{0.5}\right)\right\}. \tag{10.23}$$

where T is the load draught.

As developed, the equations should be applied to the 400 ft ship before correction to actual ship length. Only relatively small errors are incurred if the correction is applied directly to the actual ship size. The data for deriving the equations were based mainly on the BSRA series and similar forms. The equations should, as a first

Table 10.2. *Parameter ranges, Holtrop et al. [10.78]*

Ship type	*Fr* max	C_P	L/B
Tankers and bulk carriers	0.24	0.73–0.85	5.1–7.1
General cargo	0.30	0.58–0.72	5.3–8.0
Fishing vessels, tugs	0.38	0.55–0.65	3.9–6.3
Container ships, frigates	0.45	0.55–0.67	6.0–9.5

approximation, be suitable for most single-screw forms. Example 5 in Chapter 17 illustrates the use of these equations.

10.3.1.5 Holtrop and Mennen – Single-screw and Twin-screw Vessels

The regression equations developed by Holtrop *et al.* [10.78–10.81] have been used extensively in the preliminary prediction of ship resistance. The equations proposed in [10.80] and [10.81] are summarised in the following. The approximate ranges of the parameters are given in Table 10.2.

The total resistance is described as

$$R_T = R_F(1 + k_1) + R_{APP} + R_W + R_B + R_{TR} + R_A, \qquad (10.24)$$

where R_F is calculated using the ITTC1957 formula, $(1 + k_1)$ is the form factor, R_{APP} is the appendage resistance, R_W is the wave resistance, R_B is the extra resistance due to a bulbous bow, R_{TR} is the additional resistance due to transom immersion and R_A is the model-ship correlation resistance which includes such effects as hull roughness and air drag. R_{APP} and $(1 + k_1)$ are discussed in Chapters 3 and 4. This section discusses R_W, R_B, R_{TR} and R_A.

WAVE RESISTANCE R_W. In order to improve the quality of prediction of R_W, three speed ranges were used as follows:

(i) $Fr < 0.40$ obtained using Equation (10.25)
(ii) $Fr > 0.55$ obtained using Equation (10.26)
(iii) $0.40 < Fr < 0.55$ obtained by interpolation using Equation (10.27)

(i) $Fr < 0.40$

$$R_W = c_1 c_2 c_5 \nabla \rho g \exp\{m_1 Fr^d + m_4 \cos(\lambda Fr^{-2})\}, \qquad (10.25)$$

where

$$c_1 = 2223105 c_7^{3.78613} (T/B)^{1.07961} (90 - i_E)^{-1.37565}$$
$$c_2 = \exp(-1.89\sqrt{c_3})$$
$$c_3 = 0.56 A_{BT}^{1.5}/\{BT(0.31\sqrt{A_{BT}} + T_F - h_B)\}$$
$$c_5 = 1 - 0.8 A_T/(BTC_M)$$
$$c_7 = 0.229577(B/L)^{0.33333} \quad \text{when } B/L < 0.11$$
$$c_7 = B/L \quad \text{when } 0.11 < B/L < 0.25$$
$$c_7 = 0.5 - 0.0625 L/B \quad \text{when } B/L > 0.25$$
$$m_1 = 0.0140407 L/T - 1.75254\nabla^{1/3}/L - 4.79323 B/L - c_{16}$$

$$c_{16} = 8.07981 C_P - 13.8673 C_P^2 + 6.984388\, C_P^3 \quad \text{when } C_P < 0.8$$
$$c_{16} = 1.73014 - 0.7067 C_P \quad \text{when } C_P > 0.8$$
$$m_4 = c_{15} 0.4 \exp(-0.034 Fr^{-3.29})$$
$$c_{15} = -1.69385 \quad \text{when } L^3/\nabla < 512$$
$$c_{15} = -1.69385 + (L/\nabla^{1/3} - 8)/2.36 \quad \text{when } 512 < L^3/\nabla < 1726.91$$
$$c_{15} = 0 \quad \text{when } L^3/\nabla > 1726.91$$
$$d = -0.90$$

(ii) $Fr > 0.55$

$$R_W = c_{17}\, c_2\, c_5 \nabla \rho g \exp\{m_3 Fr^d + m_4 \cos(\lambda Fr^{-2})\}, \qquad (10.26)$$

where

$$c_{17} = 6919.3\, C_M^{-1.3346} (\nabla/L^3)^{2.00977} (L/B - 2)^{1.40692}$$
$$m_3 = -7.2035 (B/L)^{0.326869} (T/B)^{0.605375}$$
$$\lambda = 1.446\, C_P - 0.03 L/B \quad \text{when } L/B < 12$$
$$\lambda = 1.446\, C_P - 0.36 \quad \text{when } L/B > 12$$

(iii) $0.40 < Fr < 0.55$

$$R_W = R_{W\,(Fr=0.40)} + (10 Fr - 4)[R_{W\,(Fr=0.55)} - R_{W\,(Fr=0.40)}]/1.5, \qquad (10.27)$$

where $R_{W\,(Fr=0.40)}$ is obtained using Equation (10.25) and $R_{W\,(Fr=0.55)}$ is obtained using Equation (10.26).

The angle i_E is the half angle of entrance of the waterline at the fore end. If it is not known, the following formula can be used:

$$i_E = 1 + 89 \exp\{-(L/B)^{0.80856}(1 - C_{WP})^{0.30484}$$
$$\times (1 - C_P - 0.0225\, LCB)^{0.6367} (L_R/B)^{0.34574} (100\, \nabla/L^3)^{0.16302}\}. \qquad (10.28)$$

where LCB is LCB forward of $0.5\,L$ as a percentage of L.

If the length of run L_R is not known, it may be estimated from the following formula:

$$L_R = L_{WL}[1 - C_P + 0.06\, C_P\, LCB/(4 C_P - 1)]. \qquad (10.29a)$$

If C_M is not known, a reasonable approximation for small ships is

$$C_M = 0.78 + 0.21 C_B \text{ and for larger ships is}$$
$$C_M = 0.80 + 0.21 C_B \text{ (Molland).} \qquad (10.29b)$$

If C_{WP} is not known, a reasonable approximation for displacement craft, $0.65 < C_B < 0.80$ is

$$C_{WP} = 0.67\, C_B + 0.32. \qquad (10.29c)$$

In the preceding equations, T_A is the draught aft, T_F is the draught forward, C_M is the midship section coefficient, C_{WP} is the waterplane coefficient, C_P is the prismatic coefficient, A_T is the immersed area of transom at rest, A_{BT} is the transverse area of bulbous bow, and h_B is the centre of area of A_{BT} above the keel, Figure 10.8. It is recommended that h_B should not exceed the upper limit of $0.6\, T_F$.

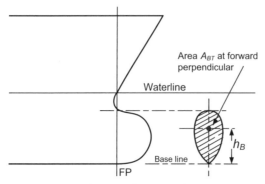

Figure 10.8. Definition of bulbous bow.

RESISTANCE DUE TO BULB R_B.

$$R_B = 0.11 \exp\left(-3P_B^{-2}\right) Fr_i^3 A_{BT}^{1.5} \rho g/(1 + Fr_i^2), \qquad (10.30)$$

where P_B is a measure of the emergence of the bow and Fr_i is the Froude number based on the immersion as follows:

$$P_B = 0.56\sqrt{A_{BT}}/(T_F - 1.5h_B) \text{ and}$$

$$Fr_i = V/\sqrt{g(T_F - h_B - 0.25\sqrt{A_{BT}}) + 0.15V^2}. \qquad (10.31)$$

RESISTANCE DUE TO TRANSOM R_{TR}.

$$R_{TR} = 0.5\rho V^2 A_T c_6, \qquad (10.32)$$

where $c_6 = 0.2(1 - 0.2Fr_T)$ when $Fr_T < 5$ or $c_6 = 0$ when $Fr_T \geq 5$. Fr_T is the Froude number based on transom immersion, as follows:

$$Fr_T = V/\sqrt{2g\,A_T/(B + BC_{WP})} \qquad (10.33)$$

MODEL-SHIP CORRELATION RESISTANCE R_A.

$$R_A = 0.5\rho\,SV^2\,C_A,$$

where S is the wetted surface area of the hull, for example, using Equations (10.83–10.86), and

$$C_A = 0.006\,(L + 100)^{-0.16} - 0.00205 + 0.003\sqrt{L/7.5}C_B^4 c_2\,(0.04 - c_4), \quad (10.34)$$

where L is in metres and

$$c_4 = T_F/L \quad \text{when } T_F/L \leq 0.04$$
$$\text{or} \quad c_4 = 0.04 \quad \text{when } T_F/L > 0.04.$$

It is noted that C_A may be higher or lower depending on the levels of hull surface roughness. The most recent analysis would suggest that the prediction for C_A can be up to 9% high, but for practical purposes use of Equation (10.34) is still recommended by Holtrop. Equation (10.34) is based on a standard roughness figure of $k_S = 150\,\mu$m. Approximate modifications to C_A for hull roughness can be made using the ITTC Bowden–Davison formula for ΔC_F, Equation (5.9), or the Townsin formula Equation (5.10).

10.3.1.6 Hollenbach – Single-screw and Twin-screw Vessels

Hollenbach [10.82] carried out a regression on the results of resistance tests on 433 models at the Vienna Ship Model Basin from 1980 to 1995. The models were of single- and twin-screw vessels. Results are presented in terms of

$$C_R = \frac{R_R}{0.5\rho\, B \cdot TV^2}.$$

It should be noted that $(B \cdot T/10)$ is used as the reference area rather than the more usual wetted surface area S. C_R was derived using the ITTC C_F line.

In addition to $L = L_{BP}$ and L_{WL} which have their usual meaning, a 'Length over surface' L_{OS} is used and is defined as follows:

- for the design draught, length between the aft end of the design waterline and the most forward point of the ship below the design waterline, for example, the fore end of a bulbous bow
- for the ballast draught, length between the aft end and the forward end of the hull at the ballast waterline, for example, the fore end of a bulbous bow would be taken into account but a rudder is not taken into account

The Froude number used in the formulae is based on the length L_{fn}, as follows:

$$\begin{aligned}
L_{fn} &= L_{OS} \quad \text{for } L_{OS}/L < 1 \\
&= L + 2/3(L_{OS} - L) \quad \text{for } 1 \leq L_{OS}/L < 1.1 \\
&= 1.0667\,L \quad \text{for } 1.1 \leq L_{OS}/L
\end{aligned}$$

Hollenbach analysed and presented the data in terms of a 'mean' value of resistance when normal constraints on the hull form will occur for design purposes and a 'minimum' resistance which might be achieved for good hull lines developed following computational and experimental investigations, and not subject to design constraints.

The coefficient C_R is generally expressed for 'mean' and 'minimum' values as

$$\begin{aligned}
C_R =\ &C_{R \cdot \text{Standard}} \cdot C_{R \cdot \text{FrKrit}} \cdot k_L (T/B)^{a1} (B/L)^{a2} (L_{OS}/L_{WL})^{a3} (L_{WL}/L)^{a4} \\
&\times (1 + (T_A - T_F)/L)^{a5} (D_P/T_A)^{a6} (1 + N_{\text{Rud}})^{a7} \\
&\times (1 + N_{\text{Brac}})^{a8} (1 + N_{\text{Boss}})^{a9} (1 + N_{\text{Thr}})^{a10}
\end{aligned} \qquad (10.35)$$

where $k_L = e_1 L^{e2}$, T_A is the draught at AP, T_F is the draught at FP, D_P is the propeller diameter, N_{Rud} is the number of rudders (1 or 2), N_{Brac} is the number of brackets (0–2), N_{Boss} is the number of bossings (0–2), N_{Thr} is the number of side thrusters (0–4).

$$\begin{aligned}
C_{R \cdot \text{Standard}} =\ &b_{11} + b_{12} Fr + b_{13} Fr^2 + C_B(b_{21} + b_{22} Fr + b_{23} Fr^2) \\
&+ C_B^2(b_{31} + b_{32} Fr + b_{33} Fr^2).
\end{aligned} \qquad (10.36)$$

$$C_{R \cdot \text{FrKrit}} = \max[1.0, (Fr/F_{r \cdot \text{Krit}})^{c1}]. \qquad (10.37)$$

where $F_{r \cdot \text{Krit}} = d_1 + d_2 C_B + d_3 C_B^2$

The formulae are valid for Froude number intervals as follows:

$$Fr_{\cdot\min} = \min(f_1, f_1 + f_2(f_3 - C_B)).$$
$$Fr_{\cdot\max} = g_1 + g_2 C_B + g_3 C_B^2.$$

The 'maximum' total resistance is given as $R_{T\cdot\max} = h_1 \cdot R_{T\cdot\text{mean}}$.

Note that for the 'minimum' resistance case, K_L and $C_{R\cdot\text{FrKrit}}$ in Equation (10.35) should be set at 1.0.

The coefficients in these equations, for 'mean' and 'minimum' resistance and single- and twin-screw vessels, are given in Table A3.7, Appendix A3.

Note that in the table of coefficients in the original paper [10.82], there was a sign error in coefficient a_3 ballast which should be $+1.1606$ *not* -1.1606.

10.3.1.7 Taylor-Gertler Series

The original tests on twin-screw model hulls were carried out by Taylor during 1910–1920 and cover the widest range of C_P, B/T and Fr yet produced. Gertler reanalysed the data as described in [10.10]. The series represents a transformation of a mathematical hull form based on an R.N. cruiser *Leviathan*. An example of a body plan for the series is shown in Figure 10.9 and the series covered the following range of speeds and hull parameters:

Speed: $V_k/\sqrt{L_f}$: 0.30–2.0 [Fr: 0.09–0.60].
B/T: 2.25, 3.00, 3.75; C_P: 0.48 to 0.86; ∇/L^3: 1.0×10^{-3} to 7.0×10^{-3} [$L/\nabla^{1/3}$: 5–10]; LCB was fixed at amidships.
Output: $C_R = \frac{R_R}{0.5\rho S V^2}$ where R_R is the residuary resistance, with C_R derived from $C_R = C_{T\text{model}} - C_{F\text{Schoenherr}}$, where $C_{F\text{Schoenherr}}$ is the Schoenherr friction line, Equation (4.10).

A typical presentation of the data, from [10.10], is shown in Figure 10.10 where, in this case, B/H is the B/T ratio.

If charts are not available, and in order to provide readily available data for design purposes, a range of data have been digitised from the charts. These are listed in Tables A3.8, A3.9, A3.10, and A3.11 in Appendix A3. Linear interpolation

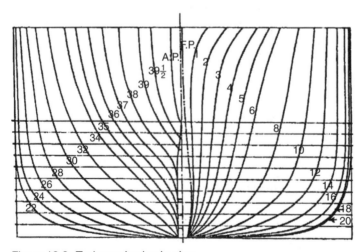

Figure 10.9. Taylor series body plan.

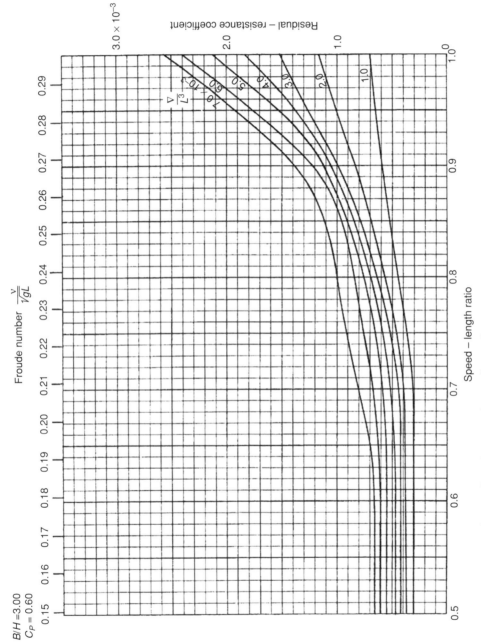

Figure 10.10. Taylor–Gertler C_R values to a base of speed.

207

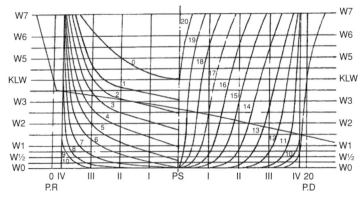

Figure 10.11. Polish series body plan: parent.

of the data in these tables should, in most cases, be satisfactory. The wetted surface area can be estimated using Equation (10.86). $C_p = C_B/C_M$ and, if C_M is not known, reasonable approximations are given in Equation (10.29b). When using the Taylor–Gertler series, the Schoenherr C_F should be used, Equation (4.10) or Equation (4.11), and it has been the practice to add a roughness allowance $\Delta C_F = 0.0004$.

10.3.1.8 Zborowski Series

A small systematic series of twin-screw models was tested in Poland [10.13]. The body plan is shown in Figure 10.11 and the series covered the following range of speed and hull parameters:

Speed: Fr: 0.25–0.35.

B/T: 2.25, 2.80, 3.35; C_B: 0.518–0.645; $L/\nabla^{1/3}$: 6.0, 6.5, 7.0; LCB was fixed at 2.25%L aft of amidships; $C_M = 0.977$.

Output: $C_{Tmodel} = \frac{R_{Tmodel}}{0.5\rho\,SV^2}$ The model length is 1.9 m and extrapolation to ship values entails the standard procedure described in Section 4.1, as follows:

$$C_{TS} = C_{TM} - (C_{FM} - C_{FS})$$

or

$$C_{TS} = C_{TM} - (1 + k)(C_{FM} - C_{FS}).$$

Values for C_{TM} for different values of speed and hull parameters are listed in Table A3.12, Appendix A3. $(1 + k)$, if used, may be estimated using the data in Chapter 4. Hull wetted surface area may be estimated using an appropriate formula, such as Equations (10.83) to (10.86), to be found in Section 10.4.

The tests did not include an aft centreline skeg and it is recommended in [10.13] that the drag of the skeg, assuming only frictional drag based on the wetted area of the skeg, should be added to the naked resistance derived above.

10.3.2 Semi-displacement Craft

10.3.2.1 Series 64

This systematic series of round bilge hull forms was tested at the David Taylor Model Basin (DTMB), West Bethesda, MD, on a wide range of hull parameters. These are described by Yeh [10.30].

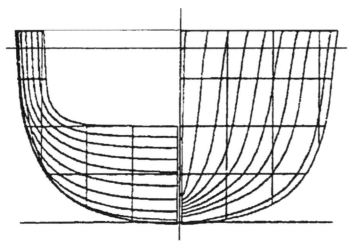

Figure 10.12. Series 64 body plan.

An example of a body plan for the series is shown in Figure 10.12. The series covered the following range of speed and hull parameters:

Speed: $V_k/\sqrt{L_f}$: 0.2–5.0 [Fr: 0.06–1.5].
B/T: 2.0, 3.0, 4.0; C_B: 0.35, 0.45, 0.55; $L/\nabla^{1/3}$: 8.0–12.4; LCB was fixed at 6.0% L aft of amidships. Data for C_R are presented in graphical and tabular form where

$$C_R = \frac{R_R}{0.5\rho S V^2}.$$

The original values for C_R were derived by subtracting Schoenherr C_F from the model total C_T. In order to provide some commonality with the NPL series, described in the next section, the Series 64 C_R values were converted to ITTC format. That is, Schoernherr C_F (Equation (4.10)) based on the model length of 3.05 m was added to the original model C_R to give model C_T. ITTC C_F Equation (4.15) was then subtracted from the model C_T to give C_R. The values for C_R (based on ITTC C_F) are given in Tables A3.13, A3.14, and A3.15, Appendix A3. The wetted surface area can be estimated using Equation (10.88). Further specific tests on Series 64 hull forms are reported in [10.31], [10.32].

The resistance of high-speed semi-displacement craft tends to be dominated by $L/\nabla^{1/3}$ ratio and a useful presentation for such craft is resistance coefficient to a base of $L/\nabla^{1/3}$ at a fixed Froude number. An early use of such an approach was by Nordström [10.27]. Examination of the data for Series 64 forms in [10.30] indicates that changes in B/T have a relatively small influence, with all C_R values lying within about 5% of a mean line. For these reasons, and to provide a practical design approach, regression analyses were carried out, [10.33]. These relate residuary resistance, C_R, solely to the $L/\nabla^{1/3}$ ratio at a number of fixed Froude numbers, Fr, for the data in [10.30]. The form of the equations is

$$C_R = a(L/\nabla^{1/3})^n. \tag{10.38}$$

Table 10.3. *Coefficients in the equation*
$1000\, C_R = a(L/\nabla^{1/3})^n$ *for Series 64,*
monohulls, $C_B = 0.35$

Fr	a	n	R^2
0.4	288	−2.33	0.934
0.5	751	−2.76	0.970
0.6	758	−2.81	0.979
0.7	279	−2.42	0.971
0.8	106	−2.06	0.925
0.9	47	−1.74	0.904
1.0	25	−1.50	0.896

The coefficients of the regressions, a and n, are given in Tables 10.3, 10.4 and 10.5.

10.3.2.2 NPL Series

This systematic series of round-bilge semi-displacement hull forms was tested at NPL, Teddington, UK, in the 1970s, Bailey [10.34]. An example of the body plan for the series is shown in Figure 10.13 and the series covered the following range of speeds and hull parameters:

Speed: $V_k/\sqrt{L_f}$: 0.8–4.1 [Fr: 0.30–1.1, Fr_∇: 0.6–3.2].

B/T: 1.7–6.7; $C_B = 0.4$ (fixed); $L/\nabla^{1/3}$: 4.5–8.3; LCB was fixed at 6.4% aft of amidships. Model length $L_{WL} = 2.54$ m.

Data for R_R/Δ and C_T for a 30.5-m ship are presented in graphical form for a range of L/B, $L/\nabla^{1/3}$ and $Fr \cdot R_R$ was derived by subtracting R_F, using the ITTC line, from the total resistance R_T.

In order to provide a more compatible presentation of the NPL data, C_R values have been calculated for the data where

$$C_R = \frac{R_R}{0.5\rho SV^2}.$$

Table 10.4. *Coefficients in the equation 1000*
$C_R = a(L/\nabla^{1/3})^n$ *for Series 64, monohulls,*
$C_B = 0.45$

Fr	a	n	R^2
0.4	36,726	−4.41	0.979
0.5	55,159	−4.61	0.989
0.6	42,184	−4.56	0.991
0.7	29,257	−4.47	0.995
0.8	27,130	−4.51	0.997
0.9	20,657	−4.46	0.996
1.0	11,644	−4.24	0.995

Table 10.5. *Coefficients in the equation 1000*
$C_R = a(L/\nabla^{1/3})^n$ *for Series 64, monohulls,*
$C_B = 0.55$

Fr	a	n	R^2
0.4	926	−2.74	0.930
0.5	1775	−3.05	0.971
0.6	1642	−3.08	0.983
0.7	1106	−2.98	0.972
0.8	783	−2.90	0.956
0.9	458	−2.73	0.941
1.0	199	−2.38	0.922

The resulting values for C_R are given in Table A3.16, Appendix A3. Wetted surface area can be estimated using an appropriate formula, such as Equation (10.88) or (10.90), to be found in Section 10.4.

10.3.2.3 NTUA Series

This systematic series of double-chine semi-displacement hull forms was developed by NTUA, Greece. These hull forms are suitable for fast semi-displacement mono-hull ferries and other such applications. A regression analysis of the resistance and trim data for the series is presented by Radojcic *et al.* [10.40]. An example of a body plan for the series is shown in Figure 10.14. The series covered the following range of speeds and hull parameters:

Speed: *Fr*: 0.3–1.1.
B/T: 3.2–6.2; $C_B = 0.34$–0.54; $L/\nabla^{1/3}$: 6.2–8.5; LCB: 12.4–14.6%L aft of amidships. Approximate mean model length is 2.35 m.

Data are presented for C_R and trim τ where

$$C_R = \frac{R_R}{0.5\rho SV^2}.$$

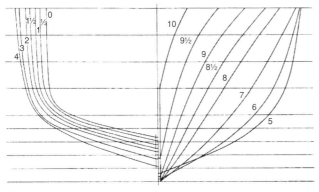

Figure 10.13. NPL series body plan: parent.

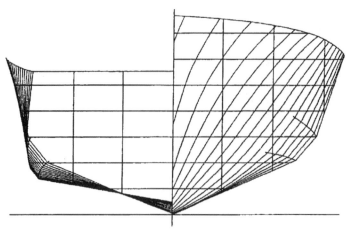

Figure 10.14. NTUA double-chine body plan.

C_R was derived by subtracting the ITTC C_{FM} from the model total resistance coefficient, C_{TM}. C_R and τ are presented as regression equations as follows:

$$C_R = \Sigma a_i \cdot x_i \quad \text{and} \quad \tau = \Sigma b_i \cdot x_i. \tag{10.39}$$

Values of the variables x_i and coefficients a_i, b_i of the regressions for C_R and trim τ are given in Tables A3.17 and A3.18, Appendix A3. Wetted surface area can be estimated using an appropriate formula, such as Equation (10.92) to be found in Section 10.4.

The results of further resistance and seakeeping experiments on the NTUA series are included in [10.41].

10.3.3 Planing Craft

The main sources of data presented are the single-chine Series 62, the Savitsky equations for planing craft and, for the lower speed range, the WUMTIA regression of hard chine forms. The WUMTIA regression is described in Section 10.3.4.2. Blount [10.94] describes the selection of hard chine or round-bilge hulls for high Froude numbers. Savitsky and Koelbel [10.95] provide an excellent review of seakeeping considerations in the design and operation of hard chine planing hulls.

10.3.3.1 Series 62

This systematic series of single chine hull forms was tested at DTMB, over a range of hull parameters. These are described by Clement and Blount [10.42]. An example of the body plan for the series is shown in Figure 10.15 and definitions of length and breadth are shown in Figure 10.16. The series covered the following range of speed and hull parameters:

Speed: Fr_∇: 1.0–3.5.
Length/breadth ratio L_p/B_{px}: 2.0, 3.06, 4.09, 5.50, 7.00.
Loading coefficient $A_P/\nabla^{2/3}$: 5.5, 7.0, 8.5. LCG aft of centroid of A_p: 0, 4, 8, 12.

Deadrise angle β: 13°. Keuning and Gerritsma [10.43] later extended the series using a deadrise angle β of 25°.

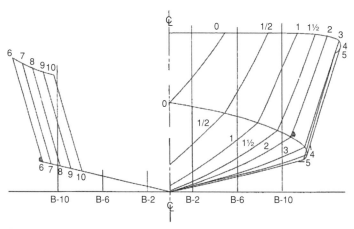

Figure 10.15. Series 62 body plan.

L_p is the projected chine length (Figure 10.16), A_P is the projected planing bottom area (for practical purposes, it can be assumed to be equivalent to the static
waterplane area), B_{px} is the maximum breadth over chines and ∇ is the displaced
volume at rest.

The data are presented in terms of total resistance per ton R_T/Δ for a 100,000 lb
displacement ship.

Radojcic [10.83] carried out a regression analysis of the Series 62 resistance data
and later updated the analysis [10.84]. Radojcic included the extension to the Series
62 by Keuning and Gerritsma [10.43], together with some of the models from Series
65 [10.45].

The resistance data are presented in terms of

$$R_T/\Delta = f[A_P/\nabla^{2/3}, L_p/B_{pa}, LCG/L_p, \beta_x].$$

at speeds of $Fr_\nabla = 1.0, 1.25, 1.50, 1.75, 2.0, 2.5, 3.0, 3.5$, where B_{pa} is the mean breadth
over chines and B_{pa} is A_p/L_p and β_x is the deadrise angle at 50% L_p.

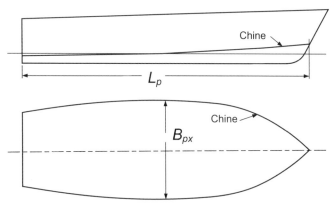

Figure 10.16. Definitions of length and breadth.

The limits of the parameters in the regression are as follows:

Loading coefficient $A_P/\nabla^{2/3}$: 4.25–9.5.
Length/beam ratio L_p/B_{pa}: 2.36–6.73.
LCG from transom $100\ LCG/L_p$: 30%–44.8%.
Deadrise angle at 50% L_p: 13°–37.4°.

Analysis of the Series 62 hull forms indicates that the ratio of maximum chine breadth to mean chine breadth B_{px}/B_{pa} varies from 1.18 to 1.22. It is suggested that a value of $B_{px}/B_{pa} = 1.21$ be used for preliminary design and powering purposes.

Regression analysis was carried out for resistance R_T/Δ, trim τ, wetted surface coefficient $S/\nabla^{2/3}$ and wetted length/chine length (length of wetted area) L_W/L_p.

The regression equations for R_T/Δ, τ, $S/\nabla^{2/3}$ and L_W/L_p all take the following form:

$$R_T/\Delta = b0 + b1 \times X1 + b2 \times X2 + b3 \times X3 + b4 \times X4 + b5 \times X5 + b6$$
$$\times\ X6 + b7 \times X7 \ldots\ldots\ldots\ldots b26 \times X26, \qquad (10.40)$$

where

$X1 = (A_P/\nabla^{2/3} - 6.875)/2.625$	$X2 = (100\ LCG/L_p - 37.4)/7.4$
$X3 = (L_p/B_{pa} - 4.545)/2.185$	$X4 = (\beta_x - 25.2)/12.2$
$X5 = X1 \times X2$	$X6 = X1 \times X3$
$X7 = X1 \times X4$	$X8 = X2 \times X3$
$X9 = X2 \times X4$	$X10 = X3 \times X4$
$X11 = X1^2$	$X12 = X2^2$
$X13 = X3^2$	$X14 = X4^2$
$X15 = X1 \times X2^2$	$X16 = X1 \times X3^2$
$X17 = X1 \times X4^2$	$X18 = X2 \times X1^2$
$X19 = X2 \times X3^2$	$X20 = X2 \times X4^2$
$X21 = X3 \times X1^2$	$X22 = X3 \times X2^2$
$X23 = X3 \times X4^2$	$X24 = X4 \times X1^2$
$X25 = X4 \times X2^2$	$X26 = X4 \times X3^2$

R_T/Δ, τ, $S/\nabla^{2/3}$ and L_W/L_p all have the same X1 to X26 values, with R_T/Δ having the 'b' coefficients b0 to b26, τ the 'a' coefficients, $S/\nabla^{2/3}$ the 'c' coefficients and L_W/L_p the 'd' coefficients. The coefficients a0–a26, b0–b26, c0–c26 and d0–d26 are given in Tables A3.19 to A3.22 in Appendix A3. These are the updated coefficients, taken from [10.84].

R_T/Δ is for a 100,000 lb displacement ship. It is more convenient to consider this as a displacement volume of $\nabla = 44.2$ m³, a displacement mass of $\Delta = 45.3$ tonnes or a displacement force of 444.4 kN.

The total resistance R_T is then calculated as

$$R_T = R_T/\Delta \times (\nabla \times \rho \times g)\ \text{kN}.$$

For ships other than the *basis* 100,000 lb (45.3 tonnes) displacement, a skin friction correction is required, as follows:

$$\text{Corrected}\ (R_T/\Delta)_{\text{corr}} = R_T/\Delta - [(C_{F\text{basis}} - C_{F\text{new}}) \times \tfrac{1}{2}\rho \times S \times V^2/1000]/\Delta,$$

and Δ_{basis} is the basis displacement of 444.4 kN. The correction will be subtracted for vessels with a length greater than the basis and added for vessels with a length less than the basis. Schoenherr C_F, Equation (4.10), or ITTC C_F, Equation (4.15), would both be suitable.

The scale ratio can be used to derive the length of the basis ship as follows:

$$\lambda = \sqrt[3]{\frac{\Delta_{new} \cdot \rho_{basis}}{\Delta_{basis} \cdot \rho_{new}}}. \tag{10.41}$$

The basis displacement $\Delta_{basis} = 45.3$ tonnes. The cube root of the largest likely change in density would lead to a correction of less than 1% and the density correction can be omitted. The scale ratio can then be written as

$$\lambda = \sqrt[3]{\frac{\Delta_{new}}{45.3}}. \tag{10.42}$$

Analysis of the Series 62 hull forms indicates that the waterline length L_{WL}, at rest, is shorter than L_p by about 1% at $L/B = 7$ up to about 2.5% at $L/B = 2$. It is suggested that an average value of 2% be used for preliminary design and powering purposes, that is:

$$L_{pnew} = L_{WL} \times 1.02$$

$$L_{pbasis} = L_{pnew}/\lambda$$

$$L_{basis} = (L_W/L_p) \times L_{pbasis}$$

$$L_{new} = (L_W/L_p) \times L_{pnew}$$

$$Re_{basis} = V \cdot L_{basis}/1.19 \times 10^{-6} \text{ and } C_{Fbasis} = f(Re)$$

$$Re_{new} = V \cdot L_{new}/1.19 \times 10^{-6} \text{and } C_{Fnew} = f(Re).$$

Example: Consider a craft with $L_{WL} = 30$ m, $\Delta = 153$ tonnes (1501 kN), travelling at 35 knots. From the regression analysis, $R_T/\Delta = 0.140$, $S = 227$ m^2 and $L_W/L_p = 0.63$ at this speed.

$$\lambda = \sqrt[3]{\frac{153}{45.3}} = 1.50.$$

$$L_{pnew} = L_{WL} \times 1.02 = 30 \times 1.02 = 30.6 \text{ m}.$$

$$L_{pbasis} = L_{pnew}/1.5 = 30.6/1.5 = 20.4 \text{ m}.$$

$$L_{basis} = (L_W/Lp) \times L_{pbasis} = 0.63 \times 20.4 = 12.85 \text{ m}.$$

$$L_{new} = (L_W/Lp) \times L_{pnew} = 0.63 \times 30.6 = 19.28 \text{ m}.$$

$$Re_{basis} = VL/v = 35 \times 0.5144 \times 12.85/1.19 \times 10^{-6} = 1.944 \times 10^8.$$

$$C_{Fbasis} = 0.075/(\log Re - 2)^2 = 0.075/(8.289 - 2)^2 = 0.001896.$$

$$Re_{new} = VL/v = 35 \times 0.5144 \times 19.28/1.19 \times 10^{-6} = 2.917 \times 10^8.$$

$$C_{Fnew} = 0.075/(\log Re - 2)^2 = 0.075/(8.465 - 2)^2 = 0.001794.$$

Skin friction correction

$$= (C_{Fbasis} - C_{Fnew}) \times \tfrac{1}{2}\rho \times S \times V^2/1000$$

$$= (0.001896 - 0.001794) \times \tfrac{1}{2} \times 1025 \times 227 \times (35 \times 0.5144)^2/1000$$

$$= 3.85 \text{ kN}.$$

$$R_T/\Delta = 0.140, \text{ and (uncorrected) } R_T = 0.140 \times 1501 = 210.14 \text{ kN}.$$

Corrected $R_T = 210.14 - 3.85 = 206.29$ kN.

It is seen that going from a 20 m/45 tonne craft up to a 30 m/153 tonne craft has led to a skin friction correction of 1.8%. It is effectively not necessary to apply the correction between about 18 m/33 tonnes up to about 23 m/67 tonnes. Applications of the Series 62 data are described in Chapter 17.

10.3.3.2 Savitsky Equations for Prismatic Planing Forms

The force developed by a planing surface and the centre at which it acts is described through equations by Savitsky [10.46].

Following flat plate tests, the following formula is put forward for the total lift (buoyant contribution and dynamic lift) acting on a flat surface with zero deadrise:

$$C_{L0} = \tau^{1.1} \left[0.0120\lambda^{1/2} + 0.0055\frac{\lambda^{2.5}}{C_V^2} \right], \tag{10.43}$$

with limits of application: $0.60 \leq C_V \leq 13; 2° \leq \tau \leq 15°$, with τ in degrees; $\lambda \leq 4.0$.

For surfaces with deadrise β, the lift coefficient requires correction to

$$C_{L\beta} = C_{L0} - 0.0065\beta C_{L0}^{0.60}, \tag{10.44}$$

with limit of application $\beta \leq 30°$, with β in degrees.

By consideration of the point of action of the buoyant contribution and the dynamic contribution, the overall position of the centre of pressure is given as

$$C_P = 0.75 - \frac{1}{5.21\left(\frac{C_V}{\lambda}\right)^2 + 2.39} = \frac{l_P}{\lambda b} = \frac{l_P}{l_m} \tag{10.45}$$

$$\text{where} \quad C_P = \frac{l_P}{\lambda b} = \frac{l_P}{l_m}$$

$$\text{with} \quad \lambda = \frac{(l_K + l_C)/2}{b} = \frac{l_m}{b}$$

$$\text{and} \quad C_V = \frac{V}{\sqrt{gb}}$$

$$\text{where} \quad C_L = \frac{\Delta}{0.5\rho b^2 V^2} \quad \text{and} \quad S = \lambda b^2 \sec\beta.$$

Further definitions, including reference to Figure 10.17, are as follows:

N: Normal bottom pressure load, buoyant and dynamic, acting perpendicular to keel through the centre of pressure.

b: Mean chine beam ($= B_{pa}$ in Series 62 definition).

D_{APP}: Appendage resistance (rudder, shafting, shaft brackets, etc.).

L_{APP}: Appendage lift.

T: Propeller thrust, along shaft line at ε to keel.

S_P: Propeller–hull interaction load, perpendicular to keel, downwards or upwards.

At the preliminary design stage, the lift and its corresponding point of action, along with the frictional forces acting, comprise the most important components

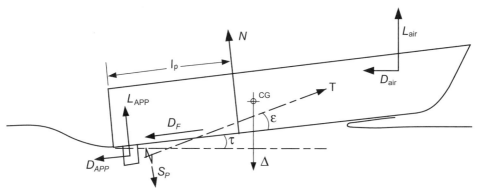

Figure 10.17. Forces on a planing craft.

in the balance of forces on a planing craft, Figure 10.17. The Savitsky equations can be used in an overall balance of forces and moments to determine the running trim angle and the thrust required. The forces acting are shown in Figure 10.17 and the general approach for deriving a balance of forces and moments is shown in Figure 10.18.

EXAMPLE USING THE SAVITSKY EQUATIONS. Consider a vessel with displacement $\Delta = 70$ tonnes, mean chine beam $b = 6.5$ m, deadrise $\beta = 20°$ and $LCG = 10.0$ m forward of transom. Speed $Vs = 45$ knots (23.15 m/s), $C_V = \frac{V}{\sqrt{gb}} = \frac{23.15}{\sqrt{9.81 \times 6.5}} = 2.899$. Required calculations are for l_m, R_T (thrust T) and trim τ.

Appendage drag and air drag are not included in this illustrative example. They can, in principle, be included in the overall balance of forces, Hadler [10.49].

It is assumed that the lines of action of the frictional drag forces and thrust act through the centre of gravity (CG), leading to the simplified force diagram shown in Figure 10.19. These forces could be applied at different lines of action and included separately in the moment balance.

Resolved forces parallel to the keel are

$$T = \Delta \sin \tau + D_F, \tag{10.46}$$

where D_F is the frictional drag with

$$D_F = \tfrac{1}{2}\rho S V^2 C_F$$

and $S = l_m \, b \sec \beta$, where β is the deadrise angle.

C_F is derived using the ITTC formula, Equation (4.15), $C_F = 0.075/(\log_{10} Re - 2)^2$.

Resolving forces perpendicular to keel,

$$N = \Delta \cos \tau. \tag{10.47}$$

Taking moments about transom at height of CG,

$$\Delta \times L_{CG} = N \times lp, \tag{10.48}$$

with $\Delta = \rho g \nabla$ (kN), $\Delta = 9.81 \times 70.0 = 686.7$ kN, and $L_{CG} = 10.0$ m, then $6867.0 = N \times lp$.

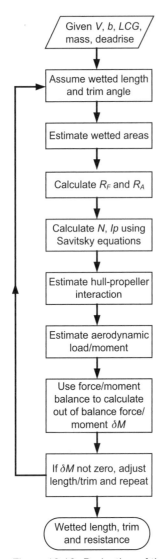

Figure 10.18. Derivation of the balance of forces and moments on a planing hull.

The Savitsky equations are used with the above force and moment balances to determine τ, l_m, l_p and N, and the resulting required thrust, T.

Following the procedure outlined in Figure 10.18, assume trim angle, τ, calculate $C_{L\beta} = \frac{\Delta}{0.5\rho b^2 V^2}$. Find C_{L0} giving the value of $C_{L\beta}$ from satisfaction of Equation (10.44). The mean wetted length, l_m, required to achieve this lift at this trim angle is found through satisfaction of the Savitsky equation for C_{L0}, Equation (10.43), where $\lambda = l_m/b$ and, in this example at 45 knots, $C_V = 2.899$.

Having derived l_m (λ), the centre of pressure l_p can be determined from Equation (10.45), as follows:

$$\frac{l_p}{l_m} = 0.75 - \frac{1}{5.21\left(\frac{C_V}{\lambda}\right)^2 + 2.39}.$$

Table 10.6. *Summary of iterative procedure using Savitsky equations*

τ	$C_{L\beta}$	C_{L0} (10.44)	λ (10.43)	l_m	l_p (10.45)	N (10.47)	δM
2.0°	0.0592	0.0898	3.8083	24.754	13.989	686.28	−2733.27
3.0°	0.0592	0.0898	2.6310	17.102	10.864	685.76	−582.98
4.0°	0.0592	0.0898	1.8701	12.155	8.301	685.03	1180.42
3.50°	0.0592	0.0898	2.2138	14.390	9.527	685.42	340.759
3.31°	0.0592	0.0898	2.3658	15.378	10.028	685.55	−7.42

The overall balance of moments on the craft: $\Delta \times L_{CG} - N \times lp = \delta M$ for $\delta M = 0$ may now be checked using the obtained values of N and lp for the assumed trim τ and derived l_m. If the forces on the craft are not in balance, a new trim angle is chosen and the calculations are repeated until $\delta M = 0$.

A summary of the calculations and iterative procedure is shown in Table 10.6.

From a cross plot or interpolation, equilibrium ($\delta M = 0$) is obtained with a trim τ of 3.31° and $l_m = 15.378$ m.

$$\text{Reynolds number } Re = Vl_m/\nu = 23.15 \times 15.378/1.19 \times 10^{-6} = 2.991 \times 10^8$$

and

$$C_F = 0.075/(\log Re - 2)^2 = 1.7884 \times 10^{-3}.$$
$$S = l_m\, b \sec \beta = 15.378 \times 6.5 \times \sec 20° = 106.37\,\text{m}^2.$$
$$D_F = \tfrac{1}{2}\rho S V^2 C_F = 0.5 \times 1025 \times 106.37 \times 23.15^2 \times 1.7884 \times 10^{-3} = 52.24\,\text{kN}.$$

Thrust required along shaft line

$$T = \Delta \sin \tau + D_F = 686.7 \sin 3.31 + 52.24 = 91.89\,\text{kN}$$
$$\text{Resistance } R_T = T \cos \tau = 91.89 \times \cos 3.31 = 91.74\,\text{kN}$$
$$\text{Effective power } P_E = R_T \times Vs = 91.74 \times 23.15 = 2123.5\,\text{kW}$$

The effects of appendages, propulsive forces and air drag can also be incorporated in the resistance estimating procedure, Hadler [10.49]. Similarly, the lines of action of the various forces can be modified as necessary.

Example applications of the Savitsky equations are included in Chapter 17.

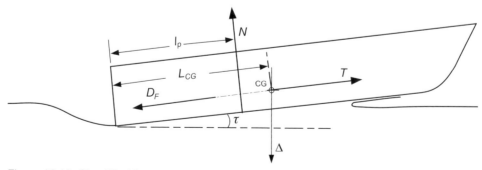

Figure 10.19. Simplified forces on a planing craft.

10.3.4 Small Craft

10.3.4.1 Oortmerssen: Small Ships

Van Oortmerssen [10.85] developed regression equations for estimating the resistance of small ships such as tugs, fishing boats, stern trawlers and pilot boats, etc., broadly in the length range from 15 m to 75 m. The objective was to provide equations that would be accurate enough for design purposes. The analysis was based on 970 data points from 93 ship models that had been tested at The Netherlands Ship Model Basin (NSMB) (now Maritime Research Institute of the Netherlands [MARIN]) in the 1960s.

Approximate limits of the data (extracted from the diagrams) are as follows:

Speed: $Fr = 0.2$–0.5.
L/B: 3.4–6.2.
LCB: $-4.4\%L$ to $+1.6\%L$ (mainly about $-1\%L$).
i_E: $15°$–$35°$ (mainly about $18°$–$30°$).

where i_E is the half-angle of entrance of the waterline. If i_E is not known, an approximation is $i_E = 120\,C_B - 50\,(0.5 < C_B < 0.7)$ (Molland)

B/T: 1.9–3.2.
C_P: 0.55–0.70 (mainly about 0.60).
C_M: 0.76–0.94 (mainly about 0.82–0.92).

If C_M is not known, an approximation for small ships is given in Equation (10.29b).

A displacement length L_D is used, defined as $L_D = 0.5(L_{BP} + L_{WL})$, with Fr, Re, LCB, C_P and C_M being based on L_D. An angle of entrance parameter is defined as

$$C_{WL} = i_E \times L_D/B.$$

L_{WL} can be used for length without incurring significant error.

The residuary resistance was derived using the ITTC1957 line for C_F. The components of the equation for residuary resistance ratio R_R/Δ are as follows:

$$\frac{R_R}{\Delta} = c_1 e^{-\frac{m}{9}Fr^{-2}} + c_2 e^{-mFr^{-2}} + c_3 e^{-mFr^{-2}}\sin Fr^{-2}$$
$$+ c_4 e^{-mFr^{-2}}\cos Fr^{-2}, \tag{10.49}$$

where

$$m = 0.14347 C_P^{-2.1976}$$
$$c_i = \{d_{i,0} + d_{i,1} \cdot LCB + d_{i,2} \cdot LCB^2 + d_{i,3} \cdot C_P + d_{i,4} \cdot C_P^2$$
$$= d_{i,5} \cdot (L_D/B) + d_{i,6} \cdot (L_D/B)^2 + d_{i,7}C_{WL} + d_{i,8}C_{WL}^2$$
$$+ d_{i,9} \cdot (B/T) + d_{i,10} \cdot (B/T)^2 + d_{i,11} \cdot C_M\} \times 10^{-3},$$

where LCB is LCB forward of $0.5\,L$ as a percentage of L and the coefficients d_i are given in Table A3.23, Appendix A3. Note that in the table of coefficients in

Table 10.7. *Trial allowances for* ΔC_F

Allowances for	ΔC_F
Roughness, all-welded hulls	0.00035
Steering resistance	0.00004
Bilge keel resistance	0.00004
Air resistance	0.00008

the original paper [10.85], there was a sign error in coefficient $d_{3,5}$ which should be +9.86873, *not* –9.86873.

The residuary resistance is calculated as

$$R_R = R_R/\Delta \times (\nabla \times \rho \times g).$$

C_F is derived using the ITTC formula.
ΔC_F allowances for trial conditions are given in Table 10.7.
Friction resistance is then derived as

$$R_F = (C_F + \Delta C_F) \times \tfrac{1}{2}\rho S V^2.$$

S can be calculated from

$$S = 3.223\nabla^{2/3} + 0.5402 L_D \nabla^{1/3}$$

which has a format similar to Equation (10.85) for larger ships.

10.3.4.2 WUMTIA: Small Craft: Round Bilge and Hard Chine

A regression analysis was carried out on chine and round-bilge hull forms which had been tested by WUMTIA at the University of Southampton, Robinson [10.86]. Over 600 hull forms had been tested by WUMTIA since 1968, including both chine and round-bilge hull forms representing vessels ranging typically from 10 m to 70 m.

Thirty test models of round-bilge generic form were used in the regression analysis and 66 of chine generic form. Tests at different displacements were also included leading to a total of 47 sets of round-bilge resistance data and 103 sets of hard chine resistance data. Separate regression coefficients were derived for the round-bilge and hard chine forms. The data were all taken from hull forms that had been optimised for their running trim characteristics at realistic operating speeds and include, for the chine hulls, the effects of change in wetted area with speed.

The analyses covered the speed range, as follows: The volume Froude number Fr_∇: 0.50–2.75, approximate length Froude number range $Fr = 0.25$–1.2, where

$$Fr_\nabla = \frac{V}{\sqrt{g\nabla^{1/3}}}, \quad Fr = \frac{V}{\sqrt{gL}} \quad \text{and} \quad Fr_\nabla = Fr \times \left(\frac{L}{\nabla^{1/3}}\right)^{0.5}. \quad (10.50)$$

The range of hull parameters L/B and $L/\nabla^{1/3}$ are shown in Figure 10.20

It is noted that for the round-bilge hulls there are few data between $L/B = 5.5$–6.5. Above $Fr_\nabla = 1.5$, the upper limit of L/B is 5.5, and it is recommended that the data for $L/B > 5.5$ be restricted to speeds $< Fr_\nabla = 1.5$.

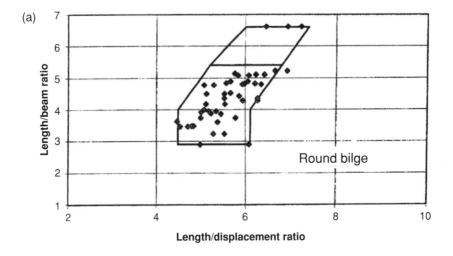

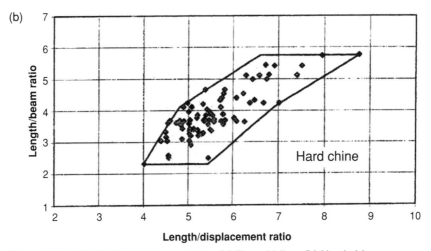

Figure 10.20. WUMTIA data boundaries: (a) Round bilge, (b) Hard chine.

The data are presented in terms of a C-Factor (C_{FAC}) which was developed by small craft designers for the prediction of power at an early design stage.

$$C_{\text{FAC}} = 30.1266 \frac{V_K}{\sqrt[4]{L}} \sqrt{\frac{\Delta}{2 P_E}}, \tag{10.51}$$

where the constant 30.1266 was introduced to conserve the value of C_{FAC}, which was originally based on imperial units. Above a Fr_{∇} of about 1.0, C_{FAC} lies typically between about 50 and 70.

Rearranging Equation (10.51),

$$P_E = 453.8 \Delta \frac{V_K^2}{\sqrt{L}} \frac{1}{C_{\text{FAC}}^2}, \tag{10.52}$$

where P_E is in kW, Δ is in tonnes, V_K is in knots and L is in metres.

The predictions for C_{FAC} are presented as regression equations.

For round-bilge hulls:

$$C_{FAC} = a_0 + a_1(L/\nabla^{1/3}) + a_2\,L/B + a_3(S/L^2)^{1/2} + a_4(L/\nabla^{1/3})^2 + a_5(L/B)^2$$
$$+ a_6(S/L^2) + a_7(L/\nabla^{1/3})^3 + a_8(L/B)^3 + a_9(S/L^2)^{3/2}. \qquad (10.53)$$

For chine hulls:

$$C_{FAC} = a_0 + a_1\,L/\nabla^{1/3} + a_2\,L/B + a_3(L/\nabla^{1/3})^2$$
$$+ a_4(L/B)^2 + a_5(L/\nabla^{1/3})^3 + a_6(L/B)^3. \qquad (10.54)$$

The wetted area S for the round-bilge and hard chine forms can be estimated using Equations (10.91) and (10.93), to be found in Section 10.4, noting that for the chine hull the wetted area is speed dependent.

The regression coefficients a_0 to a_9 in Equations (10.53) and (10.54) are given in Tables A3.24 and A3.25, Appendix A3. Note that in the table of coefficients for the hard chine hulls in the original paper [10.86], there was a sign error at $Fr_\nabla = 2.0$. The sixth term should be -0.298946, *not* $+0.298946$.

It should be noted that these regression equations, (10.53) and (10.54), tend to give slightly pessimistic predictions of power. As a result of advances in scaling techniques and the inclusion of extra model data in an updated analysis, it is recommended [10.96] that the original predictions [10.86] for the round bilge hulls be reduced on average by 4% and those for the chine hulls by 3%.

10.3.5 Multihulls

10.3.5.1 Southampton Round-Bilge Catamaran Series

This systematic series of high-speed semi-displacement catamaran hull forms was developed by the University of Southampton. These hull forms are suitable for fast semi-displacement ferries and other such applications. The results of the tests and investigations are reported in [10.50] and [10.51] and offer one of the widest parametric sets of resistance data for catamarans. Insel and Molland [10.50] also include the results of direct physical measurements of viscous resistance and wave resistance. Details of the models are given in Table 10.8. All models were tested in monohull mode and in catamaran mode at four lateral hull separations. The body plans for the series were based on extended versions of the NPL round-bilge series, Figure 10.13, and are shown in Figure 10.21. Reference [10.52] extended these data to include change in C_P.

The series covered the following range of speeds and demihull parameters:

Speed: Fr: 0.20–1.0.
Demihull parameters: C_B: 0.40 (fixed); $L/\nabla^{1/3}$: 6.3–9.5; B/T: 1.5, 2.0, 2.5; LCB: 6.4% aft; $S_C/L = 0.20, 0.30, 0.40, 0.50$, where S_C is the separation of the centrelines of the demihulls.

Insel and Molland [10.50] describe the total resistance of a catamaran as

$$C_T = (1 + \phi k)\sigma C_F + \tau C_W, \qquad (10.55)$$

Table 10.8. *Details of models: Southampton catamaran series*

Model	L(m)	L/B	B/T	$L/\nabla^{1/3}$	C_B	C_P	C_M	S (m^2)	LCB %L
3b	1.6	7.0	2.0	6.27	0.397	0.693	0.565	0.434	−6.4
4a	1.6	10.4	1.5	7.40	0.397	0.693	0.565	0.348	−6.4
4b	1.6	9.0	2.0	7.41	0.397	0.693	0.565	0.338	−6.4
4c	1.6	8.0	2.5	7.39	0.397	0.693	0.565	0.340	−6.4
5a	1.6	12.8	1.5	8.51	0.397	0.693	0.565	0.282	−6.4
5b	1.6	11.0	2.0	8.50	0.397	0.693	0.565	0.276	−6.4
5c	1.6	9.9	2.5	8.49	0.397	0.693	0.565	0.277	−6.4
6a	1.6	15.1	1.5	9.50	0.397	0.693	0.565	0.240	−6.4
6b	1.6	13.1	2.0	9.50	0.397	0.693	0.565	0.233	−6.4
6c	1.6	11.7	2.5	9.50	0.397	0.693	0.565	0.234	−6.4

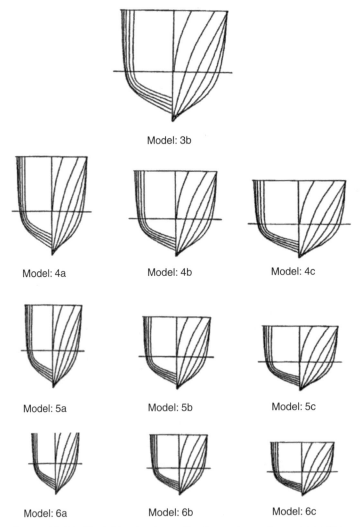

Figure 10.21. Model body plans: Southampton catamaran series.

Table 10.9. *Form factors*

	Form factors	
$L/\nabla^{1/3}$	Monohulls $(1 + k)$	Catamarans $(1 + \beta k)$
6.3	1.35	1.48
7.4	1.21	1.33
8.5	1.17	1.29
9.5	1.13	1.24

where C_F is derived from the ITTC1957 correlation line, $(1 + k)$ is the form factor for a demihull in isolation, ϕ is introduced to take account of the pressure field change around the hull, σ takes account of the velocity augmentation between the two hulls and would be calculated from an integration of local frictional resistance over the wetted surface, C_W is the wave resistance coefficient for a demihull in isolation and τ is the wave resistance interference factor.

For practical purposes, ϕ and σ were combined into a viscous interference factor β, where $(1 + \phi k)\sigma C_F$ is replaced by $(1 + \beta k)C_F$

$$\text{whence} \quad C_T = (1 + \beta k)C_F + \tau C_W, \tag{10.56}$$

noting that, for the demihull (monohull) in isolation, $\beta = 1$ and $\tau = 1$.

Data are presented in terms of C_R, where $C_R = \frac{R_R}{0.5\rho S V^2}$ and S is the static wetted area, noting that the sum of the wetted areas of both demihulls was used in the case of the catamarans. C_R was derived from

$$C_R = C_{TM} - C_{F\,\text{MITTC}}. \tag{10.57}$$

C_R data for the series are presented for the monohull and catamaran modes in Table A3.26, Appendix A3.

In applying the data, a form factor $(1 + k)$ may or may not be used. Values of $(1 + k)$ and $(1 + \beta k)$ were derived and presented in the original references [10.50] and [10.51]. These values were later revised [10.53] and the proposed form factors are given in Table 10.9.

A satisfactory fit to the monohull form factors is

$$(1 + k) = 2.76 \left(\frac{L}{\nabla^{1/3}} \right)^{-0.40}. \tag{10.58}$$

A satisfactory fit to the catamaran form factors is

$$(1 + \beta k) = 3.03 \left(\frac{L}{\nabla^{1/3}} \right)^{-0.40}. \tag{10.59}$$

It is argued by some that there is effectively little form effect on these types of hull forms and, for lack of adequate information, ITTC recommends a value $(1 + k) = 1.0$ for high-speed craft. The results reported in [10.54 to 10.58], however, indicate in a number of cases practical working values of $(1 + k)$ and $(1 + \beta k)$ up to the same order of magnitude as those in Table 10.9.

Table 10.10. *Coefficients in the equation 1000 $C_R = a(L/\nabla^{1/3})^n$ for extended NPL monohulls*

Fr	a	n	R^2
0.4	152	−1.76	0.946
0.5	2225	−3.00	0.993
0.6	1702	−2.96	0.991
0.7	896	−2.76	0.982
0.8	533	−2.58	0.982
0.9	273	−2.31	0.970
1.0	122	−1.96	0.950

In deriving the C_{TS} value for the ship, the following equations are applied.

For monohulls,

$$C_{TS} = C_{FS} + C_{Rmono} - k(C_{FM} - C_{FS}). \tag{10.60}$$

For catamarans,

$$C_{TS} = C_{FS} + C_{Rcat} - \beta k(C_{FM} - C_{FS}). \tag{10.61}$$

In Equations (10.60) and (10.61), C_{FM} is derived using the model length from which C_R was derived, in this case, 1.60 m, Table 10.8.

As discussed in Section 10.3.2.1, when describing Series 64, the resistance of high-speed semi-displacement craft tends to be dominated by $L/\nabla^{1/3}$ ratio and a useful presentation for such craft is resistance coefficient to a base of $L/\nabla^{1/3}$ at fixed Froude number. For these reasons, and to provide a practical design approach, regression analyses were carried out [10.33]. These relate residuary resistance, C_R, solely to $L/\nabla^{1/3}$ ratio at a number of fixed Froude numbers, Fr, for the monohull case. The form of the equation for the demihull (monohull) is as follows:

$$C_R = a(L/\nabla^{1/3})^n. \tag{10.62}$$

In the case of the catamaran, a residuary resistance interference factor was expressed in a similar manner, the form of the equation being:

$$\tau_R = a(L/\nabla^{1/3})^n, \tag{10.63}$$

where τ_R is the residuary resistance interference factor and is defined as the ratio of the catamaran residuary resistance to the monohull residuary resistance.

The value of τ_R was found to be dependent on speed, $L/\nabla^{1/3}$, and separation of the hulls S/L, but not to be influenced significantly by the particular hull shape. This was confirmed by comparing the interference factors for the extended NPL series [10.51] with results for a Series 64 hull form catamaran reported in [10.31], where similar trends in τ_R were observed. This is a significant outcome as it implies that the interference factors could be used in conjunction with a wider range of monohull forms, such as the Series 64 monohull forms [10.30] described in Section 10.3.2.1.

Equation (10.61) for catamarans is now written as

$$C_{TS} = C_{FS} + \tau_R C_R - \beta k(C_{FM} - C_{FS}), \tag{10.64}$$

and in this case, C_R is for the demihull (monohull).

Table 10.11. *Residuary resistance interference factor, τ_R*

Fr	S/L = 0.20		S/L = 0.30		S/L = 0.40		S/L = 0.50	
	a	n	a	n	a	n	a	n
0.4	1.862	−0.15	0.941	0.17	0.730	0.28	0.645	0.32
0.5	1.489	0.04	1.598	−0.05	0.856	0.20	0.485	0.45
0.6	2.987	−0.34	1.042	0.09	0.599	0.34	0.555	0.36
0.7	0.559	0.40	0.545	0.39	0.456	0.47	0.518	0.41
0.8	0.244	0.76	0.338	0.61	0.368	0.57	0.426	0.51
0.9	0.183	0.89	0.300	0.67	0.352	0.60	0.414	0.52
1.0	0.180	0.90	0.393	0.55	0.541	0.40	0.533	0.39

Coefficients in the equation $\tau_R = a(L/\nabla^{1/3})^n$.

The coefficients of the regressions, a and n, are given in Tables 10.10, 10.11. Wetted surface area can be estimated using Equation (10.88) or (10.90), which can be found in Section 10.4.

10.3.5.2 VWS Hard Chine Catamaran Hull Series
A series of hard chine catamarans was tested at VWS, Berlin [10.59, 10.60, 10.61]. Typical body plans, for the largest L_{WL}/b ratio, are shown in Figure 10.22; the series covered the following range of speed and hull parameters:

Speed: $Fr_\nabla = 1.0–3.5$ based on demihull ∇.
L_{WL}/b: 7.55–13.55, where b is breadth of demihull.
$L_{WL}/\nabla^{1/3}$ (demihull): 6.25–9.67.

Midship deadrise angle $\beta_M = 16°–38°$

LCB: $0.42 L_{WL}$ at $\beta_M = 38°$ to $0.38 L_{WL}$ at $\beta_M = 16°$.
Transom flap (wedge) angle δ_W: 0°–12°.

Gap ratio (clearance between demihulls) is constant at $G/L_{WL} = 0.167$. (This leads to S_C/L_{WL} values of 0.240 at $L_{WL}/b = 13.55$ up to 0.300 at $L_{WL}/b = 7.55$, where S_C is the separation of the demihull centrelines.)

Müller-Graf [10.59] describes the scope and details of the VWS hard chine catamaran series. Zips [10.60] carried out a regression analysis of the resistance data for the series, which is summarised as follows.

Non-dimensional residuary resistance ratio is expressed as R_R/Δ [R_R in Newtons, Δ in Newtons $= \nabla \times \rho \times g$]. The residuary resistance was derived using the ITTC1957 line.

The three independent hull parameters were transformed to the normalised parameters, as follows:

$$X1 = (L_{WL}/b - 10.55)/3.$$
$$X2 = (\beta_M - 27°)/11°.$$
$$X3 = \delta_W/12°.$$

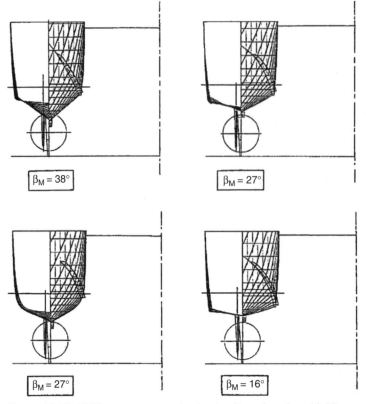

Figure 10.22. VWS catamaran series body plans: $L_{WL}/b = 13.55$.

In terms of these parameters, the length displacement ratio of the demihull is given as

$$L_{WL}/\nabla^{1/3} = 7.651877 + 1.694413 \times X1 + 0.282139 \times X1^2 - 0.052496 \times X1^2 \times X2.$$

The wetted surface area coefficient $S/\nabla^{2/3}$ (for a demihull) is given as

$$S/\nabla^{2/3} = \Sigma CS_i \times XS_i \times 10, \qquad (10.65)$$

where CS_i and XS_i are given in Table 10.12

Table 10.12. *Wetted area*
coefficients: VWS catamaran series

CS_i	XS_i
1.103767	1
0.151489	X1
0.00983	$X2^2$
−0.009085	$X1^2$
0.008195	$X1^2 \cdot X2$
−0.029385	$X1 \cdot X2^2$
0.041762	$X1^3 \cdot X2^2$

The residual resistance ratio is given as

$$R_R/\Delta = \Sigma(XR_i \times CR_i)/100, \tag{10.66}$$

where the regression parameters XR_i and the regression coefficients CR_i are given in Table A3.27, Appendix A3.

Finally, the residuary resistance is calculated as

$$R_R = R_R/\Delta \times (\nabla \times \rho \times g), \tag{10.67}$$

and, if required,

$$C_R = R_R \tfrac{1}{2}\rho SV^2.$$

The required input parameters to carry out the analysis for a particular speed are L_{WL}/b (which is transformed to X1), β_M (which is transformed to X2), δ_W (which is transformed to X3) and ∇ (or L_{WL}).

Further regression analyses of the VWS Series are presented by Müller-Graf and Radojcic [10.61], from which predictions with more accuracy may be expected.

10.3.6 Yachts

10.3.6.1 Background

For estimates of yacht resistance in the preliminary design stage, prior to towing tank or extensive CFD analysis, most designers are reliant on the Delft Systematic Yacht Hull Series (DSYHS) which is by far the most extensive published research on yacht hull performance. Data for the series have been published in [10.62] to [10.66], with the most recent update being that due to Keuning and Sonnenberg [10.67]. An example of a body plan for the series is shown in Figure 10.23.

The DSYHS consists of 50 different dedicated yacht models. Since the first publication of the series data in 1974 the series has been steadily extended to cover a large range of yacht parameters as well as an extended collection of appendage arrangements, heel angles and sea states. The first series (now known as DSYHS Series 1) consists of the parent hull (model 1) and 21 variations to this design, in 1974, of a contemporary racing yacht (models 2–22). In 1983 a new parent hull

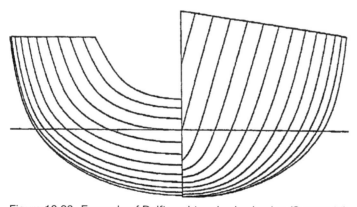

Figure 10.23. Example of Delft yacht series body plan (Sysser 44).

was introduced, this time a dedicated design was used and models 23–28 became known as Series 2. Series 3 was a very light displacement variation on Series 2 and ranges from model 29 to 40. The most recent Series 4 (models 42–50) is based on a typical 40 foot International Measurement System (IMS) design by Sparkman and Stevens (see Keuning and Sonnenberg [10.67]). Given the large number of models, the DSYHS is suitable for predicting resistance for a relatively wide range of design ratios. It is therefore used by many designers and the data are used by the Offshore Racing Congress in their Velocity Prediction Program (VPP), for the issuing of ratings to racing yachts, Offshore Racing Congress [10.68].

The DSYHS divides the resistance of the yacht's hull into a number of components and then presents a regression equation for each component, based on the towing tank tests. The sum of these components yields the total resistance for the hull. In the Delft series the resistance is treated as the sum of the resistance of the yacht in the upright condition (including appendages) and the additions (or subtractions) to this resistance when the hull is in its sailing (heeled and yawed) condition. The change in resistance due to trim, caused by crew movement and sail drive force, and added resistance in waves are also included in the more recent publications, e.g. Keuning and Sonnenberg [10.67]. These last components are not considered further here.

From Keuning and Sonnenberg [10.67], the total resistance for a hull at an angle of heel and leeway is expressed as

$$R_{\text{Total}} = R_{Fh} + R_{Rh} + R_{VK} + R_{VR} + R_{RK} + \Delta R_{Rh} + \Delta R_{RK} + R_{\text{Ind}}, \quad (10.68)$$

where

R_{Total} = total resistance of the hull, keel and rudder at an angle of heel and leeway.

R_{Fh} = frictional resistance of the hull.

R_{Rh} = residuary resistance of the hull.

R_{VK} = viscous resistance of the keel.

R_{VR} = viscous resistance of the rudder.

R_{RK} = residuary resistance of the keel.

ΔR_{Rh} = change in residuary resistance of the hull with the heel angle, ϕ.

ΔR_{RK} = change in residuary resistance of the keel with the heel angle, ϕ.

R_{Ind} = induced resistance, due to generation of side force (Fy)

at the angle of leeway, λ.

Each component is considered in turn, with equations and polynomial coefficients taken (with permission) from Keuning and Sonnenberg [10.67].

10.3.6.2 Frictional Resistance of Hull, R_{Fh}

The model test data in the Delft series have been extrapolated to a yacht length of 10.0 m. This extrapolation does not use a form factor $(1 + k)$, since the measured form factor using a Prohaska technique for the parent hulls of the series was small and there is no accepted means to calculate a form factor from the geometry of the hull. Recent measurements of the form factor of modern yacht hulls suggests that

Table 10.13. *Coefficients for polynomial: Residuary resistance of bare hull (Equation (10.72))*

Fr	0.1	0.15	0.2	0.25	0.3	0.35	0.4	0.45	0.5	0.55	0.6
a_0	−0.0014	0.0004	0.0014	0.0027	0.0056	0.0032	−0.0064	−0.0171	−0.0201	0.0495	0.0808
a_1	0.0403	−0.1808	−0.1071	0.0463	−0.8005	−0.1011	2.3095	3.4017	7.1576	1.5618	−5.3233
a_2	0.047	0.1793	0.0637	−0.1263	0.4891	−0.0813	−1.5152	−1.9862	−6.3304	−6.0661	−1.1513
a_3	−0.0227	−0.0004	0.009	0.015	0.0269	−0.0382	0.0751	0.3242	0.5829	0.8641	0.9663
a_4	−0.0119	0.0097	0.0153	0.0274	0.0519	0.032	−0.0858	−0.145	0.163	1.1702	1.6084
a_5	0.0061	0.0118	0.0011	−0.0299	−0.0313	−0.1481	−0.5349	−0.8043	−0.3966	1.761	2.7459
a_6	−0.0086	−0.0055	0.0012	0.011	0.0292	0.0837	0.1715	0.2952	0.5023	0.9176	0.8491
a_7	−0.0307	0.1721	0.1021	−0.0595	0.7314	0.0223	−2.455	−3.5284	−7.1579	−2.1191	4.7129
a_8	−0.0553	−0.1728	−0.0648	0.122	−0.3619	0.1587	1.1865	1.3575	5.2534	5.4281	1.1089

the form factor may be larger than those measured for the Delft series, however. The total viscous resistance is thus considered to be given by the frictional resistance of the hull, where

$$R_{Fh} = \tfrac{1}{2}\rho S_c V^2 C_F, \qquad (10.69)$$

where C_F is the skin friction coefficient using the ITTC1957 extrapolation line in which the Reynolds number is determined using a reference length of $0.7L_{WL}$. S_C is the wetted surface area of the hull canoe body at zero speed, which, if not known from the hydrostatic calculations, can be estimated using Equations (10.94) and (10.95), to be found in Section 10.4.

10.3.6.3 Residuary Resistance of Hull, R_{Rh}

The regression equations for residuary resistance have changed as the series has developed and more models have been tested. This may cause confusion when comparing predictions. The Delft series consists of two different residuary resistance calculation methods; one is a combination of predictions for $0.125 \leq Fr \leq 0.45$ (Equation (10.70)) and $0.475 \leq Fr \leq 0.75$ (Equation (10.71)), whilst the other method is for $0.1 \leq Fr \leq 0.6$ (Equation (10.72)). The factors a_n in Equations (10.70) and (10.72) are different coefficients for each polynomial. Equation (10.70) was first presented by Gerritsma *et al.* [10.63] with only seven polynomial factors and has subsequently been developed through the addition of further tests and regression analysis. In 1991 Equation (10.70) was updated, Gerritsma *et al.* [10.64] and a first version of Equation (10.71) was introduced. A year later the polynomial coefficients for Equation (10.70) were modified and the equation for the range $Fr = 0.475$–0.75 changed to Equation (10.71) (Gerritsma *et al.* [10.65]). Further evaluation of the testing resulted in the publication of Equation (10.72) (Keuning *et al.* [10.66]) with subsequent changes to the polynomial presented in Keuning and Sonnenberg [10.67]. The combination of Equations (10.70) and (10.71) is known as the Delft III method, with the prediction following Equation (10.72) generally referred to as Delft I, II method as the latter covers a range for which all models of the Delft series I and II have been tested. The polynomial coefficients for Equation (10.72) are given in Table 10.13.

Table 10.14. *Coefficients for polynomial: Change in residuary resistance of hull at $20°$ heel (coefficients are multiplied by 1000)*

Fr	0.25	0.30	0.35	0.40	0.45	0.50	0.55
u_0	−0.0268	0.6628	1.6433	−0.8659	−3.2715	−0.1976	1.5873
u_1	−0.0014	−0.0632	−0.2144	−0.0354	0.1372	−0.148	−0.3749
u_2	−0.0057	−0.0699	−0.164	0.2226	0.5547	−0.6593	−0.7105
u_3	0.0016	0.0069	0.0199	0.0188	0.0268	0.1862	0.2146
u_4	−0.007	0.0459	−0.054	−0.58	−1.0064	−0.7489	−0.4818
u_5	−0.0017	−0.0004	−0.0268	−0.1133	−0.2026	−0.1648	−0.1174

For $0.125 \leq Fr \leq 0.450$ (Gerritsma *et al.* [10.64]).

$$\frac{R_R}{\nabla_C \rho g} 10^3 = a_0 + a_1 C_P + a_2 LCB + a_3 \frac{B_{WL}}{T_C} + a_4 \frac{L_{WL}}{\nabla_C^{1/3}} + a_5 C_P^2 + a_6 C_P \frac{L_{WL}}{\nabla_C^{1/3}}$$

$$+ a_7 (LCB)^2 + a_8 \left(\frac{L_{WL}}{\nabla_C^{1/3}}\right)^2 + a_9 \left(\frac{L_{WL}}{\nabla_C^{1/3}}\right)^3. \tag{10.70}$$

For $0.475 \leq Fr \leq 0.75$ (Gerritsma *et al.* [10.65]),

$$\frac{R_R}{\nabla_C \rho g} 10^3 = c_0 + c_1 \frac{L_{WL}}{B_{WL}} + c_2 \frac{A_W}{\nabla_C^{2/3}} + c_3 LCB$$

$$+ c_4 \left(\frac{L_{WL}}{B_{WL}}\right)^2 + c_5 \left(\frac{L_{WL}}{B_{WL}}\right) \left(\frac{A_W}{\nabla_C^{2/3}}\right)^3. \tag{10.71}$$

For $0.1 \leq Fr \leq 0.6$ (Keuning and Sonnenberg [10.67]), with $LCB_{FPP} = $ LCB aft FP in m,

$$\frac{R_{Rh}}{\nabla_C \rho g} = a_0 + \left(a_1 \frac{LCB_{FPP}}{L_{WL}} + a_2 C_P + a_3 \frac{\nabla_C^{1/3}}{Aw} + a_4 \frac{B_{WL}}{L_{WL}}\right) \frac{\nabla_C^{1/3}}{L_{WL}}$$

$$+ \left(a_5 \frac{\nabla_C^{1/3}}{Sc} + a_6 \frac{LCB_{FPP}}{LCF_{FPP}} + a_7 \left(\frac{LCB_{FPP}}{L_{WL}}\right)^2 + a_8 C_P^2\right) \frac{\nabla_C^{1/3}}{L_{WL}}. \tag{10.72}$$

10.3.6.4 Change in Residuary Resistance of Hull with Heel, ΔR_{Rh}

When the hull heels there is a change to the distribution of displaced volume along the hull caused by the asymmetry of the heeled geometry. This results in a change to the residuary resistance of the hull. The approach taken in Keuning and Sonnenberg [10.67] is to represent this with a polynomial expression for the change in residuary resistance due to a heel angle of $20°$, for which all models in the series have been tested, namely,

$$\frac{\Delta R_{Rh\phi=20}}{\nabla_C \rho g} = u_0 + u_1 \left(\frac{L_{WL}}{B_{WL}}\right) + u_2 \left(\frac{B_{WL}}{T_C}\right) + u_3 \left(\frac{B_{WL}}{T_C}\right)^2$$

$$+ u_4 LCB + u_5 LCB^2, \tag{10.73}$$

where the coefficients are given in Table 10.14.

The residuary resistance for heel angles other than 20°, based on a smaller experimental dataset, is then calculated as

$$\Delta R_{Rh} = 6.0 \phi^{1.7} \Delta R_{Rh\phi=20},$$ (10.74)

where the heel angle is in radians.

10.3.6.5 Appendage Viscous Resistance, R_{VK}, R_{VR}

The viscous resistance of the appendages (keel and rudder) is considered to be a summation of frictional resistance and viscous pressure resistance, determined through use of a form factor. Thus,

$$R_V = R_F (1 + k)$$ (10.75)

$$R_F = \tfrac{1}{2}\rho S V^2 C_F,$$ (10.76)

where C_F is the skin friction coefficient using the ITTC1957 extrapolation line, Equation (4.15), in which the Reynolds number is determined using the average chord length of the appendage. S is the wetted surface area of the appendage.

The form factor is given as a function of the thickness/chord ratio of the sections from Hoerner [10.69], as follows:

$$(1 + k) = \left(1 + 2\left(\frac{t}{c}\right) + 60\left(\frac{t}{c}\right)^4\right).$$ (10.77)

The keel and rudder are treated in an identical manner.

A more accurate procedure may be used whereby the appendage is divided into spanwise segments, with the individual contributions to viscous resistance determined for each segment. In this case the Reynolds number for each segment is based on the local chord length of the segment. Typically, five spanwise segments are used as, for example, in the Offshore Racing Congress (ORC) velocity prediction program (VPP).

10.3.6.6 Appendage Residuary Resistance, R_{RK}

The contribution of the keel to the total residuary resistance of the yacht is estimated based on experimental tests with a variety of keels fitted beneath two hull models. A more complete description of the tests undertaken is given in Keuning and Sonnenberg [10.67]. This resistance is given by a polynomial expression, as follows:

$$\frac{R_{RK}}{\nabla_K \rho g} = A_0 + A_1\left(\frac{T}{B_{WL}}\right) + A_2\left(\frac{T_C + z_{CBK}}{\nabla_K^{1/3}}\right) + A_3\left(\frac{\nabla_C}{\nabla_K}\right),$$ (10.78)

where ∇_C is the displacement volume of the canoe body, ∇_K is the displacement volume of the keel and z_{CBK} is centre of buoyancy of keel below the waterline. T is total draught of hull plus keel.

The coefficients A_0–A_3 are given in Table 10.15.

10.3.6.7 Change in Residuary Resistance of Keel with Heel, ΔR_{RK}

The residuary resistance due to the keel of the yacht changes with heel angle as the volume of the keel is brought closer to the free surface. The interaction of the wave

Table 10.15. *Coefficients for polynomial: Residuary resistance of keel*

Fr	0.20	0.25	0.30	0.35	0.40	0.45	0.50	0.55	0.60
A_0	−0.00104	−0.0055	−0.0111	−0.00713	−0.03581	−0.0047	0.00553	0.04822	0.01021
A_1	0.00172	0.00597	0.01421	0.02632	0.08649	0.11592	0.07371	0.0066	0.14173
A_2	0.00117	0.0039	0.00069	−0.00232	0.00999	−0.00064	0.05991	0.07048	0.06409
A_3	−0.00008	−0.00009	0.00021	0.00039	0.00017	0.00035	−0.00114	−0.00035	−0.00192

produced by the hull and keel is also important. Based on experimental measurements the change in keel residuary resistance may be expressed as

$$\frac{\Delta R_{RK}}{\nabla_K \rho g} = Ch Fr^2 \phi,$$ (10.79)

where

$$Ch = H_1 \left(\frac{T_C}{T}\right) + H_2 \left(\frac{B_{WL}}{T_C}\right) + H_3 \left(\frac{T_C}{T}\right) \left(\frac{B_{WL}}{T_C}\right) + H_4 \left(\frac{L_{WL}}{\nabla_C^{1/3}}\right),$$ (10.80)

and the coefficients are given in Table 10.16.

10.3.6.8 Induced Resistance, R_{Ind}

The induced resistance of a yacht sailing at an angle of heel and leeway is that component of the resistance associated with the generation of hydrodynamic sideforce necessary to balance the sideforce produced by the rig, Fh. The induced resistance is related to the circulation around the foil and its geometry. In this formulation an 'effective' span of the foil, accounting for the presence of the free surface is used, usually referred to as the 'effective draught' of the yacht. The induced resistance is thus obtained as

$$R_{\text{Ind}} = \frac{Fh^2}{\pi T_E^2 \frac{1}{2} \rho V^2},$$ (10.81)

where Fh is the heeling force from the rig and T_E is the effective span of hull and appendages, given as

$$\frac{T_E}{T} = \left(A_1 \left(\frac{T_C}{T}\right) + A_2 \left(\frac{T_C}{T}\right)^2 + A_3 \left(\frac{B_{WL}}{T_C}\right) + A_4 TR\right)(B_0 + B_1 Fr),$$ (10.82)

where TR is the taper ratio of the keel and the polynomial coefficients are given in Table 10.17.

Table 10.16. *Coefficients for polynomial: Change in residuary resistance of keel with heel*

H_1	−3.5837
H_2	−0.0518
H_3	0.5958
H_4	0.2055

Table 10.17. *Coefficients for polynomial: Effective span of yacht*

ϕ deg.	0	10	20	30
A_1	3.7455	4.4892	3.9592	3.4891
A_2	−3.6246	−4.8454	−3.9804	−2.9577
A_3	0.0589	0.0294	0.0283	0.025
A_4	−0.0296	−0.0176	−0.0075	−0.0272
B_0	1.2306	1.4231	1.545	1.4744
B_1	−0.7256	−1.2971	−1.5622	−1.3499

10.4 Wetted Surface Area

10.4.1 Background

The wetted surface area is the area of the hull in contact with the water and is normally used to non-dimensionalise the resistance coefficients, Section 3.1.3. The static wetted area is generally used. Some change in running wetted area may occur with fast semi-displacement forms but, for practical design purposes, the static wetted area is normally used. The errors in using the static wetted area for such craft are relatively small, as discussed in the appendix to [10.97]. Planing craft will have significant changes in wetted area with change in speed, and such changes must be taken into account, see Equation (10.93) and Table 10.23, and Section 10.3.3.

If a body plan is available, the static wetted surface area may be obtained from an integration of the girths, or numerically from a CAD representation. Otherwise, approximations have to be made using empirical data. A summary is made of some empirical equations suitable for preliminary design and powering purposes.

10.4.2 Displacement Ships

The Sabit regression of BSRA Series [10.68] is as follows:

$$\frac{S}{\nabla^{2/3}} = a_0 + a_1(L/B) + a_2(B/T) - a_3 C_B. \tag{10.83}$$

The regression coefficients are given in Table 10.18

Sabit notes (discussion in [10.5]) that this equation is for the BSRA Series and vessels generally of that hull form; vessels with more 'U' form sections (such as Series 60) will have higher $\frac{S}{\nabla^{2/3}}$ for the same dimensions.

Table 10.18. *Sabit equation (10.83)*

Parameter		Coefficient	
		BSRA	Series 60
Constant	a_0	+3.371	+3.432
(L/B)	a_1	+0.296	+0.305
(B/T)	a_2	+0.437	+0.443
C_B	a_3	−0.595	−0.643

Table 10.19. *Cs values, Taylor series, Equation (10.86)*

		C_P			
$L/\nabla^{1/3}$	B/T	0.5	0.6	0.7	0.8
5.5	2.25	2.589	2.562	2.557	2.576
5.5	3.00	2.526	2.540	2.566	2.605*
5.5	3.75	2.565	2.596	2.636	2.685
6.0	2.25	2.583	2.557	2.554	2.571
6.0	3.00	2.523	2.536	2.560	2.596
6.0	3.75	2.547	2.580	2.625	2.675
7.0	2.25	2.575	2.553	2.549	2.566
7.0	3.00	2.520	2.532	2.554	2.588
7.0	3.75	2.541	2.574	2.614	2.660
8.0	2.25	2.569	2.549	2.546	2.561
8.0	3.00	2.518*	2.530	2.551	2.584
8.0	3.75	2.538	2.571	2.609	2.652
9.0	2.25	2.566	2.547	2.543	2.557
9.0	3.00	2.516*	2.528	2.544	2.581
9.0	3.75	2.536	2.568	2.606	2.649

* Extrapolated data.

Coefficients for Series 60 [10.76] are included in Table 10.18, which should also be applied to Equation (10.83).

$$\text{Denny Mumford:} \quad S = 1.7LT + \frac{\nabla}{T}. \tag{10.84}$$

$$\text{Froude:} \quad S = 3.4\nabla^{2/3} + 0.485L \cdot \nabla^{1/3} \quad \text{or} \quad \frac{S}{\nabla^{2/3}} = 3.4 + \frac{L}{2.06\nabla^{1/3}}. \tag{10.85}$$

$$\text{Taylor:} \quad S = Cs\sqrt{\nabla \cdot L}, \tag{10.86}$$

where Cs depends on C_P, B/T and $L/\nabla^{1/3}$ and values, extracted from [10.10], are given in Table 10.19.

$$C_P = C_B/C_M.$$

typical values for C_M are tankers/bulk carriers, 0.98; cargo, 0.96; container, 0.95; warship, 0.92 and, as a first approximation,

$$C_M = 0.80 + 0.21C_B. \tag{10.87}$$

10.4.3 Semi-displacement Ships, Round-Bilge Forms

The regression equation given in [10.30] for the Series 64 hull forms provides a good starting point, and its potential use in the wider sense for other round-bilge forms was investigated. The wetted area is described as

$$S = Cs\sqrt{\nabla \cdot L}, \tag{10.88}$$

where the wetted surface coefficient C_S is expressed in terms of B/T and C_B which are normally known at the preliminary design stage. It is found from the Series 64 and NPL data that, for a given C_B, C_S can be adequately described in terms of B/T and is effectively independent of the $L/\nabla^{1/3}$ ratio. The regression coefficients

Table 10.20. *Static wetted surface area regression coefficients for the derivation of C_S in Equation (10.89)*

Parameter		Coefficient
Constant	a_0	+6.554
(B/T)	a_1	−1.226
$(B/T)^2$	a_2	+0.216
C_B	a_3	−15.409
$(B/T)C_B$	a_4	+4.468
$(B/T)^2 C_B$	a_5	−0.694
C_B^2	a_6	+15.404
$(B/T)C_B^2$	a_7	−4.527
$(B/T)^2 C_B^2$	a_8	+0.655

from [10.30] have been recalculated for C_S to be in consistent (non-dimensional) units and rounded where necessary for preliminary design purposes. The form of the equation is given in Equation (10.89), and the parameters and regression coefficients are given in Table 10.20.

$$C_S = a_0 + a_1(B/T) + a_2(B/T)^2 + a_3 C_B + a_4(B/T)C_B$$
$$+ a_5(B/T)^2 C_B + a_6 C_B^2 + a_7(B/T)C_B^2 + a_8(B/T)^2 C_B^2 \qquad (10.89)$$

The equation is plotted for four values of C_B in Figure 10.24. Values for the original NPL series with $C_B = 0.40$ [10.34] have also been plotted on Figure 10.24 and it is seen that they all lie within 3% of the regression values. Values for the

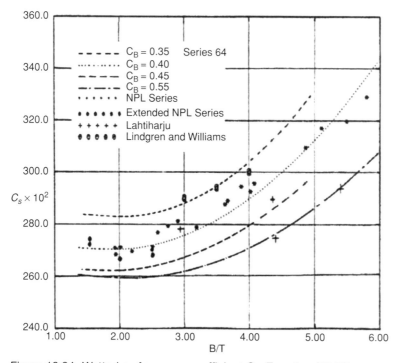

Figure 10.24. Wetted surface area coefficient C_S, Equation (10.89).

Table 10.21. *Wolfson round-bilge*
Equation (10.91)

Parameter		Coefficient
(∇)	a_1	+0.355636
(L)	a_2	+5.75893
(B)	a_3	−3.17064

extended NPL forms with $C_B = 0.4$ [10.51] also lie within 3%. The data in Lahtiharju *et al.* [10.37] (NPL basis, but changes in C_B) also fit the regressions well. The data in Lindgren and Williams [10.28] ($C_B = 0.45$) are not so good, being about 5% higher than the regression, but a docking keel is included in the wetted area.

A better fit to the NPL data ($C_B = 0.40$) is as follows, but it is restricted to the one hull form and $C_B = 0.40$:

$$C_S = 2.538 + 0.0494(B/T) + 0.01307(B/T)^2. \tag{10.90}$$

In light of the sources of the data it is recommended that the use of the equations be restricted to the B/T range 1.5–6.0 and C_B range 0.35–0.55. It is noted that there are few data at $C_B = 0.35$ for B/T higher than about 4.0, although its trend is likely to be similar to the $C_B = 0.40$ curve.

The Wolfson Unit [10.86] regression for the static wetted surface area for round-bilge hulls is as follows:

$$S = a_1(\nabla) + a_2(L) + a_3(B). \tag{10.91}$$

The regression coefficients are given in Table 10.21.

10.4.4 Semi-displacement Ships, Double-Chine Forms

The NTUA series [10.40] provides a regression for the static wetted area of hulls of double-chine form, as follows:

$$\frac{S}{\nabla^{2/3}} = a_0 + a_1(B/T)(L/\nabla^{1/3})^2 + a_2(B/T)^3 + a_3(L/\nabla^{1/3})$$
$$+ a_4(B/T)^5 + a_5(L/\nabla^{1/3})^2. \tag{10.92}$$

The regression coefficients are given in Table 10.22.

Table 10.22. *NTUA equation (10.92)*

Parameter		Coefficient
Constant	a_0	+2.400678
$(B/T)(L/\nabla^{1/3})^2$	a_1	+0.002326
$(B/T)^3$	a_2	+0.012349
$(L/\nabla^{1/3})$	a_3	+0.689826
$(B/T)^5$	a_4	−0.000120
$(L/\nabla^{1/3})^2$	a_5	−0.018380

Table 10.23. *Wolfson hard chine regression coefficients, Equation (10.93), for a range of volumetric Froude numbers*

Parameter	Fr_∇	0.5	1.0	1.5	2.0	2.5	3.0
(∇)	a_1	0.985098	0.965983	0.915863	0.860915	0.936348	0.992033
(L)	a_2	2.860999	3.229803	3.800066	4.891762	4.821847	4.818481
(B)	a_3	−1.113826	−2.060285	−4.064968	−7.594861	−9.013914	−10.85827

10.4.5 Planing Hulls, Single Chine

The Wolfson Unit [10.86] regression for running wetted surface area of hard chine hulls is as follows:

$$S = a_1(\nabla) + a_2(L) + a_3(B), \qquad (10.93)$$

noting that the wetted area is now speed dependent.

The regression coefficients for a range of volumetric Froude numbers are given in Table 10.23.

10.4.6 Yacht Forms

For the Delft series of hull forms [10.67], for the canoe body,

$$S_C = \left(1.97 + 0.171\frac{B_{WL}}{T_C}\right)\left(\frac{0.65}{C_M}\right)^{1/3}(\nabla_C L_{WL})^{1/2}, \qquad (10.94)$$

where S_C is the wetted surface area of the canoe body, T_C is the draught of the canoe body, ∇_C is the displacement (volume) of the canoe body and C_M is the midship area coefficient.

The change in viscous resistance due to heel is attributed only to the change in wetted area of the hull. This change in wetted surface area with heel angle may be approximated by

$$S_{C\phi} = S_{C(\phi=0)}\left(1 + \frac{1}{100}\left(s_0 + s_1\left(\frac{B_{WL}}{T_C}\right) + s_2\left(\frac{B_{WL}}{T_C}\right)^2 + s_3 C_M\right)\right), \qquad (10.95)$$

with coefficients given in Table 10.24.

Table 10.24. *Coefficients for polynomial: Change in wetted surface area with heel (Equation (10.95)*

ϕ	5	10	15	20	25	30	35
s_0	−4.112	−4.522	−3.291	1.85	6.51	12.334	14.648
s_1	0.054	−0.132	−0.389	−1.2	−2.305	−3.911	−5.182
s_2	−0.027	−0.077	−0.118	−0.109	−0.066	0.024	0.102
s_3	6.329	8.738	8.949	5.364	3.443	1.767	3.497

REFERENCES (CHAPTER 10)

10.1 Todd, F.H. Some further experiments on single-screw merchant ship forms – Series 60. *Transactions of the Society of Naval Architects and Marine Engineers*, Vol. 61, 1953, pp. 516–589.

10.2 Todd, F.H., Stuntz, G.R. and Pien, P.C. Series 60 – The effect upon resistance and power of variation in ship proportions. *Transactions of the Society of Naval Architects and Marine Engineers*, Vol. 65, 1957, pp. 445–589.

10.3 Lackenby, H. and Milton, D. DTMB Standard series 60. A new presentation of the resistance data for block coefficient, LCB, breadth-draught ratio and length-breadth ratio variations. *Transactions of the Royal Institution of Naval Architects*, Vol. 114, 1972, pp. 183–220.

10.4 Moor, D.I., Parker, M.N. and Pattullo, R.N.M. The BSRA methodical series – An overall presentation. Geometry of forms and variation in resistance with block coefficient and LCB. *Transactions of the Royal Institution of Naval Architects*, Vol. 103, 1961, pp. 329–440.

10.5 Lackenby, H. and Parker, M.N. The BSRA methodical series – An overall presentation: Variation of resistance with breadth-draught ratio and length-displacement ratio. *Transactions of the Royal Institution of Naval Architects*, Vol. 108, 1966, pp. 363–388.

10.6 BSRA. Methodical series experiments on single-screw ocean-going merchant ship forms. Extended and revised overall analysis. *BSRA Report NS333*, 1971.

10.7 Williams, Å. The SSPA cargo liner series propulsion. SSPA Report No. 67, 1970.

10.8 Williams, Å. The SSPA cargo liner series resistance. SSPA Report No. 66, 1969.

10.9 Roseman, D.P. (ed.) The MARAD systematic series of full-form ship models. Society of Naval Architects and Marine Engineers, 1987.

10.10 Gertler, M. A reanalysis of the original test data for the Taylor Standard Series. DTMB Report No. 806, DTMB, Washington, DC, 1954. Reprinted by SNAME, 1998.

10.11 Linblad, A.F. Experiments with models of cargo liners. *Transactions of the Royal Institution of Naval Architects*, Vol. 88, 1946, pp. 174–195.

10.12 Linblad, A.F. Some experiments with models of high-speed ships: Influence of block coefficient and longitudinal centre of buoyancy. *Transactions of the Royal Institution of Naval Architects*, Vol. 91, 1949, pp. 137–158.

10.13 Zborowski, A. Approximate method for estimation of resistance and power of twin screw merchant ships. *International Shipbuilding Progress*, Vol. 20, No. 221, January 1973, pp. 3–11.

10.14 Dawson, J. Resistance of single screw coasters. Part I, L/B = 6, *Transactions of the Institute of Engineers and Shipbuilders in Scotland*. Vol. 96, 1952–1953, pp. 313–384.

10.15 Dawson, J. Resistance and propulsion of single screw coasters. Part II, L/B = 6, *Transactions of the Institute of Engineers and Shipbuilders in Scotland*, Vol. 98, 1954–1955, pp. 49–84.

10.16 Dawson, J. Resistance and propulsion of single screw coasters. Part III, L/B = 6.5, *Transactions of the Institute of Engineers and Shipbuilders in Scotland*. Vol. 99, 1955–1956, pp. 360–441.

10.17 Dawson, J. Resistance and propulsion of single screw coasters. Part IV, L/B = 5.5. *Transactions of the Institute of Engineers and Shipbuilders in Scotland*. Vol. 102, 1958–1959, pp. 265–339.

10.18 Pattullo, R.N.M. and Thomson, G.R. The BSRA trawler series (Part I). Beam-draught and length-displacement ratio series, resistance and propulsion tests. *Transactions of the Royal Institution of Naval Architects*, Vol. 107, 1965, pp. 215–241.

10.19 Pattullo, R.N.M. The BSRA Trawler Series (Part II). Block coefficient and longitudinal centre of buoyancy variation series, resistance and propulsion tests. *Transactions of the Royal Institution of Naval Architects,* Vol. 110, 1968, pp. 151–183.

10.20 Thomson, G.R. and Pattullo, R.N.M. The BSRA Trawler Series (Part III). Block coefficient and longitudinal centre of buoyancy variation series, tests with bow and stern variations. *Transactions of the Royal Institution of Naval Architects*, Vol. 111, 1969, pp. 317–342.

10.21 Pattullo, R.N.M. The resistance and propulsion qualities of a series of stern trawlers. Variation of longitudinal position of centre of buoyancy, beam, draught and block coefficient. *Transactions of the Royal Institution of Naval Architects*, Vol. 116, 1974, pp. 347–372.

10.22 Ridgely-Nevitt, C. The resistance of trawler hull forms of 0.65 prismatic coefficient. *Transactions of the Society of Naval Architects and Marine Engineers*, Vol. 64, 1956, pp. 443–468.

10.23 Ridgely-Nevitt, C. The development of parent hulls for a high displacement-length series of trawler forms. *Transactions of the Society of Naval Architects and Marine Engineers*, Vol. 71, 1963, pp. 5–30.

10.24 Ridgely-Nevitt, C. The resistance of a high displacement-length ratio trawler series. *Transactions of the Society of Naval Architects and Marine Engineers*, Vol. 75, 1967, pp. 51–78.

10.25 Parker, M.N. and Dawson, J. Tug propulsion investigation. The effect of a buttock flow stern on bollard pull, towing and free-running performance. *Transactions of the Royal Institution of Naval Architects*, Vol. 104, 1962, pp. 237–279.

10.26 Moor, D.I. An investigation of tug propulsion. *Transactions of the Royal Institution of Naval Architects*, Vol. 105, 1963, pp. 107–152.

10.27 Nordström, H.F. Some tests with models of small vessels. Publications of the Swedish State Shipbuilding Experimental Tank. Report No. 19, 1951.

10.28 Lindgren, H. and Williams, Å. Systematic tests with small fast displacement vessels including the influence of spray strips. SSPA Report No. 65, 1969.

10.29 Beys, P.M. Series 63 – round bottom boats. Stevens Institute of Technology, Davidson Laboratory Report No. 949, 1993.

10.30 Yeh, H.Y.H. Series 64 resistance experiments on high-speed displacement forms. *Marine Technology*, July 1965, pp. 248–272.

10.31 Wellicome, J.F., Molland, A.F., Cic, J. and Taunton, D.J. Resistance experiments on a high speed displacement catamaran of Series 64 form. University of Southampton, Ship Science Report No. 106, 1999.

10.32 Karafiath, G. and Carrico, T. Series 64 parent hull displacement and static trim variations. *Proceedings of Seventh International Conference on Fast Sea Transportation, FAST'2003*, Ischia, Italy, October 2003, pp. 27–30.

10.33 Molland, A.F., Karayannis, T., Taunton, D.J. and Sarac-Williams, Y. Preliminary estimates of the dimensions, powering and seakeeping characteristics of fast ferries. *Proceedings of Eighth International Marine Design Conference, IMDC'2003*, Athens, Greece, May 2003.

10.34 Bailey, D. The NPL high speed round bilge displacement hull series. Maritime Technology Monograph No. 4, Royal Institution of Naval Architects, 1976.

10.35 Compton, R.H. Resistance of a systematic series of semi-planing transom stern hulls. *Marine Technology*, Vol. 23, No. 4, 1986, pp. 345–370.

10.36 Robson, B.L. Systematic series of high speed displacement hull forms for naval combatants. *Transactions of the Royal Institution of Naval Architects*, Vol. 130, 1988, pp. 241–259.

10.37 Lahtiharju, E., Karppinen, T., Helleraara, M. and Aitta, T. Resistance and seakeeping characteristics of fast transom stern hulls with systematically varied form. *Transactions of the Society of Naval Architects and Marine Engineers*, Vol. 99, 1991, pp. 85–118.

10.38 Gamulin, A. A displacement series of ships. *International Shipbuilding Progress*, Vol. 43, No. 434, 1996, pp. 93–107.

10.39 Radojcic, D., Princevac, M. and Rodic, T. Resistance and trim predictions for the SKLAD semi-displacement hull series. *Ocean Engineering International*, Vol. 3, No. 1, 1999, pp. 34–50.

10.40 Radojcic, D., Grigoropoulos, G.J., Rodic, T., Kuvelic, T. and Damala, D.P. The resistance and trim of semi-displacement, double-chine, transom-stern hulls. *Proceedings of Sixth International Conference on Fast Sea Transportation, FAST'2001,* Southampton, September 2001, pp. 187–195.

10.41 Grigoropoulos, G.J. and Loukakis, T.A. Resistance and seakeeping characteristics of a systematic series in the pre-planing condition (Part I). *Transactions of the Society of Naval Architects and Marine Engineers*, Vol. 110, 2002, pp. 77–113.

10.42 Clement, E.P. and Blount, D.L. Resistance tests on a systematic series of planing hull forms. *Transactions of the Society of Naval Architects and Marine Engineers*, Vol. 71, 1963, pp. 491–579.

10.43 Keuning, J.A. and Gerritsma, J. Resistance tests on a series of planing hull forms with 25° deadrise angle. *International Shipbuilding Progress*, Vol. 29, No. 337, September 1982.

10.44 Kowalyshyn, D.H. and Metcalf, B. A USCG systematic series of high-speed planing hulls. *Transactions of the Society of Naval Architects and Marine Engineers*, Vol. 114, 2006.

10.45 Hadler, J.B., Hubble, E.N. and Holling, H.D. Resistance characteristics of a systematic series of planing hull forms – Series 65. The Society of Naval Architects and Marine Engineers, Chesapeake Section, May 1974.

10.46 Savitsky, D. The hydrodynamic design of planing hulls. *Marine Technology*, Vol. 1, 1964, pp. 71–95.

10.47 Mercier, J.A. and Savitsky, D. Resistance of transom stern craft in the pre-planing regime. Davidson Laboratory, Stevens Institute of Technology, Report No. 1667, 1973.

10.48 Savitsky, D. and Ward Brown, P. Procedures for hydrodynamic evaluation of planing hulls in smooth and rough waters. *Marine Technology*, Vol. 13, No. 4, October 1976, pp. 381–400.

10.49 Hadler, J.B. The prediction of power performance of planing craft. *Transactions of the Society of Naval Architects and Marine Engineers*, Vol. 74, 1966, pp. 563–610.

10.50 Insel, M. and Molland, A.F. An investigation into the resistance components of high speed displacement catamarans. *Transactions of the Royal Institution of Naval Architects*, Vol. 134, 1992, pp. 1–20.

10.51 Molland, A.F. Wellicome, J.F. and Couser, P.R. Resistance experiments on a systematic series of high speed displacement catamaran forms: Variation of length-displacement ratio and breadth-draught ratio. *Transactions of the Royal Institution of Naval Architects*, Vol. 138, 1996, pp. 55–71.

10.52 Molland, A.F. and Lee, A.R. An investigation into the effect of prismatic coefficient on catamaran resistance. *Transactions of the Royal Institution of Naval Architects*, Vol. 139, 1997, pp. 157–165.

10.53 Couser, P.R., Molland, A.F., Armstrong, N.A. and Utama, I.K.A.P. Calm water powering predictions for high speed catamarans. *Proceedings of Fourth International Conference on Fast Sea Transportation, FAST'97*, Sydney, July 1997.

10.54 Steen, S., Rambech, H.J., Zhao, R. and Minsaas, K.J. Resistance prediction for fast displacement catamarans. *Proceedings of Fifth International Conference on Fast Sea Transportation, FAST'99*, Seattle, September 1999.

10.55 Cassella, P., Coppola, C, Lalli, F., Pensa, C., Scamardella, A. and Zotti, I. Geosim experimental results of high-speed catamaran: Co-operative investigation on resistance model tests methodology and ship-model correlation. *Proceedings of the 7th International Symposium on Practical Design of Ships and Mobile Units, PRADS'98*, The Hague, The Netherlands, September 1998.

10.56 Bruzzone, D., Cassella, P., Pensa, C., Scamardella, A. and Zotti, I. On the hydrodynamic characteristics of a high-speed catamaran with round-bilge hull: Wave resistance and wave pattern experimental tests and numerical calculations. *Proceedings of the Fourth International Conference on Fast Sea Transportation, FAST'97*, Sydney, July 1997.

10.57 Bruzzone, D., Cassella, P., Coppola, C., Russo Krauss, G. and Zotti, I. power prediction for high-speed catamarans from analysis of geosim tests and from numerical results. *Proceedings of the Fifth International Conference on Fast Sea Transportation, FAST'99*, Seattle, September 1999.

10.58 Utama, I.K.A.P. Investigation of the viscous resistance components of catamaran forms. Ph.D. thesis, University of Southampton, Department of Ship Science, 1999.

10.59 Müller-Graf, B. SUSA – The scope of the VWS hard chine catamaran hull series'89. *Proceedings of Second International Conference on Fast Sea Transportation, FAST'93*, Yokahama, 1993, pp. 223–237.

10.60 Zips, J.M. Numerical resistance prediction based on the results of the VWS hard chine catamaran hull series'89. *Proceedings of the Third International Conference on Fast Sea Transportation, FAST'95*, Lübeck-Travemünde, Germany, 1995, pp. 67–74.

10.61 Müller-Graf, B. and Radojcic, D. Resistance and propulsion characteristics of the VWS hard chine catamaran hull series'89. *Transactions of the Society of Naval Architects and Marine Engineers*, Vol. 110, 2002, pp. 1–29.

10.62 Gerritsma, J., Moeyes, G. and Onnink, R. Test results of a systematic yacht hull series. *International Shipbuilding Progress*, Vol. 25, No. 287, July 1978, pp. 163–180.

10.63 Gerritsma, J., Onnink, R. and Versluis, A. Geometry, resistance and stability of the Delft Systematic Yacht Hull Series. *7th International Symposium on Developments of Interest to Yacht Architecture 1981.* Interdijk BV, Amsterdam, 1981, pp. 27–40.

10.64 Gerritsma, J., Keuning, J. and Onnink, R. The Delft Systematic Yacht Hull (Series II) experiments. *The 10th Chesapeake Sailing Yacht Symposium, Annapolis.* The Society of Naval Architects and Marine Engineers, 1991, pp. 27–40.

10.65 Gerritsma, J., Keuning, J. and Onnink, R. Sailing yacht performance in calm water and waves. *12th International HISWA Symposium on "Yacht Design and Yacht Construction."* Delft University of Technology Press, Amsterdam, 1992, pp. 115–149.

10.66 Keuning, J., Onnink, R., Versluis, A. and Gulik, A. V. Resistance of the unappended Delft Systematic Yacht Hull Series. *14th International HISWA Symposium on "Yacht Design and Yacht Construction."* Delft University of Technology Press, Amsterdam, 1996, pp. 37–50.

10.67 Keuning, J. and Sonnenberg, U. Approximation of the hydrodynamic forces on a sailing yacht based on the 'Delft Systematic Yacht Hull Series'. *15th International HISWA Symposium on "Yacht Design and Yacht Construction"*, Amsterdam. Delft University of Technology Press, Amsterdam, 1998, pp. 99–152.

10.68 ORC VPP Documentation 2009. Offshore Racing Congress. http://www.orc.org/ (last accessed 3rd June 2010). 69 pages.

10.69 Hoerner, S.F. *Fluid-Dynamic Drag*. Published by the author, New York, 1965.

10.70 Moor, D.I. and Small, V.F. The effective horsepower of single-screw ships: Average modern attainment with particular reference to variation of C_B and *LCB*. *Transactions of the Royal Institution of Naval Architects*, Vol. 102, 1960, pp. 269–313.

10.71 Moor, D.I. Resistance and propulsion properties of some modern single screw tanker and bulk carrier forms. *Transactions of the Royal Institution of Naval Architects*, Vol. 117, 1975, pp. 201–214.

10.72 Clements, R.E. and Thompson, G.R. Model experiments on a series of 0.85 block coefficient forms. The effect on resistance and propulsive efficiency of variation in LCB. *Transactions of the Royal Institution of Naval Architects*, Vol. 116, 1974, pp. 283–317.

10.73 Clements, R.E. and Thompson, G.R. Model experiments on a series of 0.85 block coefficient forms. The effect on resistance and propulsive efficiency of variation in breadth-draught ratio and length-displacement ratio. *Transactions of the Royal Institution of Naval Architects*, Vol. 116, 1974, pp. 319–328.

10.74 Moor, D.I. and O'Connor, F.R.C. Resistance and propulsion factors of some single-screw ships at fractional draught. *Transactions of the North East Coast Institution of Engineers and Shipbuilders*, Vol. 80, 1963–1964, pp. 185–202.

10.75 Sabit, A.S. Regression analysis of the resistance results of the BSRA series. *International Shipbuilding Progress*, Vol. 18, No. 197, January 1971, pp. 3–17.

10.76 Sabit, A.S. An analysis of the Series 60 results: Part I, Analysis of forms and resistance results. *International Shipbuilding Progress*, Vol. 19, No. 211, March 1972, pp. 81–97.

10.77 Sabit, A.S. The SSPA cargo liner series regression analysis of the resistance and propulsive coefficients. *International Shipbuilding Progress*, Vol. 23, 1976, pp. 213–217.

10.78 Holtrop, J. A statistical analysis of performance test results. *International Shipbuilding Progress*, Vol. 24, No. 270, February 1977, pp. 23–28.

10.79 Holtrop, J. and Mennen, G.G.J. A statistical power prediction method. *International Shipbuilding Progress*, Vol. 25, No. 290, October 1978, pp. 253–256.

10.80 Holtrop, J. and Mennen, G.G.J. An approximate power prediction method. *International Shipbuilding Progress*, Vol. 29, No. 335, July 1982, pp. 166–170.

10.81 Holtrop, J. A statistical re-analysis of resistance and propulsion data. *International Shipbuilding Progress*, Vol. 31, 1984, pp. 272–276.

10.82 Hollenbach, K.U. Estimating resistance and propulsion for single-screw and twin-screw ships. *Ship Technology Research*, Vol. 45, Part 2, 1998, pp. 72–76.

10.83 Radojcic, D. A statistical method for calculation of resistance of the stepless planing hulls. *International Shipbuilding Progress*, Vol. 31, No. 364, December 1984, pp. 296–309.

10.84 Radojcic, D. An approximate method for calculation of resistance and trim of the planing hulls. University of Southampton, Ship Science Report No. 23, 1985.

10.85 van Oortmerssen, G. A power prediction method and its application to small ships. *International Shipbuilding Progress*, Vol. 18, No. 207, 1971, pp. 397–412.

10.86 Robinson, J.L. Performance prediction of chine and round bilge hull forms. *International Conference: Hydrodynamics of High Speed Craft*. Royal Institution of Naval Architects, London, November 1999.

10.87 Doust, D.J. Optimised trawler forms. *Transactions NECIES*, Vol. 79, 1962–1963, pp. 95–136.

10.88 Swift, P.M., Nowacki H. and Fischer, J.P. Estimation of Great Lakes bulk carrier resistance based on model test data regression. *Marine Technology*, Vol. 10, 1973, pp. 364–379.

10.89 Fairlie-Clarke, A.C. Regression analysis of ship data. *International Shipbuilding Progress*, Vol. 22, No. 251, 1975, pp. 227–250.

10.90 Molland, A.F. and Watson, M.J. The regression analysis of some ship model resistance data. University of Southampton, Ship Science Report No. 36, 1988.

10.91 Lin, C.-W., Day, W.G. and Lin W.-C. Statistical Prediction of ship's effective power using theoretical formulation and historic data. *Marine Technology*, Vol. 24, No. 3, July 1987, pp. 237–245.

10.92 Raven, H.C., Van Der Ploeg, A., Starke, A.R. and Eça, L. Towards a CFD-based prediction of ship performance – progress in predicting full-scale resistance and scale effects. *Transactions of the Royal Institution of Naval Architects*, Vol. 150, 2008, pp. 31–42.

10.93 Froude, R.E. On the 'constant' system of notation of results of experiments on models used at the Admiralty Experiment Works. *Transactions of the Royal Institution of Naval Architects*, Vol. 29, 1888, pp. 304–318.

10.94 Blount, D.L. Factors influencing the selection of a hard chine or round-bilge hull for high Froude numbers. *Proceedings of Third International Conference on Fast Sea Transportation, FAST'95*, Lübeck-Travemünde, 1995.

10.95 Savitsky, D. and Koelbel, J.G. Seakeeping considerations in design and operation of hard chine planing hulls. *The Naval Architect*. Royal Institution of Naval Architects, London, March 1979, pp. 55–59.

10.96 Correspondence with WUMTIA, University of Southampton, March 2010.

10.97 Molland, A.F., Wellicome, J.F. and Couser, P.R. Resistance experiments on a systematic series of high speed displacement catamaran forms: Variation of length-displacement ratio and breadth-draught ratio. University of Southampton, Ship Science Report No. 71, 1994.

11 Propulsor Types

11.1 Basic Requirements: Thrust and Momentum Changes

All propulsion devices operate on the principle of imparting momentum to a 'working fluid' in accordance with Newton's laws of motion:

(a) The force acting is equal to the rate of change of momentum produced.
(b) Action and reaction are equal and opposite.

Thus, the force required to produce the momentum change in the working fluid appears as a reaction force on the propulsion device, which constitutes the thrust produced by the device.

Suppose the fluid passing through the device has its speed increased from V_1 to V_2 by the device, and the mass flow per unit time through the device is $\dot{m}$, then the thrust (T) produced is given by

$$T = \text{rate of change of momentum}$$
$$= \dot{m}(V_2 - V_1). \tag{11.1}$$

The momentum change can be produced in a number of ways, leading to the evolution of a number of propulsor types.

11.2 Levels of Efficiency

The general characteristics of any propulsion device are basically as shown in Figure 11.1. The thrust equation, $T = \dot{m}\,(V_2 - V_1)$, indicates that as $V_1 \to V_2$, $T \to 0$. Thus, as the ratio (speed of advance/jet speed) $= V_1/V_2$ increases, the thrust decreases. Two limiting situations exist as follows:

(i) $V_1 = V_2$. Thrust is zero; hence, there is no useful power output ($P = TV_1$). At this condition viscous losses usually imply that there is a slight power input and, hence, at this point propulsive efficiency $\eta = 0$.
(ii) $V_1 = 0$. At this point, although the device is producing maximum thrust (usually), no useful work is being performed (i.e. $TV_1 = 0$) and, hence, again $\eta = 0$.

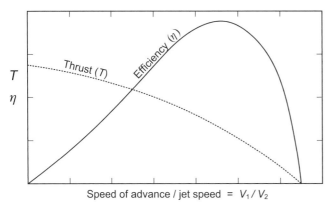

Figure 11.1. Propulsor characteristics.

Between these two conditions η reaches a maximum value for some ratio V_1/V_2. Hence, it is desirable to design the propulsion device to operate close to this condition of maximum efficiency.

11.3 Summary of Propulsor Types

The following sections provide outline summaries of the properties of the various propulsor types. Detailed performance data for the various propulsors for design purposes are given in Chapter 16.

11.3.1 Marine Propeller

A propeller accelerates a column of fluid passing through the swept disc, Figure 11.2. It is by far the most common propulsion device. It typically has 3–5 blades, depending on hull and shafting vibration frequencies, a typical boss/diameter ratio of 0.18–0.20 and a blade area ratio to suit cavitation requirements. Significant amounts of skew may be incorporated which will normally reduce levels of propeller-excited vibration and allow some increase in diameter and efficiency. The detailed characteristics of the marine propeller are described in Chapter 12. A more detailed review of the origins and development of the marine propeller may be found in Carlton [11.1]. Modifications and enhancements to the basic blade include tip rake [11.2, 11.3] and end plates [11.4].

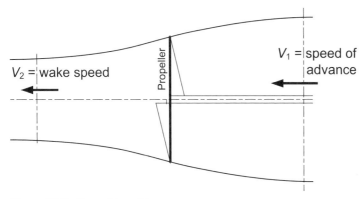

Figure 11.2. Propeller action.

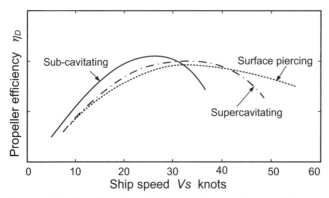

Figure 11.3. Trends in the efficiency of propellers for high-speed craft.

Specialist applications of the marine propeller include supercavitating propellers which are used when cavitation levels are such that cavitation has to be accepted, and surface piercing (partially submerged) propellers for high-speed craft. Typical trends in the efficiency and speed ranges for these propeller types are shown in Figure 11.3.

Data Sources. Published K_T–K_Q data are available for propeller series including fixed-pitch, supercavitating and surface-piercing propellers, see Chapter 16.

11.3.2 Controllable Pitch Propeller (CP propeller)

Such propellers allow the resetting of pitch for different propeller loading conditions. Hence, it is useful for vessels such as tugs, trawlers and ferries. It also provides reverse thrust. Compared with the fixed-pitch cast propeller it has a larger boss/diameter ratio of the order of 0.25. The CP propeller has increased mechanical complexity, tends to be more expensive (first cost and maintenance) than the fixed-pitch propeller, and it has a relatively small 2%–3% loss in efficiency. There may be some restriction on blade area in order to be able to reverse the blades.

Data Sources. Published series charts of K_T – K_Q for fixed-pitch propellers can be used, treating pitch as variable and allowing for a small loss in efficiency due to the increased boss size. [11.5] provides an indication of the influence of boss ratio on efficiency and [11.6] gives a description of the mechanical components of the controllable pitch propeller.

11.3.3 Ducted Propellers

11.3.3.1 Accelerating Duct
In this case the duct accelerates the flow inside the duct, Figure 11.4(a). The accelerating ducted propeller provides higher efficiency in conditions of high thrust loading, with the duct thrust augmenting the thrust of the propeller. Thus, it finds applications in vessels such as tugs when towing and trawlers when trawling. The duct can be steerable. The Kort nozzle is a proprietary brand of ducted propeller. The efficiency of ducted propellers when free-running and lightly loaded tends to be less than that of a non-ducted propeller. Rim-driven ducted thrusters have been developed where the duct forms part of the motor [11.7].

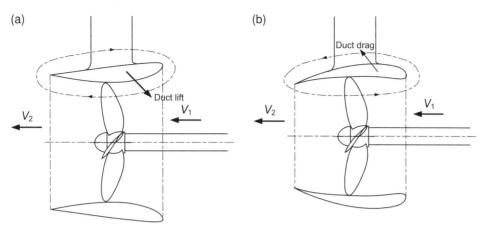

Figure 11.4. (a) Ducted propeller (accelerating). (b) Ducted propeller (decelerating).

Data Sources. Published series $K_T - K_Q$ charts for accelerating ducted propellers are available, see Chapter 16.

11.3.3.2 Decelerating Duct

In this case the duct circulation reduces the flow speed inside the duct, Figure 11.4(b). There is a loss of efficiency and thrust with this duct type. Its purpose is to increase the pressure (decrease velocity) at the propeller in order to reduce cavitation and its associated noise radiation. Its use tends to be restricted to military vessels where minimising the level of noise originating from cavitation is important.

11.3.4 Contra-Rotating Propellers

These propellers have coaxial contra-rotating shafts, Figure 11.5. The aft propeller is smaller than the upstream propeller to take account of slipstream contraction.

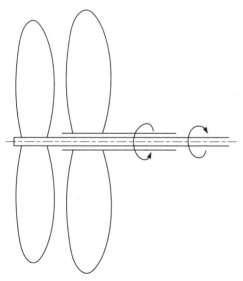

Figure 11.5. Contra-rotating propellers.

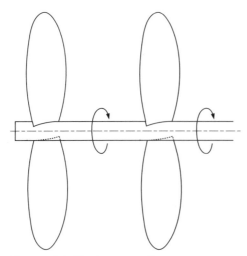

Figure 11.6. Tandem propellers.

The unit allows some recovery of rotational losses, producing a higher efficiency than the conventional propeller of the order of 5%–7%. The unit is mechanically more complex and expensive than a single propeller, including extra weight and complex gearing and sealing, together with higher maintenance costs. It has been used on torpedoes to counteract the torque reaction rotating the torpedo body. Research carried out into contra-rotating propellers includes [11.8–11.10]. Experiments have been carried out on a hybrid arrangement which comprises a combination of a conventional propeller and a downstream contra-rotating podded propeller [11.11, 11.12].

Data Sources. Some ac hoc data exist, such as [11.8–11.12], but little systematic data are available.

11.3.5 Tandem Propellers

In this design, more than one propeller is attached to the same shaft, Figure 11.6. It has been used when the thrust required to be transmitted by a shaft could not be adequately carried by one propeller on that shaft, [11.13]. This is particularly the case when there is a need to lower the risk of cavitation. An historical example is the fast naval vessel *Turbinia* [11.14].

Data Sources. Few systematic data are available, but some model test results are given in [11.13].

11.3.6 Z-Drive Units

The power is transmitted by mechanical shafting between the motor and propeller via bevel gears, Figure 11.7. There is no need for shaft brackets and space associated with conventional propellers. The propeller may be ducted. The unit is normally able to azimuth through 360°, providing directional thrust without the need for a rudder. Some efficiency losses occur through the gearing. Some systems use

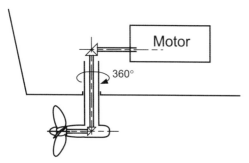

Figure 11.7. Z-Drive unit.

a propeller at both the fore and aft ends of the unit, and the propellers work in tandem.

Data Sources. Published series $K_T - K_Q$ charts for fixed-pitch propellers, with or without duct, may be used.

11.3.7 Podded Azimuthing Propellers

This it is an application of the fixed-pitch propeller and incorporates a slender high-efficiency electric drive motor housed within the pod, Figure 11.8. The propeller may be ducted. It is normally able to azimuth through 360° providing directional thrust and good manoeuvring properties. A rudder is not required. If the propeller is mounted at the leading edge of the pod, termed a puller type or tractor, a relatively clean and uniform wake is encountered, leading to less vibration, cavitation and noise. Some systems use a propeller at both the fore and aft ends of the pod; the propellers work in tandem.

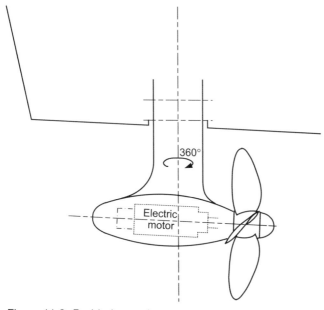

Figure 11.8. Podded propulsor.

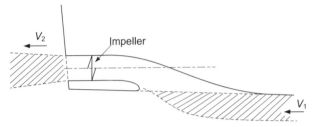

Figure 11.9. Waterjet.

The podded unit is usually associated with a pram-type stern, and the shaft may be inclined to the flow. A flap may be added at the aft end of the vertical support strut and a fin added under the pod in order to improve manoeuvring and course-keeping performance.

Data Sources. Published series $K_T - K_Q$ charts for fixed-pitch propellers can be used, together with an appropriate wake fraction and corrections for the presence of the relatively large pod and supporting strut. These aspects are discussed further in Chapter 16.

11.3.8 Waterjet Propulsion

Jet units involve drawing fluid into the hull through an intake and discharging it either above or below water (usually above) at high velocity, Figure 11.9. Various pumps may be used such as axial flow, centrifugal or piston types, but mixed axial/centrifugal pumps tend to be the most common. Waterjets have no underwater appendages which can be an advantage in some applications. For example, the safety of a shrouded propeller is attractive for small sporting and rescue craft and dive support craft. A swivelling nozzle and reversing bucket provide change of thrust direction and reverse thrust. Power losses in the pump and inlet/outlet ducting can result in low propulsive efficiency at lower speeds. It is more efficient than conventional propellers at speeds greater than about 30 knots. At lower speeds the conventional subcavitating propeller is more efficient, whilst at speeds greater than about 40–45 knots supercavitating or surface-piercing propellers may be more efficient.

Data Sources. Manufacturers' data and theoretical approaches tend to be used for design purposes, see Chapter 16.

11.3.9 Cycloidal Propeller

This is a vertical axis propeller, with the blades acting as aerofoils, Figure 11.10. Thrust can be produced in any direction and a rudder is not needed. The Voith Schneider unit is a proprietary brand of the cycloidal propeller. These propellers are commonly fitted to vessels requiring a high degree of manoeuvrability or station keeping, such as tugs and ferries. Such vessels are often double ended with a propulsion unit at each end so that the craft can be propelled directly sideways or rotated about a vertical axis without moving ahead.

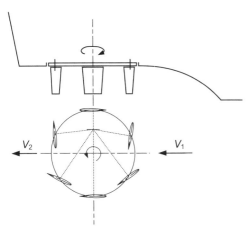

Figure 11.10. Cycloidal propeller (Voith Schneider).

Data Sources. Some published data are available in $K_T - K_Q$ form, together with manufacturers' data, see Chapter 16.

11.3.10 Paddle Wheels

Paddle wheels accelerate a surface fluid layer, Figure 11.11. They can be side or stern mounted, with fixed or feathering blades. The efficiencies achieved with feathering blades are comparable to the conventional marine propeller.

Data Sources. Some systematic performance data are available for design purposes, see Chapter 16.

11.3.11 Sails

Sails have always played a role in the propulsion of marine vessels. They currently find applications ranging from cruising and racing yachts [11.15] to the sail assist of large commercial ships [11.16, 11.17], which is discussed further in Section 11.3.16. Sails may be soft or solid and, in both cases, the sail acts like an aerofoil with the ability to progress into the wind. The forces generated, including the propulsive force, are shown in Figure 11.12. The sail generates lift (L) and drag (D) forces normal to and in the direction of the relative wind. The resultant force is F. The

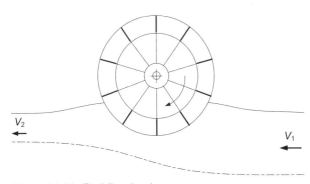

Figure 11.11. Paddle wheel.

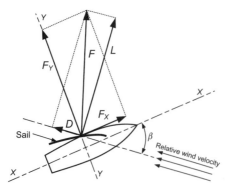

Figure 11.12. Sail forces.

resultant force can be resolved along the X and Y body axes of the boat or ship, F_X on the longitudinal ship axis and F_Y on the transverse Y axis. F_X is the driving or propulsive force.

Data Sources. Sail performance data are usually derived from wind tunnel tests. Experimental and theoretical data for soft sails are available for preliminary design purposes, see Chapter 16.

11.3.12 Oars

Rowing or sculling using oars is usually accepted as the first method of boat or ship propulsion. Typical references for estimating propulsive power when rowing include [11.18–11.21].

11.3.13 Lateral Thrust Units

Such units were originally employed as 'bow thrusters', Figure 11.13. They are now employed at the bow and stern of vessels requiring a high degree of manoeuvrability at low speeds. This includes manoeuvring in and out of port, or holding station on a dynamically positioned ship.

Data Sources. Some published data and manufacturers' data are available, see Chapter 16.

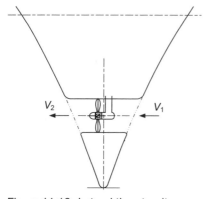

Figure 11.13. Lateral thrust unit.

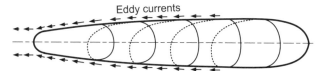

Figure 11.14. Electrolytic propulsion.

11.3.14 Other Propulsors

All of the foregoing devices are, or have been, used in service and are of proven effectiveness. There are a number of experimental systems which, although not efficient in their present form, show that there can be other ways of achieving propulsion, as discussed in the following sections.

11.3.14.1 Electrolytic Propulsion

By passing a low-frequency AC current along a solenoid immersed in an electrolyte (in this case, salt water), eddy currents induced in the electrolyte are directed aft along the coil, Figure 11.14. Thrust and efficiency tend to be low. There are no moving parts or noise, which would make such propulsion suitable for strategic applications. Research into using such propulsion for a small commercial craft is described in [11.1 and 11.22].

11.3.14.2 Ram Jets

Expanding bubbles of gas in the diffuser section of a ram jet do work on the fluid and, hence, produce momentum changes, Figure 11.15. The gas can come from either the injection of compressed air at the throat or by chemical reaction between the water and a 'fuel' of sodium or lithium pellets. The device is not self-starting and its efficiency is low. An investigation into a bubbly water ram jet is reported in [11.23].

11.3.14.3 Propulsion of Marine Life

The resistance, propulsion and propulsive efficiency of marine life have been studied over the years. It is clear that a number of marine species have desirable engineering features. The process of studying areas inspired by the actions of marine life has become known as bioinspiration [11.24]. Research has included efforts to emulate the propulsive action of fish [11.25–11.28] and changes in body shape and surface finish to minimise resistance [11.29, 11.30]. Work is continuing on the various areas

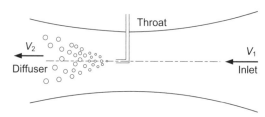

Figure 11.15. Ram jet.

of interest, and other examples of relevant research are included in [11.28, 11.31, 11.32, 11.33].

11.3.15 Propulsion-Enhancing Devices

11.3.15.1 Potential Propeller Savings
The components of the quasi-propulsive coefficient (η_D) may be written as follows:

$$\eta_D = \eta_o \times \eta_H \times \eta_R, \tag{11.2}$$

where η_H is the hull efficiency (see Chapters 8 and 16) and η_R is the relative rotative efficiency (see Chapter 16).

The efficiency η_o is the open water efficiency of the propeller and will depend on the propeller diameter (D), pitch ratio (P/D) and revolutions (rpm). Clearly, an optimum combination of these parameters is required to achieve maximum efficiency. Theory and practice indicate that, in most circumstances, an increase in diameter with commensurate changes in P/D and rpm will lead to improvements in efficiency. Propeller tip clearances will normally limit this improvement. For a fixed set of propeller parameters, η_o can be considered as being made up of

$$\eta_o = \eta_a \cdot \eta_r \cdot \eta_f, \tag{11.3}$$

where η_a is the ideal efficiency, based on axial momentum principles and allowing for a finite blade number, η_r accounts for losses due to fluid rotation induced by the propeller and η_f accounts for losses due to blade friction drag (Dyne [11.34, 11.35]). This breakdown of efficiency components is also derived using blade element-momentum theory in Chapter 15. Theory would suggest typical values of these components at moderate thrust loading as $\eta_a = 0.80$ (with a significant decrease with increase in thrust loading), $\eta_r = 0.95$ (reasonably independent of thrust loading) and $\eta_f = 0.85$ (increasing a little with increase in thrust loading), leading to $\eta_o = 0.646$. This breakdown of the components of η_o is important because it indicates where likely savings might be made, such as the use of pre- and post-swirl devices to improve η_r or surface treatment of the propeller to improve η_f.

11.3.15.2 Typical Devices
A number of devices have been developed and used to improve the overall efficiency of the propulsion arrangement. Many of the devices recover downstream rotational losses from the propeller. Some recover the energy of the propeller hub vortex. Others entail upstream preswirl ducts or fins to provide changes in the direction of the flow into the propeller. Improvements in the overall efficiency of the order of 3%–8% are claimed for such devices. Some examples are listed below:

- Twisted stern upstream of propeller [11.36]
- Twisted rudder [11.37, 11.38]
- Fins on rudder [11.39]
- Upstream preswirl duct [11.40, 11.41]
- Integrated propeller-rudder [11.42]
- Propeller boss cap fins [11.43]

11.3.16 Auxiliary Propulsion Devices

A number of devices provide propulsive power using renewable energy. The energy sources are wind, wave and solar. Devices using these sources are outlined in the following sections.

11.3.16.1 Wind
Wind-assisted propulsion can be provided by sails, rotors, kites and wind turbines. Good reviews of wind-assisted propulsion are given in [11.16] and Windtech'85 [11.44].

SAILS. Sails may be soft or rigid. Soft sails generally require complex control which may not be robust enough for large commercial vessels. Rigid sails in the form of rigid vertical aerofoil wings are attractive for commercial applications [11.44]. They can be robust in construction and controllable in operation. Prototypes, designed by Walker Wingsails, were applied successfully on a coaster in the 1980s.

ROTORS. These rely on Magnus effect and were demonstrated successfully on a cargo ship by Flettner in the 1920s. There is renewed interest in rotors; significant contributions to propulsive power have been claimed [11.45]. It may be difficult to achieve adequate robustness when rotors are applied to large commercial ships.

KITES. These have been developed over the past few years and significant contributions to power of the order of 10%–35% are estimated [11.46]. Their launching and retrieval might prove too complex and lack robustness for large commercial ships.

WIND TURBINES. These may be vertical or horizontal axis, and they were researched in some detail in the 1980s [11.44]. They are effective in practice, but require large diameters and structures to provide effective propulsion for large ships. The drive may be direct to the propeller, or to an electrical generator to supplement an electric drive.

11.3.16.2 Wave
The wave device comprises a freely flapping symmetrical foil which is driven by the ship motions of pitch and heave. With such vertical motion, the flapping foil produces a net forward propulsive force [11.28]. Very large foils, effectively impractical in size, tend to be required in order to provide any significant contribution to overall propulsive power.

11.3.16.3 Solar, Using Photovoltaic Cells
Much interest has been shown recently in this technique. Large, effectively impractical areas of panels are, however, required in order to provide any significant amounts of electricity for propulsive power at normal service speeds. Some effective applications can be found for vessels such as relatively slow-speed ferries and sight-seeing cruisers.

11.3.16.4 Auxiliary Power–Propeller Interaction
It is important to take note of the interaction between auxiliary sources of thrust, such as sails, rotors or kites, and the main propulsion engine(s), Molland and

Hawksley [11.47]. Basically, at a particular speed, the auxiliary thrust causes the propulsion main engine(s) to be offloaded and possibly to move outside its operational limits. This may be overcome by using a controllable pitch propeller or multiple engines (via a gearbox), which can be individually shut down as necessary. This also depends on whether the ship is to be run at constant speed or constant power. Such problems can be overcome at the design stage for a new ship, perhaps with added cost. Such requirements can, however, create problems if auxiliary power is to be fitted to an existing vessel.

11.3.16.5 Applications of Auxiliary Power

Whilst a number of the devices described may be impractical as far as propulsion is concerned, some, such as wind turbines and solar panels, may be used to provide supplementary power to the auxiliary generators. This will lead to a decrease in *overall* power, including propulsion and auxiliary electrical generation.

REFERENCES (CHAPTER 11)

11.1 Carlton, J.S. *Marine Propellers and Propulsion.* 2nd Edition. Butterworth-Heinemann, Oxford, UK, 2007.

11.2 Andersen, P. Tip modified propellers. *Ocean Engineering International,* Vol. 3, No. 1, 1999.

11.3 Dang, J. Improving cavitation performance with new blade sections for marine propellers. *International Shipbuilding Progress,* Vol. 51, 2004.

11.4 Dyne, G. On the principles of propellers with endplates. *Transactions of the Royal Institution of Naval Architects,* Vol. 147, 2005, pp. 213–223.

11.5 Baker, G.S. The effect of propeller boss diameter upon thrust and efficiency at given revolutions. *Transactions of the Royal Institution of Naval Architects,* Vol. 94, 1952, pp. 92–109.

11.6 Brownlie, K. *Controllable Pitch Propellers.* IMarEST, London, UK, 1998.

11.7 Abu Sharkh, S.M., Turnock, S.R. and Hughes, A.W. Design and performance of an electric tip-driven thruster. *Proceedings of the Institution of Mechanical Engineers, Part M: Journal of Engineering for the Maritime Environment,* Vol. 217, No. 3, 2003.

11.8 Glover, E.J. Contra rotating propellers for high speed cargo vessels. A theoretical design study. *Transactions North East Coast Institution of Engineers and Shipbuilders,* Vol. 83, 1966–1967, pp. 75–89.

11.9 Van Manen, J.D. and Oosterveld, M.W.C. Model tests on contra-rotating propellers. *International Shipbuilding Progress,* Vol. 15, No. 172, 1968, pp. 401–417.

11.10 Meier-Peter, H. Engineering aspects of contra-rotating propulsion systems for seagoing merchant ships. *International Shipbuilding Progress,* Vol. 20, No. 221, 1973.

11.11 Praefke, E., Richards, J. and Engelskirchen, J. Counter rotating propellers without complex shafting for a fast monohull ferry. *Proceedings of Sixth International Conference on Fast Sea Transportation, FAST'2001,* Southampton, UK, 2001.

11.12 Kim, S.E., Choi, S.H. and Veikonheimo, T. Model tests on propulsion systems for ultra large container vessels. *Proceedings of the International Offshore and Polar Engineering Conference, ISOPE-2002,* Kitakyushu, Japan, 2002.

11.13 Qin, S. and Yunde, G. Tandem propellers for high powered ships. *Transactions of the Royal Institution of Naval Architects,* Vol. 133, 1991, pp. 347–362.

11.14 Telfer, E.V. Sir Charles Parsons and the naval architect. *Transactions of the Royal Institution of Naval Architects*, Vol. 108, 1966, pp. 1–18.

11.15 Claughton, A., Wellicome, J.F. and Shenoi, R.A. (eds.) *Sailing Yacht Design*, Vol. 1 *Theory*, Vol. 2 *Practice*. The University of Southampton, Southampton, UK, 2006.

11.16 RINA. Proceedings of the Symposium on Wind Propulsion of Commercial Ships. The Royal Institution of Naval Architects, London, 1980.

11.17 Murata, M., Tsuji, M. and Watanabe, T. Aerodynamic characteristics of a 1600 Dwt sail-assisted tanker. *Transactions North East Coast Institution of Engineers and Shipbuilders*, Vol. 98, No. 3, 1982, pp. 75–90.

11.18 Alexander, F.H. The propulsive efficiency of rowing. *Transactions of the Royal Institution of Naval Architects*, Vol. 69, 1927, pp. 228–244.

11.19 Wellicome, J.F. Some hydrodynamic aspects of rowing. In *Rowing – A Scientific Approach*, ed. J.P.G. Williams and A.C. Scott. A.S. Barnes, New York, 1967.

11.20 Shaw, J.T. Rowing in ships and boats. *Transactions of the Royal Institution of Naval Architects*, Vol. 135, 1993, pp. 211–224.

11.21 Kleshnev, V. Propulsive efficiency of rowing. *Proceedings of XVII International Symposium on Biomechanics in Sports*, Perth, Australia, 1999, pp. 224–228.

11.22 Molland, A.F. (ed.) *The Maritime Engineering Reference Book*. Butterworth-Heinemann, Oxford, UK, 2008.

11.23 Mor, M. and Gany, A. Performance mapping of a bubbly water ramjet. Technical Note. *International Journal of Maritime Engineering, Transactions of the Royal Institution of Naval Architects*, Vol. 149, 2007, pp. 45–50.

11.24 Bar-Cohen, Y. Bio-mimetics – using nature to inspire human innovation. *Bioinspiration and Biomimetics*, Vol. 1, No. 1, 2006, pp. 1–12.

11.25 Gawn, R.W.L. Fish propulsion in relation to ship design. *Transactions of the Royal Institution of Naval Architects*, Vol. 92, 1950, pp. 323–332.

11.26 Streitlien, K., Triantafyllou, G.S. and Triantafyllou, M.S. Efficient foil propulsion through vortex control. *AIAA Journal*, Vol. 34, 1996, pp. 2315–2319.

11.27 Long, J.H., Schumacher, L., Livingston, N. and Kemp, M. Four flippers or two? Tetrapodal swimming with an aquatic robot. *Bioinspiration and Biomimetics*, Vol. 1, No. 1, 2006, pp. 20–29.

11.28 Bose, N. *Marine Powering Prediction and Propulsors*. The Society of Naval Architects and Marine Engineers, New York, 2008.

11.29 Fish, F.E. The myth and reality of Gray's paradox: implication of dolphin drag reduction for technology. *Bioinspiration and Biomimetics*, Vol. 1, No. 2, 2006, pp. 17–25.

11.30 Anderson, E.J., Techet, A., McGillis, W.R., Grosenbaugh, M.A. and Triantafyllou, M.S. Visualisation and analysis of boundary layer flow in live and robotic fish. *First Symposium on Turbulence and Shear Flow Phenomena*, Santa Barbara, CA, 1999, pp. 945–949.

11.31 Fish, F.E. and Rohr, J.J. Review of dolphin hydrodynamics and swimming performance. Technical Report 1801, SPAWAR Systems Center, San Diego, CA, 1999.

11.32 Triantafyllou, M.S., Triantafyllou, G.S. and Yue, D.K.P. Hydrodynamics of fish like swimming. *Annual Review of Fluid Mechanics*, Vol. 32, 2000, pp. 33–54.

11.33 Lang, T.G. *Hydrodynamic Analysis of Cetacean Performance: Whales, Dolphins and Porpoises*. University of California Press, Berkeley, CA, 1966.

11.34 Dyne, G. The efficiency of a propeller in uniform flow. *Transactions of the Royal Institution of Naval Architects*, Vol. 136, 1994, pp. 105–129.

11.35 Dyne, G. The principles of propulsion optimisation. *Transactions of the Royal Institution of Naval Architects*, Vol. 137, 1995, pp. 189–208.

11.36 Anonymous. Development of the asymmetric stern and service results. *The Naval Architect*. RINA, London, 1985, p. E181.

11.37 Molland A.F. and Turnock, S.R. *Marine Rudders and Control Surfaces*. Butterworth-Heinemann, Oxford, UK, 2007.

11.38 Anonymous. Twisted spade rudders for large fast vessels. *The Naval Architect*, RINA, London, September 2004, pp. 49–50.

11.39 Motozuna, K. and Hieda, S. Basic design of an energy saving ship. *Proceedings of Ship Costs and Energy Symposium'82*. SNAME, New York, 1982, pp. 327–353.

11.40 Anonymous. The SHI SAVER fin. Marine Power and Propulsion Supplement. *The Naval Architect*. RINA London, 2008, p. 36.

11.41 Mewis, F. Development of a novel power-saving device for full-form vessels. *HANSA International Maritime Journal*, Vol. 145, No. 11, November 2008, pp. 46–48.

11.42 Anonymous. The integrated propulsion manoeuvring system. *Ship and Boat International*, RINA, London, September/October 2008, pp. 30–32.

11.43 Atlar, M. and Patience, G. An investigation into effective boss cap designs to eliminate hub vortex cavitation. *Proceedings of the 7th International Symposium on Practical Design of Ships and Mobile Units, PRADS'98*, The Hague, 1998.

11.44 Windtech'85 *International Symposium on Windship Technology*. University of Southampton, UK, 1985.

11.45 Anonymous. Christening and launch of 'E-Ship1' in Kiel. *The Naval Architect*. RINA, London, September 2008, p. 43.

11.46 Anonymous. Skysails hails latest data. *The Naval Architect*. RINA, London, September 2008, pp. 55–57.

11.47 Molland, A.F. and Hawksley, G.J. An investigation of propeller performance and machinery applications in wind assisted ships. *Journal of Wind Engineering and Industrial Aerodynamics*, Vol. 20, 1985, pp. 143–168.

12 Propeller Characteristics

12.1 Propeller Geometry, Coefficients, Characteristics

12.1.1 Propeller Geometry

A marine propeller consists of a number of blades (2–7) mounted on a boss, Figure 12.1. Normal practice is to cast the propeller in one piece. For special applications, built-up propellers with detachable blades may be employed, such as for controllable pitch propellers or when the blades are made from composite materials.

The propeller is defined in relation to a *generator line*, sometimes referred to as the directrix, Figure 12.1. This line may be drawn at right angles to the shaft line, but more normally it is *raked*. For normal applications, blades are raked aft to provide the best clearance in the propeller aperture. For high-speed craft, the blades may be raked forward to balance bending moments due to centrifugal forces against those due to thrust loading.

Viewed from aft, the projected blade outline is not normally symmetric about the generator line but is given some *skew* or *throw round* to help clear debris and improve vibration characteristics. With skew, the blade sections meet any wake concentrations in a progressive manner, with possible reductions in vibration loading. Skew and rake generally do not have any great effect on performance.

The propeller blade is defined by a number of sections drawn through the blade, Figure 12.2. The sections lie on cylindrical surfaces coaxial with the propeller shaft. The sections are defined in relation to a pseudohelical surface defined by sweeping the generator along the shaft axis in such a way that the angle of rotation from some datum is proportional to the forward movement of the generator along the shaft axis. The intersection of the generator surface and the cylinder for a given section is thus a true helix and, when the cylinder is developed, this helix appears as a straight line. The longitudinal distance the generator moves in one complete revolution is called the *pitch* of the section, in this case the *geometric pitch*.

Although at each radius the generator sweeps out a helix, the complete generator surface is usually not a helix because of the following:

(a) The generator is raked by an amount that can vary with radius.
(b) The geometric pitch (P), Figure 12.3, is usually not constant. It normally varies with radius and is usually less at the boss than at the tip of the blade.

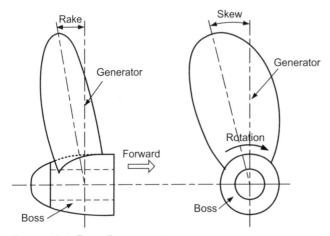

Figure 12.1. Propeller geometry.

Section shape is usually defined in relation to the generator surface at stations normal to the generator surface. Blade face and back surface heights are given above the generator surface, Figure 12.4.

The blade projected and developed section lengths are laid off around arcs, whilst the expanded lengths are laid off at fixed radii, Figure 12.5.

Typical blade sections are as follows:

(i) Simple round back sections, Figure 12.6(a). These were in common use at one time and are still used for the outermost sections of a blade and for wide-bladed propellers.

(ii) The inner (thicker) sections of most merchant propellers are normally of aerofoil shape, Figure 12.6(b), selected to give a favourable pressure distribution for avoiding cavitation and offering less drag than the equivalent round back section for the same lift. The overall application tends to amount to aerofoil sections near the root of the blade, changing gradually to round back sections

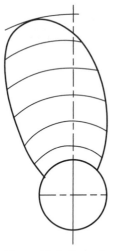

Figure 12.2. Propeller sections.

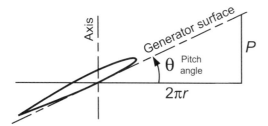

Figure 12.3. Geometric pitch.

towards the tip. Section shape is particularly important as far as cavitation inception is concerned and this is discussed in more detail in Section 12.2.

(iii) Wedge-type sections, Figure 12.6(c), are used on propellers designed for super-cavitating operation. Although not necessary for supercavitating operation, the trailing edge of the back of the section needs to be shaped in order to retain adequate performance under subcavitating operation (part loading) and operation astern.

12.1.1.1 Propeller Design Parameters

(A) Pitch, P

Various definitions are used, all being derived from the advance of some feature of the propeller during one revolution. Pitch is normally expressed non-dimensionally as a fraction of propeller diameter.

(1) Geometric pitch, Figure 12.3.

$$P/D = \text{pitch of generator surface/diameter.}$$

This is sometimes called the *face pitch* ratio. The pitch (and pitch ratio) may be constant across the blade radius. Where the propeller pitch varies radially, a mean or virtual pitch may be quoted which is an average value over the blade. It should be noted that even for constant radial pitch, the pitch angle θ will vary radially, from a large angle at the blade root to a small angle at the tip. From Figure 12.3, r is the local radius and if R is the propeller radius, then let

$$\frac{r}{R} = x.$$

In Figure 12.3, for one revolution, $2\pi r = 2\pi xR = \pi xD$. Then, at any radius r, pitch angle θ is: $\theta = \tan^{-1}(\frac{P}{\pi xD})$ and the actual local geometric pitch angle at any radius can be calculated as follows:

$$\theta = \tan^{-1}\left(\frac{P/D}{\pi x}\right). \tag{12.1}$$

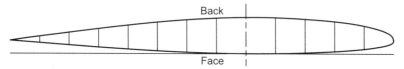

Figure 12.4. Section thickness offsets.

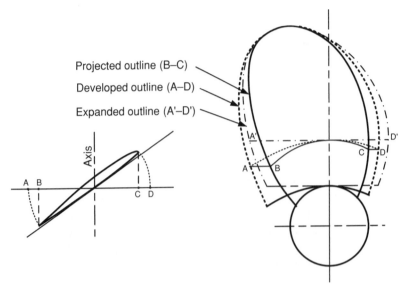

Figure 12.5. Projected, developed and expanded blade outline.

(2) Hydrodynamic pitch.

The advance at which a section will produce no thrust is the hydrodynamic pitch of that section. It is approximately the no-lift condition at that section, Figure 12.7.

(3) Effective pitch.

This is the advance for no thrust on the entire propeller, for example the point where K_T passes through zero on a $K_T - K_Q$ propeller chart, Figure 12.9.

For any given propeller type there is a fixed relationship between the various values of pitch ratio as defined above. So far as performance is concerned, the effective pitch is the critical value and the relation between mean geometric pitch and the effective pitch depends upon the pitch distribution, section type and thickness ratio.

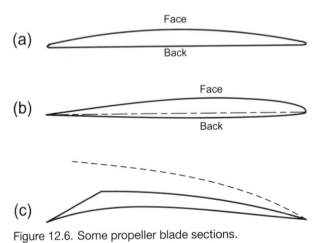

Figure 12.6. Some propeller blade sections.

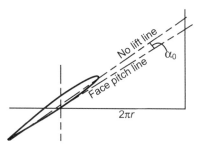

Figure 12.7. Hydrodynamic pitch.

(B) Blade area ratio (BAR)

Area ratios are defined in relation to the disc area of the propeller. Three values are in common use, Figure 12.5.

Projected blade area ratio is the projected area viewed along shaft/disc area, B–C in Figure 12.5

Developed blade area ratio (DAR) is the area enclosed by developed outline/disc area, A–D in Figure 12.5

Expanded blade area ratio (EAR) is the area enclosed by expanded outline/disc area, A′–D′ in Figure 12.5.

In each case the area taken is that outside the boss. EAR and DAR are nearly equal and either may be called BAR (simply blade area ratio). BAR values are normally chosen to avoid cavitation, as discussed in Section 12.2.

(C) Blade thickness ratio, t/D

t/D is the maximum blade thickness projected to shaft line/diameter, Figure 12.8. Typical values of t/D are $0.045 \sim 0.050$

(D) Boss/diameter ratio

The boss diameter is normally taken at the point where the generator line cuts the boss outline, Figure 12.1, and this value is used for the boss/diameter ratio. Typical values for solid propellers are $0.18\sim0.20$, whilst built-up and controllable pitch propellers are larger at about 0.25.

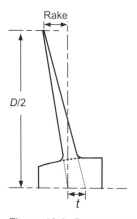

Figure 12.8. Propeller blade thickness.

12.1.2 Dimensional Analysis and Propeller Coefficients

Several systems of non-dimensional coefficients are used in propeller design work. The physical variables and their dimensions normally related by these coefficients are as follows:

Thrust,	T	$\dfrac{ML}{T^2}$ (force)
Torque,	Q	$\dfrac{ML^2}{T^2}$ (force × length)
Revs/sec,	n	$\dfrac{1}{T}$
Speed of advance,	V	$\dfrac{L}{T}$ (velocity)
Diameter,	D	L
Fluid density,	ρ	$\dfrac{M}{L^3}$ (mass/unit volume)

According to the methods of dimensional analysis [12.1], [12.2], the relationships between the quantities may be expressed as:

$$\text{Thrust} \qquad\qquad \text{Torque}$$
$$f(T, D, V, n, \rho) = 0 \qquad f(Q, D, V, n, \rho) = 0$$
$$\text{or} \quad T = f(D, V, n, \rho) \qquad \text{and} \quad Q = f(D, V, n, \rho)$$

whence

$$f_1\left[\left(\frac{T}{\rho n^2 D^4}\right), \left(\frac{V}{n \cdot D}\right)\right] = 0 \qquad f_2\left[\left(\frac{Q}{\rho n^2 D^5}\right), \left(\frac{V}{n \cdot D}\right)\right] = 0$$

$$f_3\left[\left(\frac{T}{\rho V^2 D^2}\right), \left(\frac{V}{n \cdot D}\right)\right] = 0 \qquad f_4\left[\left(\frac{Q}{\rho V^2 D^3}\right), \left(\frac{V}{n \cdot D}\right)\right] = 0$$

Commonly found systems of presentation are as follows:

(a) $K_T = T/\rho n^2 D^4$; $\quad K_Q = Q/\rho n^2 D^5$; $\quad J = V/nD$
(b) $C_T = T/\rho V^2 D^2$; $\quad C_Q = Q/\rho V^2 D^3$; $\quad J = V/nD$
(c) $\mu = n(\rho D^5/Q)^{1/2}$; $\quad \phi = V(\rho D^3/Q)^{1/2}$; $\quad \sigma = DT/2\pi Q$
(d) If a power approach is used, this yields $B_P(= NP^{1/2}/V^{2.5})$; $\quad \delta(= ND/V)$.

12.1.3 Presentation of Propeller Data

The preferred system has become K_T, K_Q, J for most purposes, Figure 12.9, since C_T, C_Q, J suffers from the disadvantage that $C_T, C_Q \to \infty$ as $V \to 0$. This renders C_T, C_Q charts useless for low-speed towing work, and meaningless at the bollard pull condition when $V = 0$.

The μ, σ, ϕ charts are convenient for certain towing duty calculations, and design charts have been developed for this purpose, Figure 16.6. For a power approach, B_P, δ charts have been used extensively in the past, although they are not suitable for low-speed or bollard conditions, Figure 16.5. It is now more usual to find data presented in terms of K_T, K_Q, J, and this is the presentation in common use, Figure 12.9.

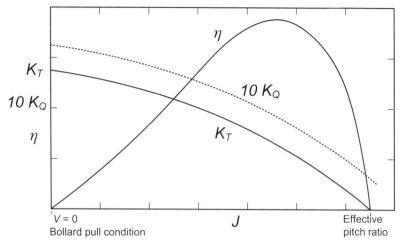

Figure 12.9. Open water $K_T - K_Q$ chart, for one pitch ratio.

As $10K_Q$ is about the same order of magnitude as K_T, this is normally plotted on the $K_T - K_Q$ chart. For a given blade area ratio, data for variation in the pitch ratio are normally shown on the same chart, Figure 16.3. Examples of the three types of chart and their applications are given in Chapter 16.

Propeller efficiency is determined as follows:

$$\eta_0 = \text{power output/power input} = \frac{TV_a}{2\pi nQ}.$$

This basic formula can be written in terms of the non-dimensional coefficients as:

$$\eta_0 = \frac{\rho n^2 D^4 K_T V_a}{2\pi n \rho n^2 D^5 K_Q}$$

or

$$\eta_0 = \frac{JK_T}{2\pi K_Q} \qquad (12.2)$$

This is a very useful formula and shows the relationship between η_0, J, K_T and K_Q.

There have been several curve fits to K_T, K_Q charts and it is clear that only K_T and K_Q need to be defined since η_0, which would be more difficult to curve fit, may be derived from Equation (12.2).

12.1.4 Measurement of Propeller Characteristics

Performance characteristics for the propeller in isolation are known as open water data. The propeller open water data may be measured using a propeller 'boat' in a test tank, Figure 12.10. In this case, unless it is a depressurised test tank such as that described in [12.3], cavitation will not occur. Alternatively, the propeller may be tested in a cavitation tunnel where open water characteristics can be measured, Figure 12.28, the pressure reduced and cavitation performance also observed, see Section 12.2.

Tests will normally entail the measurement of propeller thrust (T), torque (Q), revolutions (n) and speed (V). These may be corrected for temperature (to 15°C)

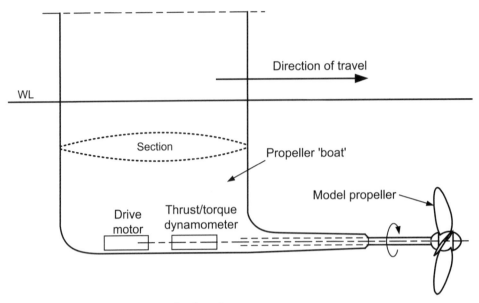

Figure 12.10. Open water propeller 'boat'.

and for tunnel wall effects (blockage) in the case of the cavitation tunnel. Finally, the results can be non-dimensionalised and plotted in terms of K_T, K_Q and η to a base of J, per Figure 12.9.

The International Towing Task Conference (ITTC) recommended procedure for propeller open water tests may be found in ITTC2002 [12.4]. The measurements made are summarised in Figure 12.11.

Other propulsors are as follows:

Ducted propellers: The tests are broadly similar to those described for the propeller, with a separate dynamometer measuring the duct thrust or drag.

Supercavitating propellers: The tests are carried out in a cavitation tunnel (Section 12.2) and are broadly similar to those described.

Surface-piercing propellers: Open water tests follow a similar pattern to those for submerged propellers, but on an inclined shaft with the propeller only partially immersed in a tank or circulating water channel. Besides rpm, speed, thrust and torque, measurements will normally also include that of the vertical force. Results of such tests are given in [12.5] and [12.6].

Podded propellers: A schematic layout of a suitable test rig is shown in Figure 12.12. It is equipped with a dynamometer close to the propeller to measure propeller thrust and torque. A separate dynamometer at the top of the unit measures the thrust of the whole unit (effectively the thrust of the propeller minus the drag of the pod and support strut). The pod/support strut can be rotated enabling tests in oblique flow to be made, such as when manoeuvring. The drive motor may be housed in the pod in line with the propeller if space allows.

Typical tests will include open water tests on the propeller alone, followed by tests on the podded unit without the propeller, then on the total podded unit including the propeller. In this way, the drag of the pod/support strut and the interference

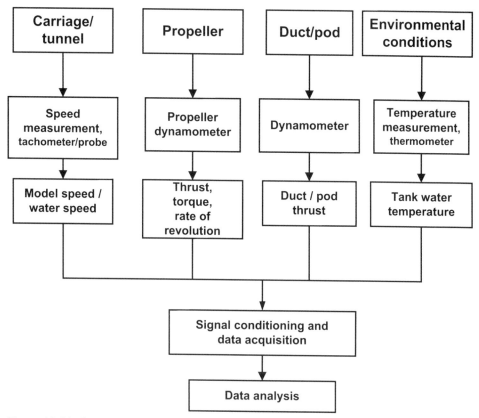

Figure 12.11. Open water test measurements.

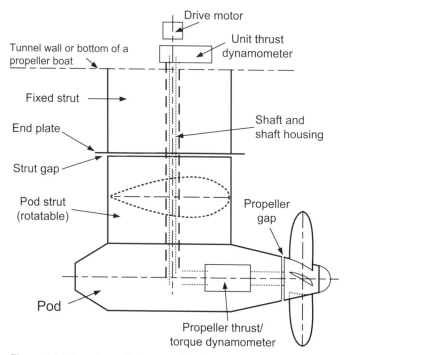

Figure 12.12. Schematic layout of podded propeller test rig.

effects of the supporting strut, pod and propeller can be determined. There are
gaps between the pod strut and the fixed strut and between the propeller and the
pod. It should be noted that gap effects between the propeller and pod can cause
erroneous thrust readings for the propeller. This can make it difficult to differen-
tiate between the total thrust of the unit, the thrust of the propeller and the net
drag of the pod/support strut. The drag of the pod/support strut in isolation can,
however, be determined by testing with the propeller removed, but this does not
provide any information on propeller-pod interference effects. If problems with
gap effects on propeller thrust are carried into the self-propulsion tests, then it
is advisable to use a torque identity analysis, as the torque is little affected by
the gap.

A recommended test procedure for podded units is described in the Appendix
of ITTC2005 [12.7]. Recommended procedures for extrapolating the propeller data
and the drag of the pod/strut support to full scale are described in ITTC2008 [12.8].
This subject is also discussed in Section 16.2.4.

> Waterjets: A special approach is necessary. The equivalent of an open water test
> with direct thrust measurements is generally not feasible. Pump jet efficiency
> will normally be derived from separate tests. Thrust will normally be determ-
> ined from momentum flux calculations using flow rate measurements. Full
> reviews and discussions of the methods employed and problems encountered
> are given in ITTC2002 [12.9] and ITTC2005 [12.10].
> Cycloidal propellers: A special approach is necessary. Results of such tests are
> given Chapter 16 and in [12.11], [12.12].
> Paddle wheels: A special approach is necessary. Results of such tests are given
> in [12.13, 12.14, 12.15].

12.2 Cavitation

12.2.1 Background

Cavitation occurs when the local fluid pressure drops to the vapour pressure of the
liquid, that is, the pressure at which the liquid vapourises. Vapour pressure depends
on temperature and the quality and content of the liquid. Cavitation can occur, in
particular, on marine propellers where peaks of low (suction) pressure can arise,
Figure 12.13. Sheet cavitation tends to occur near the nose of the blade section and
bubble cavitation tends to occur on the back.

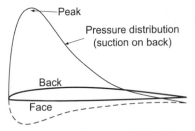

Figure 12.13. Pressure distribution on propeller blade section.

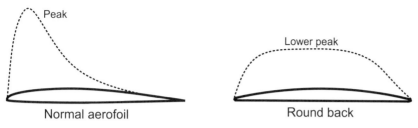

Figure 12.14. Alternative section shapes.

The magnitude (and peaks) of the pressure distribution depends on the lift coefficient (for required thrust) and on section shape and thickness. For example, compared with a normal aerofoil type section, a round back section will exhibit a lower suction peak for the same lift, Figure 12.14, although this will usually be accompanied by some increase in drag.

The effects of cavitation are as follows:

- Flow along the surface is disturbed; the effective profile properties change, causing thrust and torque reductions and decrease in efficiency,
- Possible erosion attributed to the collapse of cavitation bubbles as they move into regions of higher pressure, Figure 12.15,
- Noise as cavities collapse,
- Possible vibration, leading to blade fracture.

Thus, it is desirable to size the area of the propeller blades whereby the thrust loadings, hence the magnitude of pressure peaks, are limited in order to avoid cavitation. Also, careful choice of section shape is necessary in order to smooth out pressure peaks.

The basic physics of cavitation, and cavitation inception, is described in some detail by Carlton [12.16]. Examples of cavitation tests on propellers are included in [12.17–12.20].

It must be noted that *cavitation* should not be confused with *ventilation*. In the case of ventilation, the propeller blades are near or breaking the surface. Air is drawn down to fill the cavities in the flow at atmospheric pressure, compared with vapour pressure for cavitation. Apart from the level of pressure and the lack of erosion, the general phenomena that occur are similar to cavitation.

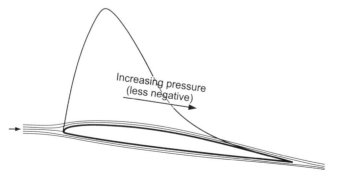

Figure 12.15. Increasing pressure as fluid flows aft.

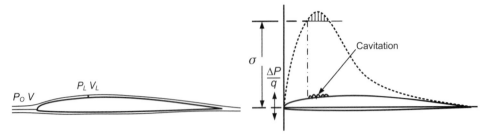

Figure 12.16. Cavitation inception.

12.2.2 Cavitation Criterion

Cavitation occurs when the local pressure P_L decreases to less than the vapour pressure, P_V, Figure 12.16.

For *NO* cavitation,

$$P_L \geq P_V.$$

If $\Delta P = P_0 - P_L$, then $\Delta P = P_0 - P_L \leq P_0 - P_V$ for *NO* cavitation, i.e. $\Delta P/q \leq (P_0 - P_V)/q$, where $q = \frac{1}{2}\rho V^2$. Hence, $\Delta P/q \leq \sigma$ for *NO* cavitation, where σ is the cavitation number and $\sigma = (P_0 - P_V)/q$, where:

P_0 is the static pressure in free stream (including atmospheric) at the point considered.

$$P_0 = P_{AT} + \rho g h.$$

h is the immersion, usually quoted to shaft axis (m)
P_{AT} is the atmospheric pressure $\cong 101 \times 10^3$ N/m²
P_V is the vapour pressure, assume for initial design purposes $\cong 3 \times 10^3$ N/m² for water.

Hence, the peaks of the pressure distribution curve $\Delta P/q$ should not exceed the cavitation number σ. $\Delta P/q$ is a function of the shape of the section and angle of attack, with $\Delta P/q \propto C_L$ for a particular section, Figure 12.17.

The cavitation number σ can be written as

$$\sigma = \frac{(P_{AT} + \rho g h - P_V)}{0.5\rho V^2} \tag{12.3}$$

and a pressure coefficient can be written as

$$C_P = \frac{(P_L - P_0)}{0.5\rho V^2}. \tag{12.4}$$

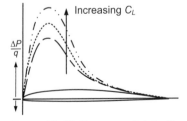

Figure 12.17. Pressure distribution change with increase in C_L.

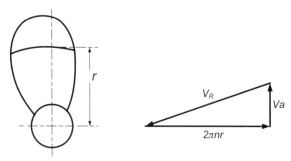

Figure 12.18. Reference velocity.

The reference velocity used (V_R) is usually the local section velocity including inflow:

$$V_R = \sqrt{Va^2 + (2\pi rn)^2} \tag{12.5}$$

where Va is the propeller advance velocity, r is the radius of propeller section considered and n is rps, Figure 12.18. For example, preliminary design criteria often consider

$$r = 0.7R = 0.7\frac{D}{2}$$

and

$$\sigma = \frac{(P_{AT} + \rho gh - P_V)}{0.5\rho V_R^2}. \tag{12.6}$$

12.2.3 Subcavitating Pressure Distributions

Marine propellers normally work in a non-uniform wake, see Chapter 8; hence, for a particular section on the propeller, the effective angle of attack will change in one revolution and may also become negative, Figure 12.19.

The subcavitating pressure distributions around the propeller sections show characteristic variations as the angle of attack changes, Figure 12.20. In each case, cavities may form at the point of minimum pressure. The physical appearance in each case is shown schematically in Figure 12.21.

Type of cavitation are as follows:

(i) Attached sheet cavitation: Attached sheet cavitation at the blade leading edge. This forms on the back for $\alpha > \alpha_i$ and on the face for $\alpha < \alpha_i$, where α_i is the ideal incidence.

(ii) Bubble cavitation: Bubble cavitation is initiated on the blade back only at the location of the pressure minimum. This can occur even at $\alpha = \alpha_i$. It can occur with sheet cavitation and a combination of face sheet cavitation with back

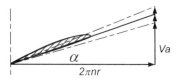

Figure 12.19. Change in blade angle of attack with change in Va.

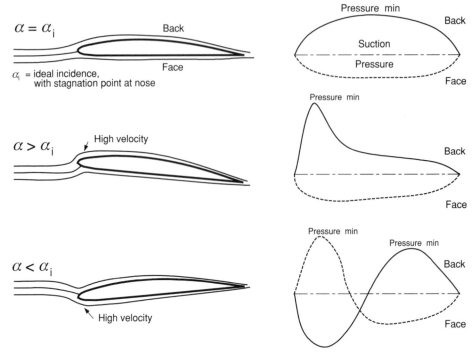

Figure 12.20. Subcavitating pressure distributions.

bubble cavitation is possible. There is the possibility of cloud cavitation down-stream of sheet cavitation. With sheet cavitation at the leading edge, the cavitation core is likely to be separated from the blade by a thin layer of fluid; hence, there is a smaller risk of erosion. With bubble cavitation, the cavity is in direct contact with the blade and erosion is likely.

(iii) Tip vortex cavitation: This is similar to the shed tip vortex on a finite lifting foil. The low pressure core of the vortex can impinge on adjacent hull structure and rudders. It can be difficult to distinguish from sheet cavitation.

(iv) Hub vortex cavitation: This depends on the convergence of the boss (hub), and can affect face or back cavitation in the blade root sections. Erosion and

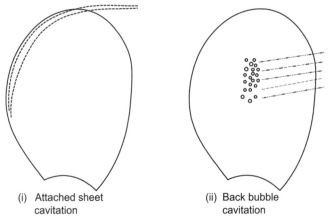

(i) Attached sheet   (ii) Back bubble
     cavitation   cavitation

Figure 12.21. Sheet and back bubble cavitation.

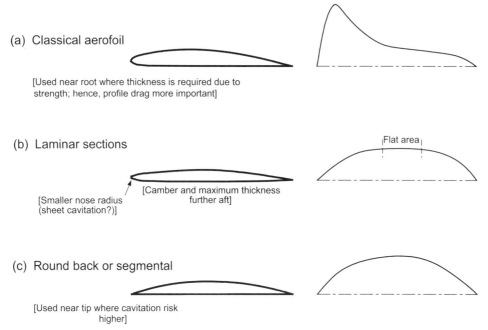

(a) Classical aerofoil

[Used near root where thickness is required due to
strength; hence, profile drag more important]

(b) Laminar sections

[Smaller nose radius   [Camber and maximum thickness
(sheet cavitation?)]   further aft]

Flat area

(c) Round back or segmental

[Used near tip where cavitation risk
higher]

Figure 12.22. Section types: typical pressure distributions at ideal incidence.

rudder damage may occur. The use of truncated cones with no boss fairing at
the trailing edge may provide a solution.

12.2.4 Propeller Section Types

These can be characterised by the pressure distributions at *ideal* incidence, Figure
12.22. It can be noted that the profile drag of laminar and round back sections is generally greater than for aerofoil type sections. Thus, the aerofoil type section tends to
be used near the root, where thickness is required for strength, changing to round
back near the tip, see also Figure 16.2.

12.2.5 Cavitation Limits

The cavitation limits for a normal propeller section can be indicated on a Gutsche
type diagram, that is, an envelope of cavitation limits, or sometimes termed a cavitation bucket which is cavitation free, Figure 12.23.

The maximum thrust (i.e. C_L) is at the intersection of the back bubble and back
sheet lines. The lines have the following forms:

Back bubble line (near α_i): $\sigma = \Delta p/q = f(C_L)$,   or $C_L = \sigma/k_1$.   (12.7)

Back sheet line (at LE): $\sigma = \Delta p/q = f(C_L - C_{Li})$,   or $C_L = \sigma/k_2 + C_{Li}$.

(12.8)

Face sheet line (at LE): $\sigma = \Delta p/q = f(C_{Li} - C_L)$,   or $C_L = C_{Li} - \sigma/k_3$.

(12.9)

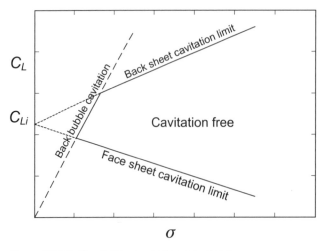

Figure 12.23. Cavitation inception envelope.

In terms of section characteristics, the limits of cavitation are given approximately by the following formulae, for sections approaching round back:

$$\text{Back bubble cavitation: } \sigma = 2/3C_L + 5/2(t/c). \qquad (12.10)$$

$$\text{Sheet cavitation: } \sigma = 0.06(C_L - C_{Li})^2/(r/c), \qquad (12.11)$$

where C_L is the operating lift coefficient, C_{Li} is the ideal design lift coefficient, t is maximum thickness, r is the nose radius, c is the section chord, and σ is based on the local relative flow speed, Equation (12.5). Typical values of the nose radius are $(r/c) = k\,(t/c)^2$, where k is given in Table 12.1.

Typical experimentally derived data attributable to Walcher, and presented in the same form as Gutsche, are shown in Figure 12.24, for changes in section thickness ratio t/l, [12.21]. This figure illustrates the main features of such data. The (vertical) width of the bucket is a measure of the tolerance of the section to cavitation-free operation, i.e. with a wider bucket, the section will be able to tolerate a much wider variation in the angle of attack without cavitating. The width and shape of the bucket will depend on section characteristics such as thickness, camber, overall shape and nose shape. For example, as seen in Figure 12.24, an increase in section thickness tends to widen the bucket and move the bucket width and shape vertically to higher values of C_L.

Table 12.1. *Values of k*

Section type	k
Round back	0.15
Joukowski	0.12
NACA 00 symmetrical	0.11
Elliptic leading edge	0.50

NACA, National Advisory Council for Aeronautics.

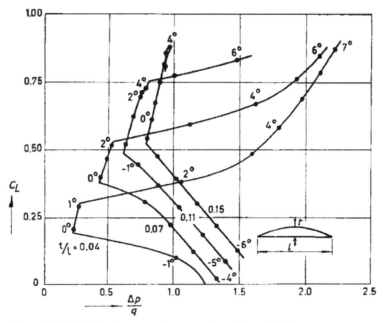

Figure 12.24. Cavitation envelopes attributable to Walcher.

It should be noted that such section information allows the detailed propeller section characteristics and choice to be investigated in some depth with respect to cavitation. For example, for a given σ and section type, the C_L to avoid cavitation (from a chart such as in Figure 12.23 or 12.24) can be retained by the choice of chord length, hence, blade area ratio. Namely, $C_L = L/\frac{1}{2}\rho c V^2$; hence, for required lift (thrust) from a spanwise thrust loading curve, and limiting value of C_L for cavitation, a suitable chord length c can be determined. This can be repeated across the blade span and a blade outline produced. A simple example applying the envelope in Figure 12.23 and Equations (12.7)–(12.11) to one elemental section is given in Chapter 17.

Some approaches use numerical methods to determine the blade surface pressures, linked to cavitation criteria, as discussed in references such as Szantyr [12.22], Carlton [12.16] and ITTC2008 [12.23]. For example, a first approximation may be achieved by applying a two-dimensional (2-D) panel method at a particular section to predict the pressure distribution, as in Figure 12.16; hence, the incidence (or C_L) when $\Delta P/q = $ local σ. This can then be used in the development of the back bubble and sheet cavitation curves in Figures 12.23 and 12.24. An example of the technique is described in [12.24], where the panel code described in [12.25] was used to predict cavitation inception on sections suitable for marine current turbines. Satisfactory correlation was obtained between the 2-D panel code and the cavitation tunnel results, Figure 12.25, although the panel code was a little conservative on the face.

12.2.6 Effects of Cavitation on Thrust and Torque

With the presence of cavitation, the flow along the surface is disturbed. The profile properties change, which causes thrust and torque reductions and a decrease

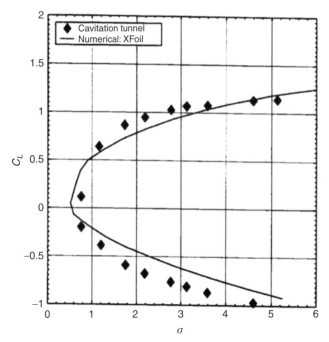

Figure 12.25. Cavitation inception envelopes for NACA 63-215 section; comparison of numerical and experimental results [12.24].

in efficiency. As the extent of the cavitation increases, there is a progressive loss of load as the suction lift on the blade is destroyed. The effects on K_T and K_Q of a decrease in cavitation number (increase in cavitation) are illustrated in Figure 12.26.

Where possible, propellers should be designed to be entirely free of cavitation (subcavitating propellers). If the cavitation number σ falls significantly, cavitation is unavoidable and, to prevent erosion, it is desirable to deliberately design so far into the back sheet cavitation zone that cavity collapse occurs well behind the blade (supercavitating propellers). As the supercavitating state is reached, the thrust, now due to face pressure only, stabilises at about half the subcavitating level. The effect on K_T is shown schematically in Figure 12.27. The effect on K_Q is similar with a subsequent loss in efficiency.

Wedge-shaped sections are often used on propellers designed for supercavitating operation, as discussed in Section 12.1.1 and shown in Figure 12.6 (c). Data for supercavitating propellers are described in Chapter 16.

12.2.7 Cavitation Tunnels

In open water propeller tests, total thrust (T) varies as the cube of scale (λ^3) and surface area (A) varies as the square of scale (λ^2). Therefore, thrust intensity (T/A) is proportional to scale. Also, atmospheric pressure is not reduced to scale. Hence, there is less thrust intensity at model scale compared with that found on working propellers at full scale. Thus, cavitation is not observed in model open water propeller tests in an experimental test tank.

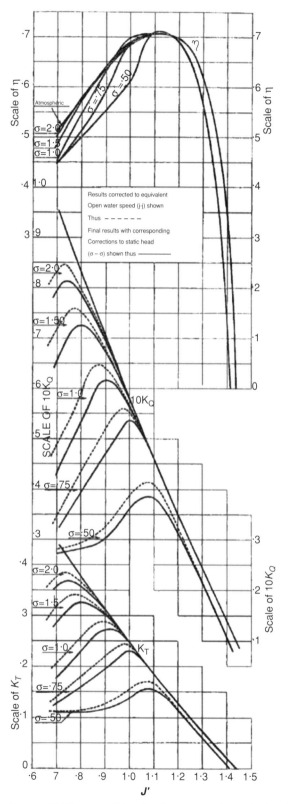

Figure 12.26. Loss of load with increase in cavitation (decrease in σ) [12.17].

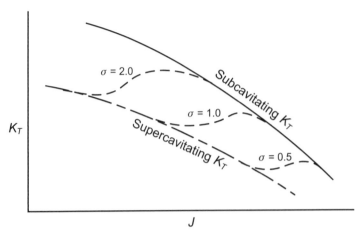

Figure 12.27. Sub- and supercavitating K_T.

A cavitation tunnel is used to simulate the correct thrust intensity at model scale, Figure 12.28. The correct cavitating conditions are reached by reducing the pressure in the tunnel. A special type of sealed test tank may also be used, which can be evacuated and the pressure reduced [12.3].

Typical cavitation tunnel working section cross-sectional dimensions vary from about 600 mm × 600 mm to 2800 mm × 1600 mm in the largest facilities. In the larger tunnels, it is possible to include a wake screen or a truncated dummy hull upstream of the propeller under test in order to simulate suitable wake distributions. In order to minimise scale effects, the propeller diameter will normally be as large as possible without incurring significant blockage effects from the tunnel walls.

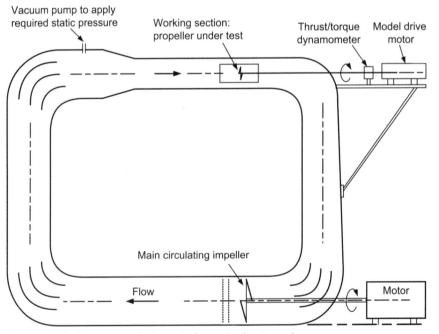

Figure 12.28. Diagrammatic layout of a cavitation tunnel.

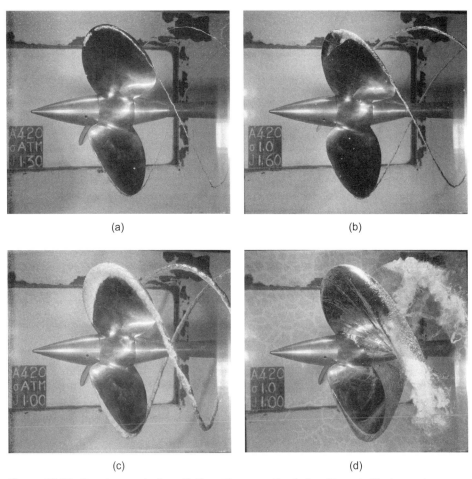

Figure 12.29. Development of cavitation; Emerson Cavitation Tunnel. Photographs courtesy of The University of Newcastle upon Tyne.

Blockage speed corrections may be applied if necessary. Water speed is usually kept as high as possible to maximise Reynolds number and minimise skin friction scale effects. The model is run at the correct J value, which then determines the propeller rpm. The pressure will be lowered as necessary to achieve the required cavitation number. The normal practice, for a given cavitation number, is to test a range of rpm at constant water speed. During each run the propeller rpm, thrust, torque, water temperature and tunnel pressure will be measured. Air content in the water will also be monitored as this can affect the onset of cavitation and affect visual flow studies. Photographs showing the development of cavitation are shown in Figure 12.29 for progressive lowering of the cavitation number. A more detailed account of cavitation and cavitation tunnels is included in Carlton [12.16].

12.2.8 Avoidance of Cavitation

It is seen from the previous sections that cavitation may be avoided by giving due attention to the blade section shape, thickness and blade area. The blade outline shape may also be modified. For example, tip offloading may be applied by a local

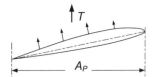

Figure 12.30. Average pressure.

reduction in pitch or a reduction in chord size near the tip. Blade skew may also improve cavitation performance [12.26]. At the preliminary design stage, however, achieving the correct blade area will be the predominant requirement.

12.2.9 Preliminary Blade Area – Cavitation Check

In the early stages of the propeller design process, the designer is primarily concerned with the selection of a suitable blade area, and hence, the choice of the most suitable standard series chart.

In general, the cavitation limit $\Delta P / \frac{1}{2} \rho V_R^2$ can be transformed, Figure 12.30, by relating the local dynamic pressure ΔP to the average difference over the blade $\bar{p}$ given by

$$\bar{p} = \frac{T}{A_P},$$ (12.12)

where T is the thrust and A_P is the projected area (viewed from aft). Thus, at cavitation inception,

$$\sigma = \tau_c,$$

where

$$\tau_c = \frac{T}{0.5 \rho A_P V_R^2}.$$ (12.13)

Such an approach is proposed by Burrill and Emmerson [12.18] for use at the preliminary design stage. Burrill and others have plotted data from cavitation tunnel and full-scale tests showing limiting τ_c values for a given cavitation number (σ), as seen in Figure 12.31. Such charts normally use a reference velocity V_R at $0.7\,R = 0.7\frac{D}{2}$ and

$$V_R = \sqrt{Va^2 + (0.7\pi nD)^2}.$$ (12.14)

Burrill provides an empirical relationship between developed area (A_D), and projected area (A_P) as follows:

$$A_P = A_D(1.067 - 0.229\,P/D),$$ (12.15)

where P/D is the pitch ratio and

$$\text{BAR} = A_D = A_P/(1.067 - 0.229\,P/D).$$ (12.16)

Empirical relationships have been developed for the various lines on the Burrill chart, Figure 12.31, as follows:

$$\text{Line (1)} \quad \tau_c = 0.21(\sigma - 0.04)^{0.46}.$$ (12.17)

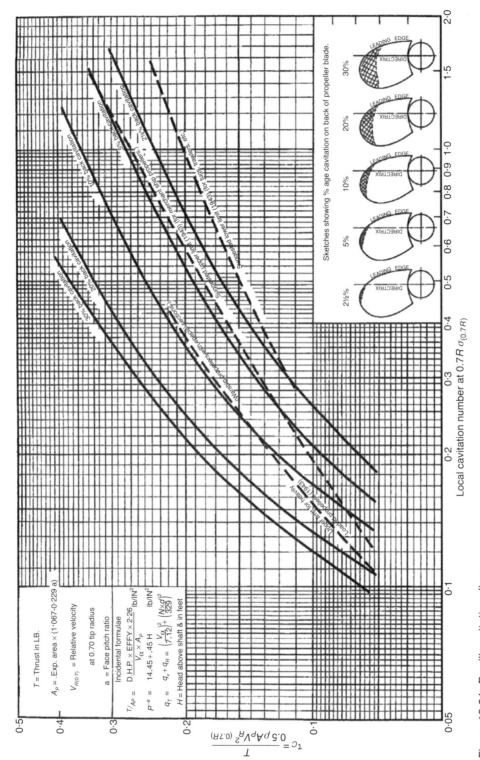

Figure 12.31. Burrill cavitation diagram.

Line (1) is a lower limit, used for heavily loaded propellers on tugs, trawlers etc.

$$\text{Line (2)} \quad \tau_c = 0.28(\sigma - 0.03)^{0.57}. \tag{12.18}$$

Line (2) is an upper limit, for merchant vessels etc., 2%–5% back cavitation, using aerofoil sections and this line is equivalent to the frequently quoted Burrill line,

$$\text{Line (3)} \quad \tau_c = 0.43(\sigma - 0.02)^{0.71}. \tag{12.19}$$

Line (3) is an upper limit used for naval vessels, fast craft etc, with 10%–15% back cavitation. This line is for uniform suction type sections and accepts a greater risk of cavitation at full power. [12.17] indicates that this line is just below the likely onset of thrust breakdown due to cavitation.

12.2.10 Example: Estimate of Blade Area

Consider a propeller with the following particulars:

$D = 4.0$ m, $P/D = 0.7$ (derived using an assumed BAR and propeller chart)
rpm $= 120$ (2.0 rps), $Va = 10$ knots ($= 5.144$ m/s)
Immersion of shaft $h = 3.0$ m
Required thrust $T = 250 \times 10^3$ N

$$V_R^2 = V_a^2 + (0.7\pi nD)^2 = 5.144^2 + (0.7\pi \times 2.0 \times 4.0)^2 = 335.97 \text{ m}^2/\text{s}^2$$
$$\sigma = (\rho gh + P_{AT} - P_V)/\tfrac{1}{2}\,\rho V_R^2$$
$$= (1025 \times 9.81 \times 3.0 + 101 \times 10^3 - 3 \times 10^3)/\tfrac{1}{2} \times 1025 \times 335.97 = 0.744.$$

Using line (2), Equation (12.18), for upper limit merchant ships, $\tau_c = 0.231$. A similar value can be determined directly from Figure 12.31. Then, $A_P = T/\tfrac{1}{2}\rho V_R^2\, \tau_c = 250 \times 10^3/\tfrac{1}{2} \times 1025 \times 335.97 \times 0.231 = 6.285$ m^2, $A_D = A_P/(1.067 - 0.229\,P/D) = 6.285/(1.067 - 0.229 \times 0.70) = 6.932$ m^2 and DAR $=$ (BAR) $= 6.932/(\pi D^2/4) = 6.932/(\pi \times 4^2/4) = 0.552$.

This would suggest the use of a B 4.55 propeller chart (BAR $= 0.550$).

If the derived P/D using the BAR $= 0.55$ chart is significantly different from the original 0.70, and the required BAR deviates significantly from the assumed BAR, then a further iteration(s) of the cavitation–blade area check may be necessary. This would be carried out using the nearest in BAR to that required.

12.3 Propeller Blade Strength Estimates

12.3.1 Background

It is normal to make propeller blades as thin as possible, in part to save expensive material and unnecessary weight and, in part, because thinner blades generally result in better performance, provided the sections are correctly chosen.

Propellers operate in a non-uniform wake flow and possibly in an unsteady flow so that blade loads are varying cyclically as the propeller rotates. Under these

Table 12.2. *Nominal propeller design
stress levels: manganese bronze*

Ship Type	Nominal mean design stress (MN/m^2)
Cargo vessels	40
Passenger vessel	41
Large naval vessels	76
Frigates/destroyers	82–89
Patrol craft	110–117

circumstances, blade failure is almost always due to fatigue, unless some accident arises (e.g. grounding) to cause loadings in excess of normal service requirements.

Two types of fatigue crack occur in practice; both originate in the blade pressure face where tensile stresses are highest. Most blades crack across the width near the boss, with the crack starting close to mid chord. Wide or skewed blades may fail by cracking inwards from the blade edge, Figure 12.32.

12.3.2 Preliminary Estimates of Blade Root Thickness

Blade design can be based on the selection of a nominal mean design stress due to the average blade loading in one revolution at steady speed, with the propeller absorbing full power. The stress level must be chosen so that stress fluctuations about this mean level do not give rise to cracking.

The normal stress level has to be chosen in relation to the following:

  (i) the degree of non-uniformity in the wake flow,
 (ii) additional loading due to ship motions,
(iii) special loadings due to backing and manoeuvring,
 (iv) the percentage of service life spent at full power,
  (v) the required propeller service life, and
 (vi) the degree of approximation of the analysis used.

In practice, the mean nominal blade stress is chosen empirically on the basis of service experience with different ship types and materials, such as those from various sources, including [12.27], quoted in Tables 12.2 and 12.3. [12.27] indicates that the allowable stresses in Table 12.3 can be increased by 10% for twin-screw vessels. The classification societies, such as [12.27], define a minimum blade thickness requirement at $0.25R$, together with blade root radius requirements.

12.3.3 Methods of Estimating Propeller Stresses

Simplified methods are available for predicting blade root stresses in which the propeller blade is treated as a simple cantilever and beam theory is applied.

Structural shell theories, using finite-element methods, may be used to predict the detailed stress distributions for the propeller blades, [12.28], [12.29], [12.30], [12.31]. These will normally be used in conjunction with computational fluid dynamics (CFD) techniques, including vortex lattice or panel methods, to determine the

Table 12.3. *Nominal propeller design stress levels for merchant ships*

Material	Nominal mean design stress (allowable) (MN/m^2)	UTS (MN/m^2)	Density (kg/m^3)
Cast iron	17	250	7200
Cast steel (low grade)	21	400	7900
Stainless steel	41	450–590	7800
Manganese bronze	39	440	8300
Nickel aluminium bronze	56	590	7600

distribution of hydrodynamic loadings on the blades. Radial cracking conditions can only be predicted by the use of such techniques. For example, vortex lattice/panel methods are required for highly skewed propellers, coupled with a finite-element stress analysis (FEA). Hydroelastic techniques [12.32] can relate the deflections from the finite-element analysis back to the CFD analysis, illustrated schematically in Figure 12.33.

Such methods provide local stresses but are computer intensive. Simple bending theories applied to the blade root section are commonly used as a final check [12.33].

12.3.4 Propeller Strength Calculations Using Simple Beam Theory

The calculation method treats the blade as a simple cantilever for which stresses can be calculated by beam theory. The method takes into account stresses due to the following:

(a) Bending moments associated with thrust and torque loading
(b) Bending moment and direct tensile loads due to centrifugal action

12.3.4.1 Bending moments due to Thrust Loading
In Figure 12.34, for a section at radius r_0, the bending moment due to thrust is as follows:

$$M_T(r_0) = \int_{r_0}^{R} (r - r_0)\frac{dT}{dr}dr. \tag{12.20}$$

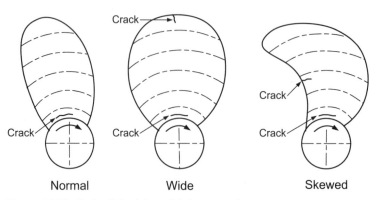

Normal   Wide   Skewed

Figure 12.32. Potential origins of fatigue cracks.

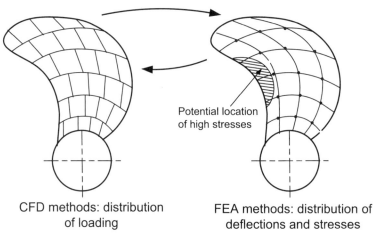

CFD methods: distribution
of loading

FEA methods: distribution of
deflections and stresses

Figure 12.33. Illustration of hydroelastic approach.

This can be rewritten as

$$M_T(r_0) = \int_{r_0}^{R} r \frac{dT}{dr} \cdot dr - r_0 T_0 \qquad (12.21)$$

$$= T_0 \bar{r} - T_0 r_0 = T_0 (\bar{r} - r_0), \qquad (12.22)$$

where T_0 is the thrust of that part of the blade outboard of r_0, and $\bar{r}$ is the centre of thrust from centreline. M_T is about an axis perpendicular to shaft centreline and blade generator.

12.3.4.2 Bending Moments due to Torque Loading
In Figure 12.35, the bending moment due to torque about an axis parallel to shaft centreline at radius r_0 is as follows:

$$M_Q = \int_{r_0}^{R} (r - r_0) \frac{dF_Q}{dr} \cdot dr$$

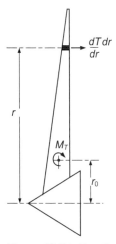

Figure 12.34. Bending moments due to thrust.

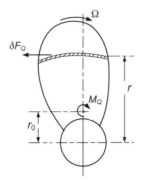

Figure 12.35. Moments due to torque loading.

but

$$\frac{dQ}{dr} = r\frac{dF_Q}{dr},$$

hence

$$M_Q = \int_{r_0}^{R} \left(1 - \frac{r_0}{r}\right)\frac{dQ}{dr} \cdot dr = Q_0 - r_0 \int_{r_0}^{R}\frac{1}{r} \cdot \frac{dQ}{dr}dr$$

$$= Q_0 - \frac{r_0}{\bar{r}}Q_0 = Q_0\left(1 - \frac{r_0}{\bar{r}}\right), \tag{12.23}$$

where Q_0 is torque due to blade outboard of r_0 and $\bar{r}$ is the centre of torque load from the centreline.

12.3.4.3 Forces and Moments due to Blade Rotation

Bending moments due to rotation arise when blades are raked, Figure 12.36. Let the tensile load $L(r)$ be the load arising due to centripetal acceleration and $A(r)$ be the

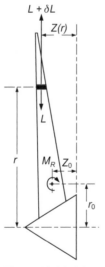

Figure 12.36. Moments due to blade rotation.

local blade cross-sectional area. The change in $L(r)$ across an element δr at radius r is given by the following:

$$\delta L = [\rho A(r)\delta r] r \,\Omega^2, \tag{12.24}$$

where ρ is the metal density, $[\rho A(r)\,\delta r]$ is the mass and

$$dL/dr = \rho \Omega^2 r A(r). \tag{12.25}$$

Since $L(r) = 0$ at the blade tip, then at $r = r_0$,

$$L(r_0) = \rho \Omega^2 \int_{r_0}^{R} r A(r) \, dr. \tag{12.26}$$

It is convenient to assume that the area A is proportional to r, and that $A(r)$ varies from $A = 0$ at the tip.

If the centre of gravity (CG) of the blade section is raked abaft the generator line by a distance $Z(r)$, then the elementary load δL from (12.24) contributes to a bending moment about the same axis as the thrust moment M_T given by

$$M_R(r_0) = \int_{r_0}^{R} [Z(r) - Z(r_0)] \frac{dL}{dr} \cdot dr = \rho \Omega^2 \int_{r_0}^{R} (Z - Z_0) r A(r) \, dr \tag{12.27}$$

or

$$M_R(r_0) = \rho \Omega^2 \int_{r_0}^{R} r Z(r) A(r) \, dr - \rho \Omega^2 Z(r_0) \int_{r_0}^{R} r A(r) dr,$$

$$M_R(r_0) = \rho \Omega^2 \left\{ \int_{r_0}^{R} r Z(r) A(r) \, dr - Z(r_0) L(r_0) \right\}$$

and

$$M_R(r_0) = \rho \Omega^2 \left\{ \int_{r_0}^{R} r Z(r) A(r) \, dr - Z(r_0) L(r_0) \right\}. \tag{12.28}$$

Equation (12.27) a can be written in a more readily useable form and, for a radius ratio $r/R = 0.2$, as follows:

$$M_{R_{0.2}} = \int_{0.2R}^{R} m(r) \cdot r \cdot \Omega^2 Z'(r) \cdot dr, \tag{12.29}$$

where Z' is $(Z - Z_0)$ and r_0 is assumed to be $0.2R$.

The centrifugal force can be written as

$$F_c = \int_{0.2R}^{R} m(r) \cdot r \cdot \Omega^2 \cdot dr, \tag{12.30}$$

where $m(r) = \rho A(r) = $ mass/unit radius.

12.3.4.4 Resolution of Bending Moments

The primary bending moments M_T, M_Q and M_R must be resolved into bending moments about the principal axes of the propeller blade section, Figure 12.37. The direction of these principal axes depends on the precise blade section shape and on

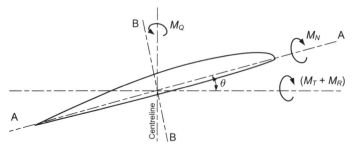

Figure 12.37. Resolution of bending moments.

the pitch angle of the section datum face at the radius r_0. Of the two principal axes shown, A–A and B–B, the section modules (I/y) is least about the axis A–A, leading to the greatest tensile stress in the middle of the blade face at P and the largest compressive stress at Q, Figure 12.38.

Applying Equation (12.1), the pitch angle is as follows:

$$\theta = \tan^{-1}\left(\frac{P/D}{\pi x}\right).$$

The significant bending moment from the blade strength point of view is thus

$$M_N = (M_T + M_R)\cos\theta + M_Q\sin\theta. \qquad (12.31)$$

This equation is used for computing the blade bending stress.

12.3.4.5 Properties of Blade Structural Section

It can be argued that the structural modulus should be obtained for a plane section A–A, Figure 12.39. In practice, cylindrical sections A′–A′ are used in defining the blade geometry and a complex drawing procedure is needed to derive plane sections.

Since pitch angles reduce as radius increases, a plane section assumes an S-shape with the nose drooping and the tail lifting, Figure 12.40.

Compared with the other approximations inherent in the simple beam theory method, the error involved in calculating the section modulus from a cylindrical section rather than a plane section is not significant. Common practice is to use cylindrical sections and to assume that the principal axis is parallel to the pitch datum line.

Typical values of I/y are as follows:

Aerofoil	$I/y = 0.095\ ct^2$.	(12.32)
Round back	$I/y = 0.112\ ct^2$.	(12.33)

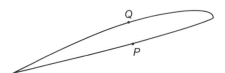

Figure 12.38. Location of largest stresses.

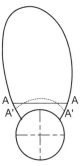

Figure 12.39. Section types.

Typical values of the area at the root are as follows:

$$A = 0.70\,ct \text{ to } 0.72\,ct, \tag{12.34}$$

where c is the chord and t is the thickness.

An approximation to the root chord ratio at $0.2R$, based on the Wageningen series, Figure 16.2 [12.34] is as follows:

$$\left(\frac{c}{D}\right)_{0.2R} = 0.416 \times \text{BAR} \times \frac{4}{Z}, \tag{12.35}$$

where Z is the number of blades.

Thickness ratio t/D at the centreline for the Wageningen series is shown in Table 12.4, together with approximate estimates of t/D at $0.2R$, $0.7R$ and $0.75R$.

Finally, the design stress $\sigma = $ direct stress $+$ bending stress, as follows:

$$\sigma = \frac{Fc}{A} + \frac{M_N}{I/y}. \tag{12.36}$$

12.3.4.6 Standard Loading Curves

Where blade element-momentum or other theoretical calculations have been performed, curves of dT/dr and dQ/dx based on these calculations may be used; see Chapter 15.

In situations where this information is not available, the following standard loading formulae provide a reasonable representation of a normal optimum load distribution [12.35]. The form of the distribution is shown in Figure 12.41.

$$\frac{dT}{dx} \quad \text{or} \quad \frac{dQ}{dx} \propto x^2\sqrt{1-x}, \tag{12.37}$$

where $x = \frac{r}{R}$.

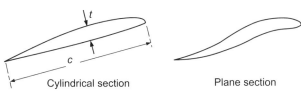

Cylindrical section   Plane section

Figure 12.40. Section shapes.

Table 12.4. *Blade thickness ratio, Wageningen series [12.34]*

Number of blades	(t/D) to centreline	$(t/D)_{0.2R}$	$(t/D)_{0.7R}$	$(t/D)_{0.75R}$
2	0.055	0.044	0.0165	0.0138
3	0.050	0.040	0.0150	0.0125
4	0.045	0.036	0.0135	0.0113
5	0.040	0.032	0.0120	0.0100

In evaluating the moments M_T and M_Q using the distribution in Equation (12.37), the following integrals are needed:

$$\int x\sqrt{1-x}\,dx = \frac{2}{15}\left(3x^2 - x - 2\right)\sqrt{1-x}. \tag{12.38}$$

$$\int x^2\sqrt{1-x}\,dx = \frac{2}{105}\left(15x^3 - 3x^2 - 4x - 8\right)\sqrt{1-x}. \tag{12.39}$$

$$\int x^3\sqrt{1-x}\,dx = \frac{2}{315}\left(35x^4 - 5x^3 - 6x^2 - 8x - 16\right)\sqrt{1-x}. \tag{12.40}$$

It may also be appropriate to assume a linear variation of blade sectional area $A(r)$ and blade rake $Z(r)$.

12.3.4.7 Propeller Strength Formulae

The following formulae are useful when using beam theory, and these may be readily inserted into Equations (12.29) and (12.30).

When the distribution of K_T and K_Q is assumed $\propto x^2\sqrt{1-x}$, then $\bar{r}$ can be derived either by numerical integration of a load distribution curve, or from Equations (12.38–12.40). When using Equations (12.38–12.40) it is found that $\bar{r}$ for thrust is $0.67R$ and $\bar{r}$ for torque is $0.57R$. Carlton [12.16] suggests values of $0.70R$ for thrust and $0.66R$ for torque, based on optimum load distributions. Based on these various values and actual load distributions, it is suggested that a value of $\bar{r} = 0.68R$ for both thrust and torque will be satisfactory for preliminary stress calculations.

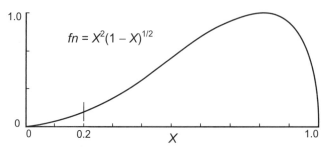

Figure 12.41. Typical spanwise load distribution.

12.3.4.8 Mass Distribution: Assumed Linear

Say M = total blade mass. Then, for an assumed boss ratio of $0.2R$,

$$M = m(r)_{0.2} \times \frac{0.8R}{2}, \qquad (12.41)$$

where $m(r)_{0.2} = (A_{0.2R} \times \rho)$ and $A_{0.2R}$ can be derived from Equation (12.34). It can then be shown that the mass distribution

$$m(r) = \frac{M}{0.32R}\left(1 - \frac{r}{R}\right) \qquad (12.42)$$

12.3.4.9 Rake Distribution: Assumed Linear

When defined from the centreline,

$$Z'(r) = \mu\left(\frac{r}{R} - 0.2\right), \qquad (12.43)$$

where μ is the tip rake to centreline, Figures 12.8 and 12.36.

12.3.4.10 High-Performance Propellers

With blades raked aft for clearance reasons, M_R and M_T are additive. Raking the blades forward reverses the sign of M_R and it is possible to use M_R to offset M_T to reduce the total bending stresses. This property is used on high-performance craft, where the amount of rake can be chosen to minimise the total bending stresses.

12.3.4.11 Accuracy of Beam Theory Strength Estimates

The beam theory strength calculation produces nominal stress values, which are an underestimate of the true blade stresses. Shell theory calculations indicate an actual stress, for a given loading, some 25% higher that the beam theory estimate. Calculations for a practical range of radial load distributions indicate that variations in loading can change estimated stress values by about 10%.

The mean stresses under the standard mean full power loading are considerably lower than the maximum blade stresses occurring in practice. For instance, trials using a strain-gauged propeller showed that, during backing and manoeuvring, a frigate propeller is subject to stresses some $3\frac{1}{2}$ times the nominal stress level, whilst the effect of non-uniform inflow into the propeller causes stress levels to vary between $\frac{1}{2}$ and $1\frac{1}{2}$ times the mean stress level. Similar stress variations in one revolution are associated with shaft inclinations to the mean flow and with ship pitch and heave motions.

The above reasons indicate why the chosen nominal design stress levels are such a small fraction of the ultimate strength of the propeller material, Table 12.3. A worked example, illustrating the estimation of propeller blade root stresses, is given in Chapter 17.

REFERENCES (CHAPTER 12)

12.1 Massey, B.S. and Ward-Smith J. *Mechanics of Fluids*. 8th Edition. Taylor and Francis, London, 2006.

12.2 Duncan, W.J., Thom, A.S. and Young, A.D. *Mechanics of Fluids*. Edward Arnold, Port Melbourne, Australia, 1974.

12.3 Noordzij, L. Some experiments on cavitation inception with propellers in the NSMB-Depressurised towing tank. *International Shipbuilding Progress*, Vol. 23, No. 265, 1976, pp. 300–306.

12.4 ITTC. Report of the Specialist Committee on Procedures for Resistance, Propulsion and Propeller Open Water Tests. Recommended procedure for Open Water Test, No. 7.5-02-03-02.1, Rev 01, 2002.

12.5 Rose, J.C. and Kruppa, F.L. Surface piercing propellers: methodical series model test results, *Proceedings of First International Conference on Fast Sea Transportation, FAST'91*, Trondheim, 1991.

12.6 Ferrando, M., Scamardella, A., Bose, N., Liu, P. and Veitch, B. Performance of a family of surface piercing propellers. *Transactions of The Royal Institution of Naval Architects*, Vol. 144, 2002, pp. 63–75.

12.7 ITTC 2005 Report of Specialist Comitteee on Azimuthing Podded Propulsors. *Proceedings of 24th ITTC*, Vol. II. Edinburgh, 2005.

12.8 ITTC Report of Specialist Comitteee on Azimuthing Podded Propulsors. *Proceedings of 25th ITTC*, Vol. II. Fukuoka, 2008.

12.9 ITTC Report of Specialist Committee on Validation of Waterjet Test Procedures. *Proceedings of 23rd ITTC*, Vol. II, Venice, 2002.

12.10 ITTC Report of Specialist Committee on Validation of Waterjet Test Procedures. *Proceedings of 24th ITTC*, Vol. II, Edinburgh, 2005.

12.11 Van Manen, J.D. Results of systematic tests with vertical axis propellers. *International Shipbuilding Progress*, Vol. 13, 1966.

12.12 Van Manen, J.D. Non-conventional propulsion devices. *International Shipbuilding Progress*, Vol. 20, No. 226, June 1973, pp. 173–193.

12.13 Volpich, H. and Bridge, I.C. Paddle wheels. Part I, Preliminary model experiments. *Transactions, Institute of Engineers and Shipbuilders in Scotland*, Vol. 98, 1954–1955, pp. 327–380.

12.14 Volpich, H. and Bridge, I.C. Paddle wheels. Part II, Systematic model experiments. *Transactions, Institute of Engineers and Shipbuilders in Scotland*, Vol. 99, 1955–1956, pp. 467–510.

12.15 Volpich, H. and Bridge, I.C. Paddle wheels. Parts IIa, III, Further model experiments and ship model correlation. *Transactions, Institute of Engineers and Shipbuilders in Scotland*, Vol. 100, 1956–1957, pp. 505–550.

12.16 Carlton, J. S. *Marine Propellers and Propulsion*. 2nd Edition. Butterworth-Heinemann, Oxford, UK, 2007.

12.17 Gawn, R.W.L and Burrill, L.C. Effect of cavitation on the performance of a series of 16 in. model propellers. *Transactions of the Royal Institution of Naval Architects*, Vol. 99, 1957, pp. 690–728.

12.18 Burrill, L.C. and Emerson, A. Propeller cavitation: Further tests on 16in. propeller models in the King's College cavitation tunnel. *Transactions North East Coast Institution of Engineers and Shipbuilders*, Vol. 79, 1962–1963, pp. 295–320.

12.19 Emerson, A. and Sinclair, L. Propeller cavitation. Systematic series of tests on five and six bladed model propellers. *Transactions of the Society of Naval Architects and Marine Engineers*, Vol. 75, 1967, pp. 224–267.

12.20 Emerson, A. and Sinclair, L. Propeller design and model experiments. *Transactions North East Coast Institution of Engineers and Shipbuilders*, Vol. 94, No. 6, 1978, pp. 199–234.

12.21 Van Manen, J.D. The choice of propeller. *Marine Technology*, SNAME, Vol. 3, No. 2, April 1966, pp. 158–171.

12.22 Szantyr, J.A. A new method for the analysis of unsteady propeller cavitation and hull surface pressures. *Transactions of the Royal Institution of Naval Architects*, Vol. 127, 1985, pp. 153–167.

12.23 ITTC. Report of Specialist Comitteee on Cavitation. *Proceedings of 25th ITTC*, Vol. II. Fukuoka, 2008.

12.24 Molland, A.F., Bahaj, A.S., Chaplin, J.R. and Batten, W.M.J. Measurements and predictions of forces, pressures and cavitation on 2-D sections sutable for marine current turbines. *Proceedings of Institution of Mechanical Engineers*, Vol. 218, Part M, 2004.

12.25 Drela, M. Xfoil: an analysis and design system for low Reynolds number aerofoils. *Conference on Low Reynolds Number Airfoil Aerodynamics*, University of Notre Dame, Notre Dame, IN, 1989.

12.26 English, J.W. Propeller skew as a means of improving cavitation performance. *Transactions of the Royal Institution of Naval Architects*, Vol. 137, 1995, pp. 53–70.

12.27 Lloyds Register. Rules and Regulations for the Classification of Ships. Part 5, Chapter 7, Propellers. 2005.

12.28 Conolly, J.E. Strength of propellers. *Transactions of the Royal Institution of Naval Architects*, Vol. 103, 1961, pp. 139–160.

12.29 Atkinson, P. The prediction of marine propeller distortion and stresses using a superparametric thick-shell finite-element method. *Transactions of the Royal Institution of Naval Architects*, Vol. 115, 1973, pp. 359–375.

12.30 Atkinson, P. A practical stress analysis procedure for marine propellers using finite elements. *75 Propeller Symposium*. SNAME, 1975.

12.31 Praefke, E. On the strength of highly skewed propellers. *Propellers/Shafting'91 Symposium*, SNAME, Virginia Beach, VA, 1991.

12.32 Atkinson, P and Glover, E. Propeller hydroelastic effects. *Propeller'88 Symposium*. SNAME, 1988, pp. 21.1–21.10.

12.33 Atkinson, P. On the choice of method for the calculation of stress in marine propellers. *Transactions of the Royal Institution of Naval Architects*, Vol. 110, 1968, pp. 447–463.

12.34 Van Lammeren, W.P.A., Van Manen, J.D. and Oosterveld, M.W.C. The Wageningen B-screw series. *Transactions of the Society of Naval Architects and Marine Engineers*, Vol. 77, 1969, pp. 269–317.

12.35 Schoenherr, K.E. Formulation of propeller blade strength. *Transactions of the Society of Naval Architects and Marine Engineers*, Vol. 71, 1963, pp. 81–119.

13 Powering Process

13.1 Selection of Marine Propulsion Machinery

The selection of propulsion machinery and plant layout will depend on design features such as space, weight and noise levels, together with overall requirements including areas of operation, running costs and maintenance. All of these factors will depend on the ship type, its function and operational patterns.

13.1.1 Selection of Machinery: Main Factors to Consider

1. Compactness and weight: Extra deadweight and space. Height may be important in ships such as ferries and offshore supply vessels which require long clear decks.
2. Initial cost.
3. Fuel consumption: Influence on running costs and bunker capacity (deadweight and space).
4. Grade of fuel (lower grade/higher viscosity, cheaper).
5. Level of emission of NOx, SOx and CO_2.
6. Noise and vibration levels: Becoming increasingly important.
7. Maintenance requirements/costs, costs of spares.
8. Rotational speed: Lower propeller speed plus larger diameter generally leads to increased efficiency.

Figure 13.1 shows a summary of the principal options for propulsion machinery arrangements and the following notes provide some detailed comments on the various propulsion plants.

13.1.2 Propulsion Plants Available

13.1.2.1 Steam Turbines

1. Relatively heavy installation including boilers. Relatively high fuel consumption but can use the lowest grade fuels.
2. Limited marine applications, but include nuclear submarines and gas carriers where the boilers may be fuelled by the boil-off from the cargo.

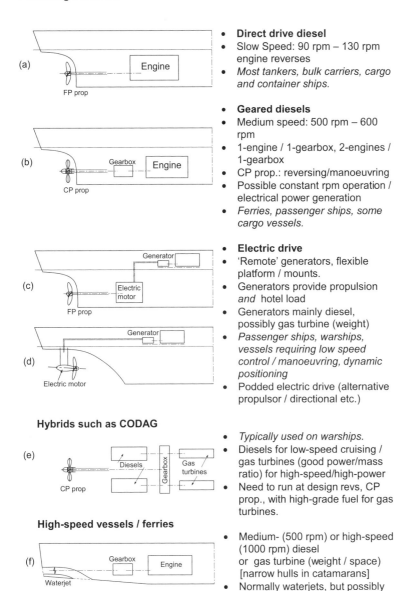

- **Direct drive diesel**
- Slow Speed: 90 rpm – 130 rpm engine reverses
- *Most tankers, bulk carriers, cargo and container ships.*

- **Geared diesels**
- Medium speed: 500 rpm – 600 rpm
- 1-engine / 1-gearbox, 2-engines / 1-gearbox
- CP prop.: reversing/manoeuvring
- Possible constant rpm operation / electrical power generation
- *Ferries, passenger ships, some cargo vessels.*

- **Electric drive**
- 'Remote' generators, flexible platform / mounts.
- Generators provide propulsion *and* hotel load
- Generators mainly diesel, possibly gas turbine (weight)
- *Passenger ships, warships, vessels requiring low speed control / manoeuvring, dynamic positioning*
- Podded electric drive (alternative propulsor / directional etc.)

Hybrids such as CODAG

- *Typically used on warships.*
- Diesels for low-speed cruising / gas turbines (good power/mass ratio) for high-speed/high-power
- Need to run at design revs, CP prop., with high-grade fuel for gas turbines.

High-speed vessels / ferries

- Medium- (500 rpm) or high-speed (1000 rpm) diesel or gas turbine (weight / space) [narrow hulls in catamarans]
- Normally waterjets, but possibly surface piercing propellers

Figure 13.1. Propulsion machinery layouts: Principal options.

13.1.2.2 Gas Turbines

1. Good power/weight ratio.
2. Use where lightness, compactness and operating flexibility (e.g. fast start-up) are important.
3. Need for high-grade fuels (expensive) and prefer to run at design revs; hence, there is generally a need for a controllable pitch propeller (or the use of a waterjet).
4. Little application to merchant/cargo ships, but applicable to warships and, more recently, to high-speed passenger/car ferries, Figure 13.1(f). Also used to generate electricity for ship services and propulsion.

Further details concerning gas turbines may be found in [13.1 and 13.2].

13.1.2.3 Diesel Engines

Diesel engines are by far the most popular form of installation for merchant ships. They have now been developed to burn heavy fuels, and engine efficiency has improved significantly over the past 20 years. Diesel engines can be divided into *slow speed* (90–130 rpm), *medium speed* (400–600 rpm) and *high speed* (1000–1800 rpm).

The following summarises the merits of the various diesels. More detailed accounts may be found in [13.1–13.4].

SLOW-SPEED DIESELS. Advantages/disadvantages

1. Fewer cylinders/low maintenance.
2. Use of lower quality fuels possible.
3. Possibility of direct drive: no gearbox, simple, reliable.
4. No gearbox losses (2%–5%).
5. Low engine revs implies low noise levels.
6. In general, heavier than medium-speed diesels, and propeller reversal for direct drive requires engine reversal.

MEDIUM-SPEED DIESELS. Advantages/disadvantages

1. Smaller, lighter, less height.
2. Choice of optimum propeller revs using a gearbox.
3. In the main, cheaper than slow-speed direct drive (even including gearbox) because of larger production for extensive land use applications.
4. Engines installed in ship in one piece, with less chance of faults or incursion of dirt.
5. Possibility of driving generator from power-take-off (PTO) shafts from gearbox.
6. Possibility of multi-engined plants: reliability, maintenance of one engine when under way, use of less than full number of engines when slow steaming.
7. Spares lighter: can be sent by air freight.
8. Better able to cope with slow-speed running.

HIGH-SPEED DIESELS. These are generally lighter than medium-speed engines, otherwise their merits are similar to the medium-speed engine.

In general, the differences between slow- and medium-speed diesels are decreasing with the choice tending to depend on application. Slow-speed/direct-drive plants are very popular for cargo ships, container ships, bulk carriers and tankers, etc., Figure 13.1(a). Geared medium-speed diesels are used for ferries, tugs, trawlers, support vessels and some passenger ships, Figure 13.1(b). High-speed engines find applications in the propulsion of fast naval craft and fast ferries. It should be noted that the use of combinations such as gas/diesel, Figure 13.1(e), and gas/gas, may be used on warships and large fast passenger/car ferries for increased flexibility and economy of operation.

13.1.2.4 Diesel (or Turbo) Electric

Diesel or gas turbine generated electricity with electric propulsor(s), Figures 13.1(c) and (d).

The merits of electric propulsion include the following:

1. Flexibility of layout, for example, the generating station can be separate from the propulsion motor(s).
2. Load diversity between ship service load and propulsion (hence, popularity for passenger ships and recent warships).
3. Economical part load running; a fixed-pitch propeller is feasible, as an electric propulsion motor can provide high torque at low revolutions.
4. Ease of control.
5. Low noise and vibration characteristics, e.g. diesel generators can be flexibly mounted/rafted, and
6. Electric podded drives are also now becoming popular, Figure 13.1(d).

Regarding electric drives, it is generally accepted that there will be some overall increase in propulsion machinery mass and some decrease in transmission efficiency between engine and propeller, that is, diesel to electricity to propeller; instead, of diesel directly to propeller.

However, the above attributes have led to the increasing use of electric drive on many passenger cruise ships, warships and other vessels such as cable and survey ships where control and dynamic positioning is important.

13.1.3 Propulsion Layouts

The principal options for propulsion machinery arrangements are summarised in Figure 13.1. It is clear that there are several alternative arrangements, but the various options are generally applied to a particular ship or group of ships types. Further detailed propulsion machinery layouts are given by Gallin *et al.* [13.5].

13.2 Propeller–Engine Matching

13.2.1 Introduction

It is important to match the propeller revolutions, torque and developed power to the safe operating limits of the installed propulsion engine. Typical power, torque and revolutions limits for a diesel engine are shown in Figure 13.2, within ABCDE.

It should be noted that the propeller pitch determines at what revolutions the propeller, and hence engine, will run. Consequently, the propeller design (pitch and revs) must be such that it is suitably matched to the installed engine.

Figure 13.3 is indicative of a typical engine-propeller matching chart. Typically it is assumed that the operator will not run the engine to higher than 90% of its continuous service rating (CSR), Figure 13.3. The marine diesel engine characteristics are usually based on the 'propeller law' which assumes $P \propto N^3$, which is acceptable for most displacement ships. The N^3 basic design power curve passes through the point A. When designing the propeller for say clean hull and calm water, it is usual to keep the actual propeller curve to the right of the engine (N^3) line, such as on line [1], to allow for the effects of future fouling and bad weather. In the case of line [1] the pitch is said to be light and if the design pitch is decreased further, the line will move to line [1a], etc.

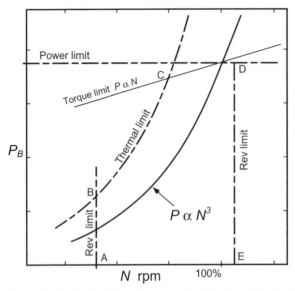

Figure 13.2. Typical diesel engine limits (within ABCDE).

As the ship fouls, or encounters heavy weather, the design curve [1] will move to the left, first to the N^3 line and then say to line [2]. It must be noted that, in the case of line [2], the maximum available power at B is now not available due to the torque limit, and the maximum operating point is at C, with a consequent decrease in power and, hence, ship speed.

Note also the upper rpm limit on the (design) line [1], say point D at 105% maximum rpm. At this point the full available power will not be absorbed in the 'clean hull-calm water' trials, and the full 'design' speed will not be achieved. In a similar manner, if the ship has a light load, such as in ballast, the design curve [1] will move to the right (e.g. towards line [1a]) and, again, the full power will not be available

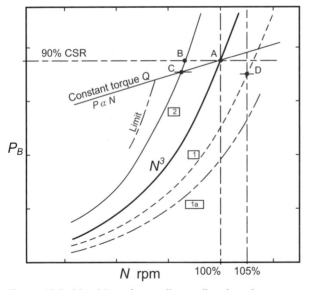

Figure 13.3. Matching of propeller to diesel engine.

due to the rpm limit and speed will be curtailed. These features must be allowed for when drawing up contractual ship design speeds (load and ballast) and trial speeds.

The basic assumption is made that $P \propto N^3$. $P \propto N^3$ is defined by the engine manufacturers as the 'Propeller Law' and they design their engines for best efficiency (e.g. fuel consumption) about this line. It does not necessarily mean that the propeller actually operates on this line, as discussed earlier.

Now $P = 2\pi n Q$ and, for constant torque (e.g. maximum torque), $P \propto n$, Figure 13.2. If it is assumed that $P(= 2\pi n Q) \propto n^3$, then $Q \propto n^2$, which implies that K_Q is constant ($K_Q = Q/\rho n^2 D^5$), $\therefore$ J is constant ($J = V/nD$); hence, if J is constant, K_T is constant ($K_T = T/\pi n^2 D^4$). If K_T is constant, then $T \propto n^2$, but if J is constant, $n^2 \propto V^2$ and $T \propto V^2$ (or $R \propto V^2$ for constant thrust deduction, t). $R \propto V^2$ is a reasonable assumption in the normal speed range for displacement craft, and, hence, $P \propto N^3$ is a reasonable assumption. It should be noted, however, that the speed index does change with different craft (e.g. high-speed craft) or if a displacement craft is overdriven, in which case $P \propto N^x$, where x will not necessarily be 3, although it will generally lie between 2.5 and 3.5.

13.2.2 Controllable Pitch Propeller (CP Propeller)

Using a controllable pitch propeller is equivalent to fitting an infinitely variable gearbox between the engine and the propeller, resulting in a range of curves for different pitch ratios, Figure 13.4.

Whilst the fixed-pitch propeller imposes a fixed relationship between revolutions and torque, say the 100% pitch line in Figure 13.4, the CP propeller gives independence between these two variables. For example, the different load requirements in Figure 13.3, curves [1], [1a] and [2], could be met with the use of a controllable pitch propeller. Typical vessels that exploit the merits of the CP propeller include tugs and trawlers where line [1] might represent the free-running condition, whilst line [2] would represent a tug towing or a trawler trawling. A decrease in

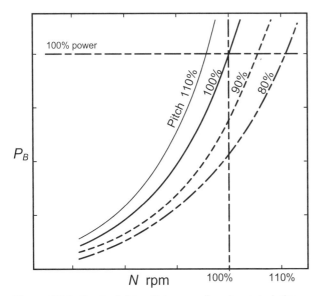

Figure 13.4. Controllable pitch propeller characteristics.

propeller pitch (with a CP propeller) will move line [2] on to the N^3 line, whilst an increase in pitch can be used to move line [1] to the N^3 line.

It should be noted that a particular pitch will be chosen as the 'design' pitch for a CP propeller, on which will be based the design calculations and blade pitch distribution, etc. When the pitch is moved away from the 'design' condition, the blades are in fact working a little off-design, but generally without significant differences in comparison with if the blades had been redesigned at each pitch ratio.

A further possible mode of operation is to run the CP propeller at constant revolutions. For example, a power take-off at constant revolutions can be used to drive a generator. This can, however, lead to inefficient operation of the engine, in particular, at reduced power, since the engine will have been designed to operate efficiently (minimum specific fuel consumption *sfc*) around the N^3 line.

Further applications of the CP propeller include ferries, offshore supply vessels and warships when good manoeuvrability or station keeping is required and when a quick response to a 'crash astern' order is required and the risk of such an order is high.

13.2.3 The Multi-Engined Plant

A multi-engine plant, Figure 13.1(b), with two medium-speed engines geared to a single propeller, offers gains in operational safety and flexibility. Some limitations in operation do, however, have to be noted. Figure 13.5 shows the case of the propeller

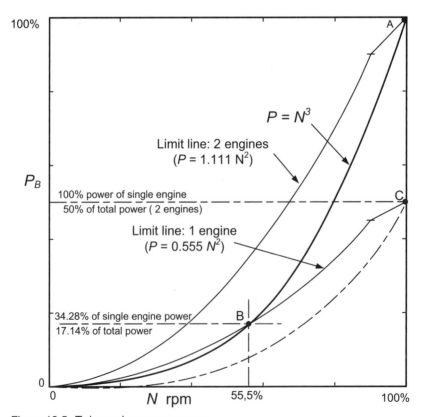

Figure 13.5. Twin-engine power curves.

designed to absorb 100% power with two engines at 100% rpm. The maximum constant torque is assumed to extend from 90% to 100% rpm with $P \propto N$, and from 0% to 90% rpm with $Q \propto N$ and $P \propto N^2$.

If one engine is disconnected, the limit line for the remaining engine cuts the propeller curve at 55.5% of full rpm, at point B, and develops only 34.28% of its full power (or 17.14% of the original combined output). If the propeller had been designed to absorb say 85% of the combined power, then the remaining engine would operate at 65.4% rpm and 47.7% power (23.7% of total). These are still very low figures and represent less than half of the available power.

If a considerable time is to be spent running on one engine then a CP propeller becomes attractive, allowing the single engine to run up to its full power and rpm at position C in Figure 13.5. An alternative would be to use a two-speed gearbox, which would also allow the single engine to run up to its full power and rpm.

The foregoing discussion is also applicable to the off-loading of a propulsion engine when some form of auxiliary propulsive power is present, such as using wind power including sails (motor sailing), wing sails, kites and wind turbines (Molland and Hawksley [13.6]).

13.3 Propeller Off-Design Performance

13.3.1 Background

It is frequently required to evaluate propeller performance at conditions other than those for which the propeller has been designed. In this case the propeller characteristics (such as diameter and pitch ratio) are already fixed and the variables are V, N, T and Q. Some examples are as follows:

Design	Performance
Tug towing	Tug free-running
Trawler free-running	Trawler trawling
Tanker/bulk carrier loaded	Tanker/bulk carrier in ballast
Service speed–load condition	Trials (or service) at light displacement
	Overload due to weather (at same speed)
	Off-loaded propeller due to auxiliary power such as wind (sails, kites, rotors)
	Estimation of ship acceleration or stopping performance

It is necessary to distinguish between the torque (or power) absorbed by the propeller at a given condition and the maximum torque (or power) that can be delivered by the propelling machinery, which depends on the type of machinery installed. Clearly, Q absorbed $\leq Q$ delivered (max), and the maximum running propeller revolutions will be such that the two are equal. At speeds less than maximum, the engine throttles must be closed such that Q delivered $= Q$ absorbed ($< Q$max). For example, the off-design propeller torque for a range of speeds can be estimated and then matched to the available (max) engine torque.

In the absence of a manufacturer's performance curves for the engine it is commonly assumed that diesel engines have constant *maximum* Q independent of n, i.e.

$P = 2\pi nQ$, and $P \propto n$ (Figure 13.2), whilst steam turbines have constant *maximum* P independent of n, i.e. $Q \propto 1/n$. Gas turbines are considerably less flexible than steam plant or diesel engines and should be run at or close to the design rpm, thus indicating the need for a CP propeller in this case or the use of fluid or electric power transmissions.

13.3.2 Off-Design Cases: Examples

These should be considered in association with the propeller design example calculations in Chapter 16.

13.3.2.1 Case 1: Speed Less Than Design Speed

Assume that D and P/D are fixed by some design condition. The ship is travelling at less than design speed and it is required to find the new delivered power and propeller revolutions.

Va and P_E are known for the new speed; hence, also T from an estimate of t, i.e.

$$P_E = R \cdot V_s, \; T = R/(1-t) = P_E/Vs/(1-t).$$

$Va = Vs(1 - w_T)$ and data for wake fraction w_T and thrust deduction factor t are given in Chapter 8.

Hence, at new speed, assume a range of revolutions

$$J = Va/nD = f(1/n).$$

$$
\begin{array}{cccc}
\text{Assumed} & & \text{Estimated} & \\
n \rightarrow & J \rightarrow & K_T \rightarrow & T'(= K_T \rho\, n^2 D^4) \\
- & - & - & - \\
- & - & - & - \quad \text{hence, } n \text{ when } T' = T \\
- & - & - & - \\
\end{array}
$$

for given n, J is calculated and K_T is read from the $K_T - K_Q$ chart at given (fixed) P/D (T' is the thrust provided by the propeller, T is the thrust required by the hull). Knowing n for the required T ($= T'$), recalculate J, hence, K_Q from the chart for given (fixed) P/D. Hence,

$$P_D = 2\pi nQ/\eta_R = 2\pi n[K_Q\rho n^2 D^5]/\eta_R \quad \text{and} \quad N = n \times 60.$$

13.3.2.2 Case 2: Increase in Resistance at Same Speed

Assume that D and P/D are fixed by some design condition. Find the power and rpm for say 20% increase in resistance (hence, T for const t) at the same speed.

Repeat Case 1 since T and Va are known. Note that in the case of overload such as this (e.g. increase in resistance at constant speed) torque will rise and care must be taken that maximum engine torque (e.g. in case of diesels and shafting) is not exceeded, i.e. $P_D = 2\pi nQ$ and torque Q for new condition $= P_D/2\pi n$.

If the maximum torque is exceeded, the throttle is closed and the ship speed decreases. A more rigorous approach, taking account of torque limits, is described in Case 3.

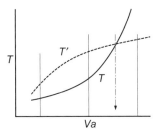

Figure 13.6. Thrust matching, noting that T' is the thrust provided by the propeller and T is the thrust required by the hull.

13.3.2.3 Case 3: Increase in Resistance at Same Speed with Torque Limit

Assume that D and P/D are fixed by some design condition. Find the power, rpm and speed for say 50% increase in resistance, hence, T. The maximum torque Qm of the diesel engine must not be exceeded. It is assumed that the thrust curve (hence, $T + 50\%$) is known over a range of speeds.

First, assume a range of speeds, say Va_1, Va_2, Va_3

For Va_1

$$
\begin{array}{ll}
\text{Assume} & \text{Estimated} \\
n \to J \;\; \to K_Q \to Q(= K_Q \rho n^2 D^5) \; K_T \to T'(= K_T \rho n^2 D^4) \\
\text{—} \quad \text{—} \quad\quad \text{—} \quad\quad \text{—} \quad\quad\quad\quad \text{—} \quad \text{—} \\
\text{—} \quad \text{—} \quad\quad \text{—} \quad\quad \text{—} \quad\quad\quad\quad \text{—} \quad \text{—} \\
\text{—} \quad \text{—} \quad\quad \text{—} \quad\quad \text{—} \quad\quad\quad\quad \text{—} \quad \text{—}
\end{array}
$$

for given n, J is calculated and K_Q read from the chart at given (fixed) P/D. rpm n increased until $Q = Qm$; hence, rpm and T' for maximum torque (Qm).

Repeat for Va_2, Va_3; hence, speed is derived at which T' matches T from the cross plot, Figure 13.6.

13.3.2.4 Case 4: Diesel Tug Maximum Thrust When Towing

Assume that propeller/engine is restricted to a maximum torque Qm.

Repeat Case 3, but only for the one towing speed; hence, maximum *total* thrust T at that speed.

$$\text{Towing thrust (pull) available} = [total\ T \times (1 - t)] - \text{tug hull resistance}$$

where $[total\ T \times (1 - t)]$ is the effective thrust T_E. Values of thrust deduction factor t for the low and zero speed (bollard) conditions are included in Chapter 8.

As discussed in Section 13.2, in order for a diesel tug (or trawler) *to develop full power when towing* (or trawling) as well as when free-running, a CP propeller or a two-speed gearbox would be required in order to increase the engine rpms/decrease torque and develop full power.

13.3.2.5 Summary

Example applications of these various off-design cases are included in Chapter 17.

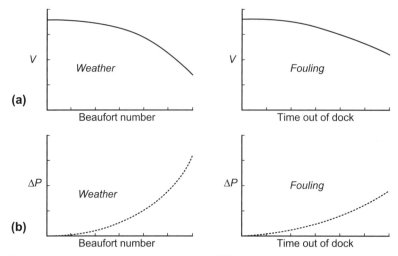

Figure 13.7. Presentation of data derived from voyage analysis.

13.4 Voyage Analysis and In-service Monitoring

13.4.1 Background

Voyage analysis entails the logging and analysis of technical data such as speed, power, propeller revolutions, displacement and weather conditions during the course of normal ship operation. The analysis is aimed at assessing the influence on propulsive power of hull roughness, fouling and wind and waves. Examples of the results of voyage analysis investigations are described in Section 3.2.3 for fouling and in Section 3.2.4 for weather.

A continuous comprehensive analysis of voyage data of ships in operation can indicate the operating efficiency and opportunities for improvement of existing ships, and lead to possible improvements which should be incorporated in future new designs.

The analysis should be designed to include quantitative assessment of the power and fuel consumption variations with speed, fouling and weather. The results of the analysis should indicate the following:

(i) The effect of weather on speed, power and fuel consumption
(ii) The effect of fouling on power, speed, fuel consumption and propeller efficiency, relative say to a time out of dock

The effects of (i) and (ii) may be presented as a speed loss for constant power, Figure 13.7(a), or the power augment required to maintain a constant speed, Figure 13.7(b), that is the cost of maintaining a scheduled speed in all weathers and hull conditions and the amount of reserve power to be installed in future tonnage, see Section 3.2.5.

A knowledge of speed loss due to weather for different ships is also necessary if weather routeing is to be employed, since routeing around a rough weather area can be economic for a ship with a high speed loss per unit wave height, or Beaufort number, and uneconomic for one with a small speed loss.

The effect of fouling will help to assess such items as the best hull finish and protection and the most economically favourable frequency of hull and propeller cleaning and docking.

The basic objectives and methodologies for voyage data analysis are described in [13.7, 13.8 and 13.9]; the basic requirements of such analyses have changed little over the years. Much pioneering work on the applications of such techniques was carried out by Aertssen [13.10, 13.11 and 13.12]. Further useful developments of the techniques are described in [13.13, 13.14]. Carlton [13.15] makes a wide-ranging review of voyage data analysis and in-service monitoring.

13.4.2 Data Required and Methods of Obtaining Data

Speed: shipborne log; noon to noon ground speed readings, sextant or GPS.

Power: measured by torsionmeter attached to shaft and $P = 2\pi n Q$, average per watch per day, or measure indicated power using BMEP, [13.1].

Revolutions: average per watch per day.

Fuel: average consumption per watch per day.

Weather: defined by wind speed (or force) and direction, and wave height and direction. Wind force and direction can be measured by instruments mounted high on the ship, or estimated by ship deck personnel as for the deck log. Wave height and direction are ideally measured by wave-recording buoys or shipborne wave recorder. It is generally impractical to measure wave properties during a normal voyage, and it is usual to assume that waves are a function of wind force and direction. Readings are normally taken each watch and averaged for day.

Fouling: time out of dock (days or weeks) recorded as an indirect measure of the deterioration of the ship's hull and propeller.

Displacement/trim: daily displacement to be recorded, normally estimated from departure displacement minus consumed fuel, stores, etc.

Normally, only whole days will be used in the analysis. Days of fog or machinery trouble will be excluded. Days during which large variations in speed, power or revolutions occur will be discarded. Draughts should be limited to a certain range, for example, between full load and 0.8 full load. Alternatively, data can be corrected to some mean draught. Data can be recorded by personnel on standard forms. Automatic data logging is likely to be employed for power, fuel, revolutions and speed. Erroneous data still have to be discarded when using this method. The data may be transmitted ashore continuously for analysis by shore-based staff, Carlton [13.15].

13.4.3 Methods of Analysis

13.4.3.1 Speed as a Base Parameter

Speed: corrected for temperature, for example, using a Reynolds number C_F type correction, as described for tank tests in Section 3.1.4.

Revolutions: corrected to some mean, using trial power/rpm relationship.

Power: corrected to some mean power (say mean for voyage) using trial power–speed curve, or assuming power to vary as V^3.

Displacement: corrected using $\Delta^{2/3}$ ratios, or tank data if available, to correct to a mean displacement.

The corrections to P_1, N_1 and Δ_1 to some standard Δ_2 and new P_2 and N_2 may be approximated as

$$P_2 = \left[\frac{\Delta_2}{\Delta_1}\right]^{2/3} P_1 \qquad (13.1)$$

and

$$N_2 = N_1 \left[\frac{P_2}{P_1}\right]^{1/3} = N_1 \left[\frac{\Delta_2}{\Delta_1}\right]^{2/9}. \qquad (13.2)$$

Effect of weather: resulting speeds can be plotted to a base of Beaufort number for ahead or cross winds.

Effect of fouling: fine weather results (Beaufort number <2.5 say) used, and plotted on a time base.

The method is 'graphical' and depends on fair curves being drawn through plotted data.

13.4.3.2 Admiralty Coefficient A_C

The Admiralty coefficient is defined as follows:

$$A_C = \frac{\Delta^{2/3} V^3}{P}. \qquad (13.3)$$

The Admiralty coefficient can be plotted to a base of time (fine weather results for fouling effects), or Beaufort number (for weather effects). Mean voyage values give a mean power increase due to fouling or weather, Figure 13.8.

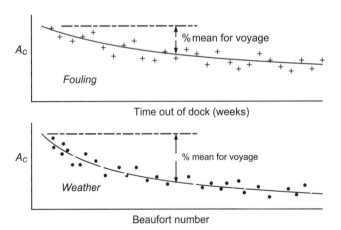

Figure 13.8. Influence of fouling and weather on Admiralty coefficient.

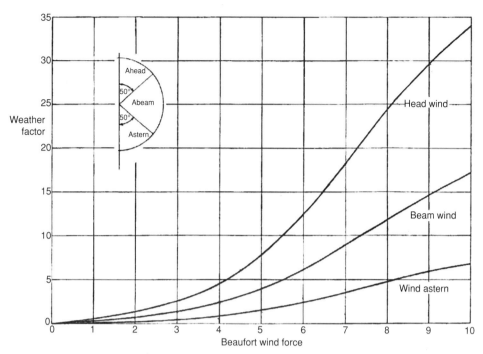

Figure 13.9. Weather factor for different headings [13.8].

If measurements of power are not available, then a fuel coefficient (F_C) may be used, as follows:

$$F_C = \frac{\Delta^{2/3} V^3}{F},$$ (13.4)

where F is the fuel consumption (tonnes) per 24 hours. This criterion has often been used by shipping companies as an overall measure of the effects of changes in engine efficiency, fouling and weather.

It should be noted that the uses of Admiralty or fuel coefficients are best confined to the derivation of trends and approximate margins. The statistical methods described in the next section place more control on the manipulation of the data and the outcomes.

13.4.3.3 Statistical Methods

If hull roughness and fouling are assumed to be a function of time out of dock, and the weather to be described as a weather factor W, then the analyses of the data might entail multiple regression analyses of the following type:

$$\frac{\Delta P}{P} = a T_D + b W + c,$$ (13.5)

where $\frac{\Delta P}{P}$ is the increase in power, T_D is the time out of dock and W is a weather factor, based say on four quadrants with a separate weighting for each quadrant, Figure 13.9. Power will be corrected to some mean displacement, a standard speed and mean revolutions, as described in the previous sections.

Calm water data can be used to determine the effects of roughness and fouling on power, based on time out of dock, and early clean smooth hull data can be used

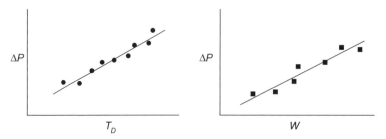

Figure 13.10. Influence on power of time out of dock (T_D) and weather (W).

to determine the effects of weather on power, based on a weather factor, as shown in Figure 13.10.

An alternative approach is to use the following:

$$\frac{P}{N^3} = a\,T_D + bW + c. \tag{13.6}$$

P/N^3 may not vary linearly with T_D, in which case an equation of the form of Equation (13.7) might be suitable, as follows:

$$\frac{P}{N^3} = a\,T_D^2 + bT_D + cW + d. \tag{13.7}$$

A further analysis of the data can include the effects of apparent slip as follows:

$$\frac{P}{N^3} = a_1 S_a + b_1, \tag{13.8}$$

where S_a is the apparent slip defined as

$$S_a = \frac{P_P N - V_S}{P_P N} = 1 - \frac{V_S}{P_P N}, \tag{13.9}$$

and P_P is the propeller pitch (m), N the revolutions/sec (rps) and V_S is the ship speed (m/s). It should be noted that this is the *apparent* slip, based on ship speed V_S, and not the *'true'* slip which is different and is based on the wake speed, Va.

Use of Equations (13.8) and (13.9) yields the increase in power due to fouling [13.8]. Manipulation of the formulae will also yield the loss of speed due to weather. Burrill [13.9] describes how the wake fraction may be derived from changes in P/N^3, Figure 13.11.

13.4.4 Limitations in Methods of Logging and Data Available

Speed: measurement of speed through the water using shipborne logs can be inaccurate because of effects such as the influence of the boundary layer or ship motions. Average speed over the ground can be measured using Global Positioning System (GPS), but this does not take account of ocean currents.

Power: torsionmeters require relatively frequent calibration. Many ships are not fitted with a torsionmeter, in which case indirect methods may be adopted such as the use of Brake Mean Effective Pressure (BMEP) [13.1] or of fuel consumption as the objective criterion.

Weather: ideally this should entail a measure of the mean wave height and direction *and* wind speed and direction. Wave recording buoys or a shipborne

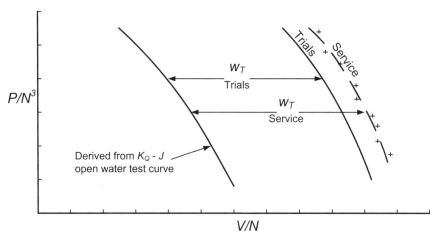

Figure 13.11. Derivation of service wake fraction.

wave recorder are not normally feasible for continuous assessment during a normal ship voyage. It is therefore possible, and often assumed, that a single weather scale can be obtained for a relative wind speed and direction, and a sea scale assumed proportional to this. This is not an unreasonable assumption since the sea disturbances are generated by the wind.

Varying draught (displacement) and trim: data have to be corrected to some mean draught by alternative methods. Alternatively, data might be limited to within certain draught limitations.

Data quality: when considering the various methods of obtaining the data for voyage analysis purposes, it is inevitable that the analysed data will show a high degree of scatter.

13.4.5 Developments in Voyage Analysis

Carlton [13.15] reviews, in some detail, the work of Townsin *et al.* [13.16], Whipps [13.17] and Bazari [13.18]. Townsin *et al.* establish a methodology for analysing the effects of roughness on the hull and propeller, Whipps develops coefficients of performance for the engine and navigation areas and Bazari applies energy-auditing principles to ship operation and design. Carlton goes on to discuss on-line data acquisition systems, continuous monitoring, integrated ship management and ship-to-shore data transmission.

13.4.6 Further Data Monitoring and Logging

Extensive engine and component condition monitoring now takes place on most ships. This may be considered as a separate process to that described for obtaining voyage data, although properties such as rpm and BMEP are complementary and will be recorded in any standard monitoring process.

A number of large ships such as bulk carriers monitor stresses in potentially high-stress areas in the hull structure. Some ships monitor motions, accelerations and slamming pressures, although such measurements are normally confined to seakeeping trials and research purposes.

REFERENCES (CHAPTER 13)

13.1 Woodyard, D.F. *Pounder's Marine Diesel Engines and Gas Turbines*. 8th Edition. Butterworth-Heinemann, Oxford, UK, 2004.

13.2 Molland, A.F. (ed.) *The Maritime Engineering Reference Book*. Butterworth-Heinemann, Oxford, UK, 2008.

13.3 Taylor, D.A. *Introduction to Marine Engineering*. Revised 2nd Edition. Butterworth-Heinemann, Oxford, UK, 1996.

13.4 Harrington, R.L. (ed.) *Marine Engineering*. Society of Naval Architects and Marine Engineers, New York, 1971.

13.5 Gallin, C., Hiersig, H. and Heiderich, O. *Ships and Their Propulsion Systems – Developments in Power Transmission*. Lohmann and Stolterfaht Gmbh, Hannover, 1983.

13.6 Molland, A.F. and Hawksley, G.J. An investigation of propeller performance and machinery applications in wind assisted ships. *Journal of Wind Engineering and Industrial Aerodynamics*, Vol. 20, 1985, pp. 143–168.

13.7 Bonebakker, J.W. The application of statistical methods to the analysis of service performance data. *Transactions of the North East Coast Institution of Engineers and Shipbuilders*, Vol. 67, 1951, pp. 277–296.

13.8 Clements, R.E. A method of analysing voyage data. *Transactions of the North East Coast Institution of Engineers and Shipbuilders*, Vol. 73, 1957, pp. 197–230.

13.9 Burrill, L.C. Propellers in action behind a ship. *Transactions of the North East Coast Institution of Engineers and Shipbuilders*, Vol. 76, 1960, pp. 25–44.

13.10 Aertssen, G. servive-performance and seakeeping trials on MV Lukuga. *Transactions of the Royal Institution of Naval Architects*, Vol. 105, 1963, pp. 293–335.

13.11 Aertssen, G. Service-performance and seakeeping trials on MV Jordaens. *Transactions of the Royal Institution of Naval Architects*, Vol. 108, 1966, pp. 305–343.

13.12 Aertssen, G. and Van Sluys, M.F. Service-performance and seakeeping trials on a large containership. *Transactions of the Royal Institution of Naval Architects*, Vol. 114, 1972, pp. 429–447.

13.13 Berlekom, Van W.B., Trägårdh, P. and Dellhag, A. Large tankers – Wind coefficients and speed loss due to wind and sea. *Transactions of the Royal Institution of Naval Architects*, Vol. 117, 1975, pp. 41–58.

13.14 Townsin, R.L., Moss, B., Wynne, J.B. and Whyte, I.M. Monitoring the speed performance of ships. *Transactions of the North East Coast Institution of Engineers and Shipbuilders*, Vol. 91, 1975, pp. 159–178.

13.15 Carlton, J.S. *Marine Propellers and Propulsion*. 2nd Edition. Butterworth-Heinemann, Oxford, UK, 2007.

13.16 Townsin, R.L., Spencer, D.S., Mosaad, M. and Patience, G. Rough propeller penalties. *Transactions of the Society of Naval Architects and Marine Engineers*, Vol. 93, 1985, pp. 165–187.

13.17 Whipps, S.L. On-line ship performance monitoring system: operational experience and design requirements. *Transactions IMarEST*, Vol. 98, Paper 8, 1985.

13.18 Bazari, Z. Ship energy performance benchmarking/rating: Methodology and application. *Transactions of the World Maritime Technology Conference International Co-operation on Marine Engineering Systems, (ICMES 2006)*, London, 2006.

14 Hull Form Design

14.1 General

14.1.1 Introduction

The hydrodynamic behaviour of the hull over the total speed range may be separated into three broad categories as displacement, semi-displacement and planing. The approximate speed range of each of these categories is shown in Figure 14.1. Considering the hydrodynamic behaviour of each, the displacement craft is supported entirely by buoyant forces, the semi-displacement craft is supported by a mixture of buoyant and dynamic lift forces whilst, when planing, the hull is supported entirely by dynamic lift. The basic development of the hull form will be different for each of these categories.

This chapter concentrates on a discussion of displacement craft, with some comments on semi-displacement craft. Further comments and discussion of semi-displacement and planing craft are given in Chapters 3 and 10.

14.1.2 Background

The underwater hull form is designed such that it displaces a prescribed volume of water ∇, and its principal dimensions are chosen such that

$$\nabla = L \times B \times T \times C_B, \tag{14.1}$$

where ∇ is the volume of displacement (m^3), L, B and T are the ship length, breadth and draught (m) and C_B is the block coefficient.

In theory, with no limits on the dimensions, there are an infinite number of combinations of L, B, T and C_B that would satisfy Equation (14.1). In practice, there are many objectives and constraints which limit the range of choice of the dimensions. These include physical limits on length due to harbours, docks and docking, on breadth due to harbour and canal restrictions and on draught due to operational water depth. Combinations of the dimensions are constrained by operational requirements and efficiency. These include combinations to achieve low calm water resistance and powering, hence fuel consumption, combinations to behave well in a seaway and the ability to maintain speed with no slamming, breadth to achieve

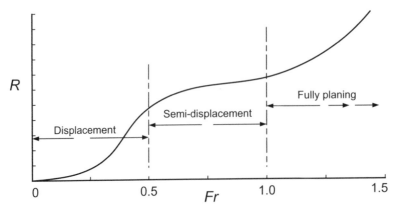

Figure 14.1. Approximate speed ranges for displacement, semi-displacement and planing craft.

adequate stability together with the cost of construction which will influence operational costs. In practice, different combinations of L, B, T and C_B will generally evolve to meet best the requirements of alternative ship types. It should be noted that the 'optimum' choice of dimensions will relate to one, say design, speed and it is unlikely that the hull form will be the optimum at all speeds.

The next section discusses the choice of suitable hull form parameters subject to these various, and sometimes conflicting, constraints and requirements.

14.1.3 Choice of Main Hull Parameters

This section discusses the main hull parameters that influence performance and the typical requirements that should be taken into account when considering the choice of these parameters.

14.1.3.1 Length-Displacement Ratio, $L/\nabla^{1/3}$

The length-displacement ratio (or slenderness ratio) usually has an important influence on hull resistance for most ship types. With increasing $L/\nabla^{1/3}$ for constant displacement, the residuary resistance R_R decreases; the effect is more important as speed increases. With constant displacement ∇ and draught T, wetted surface area and frictional resistance R_F tend to increase with increase in length (being greater than the decrease in C_F due to increase in Re) with net increase in R_F, the opposite effect from the residuary resistance. Hence, there is the possibility of an optimum L where total resistance R_T is minimum, Figure 14.2. This may be termed the optimum 'hydrodynamic' length. Results of standard series tests, such as the BSRA series [14.1] indicate the presence of an optimum $L/\nabla^{1/3}$. The influence of $L/\nabla^{1/3}$ is also illustrated by the Taylor series [14.2], using the summarised Taylor–Gertler data in Tables A3.8–A3.11, Appendix A3. The range of $L/\nabla^{1/3}$ is typically 5.5–7.0 for cargo vessels, 5.5–6.5 for tankers and bulk carriers, 7.0–8.0 for passenger ships and 6.0–9.0 for semi-displacement craft.

14.1.3.2 Length/Breadth Ratio, L/B

With increase in $L/\nabla^{1/3}$, and other parameters held constant, L/B increases. With the effect of an increase in L leading to a decrease in R_R, large L/B is favourable for faster ships. For most commercial ships, length is the most expensive dimension as

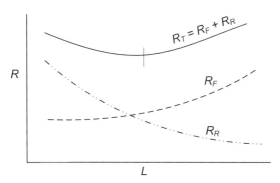

Figure 14.2. 'Hydrodynamic' optimum length.

far as construction costs are concerned. Hence, whilst an increase in L/B will lead to a decrease in specific resistance, power and fuel costs, there will be an increase in capital costs of construction. The sum of the capital and fuel costs leads to what may be termed the optimum 'economic' length, Figure 14.3, which is likely to be different from (usually smaller than) the 'hydrodynamic' optimum. It should also be noted that a longer ship will normally provide a better seakeeping performance.

The range of L/B is typically 6.0–7.0 for cargo vessels, 5.5–6.5 for tankers and bulk carriers, 6.0–8.0 for passenger ships and 5.0–7.0 for semi-displacement craft.

14.1.3.3 Breadth/Draught Ratio, *B/T*

Wave resistance increases with increase in B/T as displacement is brought nearer to the surface. Results of standard series tests, for example, the British Ship Research Associatin (BSRA) series [14.1], indicate such an increase in resistance with increase in B/T. This might, however, conflict with a need to improve transverse stability, which would require an increase in B and B/T. The influence of B/T is also clearly illustrated by the Taylor series [14.2], using the summarised Taylor–Gertler data in Tables A3.8–A3.11, Appendix A3. A typical average B/T for a cargo vessel is about 2.5, with values for stability-sensitive vessels such as ferries and passenger ships rising to as much as 5.0.

14.1.3.4 Longitudinal Centre of Buoyancy, *LCB*

LCB is normally expressed as a percentage of length from amidships. The afterbody of a symmetrical hull (symmetrical fore and aft with $LCB = 0\%L$) produces less wavemaking resistance than the forebody, due to boundary layer suppression of the

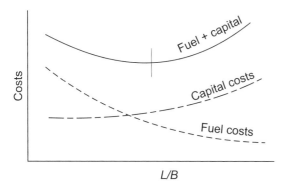

Figure 14.3. 'Economic' optimum length.

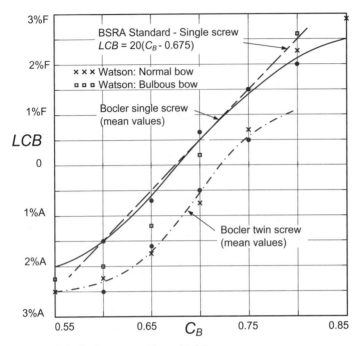

Figure 14.4. Optimum position of *LCB*.

afterbody waves. By moving *LCB* aft, the wavemaking of the forebody decreases more than the increase in the afterbody, although the pressure resistance of the afterbody will increase. The pressure resistance of fine forms (low C_P) is low; hence, *LCB* can be moved aft to advantage. The ultimate limitation will be due to pressure drag and propulsion implications. Conversely, the optimum *LCB* (or optimum range of *LCB*) will move forward for fuller ships. The typical position of *LCB* for a range of C_B is shown in Figure 14.4 which is based on data from various sources, including mean values from the early work of Bocler [14.3] and data from Watson [14.4]. It is seen in Figure 14.4 that, for single-screw vessels, the *LCB* varies typically from about 2%*L* aft of amidships for faster finer vessels to about 2%*L* to 2.5%*L* forward for slower full form vessels. Bocler's twin-screw values are about 1% aft of the single-screw values. This is broadly due to the fact that the twin-screw vessel is not as constrained as a single-screw vessel regarding the need to achieve a good flow into the propeller. It should also be noted that these optimum *LCB* values are generally associated with a particular speed range, normally one that relates C_B to Fr, such as Equation (14.2). For example, the data of Bocler [14.3] and others would indicate that the *LCB* of overdriven coasters should be about 0.5%*L* further forward than that for single-screw cargo vessels.

It should be noted that, in general, the optimum position, or optimum range, of *LCB* will change for different hull parameters and, for example, with the addition of a bulbous bow. For example, the Watson data would indicate that the *LCB* for a vessel with a bulbous bow is about 0.5%*L* forward of the *LCB* for a vessel with a normal bow, Figure 14.4. However, whilst the *LCB* data and lines in Figure 14.4 show suggested mean values for minimum resistance, there is some freedom in the position of *LCB* (say ± 0.5%*L*) without having a significant impact on

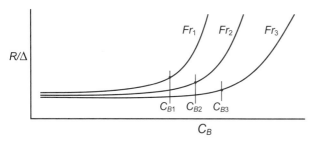

Figure 14.5. Hydrodynamic boundary, or economic, speed.

the resistance. For this reason, a suitable approach at the design stage is to use an average value for LCB from the data in Figure 14.4.

The hydrodynamic characteristics discussed may be modified by the practical requirements of a particular location of LCG, and its relation to LCB, or required limits on trim. Such practical design requirements are discussed by Watson [14.4].

14.1.3.5 Block Coefficient, C_B

C_B defines the overall fullness of the design, as described by Equation (14.1) and will have been derived in the basic design process. This is likely to have entailed the use of empirical formulae such as Equation (14.2), variations of which can be found in [14.5] and [14.6].

$$C_B = 1.23 - 2.41 \times Fr. \tag{14.2}$$

This is sometimes termed the hydrodynamic boundary, or economic, speed and can be found from standard series data, such as for the BSRA series in Figure 10.3. For each speed, the hydrodynamic boundary C_B is taken to be where the resistance curve starts to increase rapidly, Figure 14.5. A relationship, such as Equation (14.2), can then be established.

The hydrodynamic performance of the hull form is described better by the midship and prismatic coefficients, C_M and C_P.

14.1.3.6 Midship Coefficient, C_M

$C_M = C_B/C_P$, and C_B should remain constant to preserve the design displacement, Equation (14.1). A fuller C_M will lead to a smaller C_P. This may also give rise to a decrease in resistance, but this is limited since the transition between amidships and the ends of the ship has to be gradual.

14.1.3.7 Prismatic Coefficient, C_P

An increase in C_P leads to a decrease in C_M, whilst retaining the same C_B and ∇. The displacement is shifted from amidships towards the ends. The bow and stern waves change, and interference effects change the wavemaking, as discussed in Section 3.1.5. In general, fine ends are favourable at low speeds, whilst at higher speeds fuller ends may be favourable, Figure 14.6. Thus, the C_P will *increase* at higher speeds, such as the trend shown in Figure 14.7, based on data for the Taylor series, [14.2]. For the lower speed range, Fr up to about 0.28, C_P is generally limited by the hydrodynamic boundary speed. The overall variation in C_P with Fr is shown in Figure 14.7.

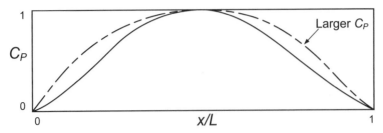

Figure 14.6. Typical sectional area curves.

14.1.3.8 Sectional Area Curve

The influence of the sectional area curve (SAC) depends on the size and distribution of C_P, discussed earlier, and with similar influences on performance.

The fore end of the SAC may be adjusted whereby some wave cancellation may be achieved. The objectives are to place the maximum curvature under the first bow wave crest and the maximum SAC slope under the bow wave trough, $\lambda/2$ from the fore end, where λ is the length of the wave and $\lambda/L = 2\pi Fr^2$. The concept is shown in Figure 14.8. The suitable location of the SAC maximum slope, based on wave length theory and experiment, is shown in Figure 14.9. It is noted that the theoretical values are aft of the best location derived from experiments.

14.1.4 Choice of Hull Shape

It is useful to consider the hull shape in terms of horizontal waterlines and vertical sections, Figure 14.10.

The midship shape, and area, will result from the choice of C_M and C_P for hydrodynamic reasons and for practical hold shapes, Figure 14.11. A small bilge radius and large C_M (≈ 0.98) tends to be used for large tankers and bulk carriers, maximising tank space and leading to a 'box type' vessel, which is also easier to construct. There may be a practical incentive to increase C_M for a container ship as far as is hydrodynamically reasonable, in order to provide the best hold shape for containers. A small rise of floor (ROF) may be employed, which aids drainage and pumping in double-bottom tanks and may offer some improvement in directional stability.

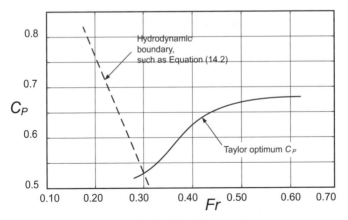

Figure 14.7. Variation in design C_P with speed.

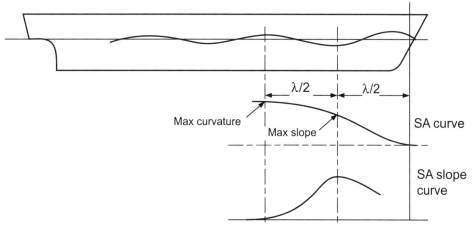

Figure 14.8. Suitable location of maximum slope of SAC.

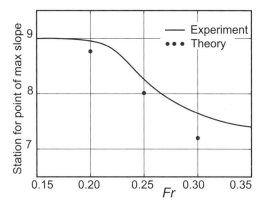

Figure 14.9. Variation of SAC maximum slope with speed.

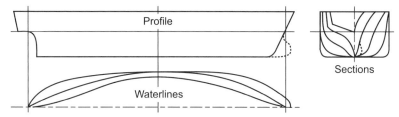

Figure 14.10. Horizontal waterlines and transverse vertical sections.

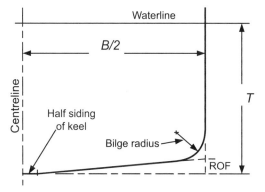

Figure 14.11. Midship section.

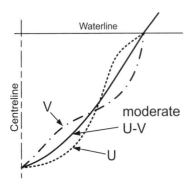

Figure 14.12. Alternative section shapes, with same underwater sectional area and same waterline breadth.

As one moves away from amidships, it should be appreciated that, fundamentally, there is an infinite number of alternative section shapes that would provide the correct underwater sectional area, hence correct underwater volume. Examples of three such alternatives are shown in Figure 14.12. The two extremes are often termed 'U'-type sections and 'V'-type sections.

In Figure 14.12, the waterline breadth has been held constant. If the design process is demanding extra initial stability then, from a hull design point of view, the simplest way is to provide more breadth B and, possibly, to decrease draught T, i.e.

$$GM = KB + BM - KG$$

and

$$BM = J_{XX}/\nabla = f[L \cdot B^3 / L \cdot B \cdot T \cdot C_B] = f[B^2 / T \cdot C_B],$$

noting that, for constant C_B, the change in metacentric height GM is a function of $[B^2 / T]$.

The approach, therefore, is to increase the waterline breadth but maintain the same underwater transverse sectional area, hence displacement, Figure 14.13. This procedure, as a consequence, tends to reshape the sections from a 'U' form to a more 'V' form. Such a procedure has been applied to passenger ships and the aft end of twin-screw car ferries, where an increase in breadth for car lanes may be required and/or higher stability may be sought.

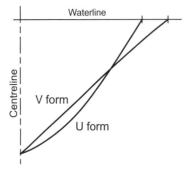

Figure 14.13. Alternative section shapes, with same underwater sectional area but change in waterline breadth.

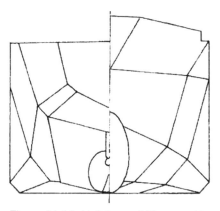

Figure 14.14. Hull form of *Pioneer* ship.

A number of straight framed ships (rather than using curved frames) have been proposed and investigated over the years. This has generally been carried out in order to achieve a more production-friendly design and/or to provide a hold shape that is more suitable for box-type cargoes such as pallets and containers. Such investigations go back to the period of the First World War, [14.7].

Blohm and Voss Shipbuilders developed the straight framed *Pioneer* ship in the 1960s, with a view to significantly reducing ship production costs. The hull form is built up from straight lines, with a number of knuckles, and the hull structure is comprised of a number of flat panels, Figure 14.14. Compared with preliminary estimates, the extra time taken for fairing the flat panels and the forming/joining of knuckle joints in the transverse frames, tended to negate some of the production cost savings.

Johnson [14.8] investigated the hydrodynamic consequences of adopting straight framed hull shapes. Model resistance and propulsion tests were carried out on the four hull shapes shown in Figure 14.15, which follow an increasing degree of simplification. The block coefficient was held constant at $C_B = 0.71$. The basic concept was to form the knuckle lines to follow the streamlines that had been mapped on the conventionally shaped parent model, A71. In addition, many of the resulting plate shapes could be achieved by two-dimensional rolling.

Resistance and propulsion tests were carried out on the four models. At the approximate design speed, relative to parent model A71, model B71 gave a reduction in resistance of 2.9%, whilst models C71 and D71 gave increases in resistance of 5.3% and 50.3%. The results for C71 indicate the penalty for adopting a flat bottom aft, and for model D71 the penalty for adopting very simplified sections. Relative to the parent model, A71, propulsive power, including propeller efficiency was –4.7% for B71, −1.5% for C71 and +39.8% for D71.

Wake patterns were also measured for models B71, C71 and D71 to help understand the changes in propulsive efficiency.

Tests were also carried out on a vessel with a block coefficient of 0.82. The first model was a conventionally shaped parent and, the second, a very simplified model with straight frames for fabrication purposes. The resistance results for the straight framed model were 19% worse than the parent, but there was relatively little change in the propulsive efficiency.

Overall, the results of these tests showed that it is possible to construct ship forms with straight sections and yet still get improvements in resistance and self

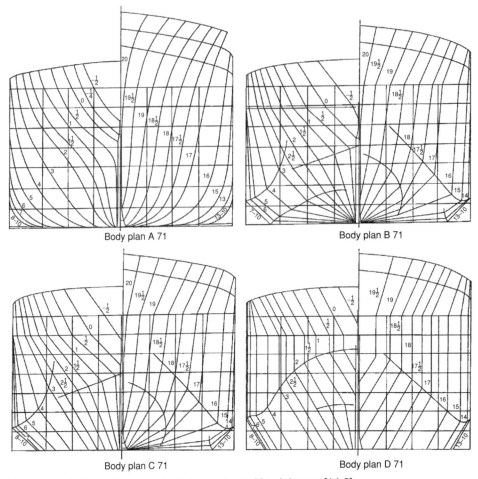

Figure 14.15. Straight framed hull shapes tested by Johnson [14.8].

propulsion in still water. This was found to hold, however, on the condition that the knuckle lines follow the stream flow.

Silverleaf and Dawson [14.9] provide a good overview of the fundamentals of hydrodynamic hull design. A wide discussion of hull form design is offered in Schneekluth and Bertram [14.6].

14.2 Fore End

14.2.1 Basic Requirements of Fore End Design

There are two requirements of fore end design:

(i) Determine the influence on hull resistance in various conditions of loading
(ii) Take note of the influence on seakeeping and manoeuvring performance.

The shape of the sections at the fore end can be considered in association with the half angle of entrance of the design waterline, $1/2\,\alpha_E$, Figure 14.16. With a constant sectional area curve, $1/2\,\alpha_E$ governs the form of the forebody sections,

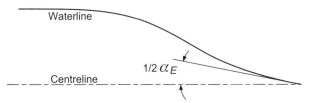

Figure 14.16. Definition of half-angle of entrance $1/2\,\alpha_E$.

that is low $1/2\,\alpha_E$ leads to a 'U' form and high $1/2\,\alpha_E$ leads to a 'V' form. 'V' forms tend to move displacement nearer the surface and to produce more wavemaking. At the same time, vessels such as container ships, looking for extra breadth forward to accommodate more containers on deck, might be forced towards 'V' sections. The effect of $1/2\,\alpha_E$ depends on speed. With a large $1/2\,\alpha_E$ there is high resistance at low speeds whilst at high speed a contrary effect may exist, such as in the case of overpowered or 'overdriven' coasters. With a relatively low C_P and high speeds, a small $1/2\,\alpha_E$ is preferable, yielding 'U' sections and lower wavemaking. This may be tempered by the fact that 'U' forms tend to be more susceptible to slamming. The effect of forebody shape on ship motions and wetness is discussed in [14.10], [14.11] and [14.12]. Moderate 'U-V' forms may provide a suitable compromise. Typical values of $1/2\,\alpha_E$ for displacement vessels are shown in Table 14.1.

14.2.2 Bulbous Bows

Bulbous bows can be employed to reduce the hull resistance of ships. Their role in the case of finer faster vessels tends to entail the reduction of wavemaking resistance whilst, in the case of slower fuller ships, the role tends to entail the reduction of viscous resistance. The resistance reduction due to a bulb for a full form slow ship can exceed the wave resistance alone. For full form slower ships the bulbous bow tends to show most benefit in the ballast condition. It should also be noted that a bulb tends to realign the flow around the fore end, but this is carried downstream and the bulb is also found to influence the values of wake fraction, thrust deduction factor and hull efficiency [14.13].

The application of a bulbous bow entails the following two steps:

(i) Decide whether a bulb is likely to be beneficial, which will depend on parameters such as ship type, speed and block coefficient
(ii) Determine the actual required characteristics and design of the bulb.

Table 14.1. *Typical values of half-angle*
of entrance: displacement ships

C_B	$1/2\,\alpha_E$ (deg)
0.55	8
0.60	10
0.70	20
0.80	35

The benefits of using a bulb are likely to depend on the existing basic components of resistance, namely the proportions of wave and viscous resistance. The longitudinal position of the bulb causes a wave phase difference whilst its volume is related to wave amplitude. At low speeds, where wavemaking is small, the increase in skin friction resistance arising from the increase in wetted area due to the bulb is likely to cancel any reductions in resistance. At higher speeds, a bulb can improve the flow around the hull and reduce the friction drag, as deduced by Steele and Pearce [14.14] from tests on models with normal and bulbous bows.

Also, bulb cancellation effects are likely to be speed dependent because the wave length (and position of the wave) changes with speed, whereas the position of the bulb (pressure source) is fixed. The early work of Froude around 1890 and that of Taylor around 1907 should be acknowledged; both recognised the possible benefits of bulbous bows. The earliest theoretical work on the effectiveness of bulbous bows was carried out by Wigley [14.15]. Ferguson and Dand [14.16] provide a fundamental study of hull and bulbous bow interaction.

Sources providing guidance on the suitability of fitting a bulbous bow include the work of BSRA [14.1], the classical work of Kracht [14.13] and the regression work of Holtrop [14.17]. The BSRA results are included in Figure 10.6 in Chapter 10. The data are for the loaded condition and are likely to be suitable for many merchant ships such as cargo and container ships, tankers and bulk carriers and the like. It is interesting to note from Figure 10.6 that the largest reductions occur at lower C_B and higher speeds, with reductions up to 20% being realised. For higher C_B and lower speeds, the reductions are generally much smaller. However, for slower full form ships, significant benefits can be achieved in the ballast condition and, for this reason, most full form vessels such as tankers and bulk carriers, which travel for significant periods in the ballast condition, are normally fitted with a bulbous bow. Reductions in resistance in the ballast condition of up to 15% have been reported for such vessels [14.18].

The regression analysis of Holtrop [14.17] includes an estimate of the influence of a bulbous bow. This is included as Equation (10.30) in Chapter 10. Holtrop, in his discussion to [14.13] indicates that, for a test case, his approach produces broadly similar results to those in [14.13].

Moor [14.19] presents useful experimental data from tests on a series of bulbous (ram) bows with a progressive increase in size. Guidance is given on choice of bow, which depends on load and/or ballast conditions and speed.

When considering the actual required characteristics of the bulb, the work of Kracht [14.13] provides a good starting point. Kracht defines three types of bulb as the Δ-Type, the O-Type and the ∇-Type, Figure 14.17. Broad applications of these three types are summarised as follows:

Δ-Type: Suitable for ships with large draught variations and U-type forward sections. The effect of the bulb decreases with increasing draught and vice versa. There is a danger of slamming at decreased draught.
O-Type: Suitable for both full and finer form ships, fits well into U- and V-type sections and offers space for sonar and sensing equipment. It is less susceptable to slamming.
∇-Type: It is easily faired into V-shaped forward sections and has, in general, a good seakeeping performance.

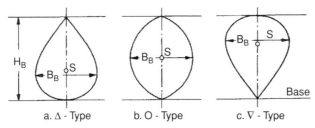

Figure 14.17. Bulb types.

In all cases, the bulb should not emerge in the ballast condition beyond point B in Figure 14.18.

Six parameters used to describe the geometry of the bulb are as follows: Figure 14.18:

Length parameter: $C_{LPR} = L_{PR}/L_{BP}$, where L_{PR} is the protruding length of the bulb.

Breadth parameter: $C_{BB} = B_B/B$, where B_B is the maximum breadth of the bulb at the forward perpendicular (FP) and B is the ship breadth

Depth parameter: $C_{ZB} = Z_B/T_{FP}$, where Z_B is the height of the forward most point of the bulb and T_{FP} is the draught at the forward perpendicular.

Cross-section parameter: $C_{ABT} = A_{BT}/A_X$, where A_{BT} is the cross-sectional area of the bulb at the FP and A_X is the midship section area.

Lateral parameter: $C_{ABL} = A_{BL}/A_X$, where A_{BL} is the area of the ram bow in the longitudinal plane and A_X is the midship section area.

Volume parameter: $C_{\nabla PR} = \nabla_{PR}/\nabla$, where ∇_{PR} is the nominal bulb volume and ∇ is the ship volumetric displacement.

Kracht suggests that the length, cross-section and volume parameters are the most important.

In order to describe the characteristics and benefits of the bulbous bow, Kracht uses a residual power reduction coefficient, $\Delta C_{P\nabla R}$, which is a measure of the percentage reduction in power using a bulb compared with a normal bow, a larger value representing a larger reduction in power. The data were derived from an analysis of routine test results in two German test tanks. Examples of $\Delta C_{P\nabla R}$ for $C_B = 0.71$ over a range of Froude numbers $Fr(F_N$ in diagram) are shown in Figures 14.19–14.23 for C_{LPR}, C_{BB}, C_{ABT}, C_{ABL} and $C_{\nabla PR}$.

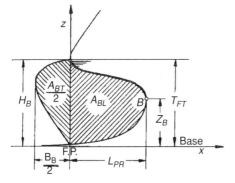

Figure 14.18. Definitions of bulb dimensions.

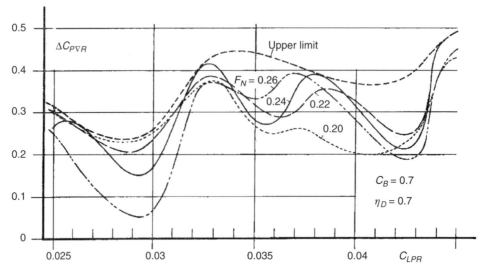

Figure 14.19. Residual power reduction coefficient as a function of C_{LPR}.

Use of the data allows combinations of bulb characteristics to be chosen to maximise the savings in power (maximum $\Delta C_{P\nabla R}$). For example, assume a speed of $Fr = 0.26$, and assume a design requirement of $L_{PR}/L_{BP} < 3.5\%$. If $L_{PR}/L_{BP} < 0.035$, then from Figure 14.19 the maximum $\Delta C_{P\nabla R}$ at $Fr = 0.26$ is 0.38 at $L_{PR}/L_{BP} = 0.033$ (3.3%). From Figure 14.20 at $\Delta C_{P\nabla R} = 0.38$, a suitable breadth coefficient $C_{BB} = 0.155$ (15.5%) and from Figure 14.21 a suitable cross-section coefficient $C_{ABT} = 0.12$ (12%). Suitable values for C_{ABL} and $C_{\nabla PR}$ can be found in a similar manner.

The data and methodology of Kracht have been applied to high-speed fine form ships by Hoyle *et al.* [14.20]. A series of bulb forms were developed and analysed using numerical and experimental methods, Figure 14.24. The use of the design charts is illustrated and the derivation of charts for other block coefficients is described. The Kracht design charts produced acceptable, but not optimum, initial

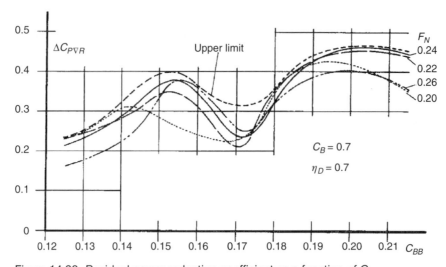

Figure 14.20. Residual power reduction coefficient as a function of C_{BB}.

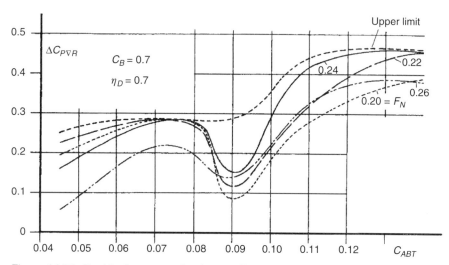

Figure 14.21. Residual power reduction coefficient as a function of C_{ABT}.

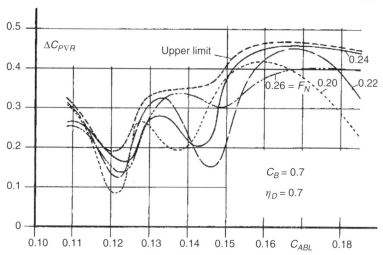

Figure 14.22. Residual power reduction coefficient as a function of C_{ABL}.

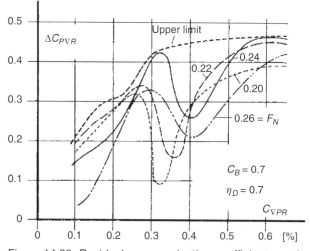

Figure 14.23. Residual power reduction coefficient as a function of $C_{\nabla PR}$.

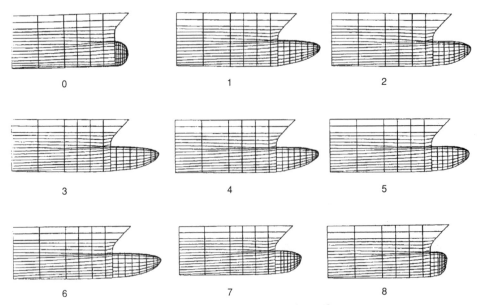

Figure 14.24. Bulb designs investigated by Hoyle *et al.* [14.20].

designs; increases in bulb breadth and volume tended to lower the resistance further. The decreases in resistance due to the bulbs varied with speed. Bulb 8 showed the worst results, whilst Bulb 0 showed reasonable reductions, although bettered over much of the speed range by Bulbs 4 and 6. The numerical methods employed provided an accurate relative resistance ranking of the bulbous bow configurations. This demonstrated the potential future use of numerical methods for such investigations.

14.2.3 Seakeeping

In general, a bulbous bow does not significantly affect ship motions or seakeeping characteristics [14.13], [14.18] and [14.21], and the bulb can be designed for the calm water condition. It is, however, recommended in [14.21] that it is prudent to avoid extremely large bulbs, which tend to lose their calm water benefits in a seaway.

14.2.4 Cavitation

Cavitation can occur over the fore end of bulbous bows of fast vessels. The use of elliptical horizontal sections at the fore end can help delay the onset of cavitation.

14.3 Aft End

14.3.1 Basic Requirements of Aft End Design

There are four requirements of aft end design:

(i) The basic aft end shape should minimise the likelihood of flow separation and its influence on hull resistance and the performance of the propulsor.

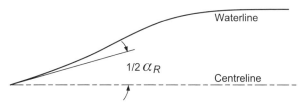

Figure 14.25. Definition of half-angle of run $1/2\alpha_R$.

(ii) The shape should ideally be such that it produces a uniform wake in way of the propulsor(s).
(iii) The aft end should suit the practical and efficient arrangement of propulsors, shaft brackets or bossings and rudders.
(iv) There should be adequate clearances between propulsor(s) and the adjacent structure such as hull, sternframe and rudder.

Resistance tests and flow visualisation studies are used to measure the effectiveness of the hull shape. Wake surveys (see Chapter 8) are used to assess the distribution of the wake and degree of non-uniformity. These provide a measure of the likely variation in propeller thrust loading and the possibility of propeller-excited vibration.

The shape of the sections at the aft end can be considered in association with the half angle of run, $1/2\alpha_R$, Figure 14.25. Large $1/2\alpha_R$ leads to 'V' sections aft and less resistance, and is typically applied to twin-screw vessels. Smaller $1/2\alpha_R$ with moderate 'U' sections is normally applied to single-screw vessels, in general leading to an increase in resistance. This is generally offset by an increase in propulsive efficiency. For example, Figure 8.4 in Chapter 8 illustrates the influence on wake distribution when moving from what is effectively a 'V' section stern to a 'U' shape and then to a bulbous stern, as shown in Figure 14.26. With the 'U' and bulbous sterns, the lines of constant wake become almost concentric, leading to decreases in propeller force variation and vibration, and a likely improvement in overall efficiency.

Excellent insights into the performance of different aft end section shapes and arrangements are provided in [14.22], [14.23] and [14.24]. Resistance and propulsion tests were carried out on vessels with $C_B = 0.65$ [14.22] and $C_B = 0.80$ [14.23] and with aft end section shapes representing the parent (moderate U-V form), U

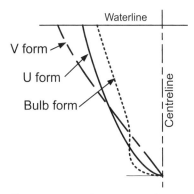

Figure 14.26. 'V', 'U' and bulbous sterns.

Table 14.2. *Effect of bulb on resistance and propulsion*

Stern type	ⓒ	η_D	ⓒ / η_D
Parent	1.012	0.643	1.574
U type	1.041	0.674	1.545
Bulb	1.042	0.662	1.574
Concentric	1.000	0.654	1.529

form, a bulb (bulbous near base line) and concentric bulb (concentric with shaft line). Some results extracted from [14.23] for $C_B = 0.80$ and $Fr = 0.21$, are given in Table 14.2. This table broadly demonstrates the phenomena already described in terms of changes in resistance and propulsive efficiency. Namely, in general terms, the resistance increases when moving to U-shaped sections, but there is a small improvement in propulsive efficiency and a net decrease in delivered power P_D ($= f \, ⓒ/ \, \eta_D$). This is true also for the concentric form. There is also a more uniform wake distribution for the U and concentric bulb forms. These results serve to demonstrate the need to consider both resistance and propulsion effects when designing the aft end.

For single-screw vessels, the aft end profile generally evolves from the requirements of draught, propeller diameter, rudder location and clearances, Figure 14.27. Draught issues will include the ballast condition and adequate propeller immersion. Rudder location and its influence on propulsion is discussed by Molland and Turnock [14.25]. Suitable propeller clearances (in particular, to avoid propeller vibration) can be obtained from the recommendations of classification societies, such as [14.26]. For preliminary design purposes, a minimum propeller tip clearance of 20%D can be used ('a' in Figure 14.27). Most vessels now incorporate a transom stern which increases deck area, providing more space for mooring equipment, or allowing the deckhouse to be moved aft, or containers to be stowed aft, whilst at the same time generally lowering the cost of construction.

For twin-screw vessels, conventional V-type sections have generally been adopted, with the propeller shafts supported in bossings or on shaft brackets, Figure 14.28. Again, a transom stern is employed. For some faster twin-screw forms, such as warships, a stern wedge (or flap) over the breadth of the transom may be

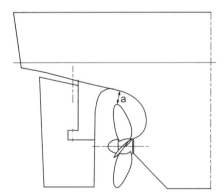

Figure 14.27. Aft end profile, single screw.

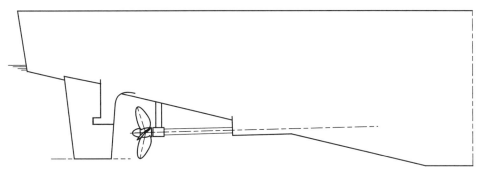

Figure 14.28. Aft end arrangement, twin screw.

employed which deflects the flow downward as it leaves the transom, providing a trim correction and resistance reduction, [14.27].

More recent investigations, mainly for large container ships requiring very large propulsive power, have considered twin screws with twin-skeg forms [14.28]. The stern is broadly pram shape, with the skegs suitably attached, Figure 14.29. Satisfactory overall resistance and propulsion properties have been reported for these arrangements, although first and running costs are likely to be higher than for an equivalent single-screw installation.

14.3.2 Stern Hull Geometry to Suit Podded Units

When podded propulsors are employed, a pram-type stern can be adopted. Because there are no bossings or shafting upstream, a tractor (pulling) unit is then working in a relatively undisturbed wake.

The pram-type stern promotes buttock flow which, if the hull lines are designed appropriately, can lead to a decrease in hull resistance [14.29]. It is generally accepted that, in order to avoid flow separation, the slope of the pram stern should not be more than about 15°. A useful investigation into pram stern slope was carried out by Tregde [14.30]. He estimated the limiting slope using an inverse design method

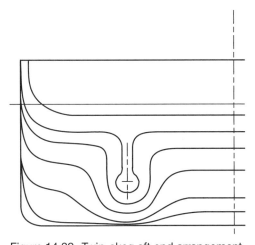

Figure 14.29. Twin-skeg aft end arrangement.

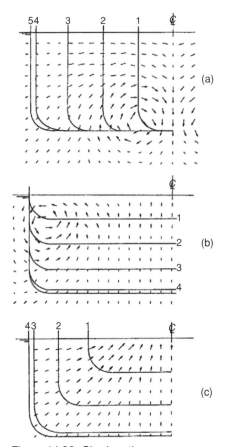

Figure 14.30. Shed vortices.

and the principle of Stratford flow [14.31], where a pressure and, hence, velocity distribution is prescribed which just precludes the onset of separation.

When considering the overall shape, it should be noted that a steady change in waterline and buttock slope should be adopted in order to avoid shed vortices, with consequent increase in resistance. This can be seen from the tuft study results in Figure 14.30 [14.32], where (a) is waterline flow, (b) is buttock flow and (c) provides a good compromise with the absence of vortices.

Research has shown [14.33] that the optimum longitudinal pod inclination is about the same as that for the corresponding buttock line. For good propeller efficiency, the pod should be located at a minimum distance of 5%L from the transom.

Ukon *et al.* [14.29] investigated the propulsive performance of podded units for single-screw vessels with a conventional stern hull, buttock flow stern and a stern bulb hull form. The buttock flow stern was found to have the lowest resistance and effective power requirement. The wake fraction and thrust deduction factor for the buttock flow stern were low, leading to a low hull efficiency of 1.031. The wake gain for the stern bulb hull led to the highest hull efficiency (1.304) and overall propulsive efficiency, and the lowest overall delivered power requirement. The paper concludes that (for single-screw vessels) the bulb stern is a promising option.

The seakeeping behaviour of a pram stern has to be taken into account. If the stern surfaces are too flat, this can give rise to slamming in a following sea. In [14.33]

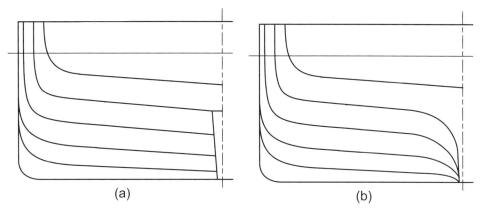

Figure 14.31. Pram stern.

it is proposed that the transverse slope of a section relative to the still waterplane should be greater than about 5°. In order to provide directional stability, a skeg will normally be incorporated in the pram stern. This may be incorporated as a separate fabrication, Figure 14.31 (a), or shaped to form part of the hull, Figure 14.31 (b).

14.3.3 Shallow Draught Vessels

Some tankers with a draught, and hence propeller diameter, limitation have been designed with twin screws. This is technically viable and acceptable, but will generally lead to higher build and operational costs. Other shallow draught tankers have been fitted with ducted propellers with successful results [14.34], [14.35]. The restricted propeller diameter leads to higher thrust loadings, which is where the ducted propeller can be helpful, with the duct augmenting the thrust of the propeller, see Section 11.3.3.

Shallow draught vessels such as those found on inland waterways have successfully employed tunnel sterns, Figure 14.32. As a larger propeller diameter will normally improve the efficiency, the use of a tunnel allows some increase in diameter. Care must be taken to ensure that there is adequate immersion of the propeller, and adequate vertical tunnel outboard of the propeller to preclude ventilation of the propeller around the side of the hull. A combination of a ducted propeller within a partial tunnel has also been employed. Tunnels can also be applied to single-screw vessels using similar approaches.

Some discussion on the use of tunnels is included in Carlton [14.36]. For smaller craft, a useful source of information on tunnels for such craft may be found in Harbaugh and Blount [14.37].

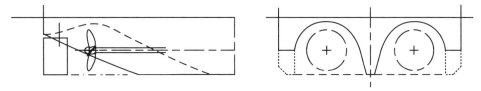

Figure 14.32. Shallow draught vessel with tunnel stern.

14.4 Computational Fluid Dynamics Methods Applied to Hull Form Design

Until recent years, hull form development has been mainly carried out using experimental techniques. Initially, this concerned the measurement of model total resistance and its extrapolation to full scale, as discussed in Chapter 4. Since the 1960s, much experimental effort has been directed at measuring the individual components of hull resistance, allowing a better insight into why changes in hull form lead to changes in resistance. This is discussed in Chapter 7.

Theoretical work has been carried out over many years, including that of Havelock and Kelvin, but it was the advent of the modern computer and numerical computational methods that allowed extensive investigations into the flow over the hull and the influence of hull form changes on the flow. Computational fluid dynamics (CFD) has not yet replaced the experimental approach, but can be used very successfully with experiments in a complementary manner. In particular, CFD predictions can be used in planning experiments and indicating potential areas of investigation. At the same time, good quality experimental data, particularly those relating to the individual resistance components, are used to validate CFD predictions. Rapid progress is being made towards developing computational methods that offer very realistic predictions both at model and full scale [14.38].

Further discussion of the applications of CFD approaches to hull design, wake and propeller design are included in Chapters 8, 9 and 15.

Examples where hull forms have been developed using a mixture of CFD and experiments are provided [14.39] and [14.40]. Other examples of the use of CFD and experiments in hull form design and interaction with the propeller may be found in [14.41], [14.42] and [14.43].

REFERENCES (CHAPTER 14)

14.1 BSRA. Methodical series experiments on single-screw ocean-going merchant ship forms. Extended and revised overall analysis. BSRA Report NS333, 1971.

14.2 Gertler, M. A reanalysis of the original test data for the Taylor standard series. David Taylor Model Basin Report No. 806. DTMB, Washington, DC, 1954. Reprinted by Society of Naval Architects and Marine Engineers, 1998.

14.3 Bocler, H. The position of the longitudinal centre of buoyancy for minimum resistance. *Transactions of the Institute of Engineers and Shipbuilders in Scotland.* Vol. 97, 1953–1954, pp. 11–63.

14.4 Watson, D.G.M. *Practical Ship Design.* Elsevier Science, Oxford, UK, 1998.

14.5 Molland, A.F. (ed.) *Maritime Engineering Reference Book.* Butterworth-Heinemann, Oxford, UK, 2008.

14.6 Schneekluth, H. and Bertram, V. *Ship Design for Efficiency and Economy.* 2nd Edition. Butterworth-Heinemann, Oxford, UK, 1998.

14.7 McEntee, W. Cargo ship lines on simple form. *Transactions of the Society of Naval Architects and Marine Engineers,* Vol. 25, 1917.

14.8 Johnson, N.V. Experiments with straight framed ships. *Transactions of the Royal Institution of Naval Architects,* Vol. 106, 1964, pp. 197–211.

14.9 Silverleaf, A. and Dawson, J. Hydrodynamic design of merchant ships for high speed operation. *Transactions of the Royal Institution of Naval Architects,* Vol. 109, 1967, pp. 167–196.

14.10 Swaan, W.A. and Vossers, G. The effect of forebody section shape on ship behaviour in waves. *Transactions of the Royal Institution of Naval Architects*, Vol. 103, 1961, pp. 297–328.

14.11 Ewing, J.A. The effect of speed, forebody shape and weight distribution on ship motions. *Transactions of the Royal Institution of Naval Architects*, Vol. 109, 1967, pp. 337–346.

14.12 Lloyd, A.R.J.M., Salsich, J.O. and Zseleczky, J.J. The effect of bow shape on deck wetness in heads seas. *Transactions of the Royal Institution of Naval Architects*, Vol. 128, 1986, pp. 9–25.

14.13 Kracht, A.M. Design of bulbous bows. *Transactions of the Society of Naval Architects and Marine Engineers*, Vol. 86, 1978, pp. 197–217.

14.14 Steele, B.N. and Pearce, G.B. Experimental determination of the distribution of skin friction on a model of a high speed liner. *Transactions of the Royal Institution of Naval Architects*, Vol. 110, 1968, pp. 79–100.

14.15 Wigley, W.C.S. The theory of the bulbous bow and its practical application. *Transactions of the North East Coast Institution of Engineers and Shipbuilders*, Vol. 52, 1935–1936.

14.16 Ferguson, A.M. and Dand, I.W. Hull and bulbous bow interaction. *Transactions of the Royal Institution of Naval Architects*, Vol. 112, 1970, pp. 421–441.

14.17 Holtrop, J. A statistical re-analysis of resistance and propulsion data. *International Shipbuilding Progress*, Vol. 31, 1984, pp. 272–276.

14.18 Lewis, E.V. (ed.). *Principles of Naval Architecture*. The Society of Naval Architects and Marine Engineers, New York, 1989.

14.19 Moor, D.I. Resistance and propulsion properties of some modern single screw tanker and bulk carrier forms. *Transactions of the Royal Institution of Naval Architects*, Vol. 117, 1975, pp. 201–204.

14.20 Hoyle, J.W., Cheng, B.H., Hays, B., Johnson, B. and Nehrling, B. A bulbous bow design methodology for high-speed ships. *Transactions of the Society of Naval Architects and Marine Engineers*, Vol. 94, 1986, pp. 31–56.

14.21 Blume, P. and Kracht, A.M. Prediction of the behaviour and propulsive performance of ships with bulbous bows in waves. *Transactions of the Society of Naval Architects and Marine Engineers*, Vol. 93, 1985, pp. 79–94.

14.22 Thomson, G.R. and White, G.P. Model experiments with stern variations of a 0.65 block coefficient form. *Transactions of the Royal Institution of Naval Architects*, Vol. 111, 1969, pp. 299–316.

14.23 Dawson, J. and Thomson, G.R. Model experiments with stern variations of a 0.80 block coefficient form. *Transactions of the Royal Institution of Naval Architects*, Vol. 111, 1969, pp. 507–524.

14.24 Thomson, G.R. and Pattullo, R.N.M. The BSRA Trawler Series (Part III). Block coefficient and longitudinal centre of buoyancy variation series, tests with bow and stern variations. *Transactions of the Royal Institution of Naval Architects*, Vol. 111, 1969, pp. 317–342.

14.25 Molland, A.F. and Turnock, S.R. *Marine Rudders and Control Surfaces*. Butterworth-Heinemann, Oxford, UK, 2007.

14.26 Lloyd's Register. *Rules and Regulations for the Classification of Ships*. Part 3, Chapter 6. July 2005.

14.27 Kariafiath, G., Gusanelli, D. and Lin, C.W. Stern wedges and stern flaps for improved powering – US Navy experience. *Transactions of the Society of Naval Architects and Marine Engineers*, Vol. 107, 1999, pp. 67–99.

14.28 Kim, J., Park, I.-R., Van, S.-H., and Park, N.-J. Numerical computation for the comparison of stern flows around various twin skegs. *Journal of Ship and Ocean Technology*, Vol. 10, No. 2, 2006.

14.29 Ukon, Y., Sasaki, N, Fujisawa, J. and Nishimura, E. The propulsive performance of podded propulsion ships with different shape of stern hull. *Second*

International Conference on Technological Advances in Podded Propulsion, T-POD. University of Brest, France, 2006.

14.30 Tregde, V. Aspects of ship design; Optimisation of aft hull with inverse geometry design. Dr.Ing. thesis, Department of Marine Hydrodynamics, University of Science and Technology, Trondheim, 2004.

14.31 Stratford, B.S. The prediction of separation of the turbulent boundary layer. *Journal of Fluid Mechanics*, Vol. 5, No. 17, 1959, pp. 1–16.

14.32 Muntjewert, J.J. and Oosterveld, M.W.C. Fuel efficiency through hull form and propulsion research – a review of recent MARIN activities. *Transactions of the Society of Naval Architects and Marine Engineers*, Vol. 95, 1987, pp. 167–181.

14.33 Bertaglia, G., Serra, A. and Lavini, G. Pod propellers with 5 and 6 blades. *Proceedings of International Conference on Ship and Shipping Research, NAV'2003*, Palermo, Italy, 2003.

14.34 Flising, A. Ducted propeller installation on a 130,000 TDW tanker – A research and development project. *RINA Symposium on Ducted Propellers*. RINA, London, 1973.

14.35 Andersen, O. and Tani, M. Experience with SS *Golar Nichu*. *RINA Symposium on Ducted Propellers*. RINA, London, 1973.

14.36 Carlton, J.S. *Marine Propellers and Propulsion*. 2nd Edition. Butterworth-Heinemann, Oxford, UK, 2007.

14.37 Harbaugh, K.H. and Blount, D.L. An experimental study of a high performance tunnel hull craft. Paper H, *Society of Naval Architects and Marine Engineers*, Spring Meeting, 1973.

14.38 Raven, H.C., Van Der Ploeg, A., Starke, A.R. and Eça, L. Towards a CFD- based prediction of ship performance – progress in predicting full-scale resistance and scale effects. *Transactions of the Royal Institution of Naval Architects*, Vol. 150, 2008, pp. 31–42.

14.39 Hämäläinen, R. and Van Heerd, J. Hydrodynamic development for a large fast monohull passenger ferry. *Transactions of the Society of Naval Architects and Marine Engineers*, Vol. 106, 1998, pp. 413–441.

14.40 Valkhof, H.H., Hoekstra, M. and Andersen, J.E. Model tests and CFD in hull form optimisation. *Transactions of the Society of Naval Architects and Marine Engineers*, Vol. 106, 1998, pp. 391–412.

14.41 Tzabiras, G.D. A numerical study of additive bulb effects on the resistance and self-propulsion characteristics of a full form ship. *Ship Technology Research*, Vol. 44, 1997.

14.42 Turnock, S.R., Phillips, A.B. and Furlong, M. URANS simulations of static drift and dynamic manoeuvres of the KVLCC2 Tanker. *Proceedings of the SIMMAN International Manoeuvring Workshop*. Copenhagen, April 2008.

14.43 Larsson, L. and Raven, H.C. Principles of Naval Architecture: Ship Resistance and Flow. The Society of Naval Architects and Marine Engineers, New York, 2010.

15 Numerical Methods for Propeller Analysis

15.1 Introduction

Ship powering relies on a reliable estimate of the relationship between the shaft torque applied and the net thrust generated by a propulsor acting in the presence of a hull. The propeller provides the main means for ship propulsion. This chapter considers numerical methods for propeller analysis and the hierarchy of the possible methods from the elementary through to those that apply the most recent computational fluid dynamics techniques. It concentrates on the blade element momentum approach as the method best suited to gaining an understanding of the physical performance of propeller action. Further sections examine the influence of oblique flow and tangential wake, the design of wake-adapted propellers and finally the assessment of cavitation risk and effects.

Although other propulsors can be used, Chapter 11, the methods of determining their performance have many similarities to those applied to the conventional ship propeller and so will not be explicitly covered. The main details of the computational fluid dynamic (CFD) based approaches are covered in Chapter 9 as are the methods whereby coupled self-propulsion calculations can be applied, Section 9.6.

Further details of potential-based numerical analysis of propellers are covered by Breslin and Anderson [15.1], and Carlton [15.2] gives a good overview.

15.2 Historical Development of Numerical Methods

From the start of mechanically based propulsion, there was an awareness of the need to match propeller design to the requirement of a specific ship design. The key developments are summarised, based on [15.2, 15.3], as follows.

Rankine [15.4], considering fluid momentum, found the ideal efficiency of a propeller acting as an actuator disc. The rotor is represented as a disc capable of sustaining a pressure difference between its two sides and imparting linear momentum to the fluid that passes through it. The mechanism of thrust generation requires the evaluation of the mass flow through a stream tube bounded by the disc. Froude [15.5], in his momentum theory, allowed the propeller to impart a rotational velocity to the slipstream.

In 1878 William Froude [15.6] developed the theory of how a propeller section, or blade element, could develop the force applied to the fluid. It was not until the work of Betz [15.7] in 1919, and later Goldstein [15.8] in 1929 employing Prandtl's [15.9] lifting line theory, that it was shown that optimum propellers could be designed. This approach is successful for high-aspect ratio blades more suited to aircraft. For the low-aspect ratio blades widely used for marine propellers, this assumption is not valid. It was not until 1952, when Lerbs [15.10] published his paper on the extension of Goldstein's lifting line theory for propellers with arbitrary radial distributions of circulation in both uniform and radially varying inflow, that, at last, marine propellers could be modelled with some degree of accuracy. Although its acceptance was slow, it still is, even today, universally accepted as a good procedure for establishing the principal characteristics of the propeller at an early design stage.

The onset of digital computers allowed the practical implementation of numerical lifting surface methods. This allowed the influence of skew and the radial distribution of circulation to be modelled, Sparenberg [15.11]. There was a rapid development of techniques based on lifting surfaces [15.12–15.15] which were then further refined as computer power increased [15.16, 15.17].

The above methods, although suitable for design purposes, provided limited information on the section flow. Hess and Valarezo [15.18] developed a boundary element method (BEM) or surface panel code that allowed the full geometry of the propeller to be modelled, and this approach, or related ones, has been widely adopted.

At a similar time early work was being undertaken into the use of Reynolds averaged Navier–Stokes (RANS) codes for propeller analysis. For example, Kim and Stern [15.19] showed the possibilities of such analysis for a simplified propeller geometry. The work of authors such as Uto [15.20] and Stanier [15.21–15.23] provided solutions for realistic propeller geometries with detailed flow features. Chen and Stern [15.24] undertook unsteady viscous computations and investigated their applicability, although they obtained poor results due to the limited mesh size feasible at the time. Maksoud *et al.* [15.25, 15.26] performed unsteady calculations for a propeller operating in the wake of a ship using a non-matching multiblock scheme.

Finally, the ability to deal with large-scale unsteadiness through the availability of massive computational power has allowed propeller analysis to be extended to extreme off-design conditions. The application of large eddy simulations (LES) to propeller flows such as crash back manoeuvres is the current state of the art. Notable examples are found in the publications of Jessup [15.27] and Bensow and Liefvendahl [15.28] along with the triennial review of the International Towing Tank Conference (ITTC) Propulsion Committee [15.29].

15.3 Hierarchy of Methods

Table 15.1, developed from Phillips *et al.* [15.30], classifies the various approaches in increasing order of physical and temporal accuracy. A simplified computational cost measure is also included. This represents an estimate of the relative cost of each technique normalised to the baseline blade element-momentum theory (BEMT) which has a cost of one. As can be seen, the hierarchy reflects the historical development, as well as the progressively more expensive computational cost.

Table 15.1. *Numerical methods for modelling propellers*

Method	Description	Cost
Momentum theory	The propeller is modelled as an actuator disc over which there is an instantaneous pressure change, resulting in a thrust acting at the disc. The thrust, torque and delivered power are attributed to changes in the fluid velocity within the slipstream surrounding the disc, Rankine [15.4], Froude [15.5]	<1
Blade element theory	The forces and moments acting on the blade are derived from a number of independent sections represented as two-dimensional aerofoils at an angle of attack to the fluid flow. Lift and drag information for the sections must be provided a priori and the induced velocities in the fluid due to the action of the propeller are not accounted for, Froude [15.6].	<1
Blade element-momentum theory	By combining momentum theory with blade element theory, the induced velocity field can be found around the two-dimensional sections, Burrill [15.42], Eckhardt and Morgan [15.40], O'Brien [15.41]. Corrections have been presented to account for the finite number of blades and strong curvature effects.	1
Lifting line method	The propeller blades are represented by lifting lines, which have a varying circulation as a function of radius. This approach is unable to capture stall behaviour, Lerbs [15.10].	~ 10
Lifting surface method	The propeller blade is represented as an infinitely thin surface fitted to the blade camber line. A distribution of vorticity is applied in the spanwise and chordwise directions, Pien [15.12].	$\sim 10^2$
Panel method	Panel methods extend the lifting surface method to account for blade thickness and the hub by representing the surface of the blade by a finite number of vortex panels, Kerwin [15.13].	$\sim 10^3$
Reynolds averaged Navier–stokes	Full three-dimensional viscous flow field modelled using a finite volume or finite-element approach to solve the averaged flow field, Stanier [15.21–15.23], Adbel-Maksoud et al. [15.25, 15.26].	$\sim 10^6$
Large eddy simulation	Bensow and Liefvendahl [15.28]	$\sim 10^8$

Automated design optimisation techniques rely on the ability to evaluate multiple designs within a reasonable time frame and at an appropriate cost. The design goals of a propeller optimisation seek to minimise required power for delivered thrust with a sufficiently strong propeller that avoids cavitation erosion at design and off-design conditions [15.2, 15.31]. The physical fidelity of the simulation can be traded against the computational cost if suitable empiricism can be included in interpreting the results of the analysis. For instance, as viscous effects often only have limited influence at design, an estimate of skin friction can be included with a potential-based surface panel method alongside a cavitation check based on not going beyond a certain minimum surface pressure to select an optimum propeller.

15.4 Guidance Notes on the Application of Techniques

15.4.1 Blade Element-Momentum Theory

As will be shown in Section 15.5, for concept propeller design the blade element-momentum theory in its various manifestations provides a very rapid technique for achieving suitable combinations of chord and pitch for given two-dimensional sectional data. The resultant load distributions can be used with the one-dimensional beam theory–based propeller strength calculations in Section 12.3 to determine the required blade section thickness and cavitation inception envelopes, Figures 12.23 and 12.24, to assess cavitation risk.

Although the method relies on empirical or derived two-dimensional section data, this is also one of its strengths as it allows tuned performance to account for the influence of sectional thickness, chord-based Reynolds number, and viscous-induced effects such as stall.

15.4.2 Lifting Line Theories

In these methods each blade section is represented by a single-line vortex whose strength varies from section to section [15.9, 15.10]. A trailing vortex sheet, behind each blade, is typically forced to follow a suitable helical surface. As a result, there is no detail as to the likely based variation in chordwise loading or location of the centre of effort. Such an approach is more suited to high aspect ratio blades which are lightly loaded, e.g. $J > 0.25$, and not those with significant skew.

15.4.3 Surface Panel Methods

The surface panel method may be considered as the workhorse computational tool for detailed propeller design that, if used appropriately, can predict propeller performance with a high degree of confidence. Difficulties arise in more extreme designs or where significant cavitation is expected. The typical process of applying this method requires a series of steps to develop the full surface geometry. Figure 15.1 illustrates such a process for solving a propeller design using the Palisupan surface panel code, Turnock [15.32]. In this process, a table of propeller section offsets, chord, pitch, rake, skew and thickness is processed to generate a series of sections, each consisting of a set of Cartesian coordinate nodes [15.3]. A bicubic spline interpolation is used to subdivide the complete blade surface into a map of Nt chordwise and Ns spanwise panels. The use of appropriate clustering functions allows the panels to be clustered near the leading and trailing edges as well as the tip. The quality of the numerical solution will strongly depend on selection of the appropriate number of panels for a given geometry. As the outboard propeller sections are thin ($t/c \sim 6\%$), a large number of panels are required around each section (50+) in order to avoid numerical problems at the trailing edge. Similarly, the aspect ratio of panels should typically not exceed three, so the spanwise number of panels should be selected to keep the panel aspect ratio below this threshold.

The most complex part of the process is the selection of the appropriate shape of the strip of wake panels that trail behind from each pair of trailing edge panels. Numerically, the surface panel method requires the trailing wake to follow a stream surface. However, the shape of this stream surface is not known *a priori*. What is known is that, as the wake trails behind the propeller, the race contracts and the conservation of angular momentum requires the local pitch to reduce. In the far wake, the vorticity associated with all the wake panels will have coalesced into a single tip vortex and a contra-rotating hub vortex for each blade. The stability

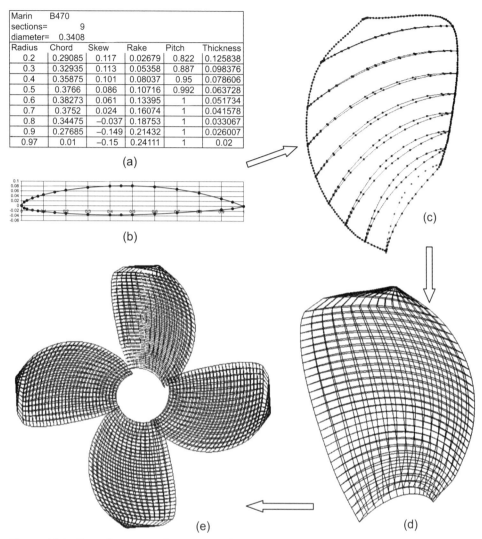

Marin	B470				
sections=	9				
diameter=	0.3408				
Radius	Chord	Skew	Rake	Pitch	Thickness
0.2	0.29085	0.117	0.02679	0.822	0.125838
0.3	0.32935	0.113	0.05358	0.887	0.098376
0.4	0.35875	0.101	0.08037	0.95	0.078606
0.5	0.3766	0.086	0.10716	0.992	0.063728
0.6	0.38273	0.061	0.13395	1	0.051734
0.7	0.3752	0.024	0.16074	1	0.041578
0.8	0.34475	−0.037	0.18753	1	0.033067
0.9	0.27685	−0.149	0.21432	1	0.026007
0.97	0.01	−0.15	0.24111	1	0.02

(a)

(b)

(c)

(d)

(e)

Figure 15.1. Propeller generation process for a surface panel code.

of numerical schemes that attempt to model such a process is always in question. As a result, practical techniques will make an assumption of the expected wake pitch and race contraction variation based on the expected propeller thrust loading [15.3, 15.33]. A practical way of doing this is to use blade element-momentum theory along with the wake contraction expressions of Gutsche [15.34]. A simpler approach is often adopted based on defining an average wake pitch that usually is chosen to be a suitable value between the geometric pitch and the far wake hydrodynamic pitch. It is worth noting that altering this wake pitch value can shift the thrust and torque values up or down for a given J. It is found that the propeller wake needs to be panelled for 5 to 10 diameters downstream and, as a result, the number of wake panels can be an order of magnitude higher than the number of blade panels.

The influence of the hub is required to ensure the correct circulation at the blade root. The panels on this hub are usually best aligned with the local geometric pitch of the propeller. The generation of the panels in this region, as shown in Figure 15.1, requires a suitable geometrical transformation to ensure orthogonality.

Once the propeller, hub and wake have had a panel geometry created, the numerical application of a surface panel method is straightforward. For a steady flow, the boundary condition on the propeller blade surface requires zero normal velocity based on the resultant velocity of the free stream and blade rotational speed. Rotational symmetry can used so that the problem is only solved for one of multiple blades and a segment of hub. It is good practice to investigate the sensitivity of the resultant propeller thrust and torque to the number of panels on the blade, hub and in the wake. An advantage of surface panel methods is that the blade loading can be applied directly to three-dimensional finite-element analysis based structural codes, as mentioned in Chapter 12, Section 12.3.3.

Unsteady versions of panel codes can be used to investigate the behaviour of a propeller in a hull wake or even to deal with the complex flow found in surface-piercing propellers [15.35].

15.4.4 Reynolds Averaged Navier–Stokes

As detailed in Chapter 9, it is not the intention in this book to give the full details of the complexity associated with applying RANS-based CFD. However, there are a number of practical aspects of using CFD flow solvers for ship propellers that are worth noting.

In addition to defining precisely the surface geometry of the propeller and hub, CFD codes will require a suitably created mesh of elements that fill the space within the solution domain. It is the definition of this domain that is particularly complex for a rotating propeller. The quality of the mesh and whether it suitably captures all the necessary flow features will determine the accuracy of the solution. In the case of the propeller, the viscous wake and its downstream propagation, along with the tip vortex, has a strong influence on the accurate prediction of forces [15.3, 15.36].

The mesh around the blade needs to be chosen to match the selected turbulence model and to expand at a suitable rate in the surface normal direction. Typically, at least 10 cells will be required within the turbulent boundary layer thickness which should have a first cell thickness of between 30 and 250 for 'law of the wall' turbulence models or <3 for those which capture the sublayer directly. A similar approach should be chosen for the hub. It should be noted that for many real hubs there may be a flow separation zone. The accurate capture of this behaviour can be important in determining an accurate prediction of propeller thrust, although it is worth noting that better design to avoid separation will improve the efficiency of the propeller.

For a steady and open flow condition only a single blade requires modelling with the interface faces selected as periodic boundary conditions. Typical domain size would extend to at least two diameters in a radial direction and in the upstream direction, whereas 5 to 10 diameters would be appropriate in the downstream direction.

Ideally, the mesh will consist of hexahedral cells, one of whose principal axes is aligned with the flow. Thus, a mesh which follows an approximately helical structure is more likely to avoid problems with numerical diffusion. The off-body features of the viscous wake and the vorticity sheet roll up into the tip and hub vortices are much more difficult to capture. A recommended approach, developed by Pashias [15.3] and refined by Phillips [15.37], uses the vortex identification technique, Vortfind of Pemberton *et al.* [15.38], to first run a coarse mesh that identifies the tip vortex near to the blade, predict its track and then generate a refined mesh suited to capturing a vortex for a suitable distance downstream. Resolving on this finer mesh, and progressively repeating the process as necessary, allows the wake structure to be maintained for significant distances ($>10D$) downstream.

In selecting the turbulence closure model it is worth considering that the behaviour of turbulent strain is likely to be anisotropic within the vortex core. The correct modelling of this behaviour will be important in controlling the accuracy with which the vortex core pressure is predicted and hence the likelihood of cavitation prediction in multiphase calculations.

15.5 Blade Element-Momentum Theory

The combination of axial momentum theory and analysis of section, or blade element, performance is used to derive a rapid and, with appropriate empirical correction factors, a powerful propeller analysis tool suitable for overall shape optimisation [15.39–15.41]. The approach combines the two initial strands of propeller analysis with the blade element, identifying the developed forces for a given flow incidence at a given section with the necessary momentum changes needed to generate those forces. This is illustrated in Figure 15.2(a) where the blade element approach provides information on the action of the blade element but not the momentum changes (induced velocities a, a') whilst the momentum approach, Figure 15.2(b), provides information on the momentum changes (a, a') but not the actual action of the blade element. The problem can be solved by combining the theories in such a way that that part of the propeller between radius r and ($r + \delta r$) is analysed by matching forces generated by the blade elements, as two-dimensional lifting foils, to the momentum changes occurring in the fluid flowing through the propeller disc between these radii.

15.5.1 Momentum Theory

Simple actuator disc theory shows that the increment of axial velocity at the disc is half that which occurs downstream. It can be shown that the same result is true

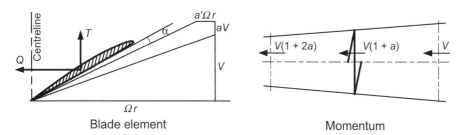

Figure 15.2. Blade element and momentum representations of propeller action.

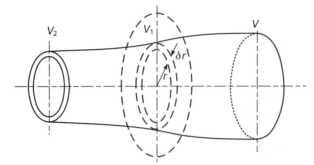

Figure 15.3. Annulus breakdown of momentum through propeller disc.

of the angular momentum change. Figure 15.3 illustrates the changes to an annular stream tube as it passes through an actuator disc. An actuator disc, defined as having no thickness, is porous so that flow passes through it, but yet develops a pressure increase due to work being done on the fluid.

The relative axial velocity at disc $V_1 = V(1+a)$, where a is the axial inflow factor. Similarly, if the angular velocity relative to the blades forward of the propeller is Ω, then the angular velocity relative to the blades at disc is $\Omega(1 - a')$, where a' is the circumferential inflow factor.

Consider the flow along an annulus of radius r and thickness δr at the propeller disc of an infinitely bladed propeller. The thrust and torque on the corresponding section of the propeller can be obtained from the momentum changes occurring as the fluid flow downstream of the annulus. Definitions are as follows:

Speed of advance of propeller $= V$.
Angular velocity of propeller $= \Omega$.
Disc radius $= R$.
Axial velocity at disc $V_1 = V(1 + a)$, where a is the axial inflow factor.
Axial velocity in wake $V_2 = V(1 + 2a)$.
Fluid angular velocity at disc $\omega_1 = a'\Omega$, where a' is the circumferential inflow factor.
Fluid angular velocity in wake $\omega_2 = 2a'\Omega$.

The mass flow rate through annulus is $2\pi r \delta r \rho V(1 + a)$ and the thrust on the annular disc will be equal to the axial rate of momentum change, as follows:

$$\delta T = 2\pi r \delta r \rho V (1 + a)(V_2 - V) = 2\pi r \delta r \rho V^2 (1 + a) 2a \quad \text{as } V_2 = V(1 + 2a).$$

The torque on element is the angular momentum change (or moment of momentum change), as follows:

$$\delta Q = 2\pi r \delta r \rho V(1 + a) r^2 \omega_2 = 2\pi r \delta r \rho V(1 + a) r^2 2a'\Omega.$$

Thus, the thrust and torque loadings per unit span on the propulsor are as follows:

$$\frac{dT}{dr} = 4\pi \rho r V^2 a(1 + a) \tag{15.1}$$

and

$$\frac{dQ}{dr} = 4\pi \rho r^3 \Omega V a'(1 + a). \tag{15.2}$$

15.5.1.1 Correction for Finite Number of Blades

With a finite number of blades, flow conditions will not be circumferentially uniform and the average inflow factors will differ from those at the blades. An averaging factor, K, called the Goldstein factor, can be introduced and Equations (15.1) and (15.2) can be rewritten as follows:

$$\frac{dT}{dr} = 4\pi\rho r V^2 Ka(1+a) \tag{15.3}$$

and

$$\frac{dQ}{dr} = 4\pi\rho r^3 \Omega\, V Ka'(1+a), \tag{15.4}$$

where a and a' are now values at the blade location. Lifting line theory can be used to calculate K and charts are available for propellers with 2–7 blades, as discussed in the next section.

The local section efficiency η can be obtained from these equations as follows:

$$\eta = \frac{P_E}{P_D} = \frac{TV}{2\pi n Q} = \frac{TV}{\Omega Q} \quad \text{and} \quad \eta = \frac{V\dfrac{dT}{dr}}{\Omega\dfrac{dQ}{dr}} = \left(\frac{V}{r\Omega}\right)^2 \frac{a}{a'}. \tag{15.5}$$

These basic momentum equations can be put into a non-dimensional form as follows:

Write $r = xR$, R = disc radius, D = disc diameter and $\Omega = 2\pi n$, n = rps.

$$dT = \rho\mathrm{n}^2 D^4\, dK_T \quad dr = R\, dx = \frac{D}{2}dx$$

$$dQ = \rho\mathrm{n}^2 D^5 dK_Q \quad J = \frac{V}{nD},$$

whence Equation (15.3) becomes

$$\frac{dK_T}{dx} = \pi J^2 x Ka(1+a), \tag{15.6}$$

Equation (15.4) becomes

$$\frac{dK_Q}{dx} = \frac{1}{2}\pi^2 J x^3 Ka'(1+a) \tag{15.7}$$

and Equation (15.5) becomes

$$\eta = \left(\frac{J}{\pi x}\right)^2 \frac{a}{a'}. \tag{15.8}$$

15.5.2 Goldstein K Factors [15.8]

Goldstein analysed the flow induced by a system of constant pitch helical surfaces of infinite length and produced a method of computing average momentum flux, as compared with infinite blades, in terms of the fluid velocities on the surface of the sheets in way of the blades.

Several authors have published calculated values of Goldstein K factors for sheets with 2–7 blades [15.39–15.41]. Figure 15.4 illustrates typical charts for three and four blade propellers. Widely differing values can be seen for different radii x and $\lambda_i = x \tan\phi$, where ϕ is the local section hydrodynamic pitch angle, Figure 15.5.

It should be noted that, in some theories, such as the theory of Burrill [15.42], the slipstream contraction is allowed for and separate Goldstein factors applied at the disc *and* downstream.

A suitable functional relationship for K, due to Wellicome, is given by:

$$K = \frac{2}{\pi} \cos^{-1}\left(\frac{\cosh(xF)}{\cosh(F)}\right)$$

where $F = \frac{Z}{2x \tan\phi} - \frac{1}{2}$ for $F \leq 85$, otherwise $K = 1$, and Z is the number of blades.

15.5.3 Blade Element Equations

A velocity vector diagram including the inflow velocity components induced by the propeller action is shown in Figure 15.5. The axial inflow increases the relative fluid velocity whilst the circumferential inflow reduces the *relative* velocity, since the angular velocity produced in the fluid is in the same sense as the blade rotation.

Using two-dimensional section data the spanwise lift and drag forces on the blade can be expressed as follows:

$$\frac{dL}{dr} = \tfrac{1}{2}\rho Z c U^2 C_L(\alpha) \tag{15.9}$$

$$\frac{dD}{dr} = \tfrac{1}{2}\rho Z c U^2 C_D(\alpha). \tag{15.10}$$

where Z is the number of blades, c is the blade chord and lift and drag coefficients C_L and C_D depend on angle of attack α. From Equations (15.9) and (15.10) $\tan\gamma = \frac{C_D(\alpha)}{C_L(\alpha)}$ and from the vector diagram,

$$\tan\psi = \frac{V}{\Omega r} = \frac{J}{\pi x} \tag{15.11}$$

and

$$\tan\phi = \frac{V(1+a)}{\Omega r(1-a')} = \frac{1+a}{1-a'} \cdot \tan\psi. \tag{15.12}$$

The local section pitch P is the sum of the induced flow angle ϕ and the effective angle of attack α, Figure 15.6, and $2\pi r = 2\pi xR = \pi xD$. Hence,

$$\tan(\phi + \alpha) = \frac{P}{2\pi r} = \frac{P}{2\pi xR} = \frac{P}{\pi xD} = \left(\frac{P/D}{\pi x}\right). \tag{15.13}$$

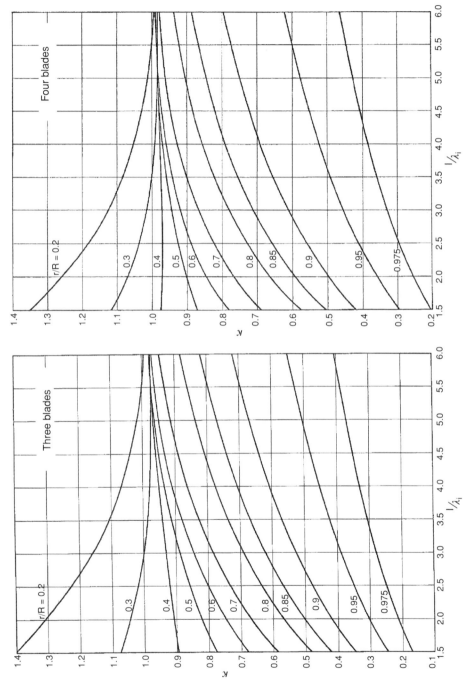

Figure 15.4. Goldstein K factors for three- and four-bladed propellers [15.40].

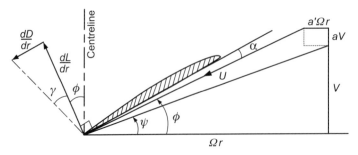

Figure 15.5. Blade element diagram.

The section lift and drag can be resolved to give the section thrust and torque as follows:

$$\frac{dT}{dr} = \frac{dL}{dr} \cdot \cos\phi - \frac{dD}{dr} \cdot \sin\phi = \frac{dL}{dr} \cdot \cos\phi(1 - \tan\phi\tan\gamma) \qquad (15.14)$$

$$\frac{dQ}{dr} = r\left(\frac{dL}{dr} \cdot \sin\phi + \frac{dD}{dr} \cdot \cos\phi\right) = r\frac{dL}{dr}\cos\phi(\tan\phi + \tan\gamma). \qquad (15.15)$$

From the velocity diagram

$$U = r\Omega(1 - a')\sec\phi$$
$$= \pi n D x(1 - a')\sec\phi.$$

Combining Equations (15.14) and (15.9) then,

$$2\rho n^2 D^3 \frac{dK_T}{dx} = \frac{1}{2}\rho Z c C_L \cdot \pi^2 n^2 D^2 x^2 (1 - a')^2 \sec\phi(1 - \tan\phi \cdot \tan\gamma)$$

$$dr = \frac{D}{2}dx$$

$$\therefore \quad \frac{dK_T}{dx} = \frac{\pi^2}{4}\left(\frac{Zc}{D}\right)C_L \cdot x^2(1 - a')^2 \sec\phi(1 - \tan\phi \cdot \tan\gamma). \qquad (15.16)$$

Similarly, Equation (15.15) becomes

$$\frac{dK_Q}{dx} = \frac{\pi^2}{8}\left(\frac{Zc}{D}\right)C_L \cdot x^3(1 - a')^2 \sec\phi(\tan\phi + \tan\gamma). \qquad (15.17)$$

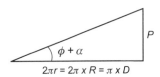

Figure 15.6. Pitch angle.

Equations (15.14) and (15.15) can be combined into an alternative equation for local efficiency, as follows:

$$\eta = \frac{V\dfrac{dT}{dr}}{\Omega\dfrac{dQ}{dr}} = \frac{V}{r\Omega} \cdot \frac{1 - \tan\phi\tan\gamma}{\tan\phi + \tan\gamma} = \frac{\tan\psi}{\tan(\phi + \gamma)}. \tag{15.18}$$

This equation can be expressed in various forms, as follows:

$$\eta = \frac{\tan\psi}{\tan\phi} \cdot \frac{\tan\phi}{\tan(\phi + \gamma)} = \frac{1 - a'}{1 + a} \cdot \frac{\tan\phi}{\tan(\phi + \gamma)} \tag{15.19}$$

or

$$\eta = \eta_a \times \eta_r \times \eta_f, \tag{15.20}$$

where $\eta_a = \frac{1}{1+a}$ ideal, per actuator disc theory (Froude efficiency), $\eta_r = (1 - a')$, the rotational loss factor and $\eta_f = \frac{\tan\phi}{\tan(\phi+\gamma)}$ the blade friction drag loss factor. Typical values are $\eta_a = 0.80$, $\eta_r = 0.95$ and $\eta_f = 0.90$, giving an overall $\eta = 0.8 \times 0.95 \times 0.90 = 0.68$.

This particular breakdown of the components of open water efficiency is discussed further in Section 11.3.15, when considering propeller efficiency improvements and energy savings.

15.5.4 Inflow Factors Derived from Section Efficiency

There are two independently derived equations for η and the combination of these is fundamental to the solution of the blade element-momentum theory. From momentum theory, Equation (15.8),

$$\eta = \frac{a}{a'} \cdot \tan^2\psi$$

From blade element theory, Equation (15.18),

$$\eta = \frac{\tan\psi}{\tan(\phi + \gamma)}$$

and ideal efficiency

$$\eta_i = \frac{\tan\psi}{\tan\phi} = \frac{(1 - a')}{(1 + a)} \quad (C_D = 0; \quad \gamma = 0)$$

Hence,

$$(1 + a) = \frac{(1 - a')}{\eta_i}$$

and

$$a' = 1 - \eta_i(1 + a). \tag{15.21}$$

Also, from Equation (15.8),

$$a' = \frac{a\tan^2\psi}{\eta}, \tag{15.22}$$

equating Equations (15.21) and (15.22), as follows:

$$\frac{a \tan^2 \psi}{\eta} = 1 - \eta_i(1 + a)$$

$$= 1 - \eta_i - a\eta_i$$

$$a\left(\eta_i + \frac{\tan^2 \psi}{\eta}\right) = 1 - \eta_i$$

and finally,

$$a = \frac{1 - \eta_i}{\eta_i + \dfrac{1}{\eta}\tan^2 \psi}. \tag{15.23}$$

The overall design procedure, and the derivation of a, a', dK_T/dx and η, will normally be an iterative process, typically following the steps shown in Figure 15.7.

Since C_L and dK_T/dx are not very dependent on drag, a common approach is initially to assume $C_D = 0$, hence $\gamma = 0$ and $\eta = \eta_i$ for a first iteration, and an initial solution for a can be obtained from Equation (15.23). This can be used to derive dK_T/dx from Equation (15.6), hence, C_L from Equation (15.16). C_D can then be introduced to yield $\gamma \ (= \tan^{-1} C_D/C_L)$ and the actual η. The process is then repeated until convergence for η is achieved.

The process described so far is for an element of the propeller span dr at radius r $(x = r/R)$. This is repeated over the range of x values $(0.20 \rightarrow 1.0)$ and total values of K_T, K_Q and η obtained by quadrature.

15.5.5 Typical Distributions of a, a' and dK_T/dx

The radial variation of the inflow factors a and a' has the general form shown in Figure 15.8. Typical values are $a = 0.3$–0.4 and $a' = 0.02$–0.04.

The actual thrust loading (dK_T/dx) curve exhibits the general form shown in Figure 15.9, noting that the presence of the hub holds up the thrust generated towards the root and that, for reasons of structural strength, open water propellers tend to have less thrust per unit span towards the tip.

15.5.6 Section Design Parameters

The design of two-dimensional sections for propellers has many similarities to the design of those for lifting surfaces such as rudders or aircraft wings. The main aspects of the flow regime and associated section performance that require understanding are the following:

(i) The local chord-based Reynolds number.
(ii) The required local camber.
(iii) The necessary strength required at a given section, for example, the second moment area and associated strength of the material used for construction; this often influences the selected thickness, camber and choice of section type.
(iv) Likelihood of different types of cavitation.
(v) Sensitivity to imperfections such as roughness and fouling.

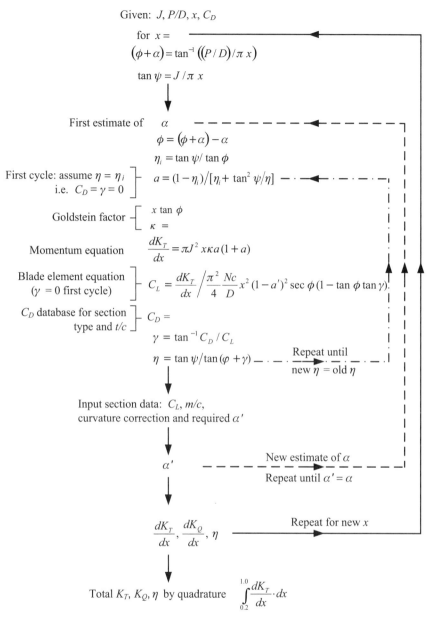

Given: $J, P/D, x, C_D$

for $x =$

$(\phi + \alpha) = \tan^{-1}\left((P/D)/\pi x\right)$

$\tan\psi = J/\pi x$

First estimate of $\quad \alpha$

$\phi = (\phi + \alpha) - \alpha$

$\eta_i = \tan\psi/\tan\phi$

First cycle: assume $\eta = \eta_i$ $\quad a = (1 - \eta_i)/[\eta_i + \tan^2\psi/\eta]$
i.e. $C_D = \gamma = 0$

Goldstein factor $\quad \kappa = \dfrac{x\tan\phi}{}$

Momentum equation $\quad \dfrac{dK_T}{dx} = \pi J^2 x\kappa a(1 + a)$

Blade element equation $\quad C_L = \dfrac{dK_T}{dx} \bigg/ \dfrac{\pi^2}{4}\dfrac{Nc}{D}x^2(1 - a')^2 \sec\phi(1 - \tan\phi\tan\gamma)$
($\gamma = 0$ first cycle)

C_D database for section $\quad C_D =$
type and t/c

$\gamma = \tan^{-1} C_D/C_L$

$\eta = \tan\psi/\tan(\varphi + \gamma)$ — · — Repeat until
new η = old η

Input section data: $C_L, m/c,$
curvature correction and required α'

α' — — — New estimate of α
Repeat until $\alpha' = \alpha$

$\dfrac{dK_T}{dx}, \dfrac{dK_Q}{dx}, \eta$ $\quad$ Repeat for new x

Total K_T, K_Q, η by quadrature $\quad \displaystyle\int_{0.2}^{1.0}\dfrac{dK_T}{dx}\cdot dx$

Figure 15.7. Flow algorithm for blade element-momentum theory.

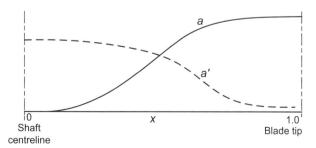

0   x 1.0
Shaft   Blade tip
centreline

Figure 15.8. Distributions of a, a' on a typical propeller.

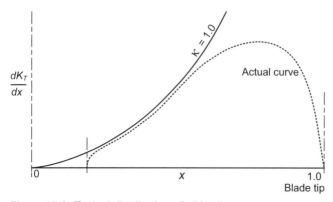

Figure 15.9. Typical distribution of dK_T/dx.

Most of these factors, and how they influence the selection of an appropriate section and the design of a specific section, need to be borne in mind when considering section design. Section selection is discussed in Carlton [15.2] and in Molland and Turnock [15.43]. An excellent analysis tool, Xfoil, developed by Drela [15.44], is suitable for foil section design and optimisation.

15.5.7 Lifting Surface Flow Curvature Effects

The velocity diagram used in the blade element analysis applies essentially at about blade mid chord of the relatively wide blade (large c/D and BAR) employed for marine propellers. As the flow progresses downstream, the inflow velocities (and, hence, local a and a' factors) are continuously changing, as shown in Figure 15.10. This is particularly true of a' which varies rapidly near the propeller. The result is a progressive rotation of the local flow vector as the fluid passes the propeller blade so that the propeller is working in a curving flow, as shown schematically in Figure 15.11.

Because of the flow curvature, the effective camber of the section, Figure 15.12, is reduced and this results in a loss of lift at a fixed angle of attack compared with the two-dimensional section performance in a straight flow, Figure 15.13. This loss in lift can be compensated for by either an increase in camber or a change of angle of attack (i.e. in local pitch), compared with the values from two-dimensional section data. Within the limits imposed by cavitation criteria and possible manufacturing constraints, camber and angle of attack can be considered as interchangeable.

Detailed two-dimensional section data, see for example [15.45], can be used for precise information, but the following values give a broad guide:

$$\text{(i)} \qquad \frac{dC_L}{d\alpha} = 0.10 \text{ per degree} \qquad (15.24)$$

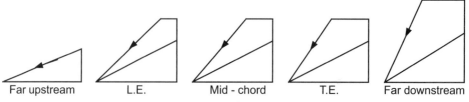

Figure 15.10. Effect of flow acceleration on effective incidence.

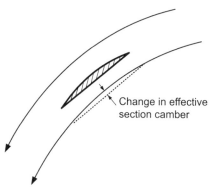

Figure 15.11. Change of flow direction with section camber.

is a suitable approximation for most sections.

$$(ii) \quad \frac{dC_L}{d(m/c)} = \begin{array}{l} 9 \text{ NACA a} = 0.5 \text{ mean line} \\ 11 \text{ NACA a} = 0.7 \text{ mean line} \\ 12 \text{ Round back sections} \end{array} \quad (15.25)$$

where m/c is the maximum camber/chord.

15.5.8 Calculations of Curvature Corrections

The earliest curvature corrections were empirical and involved parameters chosen to bring calculated thrust loadings into line with open water test data [15.39, 15.42].

Common practice is to use curvature corrections based on a theoretical model devised by Ludweig and Ginzel [15.46] or further variants of this type of model [15.47]. This model represents the earliest use of a vortex lattice model to represent a marine propeller blade. Curvature corrections derived from this model have been published in various papers. One commonly used correction diagram is that given in Eckhardt and Morgan [15.40], which gives a two-stage correction based on the assumption that the blade sections were designed for $\alpha = 0$ (or to run at $\alpha = \alpha_i$).

Figure 15.14 shows an example of a Ludweig–Ginzel chart that gives the camber correction as a function of blade area ratio, where $x = r/R$ and $\lambda_i = x \tan \phi$. Restrictions on the Ludweig–Ginzel model were as follows:

1. The blade outline was elliptical and symmetric about mid chord.
2. The chordwise loading was constant ($\Delta P =$ constant) corresponding to a NACA $a = 1.0$ mean line.
3. The radial load distribution was that for open water optimum loading.
4. The calculations were restricted to the induced velocities and rate of change of induced velocities at midchord.

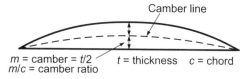

Figure 15.12. Definition of camber.

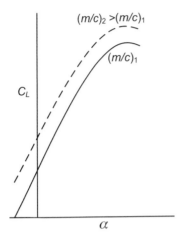

Figure 15.13. Change in lift with flow-induced curvature and reduced camber.

Less restrictive curvature corrections have been developed, such as those by Morgan *et al.* [15.47].

The Ludweig–Ginzel approach may be summarised as follows. The required camber of a propeller at $\alpha = 0$ is obtained as follows:

$$\frac{m}{c} = k_1 k_2 \cdot \frac{m_o}{c}, \tag{15.26}$$

where m_o/c is the camber required to produce the same C_L with straight two-dimensional flow. For example, consider a section at $x = 0.70$ of a propeller

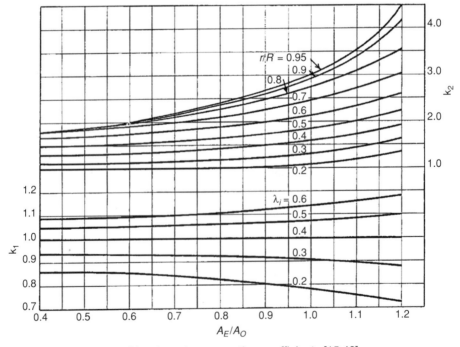

Figure 15.14. Ludweig–Ginzel camber correction coefficients [15.40].

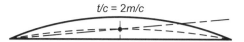

Figure 15.15. Use of camber to give a flat back.

with $A_D = 0.920$, producing $C_L = 0.145$ at $\phi = 23.45°$.

$$\lambda_i = x \tan \phi = 0.304, \qquad k_1 = 0.92 \qquad \text{and } k_2 = 2.3,$$

using Equation (15.25) for a round back (RB) section, $\frac{m_o}{c} = \frac{0.145}{12} = 0.0121$ then at $\alpha = 0$, $\frac{m}{c} = 0.92 \times 2.3 \times 0.0121 = 2.12 \times 0.0121 = 0.0256$ in curved flow. Assume $\alpha = 1°$, then $dC_L = 0.1 \times 1.0 = 0.10$ using Equation (15.24) in straight flow. The equivalent camber reduction is $\delta(m/c) = \frac{0.10}{12} = 0.0084$. Hence, the camber required for operation at $\alpha = 1°$ is $m/c = 0.0256 - 0.0084 = 0.0172$.

This information can be used to designate a round back section whose thickness is $t/c = 2m/c$, as shown in Figure 15.15.

In this particular case, camber would be provided by a flat-faced RB section of thickness $t/c = 0.0172 \times 2 = 0.0344$ which would be about right at this radius from the strength viewpoint.

The curvature correction applied directly as just a camber correction can lead to hollow-faced sections, Figure 15.16. These have the drawback of being difficult to manufacture and often have poor astern performance.

15.5.9 Algorithm for Blade Element-Momentum Theory

Figure 15.7 shows a flow chart for the application of blade element-momentum theory for assessing propeller performance. The overall flow path lends itself to a computer-based process. The Goldstein factors, Figure 15.4, and the Ludweig–Ginzel curvature corrections, Figure 15.14, can be incorporated as curve fits or lookup tables. The drag coefficient can be obtained using Figure 15.17 which shows data derived from Hill [15.39] and Burrill [15.42]. It can be noted that working angles of attack for a blade section are typically up to about $2°$ and, over the main working portion of the blade span, the drag coefficient is of the order of $C_D = 0.008$–0.015.

Such an algorithm can be used as part of an optimisation process by supplying the necessary outer loops that supply test values of overall advance ratio J and P/D ratio for each radius. Progressively more sophisticated approaches can be adopted. For example, a constant value of C_D and lift–curve slope can be used, or a look–up table that supplies the two-dimensional values for the local angle of attack and sectional Reynolds number. Such a modification allows the influence of stall and other viscous-related effects to be included within the optimisation process.

Section chord size c (hence BAR) can be chosen for a required local C_L (based on cavitation number) to avoid cavitation. The drag coefficient can be systematically varied to investigate the influence of thickness or roughness on efficiency or the

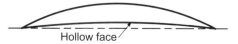

Figure 15.16. Example of a hollow face.

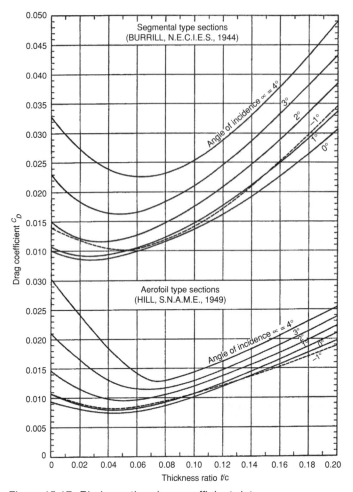

Figure 15.17. Blade section drag coefficient data.

likely influence of scale effect. The radial load distributions can be investigated, such as off-loading the tip by reducing P/D to avoid cavitation and the overall influence on efficiency. The algorithm has been used to derive the velocities a and a' induced by the propeller across the blade span and to predict the direction and velocity of flow onto a rudder downstream of a propeller [15.48].

The influence of a real ship wake can be included through the use of a modified inflow speed at each radius based on $w_{T'}$, as explained in Chapter 8. This allows a wake-adapted propeller to be found directly using the blade element-momentum theory algorithm.

15.6 Propeller Wake Adaption

15.6.1 Background

All the necessary elements of a simplified method of designing a marine propeller have been considered, using the blade element-momentum equations, the Goldstein K factor tip loss correction and the Ludweig–Ginzel curvature correction. Provided

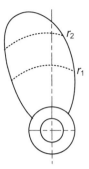

Figure 15.18. Two sample sections efficiency compared at two radii.

the blades are not too close to cavitation inception, when more rigorous methods are needed, and provided the blockage due to blade thickness can be ignored (which basically implies a small number of blades), these methods are sufficiently accurate for normal design purposes.

One missing parameter is the type of radial load distribution to choose. In most cases, the design is for a propeller to work in a non-uniform wake field behind a ship. The propeller is to be designed for the circumferential average flow conditions at each radius in order to achieve the best overall quasi-propulsive coefficient to minimise the delivered power requirement P_D for a specified effective power P_E.

15.6.2 Optimum Spanwise Loading

From the basic equation (see Section 16.1),

$$\eta_D = \eta_o \times \eta_R \times \eta_H = \eta_b \times \eta_H,$$

where $\eta_b = \eta_o \times \eta_R$ is the efficiency behind the ship. It can be argued using a comparison of load at two radii, as shown in Figure 15.18, as follows:

1. By altering the pitch distribution the load can be redistributed from radius r_1 to radius r_2 or vice versa to achieve the desired thrust T.
2. If the local efficiency $\eta_{D1} > \eta_{D2}$ (i.e. position 2 is more heavily loaded than position 1), then a transfer of load from r_1 to r_2 will reduce the overall η_D whilst a transfer from r_2 to r_1 will improve η_D. (Note, $\eta_H = (1 - t)/(1 - w_T)$, $w_{T1} > w_{T2}$ and $\eta_{H1} > \eta_{H2}$).
3. If any pair of radii can be found for which $\eta_{D1} \neq \eta_{D2}$, then a change of pitch distribution can improve η_D overall and hence, as it stands, the propeller will not be optimum.
4. The only case where no improvement can be made is when $\eta_D =$ constant from boss to tip.

Thus the propeller is optimum if $\eta_D = \eta_b \cdot \eta_H =$ constant.

In terms of thrust deduction fraction t and average (circumferential) wake fraction,

$$\eta_b \cdot \frac{1 - t}{1 - w_T} = \text{constant.} \tag{15.27}$$

This equation is the Van Manen optimum loading criterion. In the case of a propeller designed for open water $t = w_T = 0$ and then $\eta = $ constant (for all radii).

The above analysis is a simplification of the situation since it ignores the fact that a change of radial loading also produces a change of local efficiency. In other words, a transfer of load from r_1 to r_2 could reverse the inequality $\eta_{D1} < \eta_{D2}$. A more rigorous argument attributable to Betz results in the following criterion:

$$\eta_o \sqrt{\eta_H} = \text{constant.} \tag{15.28}$$

For a lightly loaded propeller, $\eta = \frac{1-a'}{1+a}$ (when $\gamma = 0$)

or

$$\eta^2 = \frac{(1-a')^2}{(1+a)^2} = \frac{1-2a'}{1+2a}$$

if a and a' are small, so that Equation (15.28) can be written in an alternative form, as follows:

$$\frac{1-2a'}{1+2a} \cdot \frac{1-t}{1-W_T} = \text{constant.} \tag{15.29}$$

Both forms can be found in published papers and are referred to either as the Betz condition or as the Lerbs condition. In practical operation, w_T can be measured, but local values of t cannot be easily assessed. Lerbs recommends taking t constant and applying Equation (15.28) in the following form:

$$\eta_0 \propto \sqrt{1 - w_T}. \tag{15.30}$$

Van Manen prefers to assume, with some theoretical justification, that $(1 - t) \propto (1 - w_T)^{1/4}$, in which case Equation (15.27) reduces to

$$\eta \propto (1 - w_T)^{3/4}. \tag{15.31}$$

The differences between pitch distributions and overall performance of propellers designed according to either of these two criteria are generally small. Pitch distributions of propellers designed to fit a normal radial mean wake variation (see Chapter 8) show a 10%–20% pitch reduction towards the boss. The open water optimum is nearly constant pitch as shown schematically in Figure 15.19. A propeller which has a pitch distribution calculated as optimum for a particular wake is said to be a *wake-adapted* propeller. Wake-adapted propellers should give $\eta_R > 1$, as they perform better behind the ship than in open water.

It should be noted that in order to suppress the tip vortex and hence to reduce propeller noise, the pitch of naval propellers is frequently reduced below the optimum design at the blade tip. Also, tip loadings for commercial propellers may be reduced to reduce blade stresses and to reduce cavitation and propeller-excited vibration. Off-loading the tip, for example by reducing P/D locally or by changing the blade shape, Figure 15.20, leads to a redistribution of load as shown schematically in Figure 15.21.

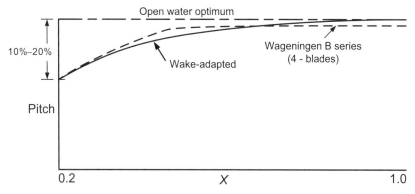

Figure 15.19. Typical pitch distributions.

A worked example for a wake-adapted propeller is included as Example 17 in Chapter 17.

15.6.3 Optimum Diameters with Wake-Adapted Propellers

Circumferential average wake values are higher near the boss and, hence, average η_H values increase as the propeller diameter is reduced. An optimum $\eta_D = \eta_b \eta_H$ is required and, for slight reductions in diameter, a gain in η_H more than offsets a small loss in η_b. Thus, the wake-adapted optimum diameter is *less* than the open water optimum as derived from design charts.

On the basis of optimum diameter calculations from blade element-momentum theory, Burrill [15.49] recommended that diameters computed from open water charts should be reduced as follows: single-screw merchant ships, 8% less (5% less, BSRA recommendation); twin-screw, 3% less; planing hull types, no reduction.

These figures reflect the degree of non-uniformity in each type of wake distribution. Pitch should be increased by about the same percentage that the diameter is reduced.

15.7 Effect of Tangential Wake

The origins of tangential wake are described in Chapter 8, Section 8.9. For propellers which operate in a non-zero tangential wake the blade element diagram can

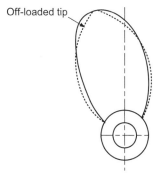

Figure 15.20. Change in shape to reduce propeller tip loading.

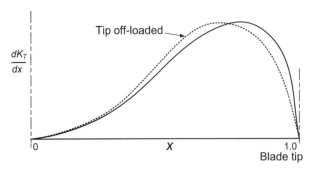

Figure 15.21. Redistribution of load with propeller tip off-loading.

be suitably modified. Figure 15.22 includes the necessary modifications. It can be seen that, depending on the sign of the tangential wake ($\pm a''$), the effective incidence will be either reduced or increased. The analysis can be correspondingly modified as follows. U_τ is taken to be the wake fraction in the plane of propeller and tangential to the radial direction, Figure 15.23. This is equivalent to a local rotation of wake.

$$\text{Tangential wake velocity } V_T = U_\tau V_S = \frac{U_\tau Va}{(1 - w_T)}$$

$$= r \, \Omega \, a'' \text{ say}$$

hence, the wake rotation factor

$$a'' = \frac{U_\tau}{(1 - w_T)} \frac{V_a}{r\Omega} = \frac{U_\tau}{(1 - w_T)} \tan \psi$$

that is a correction to $\tan \psi \cdot a''$ accounts for tangential wake in Figure 15.22.

The momentum equations remain unchanged, whereas the blade element equations now include the additional tangential component with the appropriate sign. Hence, the blade element efficiency Equation (15.18) becomes

$$\eta = \frac{\tan \psi}{\tan(\phi + \gamma)} = \frac{\tan \psi}{\tan \phi} \cdot \frac{\tan \phi}{\tan(\phi + \gamma)} = \frac{1 - a'' - a'}{1 + a} \cdot \frac{\tan \phi}{\tan(\phi + \gamma)} \qquad (15.32)$$

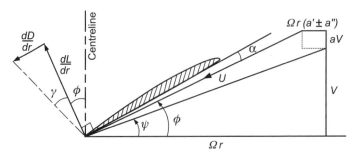

Figure 15.22. Blade velocity diagram including tangential wake.

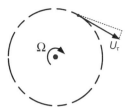

Figure 15.23. Definition of U_τ.

This replaces the earlier equation. The calculation of performance proceeds as for the non-rotating case. For example, for a downgoing blade (with upward flow), the blade element equations change as follows:

$$\eta_i = \frac{1 - a' + a''}{1 + a}, \quad \text{and} \quad \eta = \frac{a}{a'} \tan^2 \psi$$

giving

$$a = \frac{1 - \eta_i + a''}{\eta_i + \dfrac{1}{\eta} \tan^2 \psi}$$

and, neglecting friction

$$\frac{dK_T}{dx} = \pi x K J^2 a (1 + a) = \frac{\pi x}{2} \left(\frac{C \cdot Z}{D} \right) (1 - a' + a'') \sec \phi \, C_L$$

where

$$\tan \phi = \frac{1 + a}{1 - a' + a''} \tan \psi.$$

For an upgoing blade, the sign would change to $- a''$.

15.8 Examples Using Blade Element-Momentum Theory

15.8.1 Approximate Formulae

The following approximate formulae, based on one section only, are useful for preliminary calculations. Simple estimates can be made using one representative propeller section only, for example, at $x = 0.70$, and estimating the overall performance from the data at this section using a standard approximation to the $\frac{dK_T}{dx}$ and $\frac{dK_Q}{dx}$ curves.

Such estimates frequently assume that $\frac{dK_T}{dx} \propto x^2 \sqrt{1 - x}$ and that $\frac{dK_Q}{dx}$ also varies in this way.

The section $x = 0.70$ is chosen as the basis. On this basis and assuming a normal boss radius ($x = 0.20$), it is found that the propeller coefficients are as follows:

$$K_T = 0.559 \left(\frac{dK_T}{dx} \right)_{x=0.7} \tag{15.33}$$

and

$$K_Q = 0.559 \left(\frac{dK_Q}{dx} \right)_{x=0.7}. \tag{15.34}$$

Propeller outlines vary somewhat, but a typical blade width at $x = 0.70$ is as follows:

$$\left(\frac{Z \cdot c}{D}\right)_{x=0.7} = 2.2 A_D \pm 5\%, \tag{15.35}$$

where Z is the number of blades and A_D is the developed blade area ratio $\approx$ Blade area ratio (BAR).

15.8.2 Example 1

An example of an approximate preliminary performance estimate (excluding detailed section design) follows.

Consider section $x = 0.70$ on a propeller with $P/D = 1.0$ operating at $J = 0.70$ and $\alpha = 1.0°$.

From Equation (15.11),

$$\tan \psi = \frac{J}{\pi x} = 0.3183.$$

From vector diagram,

$$\tan(\phi + \alpha) = \frac{P/D}{\pi x} = 0.4547,$$

$$\therefore \quad \phi + \alpha = 24.45°$$

$$\phi = 23.45°.$$

Assumed section data are as follows:

$C_L = 0.145$ (chosen to suit cavitation number).
$C_D = 0.010$ (from section data).
$\alpha = 1.0°$ (to suit required C_L).
From $\tan \gamma = C_D/C_L$, $\gamma = 3.95°$.
From Equation (15.18), $\eta_f = \frac{\tan \phi}{\tan(\phi + \gamma)} = 0.837$.
$\eta_i = \frac{1-a'}{1+a} = \frac{\tan \psi}{\tan(\phi + \gamma)} = 0.734$ \quad (with $\gamma = 0$)
and overall efficiency $\eta = \eta_i \times \eta_f = 0.614$.
From Equation (15.23), $a = \frac{1-\eta_i}{\eta_i + \tan^2 \psi/\eta} = 0.296$
and $a' = \frac{a}{\eta} \cdot \tan^2 \psi = 0.0488$.
$\lambda_i = x \tan \phi = x \tan 23.45 = 0.304$.

From the Goldstein chart for a three-bladed propeller, Figure 15.4(a), $K = 0.82$.
From Equation (15.6), $\frac{dK_T}{dx} = \pi J^2 x K a(1 + a) = 0.339$.
From Equation (15.33) estimated overall $K_T = 0.559 \times 0.339 = 0.190$.
From Equation (15.7), $\frac{dK_Q}{dx} = \frac{\pi^2}{2} J x^3 K a'(1 + a) = 0.0615$.
From Equation (15.34), estimated overall $K_Q = 0.559 \times 0.0615 = 0.0344$.
From Equation (15.16)

$$\frac{dK_T}{dx} = \frac{\pi^2}{4}\left(\frac{Zc}{D}\right) C_L x^2 (1 - a')^2 \sec \phi (1 - \tan \phi \tan \gamma).$$

Hence, $Zc/D = 2.021$

From Equation (15.35), estimated $A_D(=\text{BAR}) = 0.920$.

In summary, the estimate indicates the following performance from a propeller with $A_D = 0.92$:

$$P/D = 1.0 \text{ at } J = 0.70 : K_T = 0.190, \ K_Q = 0.0344, \ \eta = 0.614.$$

This can be compared with the Gawn series $A_D = 0.92$ (interpolating for BAR $= 0.80$–0.95), $K_T = 0.188$, $K_Q = 0.034$, $\eta = 0.630$.

15.8.3 Example 2

These calculations exclude detailed section design: given J, P/D or η_i, C_D. Design for given α, e.g. if given P/D,

$$(\phi + \alpha) = \tan^{-1} \frac{P/D}{\pi x}.$$

$$\phi = (\phi + \alpha) - \alpha.$$

hence $\tan \phi$ and $\tan \psi = J/\pi x$.

$$\eta_i = \tan \psi / \tan \phi.$$

If given η_i,

$$\tan \phi = \tan \psi / \eta_i.$$

$$\phi = \tan^{-1}(\tan \psi / \eta_i) \ (\phi + \alpha) = \phi + \alpha \quad \text{and} \quad \frac{P/D}{\pi x} = \tan^{-1}(\phi + \alpha),$$

hence, P/D for required η_i.

EXAMPLE. Given the data below (for $x = 0.70$) calculate dK_T/dx, C_L and the overall section efficiency at radius $x = 0.70$ for a propeller designed to operate at $J = 0.65$, $P/D = 0.85$, $Z \cdot c/D = 1.50$, $\alpha = 0.50^0$, $C_D = 0.010$,

$$\tan(\phi + \alpha) = \frac{P/D}{\pi x} = \frac{0.85}{\pi \times 0.7} = 0.3865,$$

$$(\phi + \alpha) = 21.13^\circ.$$

$\alpha = 0.50^\circ$ given

$$\text{Then } \phi = 20.63^\circ \quad \{\tan \phi = 0.376\}$$

$$\tan \psi = J/\pi x = 0.65/\pi \times 0.70 = 0.296.$$

$$\eta_i = \tan \psi / \tan \phi = \frac{0.296}{0.376} = 0.786.$$

Using Equation (15.23),

$$a = \frac{1 - \eta_i}{\eta_i + \tan^2 \psi / \eta}.$$

Blade drag loss has a relatively small influence on the working values of C_L and α, hence initially assume $\gamma = 0$, $(C_D = 0)$ and $\eta = \eta_i$. Hence, the first estimate of a is as follows:

$$a = (1 - 0.786)/\left(0.786 + \frac{0.296^2}{0.786}\right) = 0.2384 \quad \{0.2319\}$$

$\lambda_i = x \tan\phi = 0.7 \times 0.376 = 0.263$ and $1/\lambda_i = 3.8$.

From the Goldstein chart for four blades, Figure 15.4(b), $K = 0.92$, then

$$\frac{dK_T}{dx} = \pi J^2 x K a (1 + a) = 0.2524 \quad \{0.2441\}$$

also

$$\frac{dK_T}{dx} = \frac{\pi^2}{4}\left(\frac{Zc}{D}\right) C_L x^2 (1 - a')^2 \phi \quad (\text{assuming } \gamma = 0)$$

$$[\text{and } (1 - a') = \eta_i(1 + a) = 0.9734] \quad \{0.9683\},$$

whence

$$C_L = 0.1375 \quad \{0.1343\}$$

$$\tan\gamma = \frac{C_D}{C_L} = \frac{0.01}{0.1375} = 0.0727 \quad \{0.07446\}$$

and

$$\gamma = 4.160° \quad \{4.258°\}$$

As an approximate correction for blade drag write:

$$(\text{see Equation (15.16)}) \quad \frac{dK_T}{dx} = 0.2524 \times (1 - \tan\phi\tan\gamma) = 0.2455$$

and overall

$$\eta = \frac{\tan\psi}{\tan(\phi + \gamma)} = 0.641 \quad \{0.638\}$$

or iterate with a new value of η, hence, new a, and correct $\frac{dK_T}{dx} = 0.2441$ etc. The second iteration is shown in {braces}.

15.8.4 Example 3

The previous calculations are extended to include section design and derivation of required α.

Estimate K_T, K_Q, η at $J = 1.20$ for a three-bladed propeller with $A_D = 0.95$, $P/D = 1.40$. Approximate geometric data at $x = 0.70$, from Equation (15.35):

$$\frac{Zc}{D} = 2.09 \text{ (i.e. } = 2.2 \times A_D).$$

Assume section camber $m/c = 0.0175$ (i.e. $t/c = 3\frac{1}{2}\%$ for round back or segmental section) and assume $C_D = 0.008$ for this section thickness/type.

$$(\phi + \alpha) = \tan^{-1}\frac{P/D}{\pi x} = 32.48°$$

at

$$J = 1.20, \ \tan\psi = J/\pi x = 0.5457.$$

Working values of C_L and α are affected very little by blade drag, so initially assume $\gamma = 0$.

The calculations are carried out per Table 15.2. α' is the incidence required to produce the C_L value. $(m/c)_{\alpha=0}$ is derived from Equation (15.25) and using a Ludweig–Ginzel curvature correction and noting assumed $m/c = 0.0175$. K comes from the Goldstein chart.

In general, two to three iterations are all that are needed. From the third estimate in Table 15.2:

$$\tan\gamma = \frac{C_D}{C_L} = 0.008/0.0999 = 0.080$$

$$\therefore \qquad \gamma = 4.58°.$$

From Equation (15.18) the section efficiency can be computed as follows:

$$\eta = \tan\psi/\tan(\phi + \gamma) = \frac{0.5457}{\tan(32.22 + 4.58)} = 0.729.$$

From Equation (15.16), the effect of drag is to reduce $\frac{dK_T}{dx}$ by a factor $(1 - \tan\phi\tan\gamma)$ from which the final estimate of $\frac{dK_T}{dx}$ is as follows:

$$\frac{dK_T}{dx} = 0.2760 \ (1 - \tan(32.22)\tan(4.58)) = 0.2621$$

and $K_T = 0.559 \times 0.2621 = 0.147$ and, from the general equation, $K_Q = \frac{JK_T}{2\pi\eta} = 0.0384$, or repeat the whole cycle using η in Equation (15.23) to derive updated a (including drag), hence $\frac{dK_T}{dx}$. The method is shown in the full analysis path in Figure 15.7.

In this and the previous examples, for a more reliable estimate of total K_T and K_Q, the whole procedure should be repeated at each radius and overall values should be derived by quadrature.

A calculation to derive α' in the first iteration, with assumed $m/c = 0.0175$, is as follows:

First iteration: required $C_L = 0.1074$.
Assuming $\frac{dC_L}{d(m/c)} = 12$, Equation (15.25).
Camber required (for $C_L = 0.1074$ and $\lambda_i = x\tan\phi = 0.446$) is as follows:

$$(m/c)_{\alpha=0} = \frac{C_L}{12} \times k_1 \times k_2 = \frac{0.1074}{12} \times 1.03 \times 2.3 = 0.0212,$$
$$(k_1, k_2 \text{ from Figure 15.14}),$$

but actual camber $= 0.0175$.
Then, camber 'deficit' $= 0.0212 - 0.0175 = 0.0037$.

Table 15.2. *Iterations for α*

			Equation No.							
			15.18	15.23			15.6	15.16	15.25	15.24
Item	α	ϕ	η_i	a	$x \tan \phi$	K	$\frac{dK_T}{dx}$	C_L	$(m/c)_{\alpha=0}$	α'
1st estimate	0	32.48	.8572	.1185	.446	.705	.2959	.1074	.0212	0.440
2nd estimate	0.5°	31.98	.8745	.1033	.437	.711	.2567	.0926	.0183	0.09°
3rd estimate	0.26°	32.22	.8659	.1108	.441	.708	.2760	.0999	.0197	0.27°

The deficit is required to be made up by the incidence 'deficit' of lift $(C_L) = 0.0037 \times 12 = 0.044$. From Equation (15.24) $\frac{dC_L}{d\alpha} = 0.1$, and the incidence required $\alpha' = 0.044/0.1 = 0.44°$ (see first row of Table 15.2).

Further example applications of blade element-momentum theory are given in Chapter 17, Example Application 23.

REFERENCES (CHAPTER 15)

15.1 Breslin, J.P. and Anderson, P. *Hydrodynamics of Ship Propellers*. Cambridge Ocean Technology Series, Cambridge University Press, Cambridge, UK, 1996.

15.2 Carlton, J.S. *Marine Propellers and Propulsion*. 2nd Edition, Butterworth-Heinemann, Oxford, UK, 2007.

15.3 Pashias, C. Propeller tip vortex capture using adaptive grid refinement with vortex identification. PhD thesis, University of Southampton, 2005.

15.4 Rankine, W.J.M. On the mechanical principles of the action of propellers. *Transactions of the Institution of Naval Architects*, Vol. 6, 1865, pp. 13–39.

15.5 Froude, R.E. On the part played in propulsion by differences in fluid pressure. *Transactions of the Royal Institution of Naval Architects*, Vol. 30, 1889, pp. 390–405.

15.6 Froude, W. On the elementary relation between pitch, slip and propulsive efficiency. *Transactions of the Institution of Naval Architects*, Vol. 19, 1878, pp. 47–65.

15.7 Betz, A. Schraubenpropeller mit geringstem Energieverlust. *K. Ges. Wiss, Gottingen Nachr. Math.-Phys.*, 1919, pp. 193–217.

15.8 Goldstein, S. On the vortex theory of screw propellers. *Proceedings of the Royal Society, London Series A*, Vol. 123, 1929, pp. 440–465.

15.9 Prandtl, L. Application of modern hydrodynamics to aeronautics. *NACA Annual Report*, 7th, 1921, pp. 157–215.

15.10 Lerbs, H.W. Moderately loaded propellers with a finite number of blades and an arbitrary distribution of circulation. *Transactions of the Society of Naval Architects and Marine Engineers*, Vol. 60, 1952, pp. 73–123.

15.11 Sparenberg, J.A. Application of lifting surface theory to ship screws. *Proceedings of the Koninhlijke Nederlandse Akademie van Wetenschappen, Series B, Physical Sciences*, Vol. 62, No. 5, 1959, pp. 286–298.

15.12 Pien, P.C. The calculation of marine propellers based on lifting surface theory. *Journal of Ship Research*, Vol. 5, No. 2, 1961, pp. 1–14.

15.13 Kerwin, J.E. The solution of propeller lifting surface problems by vortex lattice methods. Report, Department Ocean Engineering, MIT, 1979.

15.14 van Manen J.D. and Bakker A.R. Numerical results of Sparenberg's lifting surface theory of ship screws. *Proceedings of 4th Symposium on Naval Hydrodynamics*, Washington, DC, 1962, pp. 63–77.

15.15 English J.W. The application of a simplified lifting surface technique to the design of marine propellers. National Physical Laboratory, Ship Division Report, 1962.

15.16 Brockett, T.E. Lifting surface hydrodynamics for design of rotating blades. *Proceedings of SNAME Propellers '81 Symposium*, Virginia, 1981.

15.17 Greeley, D.S. and Kerwin, J.E. Numerical methods for propeller design and analysis in steady flow. *Transactions of the Society of Naval Architects and Marine Engineers*, Vol. 90, 1982, pp. 415–453.

15.18 Hess, J.L. and Valarezo, W.O. Calculation of steady flow about propellers by means of a surface panel method. AIAA Paper No. 85-0283, 1985.

15.19 Kim, H.T. and Stern, F. Viscous flow around a propeller-shaft configuration with infinite pitch rectangular blades. *Journal of Propulsion and Power*, Vol. 6, No. 4, 1990, pp. 434–444.

15.20 Uto, S. Computation of incompressible viscous flow around a marine propeller. *Journal of Society of Naval Architects of Japan*, Vol. 172, 1992, pp. 213–224.

15.21 Stanier, M.J. Design and evaluation of new propeller blade section, *2nd International STG Symposium on Propulsors and Cavitation*, Hamburg, Germany, 1992.

15.22 Stanier, M.J. Investigation into propeller skew using a 'RANS' code. Part 1: Model scale. *International Shipbuilding Progress*, Vol. 45, No. 443, 1998, pp. 237–251.

15.23 Stanier, M.J. Investigation into propeller skew using a 'RANS' code. Part 2: Scale effects. *International Shipbuilding Progress*, Vol. 45, No. 443, 1998, pp. 253–265.

15.24 Chen, B. and Stern, F. RANS simulation of marine-propulsor P4119 at design condition, *22nd ITTC Propulsion Committee Propeller RANS/PANEL Method Workshop*, Grenoble, April 1998.

15.25 Maksoud, M., Menter, F.R. and Wuttke, H. Numerical computation of the viscous flow around Series 60 C_B=0.6 ship with rotating propeller. *Proceedings of 3rd Osaka Colloquium on Advanced CFD Applications to Ship Flow and Hull Form Design*, Osaka, Japan, 1998, pp. 25–50.

15.26 Maksoud, M., Menter F., Wuttke, H. Viscous flow simulations for conventional and high-skew marine propellers. *Ship Technology Research*, Vol. 45, 1998.

15.27 Jessup, S. Experimental data for RANS calculations and comparisons (DTMB4119. *22nd ITTC Propulsion Committee, Propeller RANS/Panel Method Workshop*, Grenoble, France, April 1998.

15.28 Bensow, R.E. and Liefvendahl, M. Implicit and explicit subgrid modelling in LES applied to a marine propeller. *38th Fluid Dynamics Conference*, AIAA Paper, 2008-4144, 2008.

15.29 Kim, K., Turnock, S.R., Ando, J., Becchi, P., Minchev, A., Semionicheva, E.Y., Van, S.H., Zhou, W.X. and Korkut, E. The Propulsion Committee: final report and recommendations. *The 25th International Towing Tank Conference*, Fukuoka, Japan, 2008.

15.30 Phillips, A.B., Turnock, S.R. and Furlong, M.E. Evaluation of manoeuvring coefficients of a self-propelled ship using a blade element momentum propeller model coupled to a Reynolds Averaged Navier Stokes flow solver. *Ocean Engineering*, Vol. 36, 2009, pp. 1217–1225.

15.31 Liu, Z. and Young, Y.L. Utilization of bend-twist coupling for performance enhancement of composite marine propellers, *Journal of Fluids and Structures*, 25(6), 2009, pp. 1102–1116.

15.32 Turnock, S.R. Technical manual and user guide for the surface panel code: PALISUPAN. University of Southampton, Ship Science Report No. 100, 66 p., 1997.

15.33 Turnock, S.R. Prediction of ship rudder-propeller interaction using parallel computations and wind tunnel measurements. PhD thesis, University of Southampton, 1993.

15.34 Gutsche, F, Die induction der axialen strahlzusatgescnwindigheit in der umgebung der shcraubenebene. *Schiffstecknik*, No. 12/13, 1955.

15.35 Young, Y.L. and Kinnas, S.A. Analysis of supercavitating and surface-piercing propeller flows via BEM. *Computational Mechanics*, Vol. 32, 2003, pp. 269–280.

15.36 Turnock, S.R., Pashias, C. and Rogers, E. Flow feature identification for capture of propeller tip vortex evolution. *Proceedings of the 26th Symposium on Naval Hydrodynamics*. Rome, Italy, INSEAN Italian Ship Model Basin / Office of Naval Research, 2006, pp. 223–240.

15.37 Phillips, A.B. Simulations of a self propelled autonomous underwater vehicle. Ph.D. thesis, University of Southampton, 2010.

15.38 Pemberton, R.J., Turnock, S.R., Dodd, T.J. and Rogers, E. A novel method for identifying vortical structures. *Journal of Fluids and Structures*, Vol. 16, No. 8, 2002, pp. 1051–1057.

15.39 Hill, J.G. The design of propellers. *Transactions of the Society of Naval Architects and Marine Engineers*, Vol. 57, 1949, pp. 143–192.

15.40 Eckhardt, M.K. and Morgan, W.B. A propeller design method. *Transactions of the Society of Naval Architects and Marine Engineers*, Vol. 63, 1955, pp. 325–374.

15.41 O'Brien, T.P. *The Design of Marine Screw Propellers*. Hutchinson & Co., London, 1967.

15.42 Burrill, L.C. Calculation of marine propeller performance characteristics. *Transactions of North East Coast Institution of Engineers and Shipbuilders*, Vol. 60, 1944.

15.43 Molland, A.F., and Turnock, S.R. *Marine Rudders and Control Surfaces*. Butterworth-Heinemann, Oxford, UK, 2007.

15.44 Drela, M. Xfoil: an analysis and design system for low Reynolds number aerofoils. *Conference on Low Reynolds Number Airfoil Aerodynamics*. University of Notre Dame, Indiana, 1989.

15.45 Abbott, I.H. and Von Doenhoff, A.E. *Theory of Wing Sections*. Dover Publications, New York, 1958.

15.46 Ginzel, G.I. Theory of the broad–bladed propeller, ARC, Current Papers, No. 208, Her Majesty's Stationery Office, London, 1955.

15.47 Morgan, W.B., Silovic, V. and Denny, S.B. Propeller lifting surface corrections. *Transactions of the Society of Naval Architects and Marine Engineers*, Vol. 76, 1968, pp. 309–347.

15.48 Molland, A.F. and Turnock, S.R. A compact computational method for predicting forces on a rudder in a propeller slipstream. *Transactions of the Royal Institution of Naval Architects*. Vol. 138, 1996, pp. 227–244.

15.49 Burrill, L.C. The optimum diameter of marine propellers: A new design approach. *Transactions of North East Coast Institution of Engineers and Shipbuilders*, Vol. 72, 1955, pp. 57–82.

16 Propulsor Design Data

16.1 Introduction

16.1.1 General

The methods of presenting propeller data are described in Section 12.1.3. A summary of the principal propulsor types is given in Chapter 11. It is important to note that different propulsors are employed for different overall design and operational requirements. For example, a comparison of different propulsors based solely on efficiency is shown in Figure 16.1, [16.1]. This does not, however, take account of other properties such as the excellent manoeuvring capabilities of the vertical axis propeller, the mechanical complexities of the highly efficient contra-rotating propeller or the restriction of the higher efficiency of the ducted propeller to higher thrust loadings.

As described in Chapter 2, the propeller quasi-propulsive coefficient η_D can be written as follows:

$$\eta_D = \eta_O \times \eta_H \times \eta_R, \tag{16.1}$$

where η_O is the propeller open water efficiency, and η_H is the hull efficiency, defined as follows:

$$\eta_H = \frac{(1-t)}{(1-w_T)}, \tag{16.2}$$

where t is the thrust deduction factor, and w_T is the wake fraction. η_R is the relative rotative efficiency. Data for the components of η_H and η_R are included in Section 16.3.

Section 16.2 describes design data and data sources for η_O. The propulsors have been divided into a number of categories. Sources of data for the various categories are described. Examples and applications of the data are provided where appropriate.

16.1.2 Number of Propeller Blades

An early design decision concerns the choice of the number of blades. The number of blades is governed mainly by the effects of propeller-excited vibration and, in

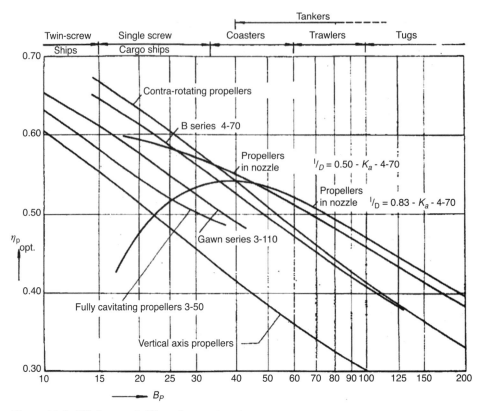

Figure 16.1. Efficiency of different propulsor types.

particular, vibration frequencies. Such excitation occurs due to the non-uniform nature of the wake field, see Chapter 8. Propeller forces are transmitted to the hull through bearing forces via the stern bearing, and hull surface forces which are transmitted from the pressure field that rotates with the propeller. Exciting frequencies arising are the blade passage frequency (rpm) and multiples of blade number (rpm $\times Z$). For example, a four-bladed propeller running at 120 rpm would excite at 120 cpm, 480 cpm, 960 cpm, etc. For a propeller with an even number of blades, the most important periodic loads are T and Q which would excite shaft vibration or torsional hull vibration. For a propeller with an odd number of blades, the vertical and horizontal forces and moments, F_V, M_V, F_H and M_H, will be dominant, leading to vertical or horizontal hull vibration. Changing the number of blades can therefore cure one problem but create another. Estimates will normally be made of hull vibration frequencies (vertical, horizontal and torsional) and propeller-rpm (and multiples) chosen to avoid these frequencies. A more detailed discussion of propeller-excited vibration can be found in [16.2–16.4]. From the point of view of propeller efficiency at the design stage, changes in blade number do not lead to large changes in open water efficiency η_0. Four blades are the most common, and changing to three blades would typically lead to an increase in efficiency of about 3% for optimum diameter (1% for non-optimum diameter), whilst changing to five blades would typically lead to a reduction in efficiency of about 1% [16.5]. The magnitude of such efficiency changes can be estimated using, say, the Wageningen Series data for two to seven blades, Section 16.2.1.

16.2 Propulsor Data

16.2.1 Propellers

16.2.1.1 Data

Tests on series of propellers have been carried out over a number of years. In such tests, systematic changes in P/D and BAR are carried out and the performance characteristics of the propeller are measured. The results of standard series tests provide an excellent source of data for propeller design and analysis, comparison with other propellers and benchmark data for computational fluid dynamics (CFD) and numerical analyses.

The principal standard series of propeller data, for fixed pitch, fully submerged, non-cavitating propellers, are summarised as follows: Wageningen B series [16.6], Gawn series [16.7], Au series [16.8], Ma series [16.9], KCA series [16.10], KCD series [16.11, 16.12], Meridian series [16.13]. The Wageningen and Gawn series are discussed in more detail. All of the series are described in some detail by Carlton [16.14].

(I) WAGENINGEN SERIES. The Wageningen series has two to seven blades, BAR = 0.3–1.05 and $P/D = 0.60$–1.40. The general blade outline of the Wagengingen B series, for four blades, is shown in Figure 16.2. The full geometry of the propellers is included in [16.14]. Typical applications include most merchant ship types. Figures 16.3 and 16.4 give examples of Wageningen $K_T - K_Q$ charts for the B4.40 and B4.70 propellers. Examples of $Bp - \delta$ and $\mu - \sigma - \phi$ charts are given in Figures 16.5 and 16.6.

As discussed in Section 12.1.3, the $\mu - \sigma - \phi$ chart is designed for towing calculations and the $Bp - \delta$ chart is not applicable to low- or zero-speed work. The $K_T - K_Q$ chart covers all speeds and is more readily curve-fitted or digitised for computational calculations. Consequently, it has become the most practical and popular presentation in current use.

Polynomials have been fitted to the Wageningen $K_T - K_Q$ data [16.15], Equations (16.3) and (16.4). These basic equations are for a Reynolds number Re of 2×10^6. Further equations, (16.5) and (16.6), allow corrections for Re between 2×10^6 and 2×10^9. The coefficients of the polynomials, together with the ΔK_T and ΔK_Q corrections for Reynolds number, are listed in Appendix A4, Tables A4.1 and A4.2.

$$K_T = \sum_{n=1}^{39} C_n (J)^{S_n} (P/D)^{t_n} (A_E/A_0)^{u_n} (z)^{v_n}. \tag{16.3}$$

$$K_Q = \sum_{n=1}^{47} C_n (J)^{S_n} (P/D)^{t_n} (A_E/A_0)^{u_n} (z)^{v_n}, \tag{16.4}$$

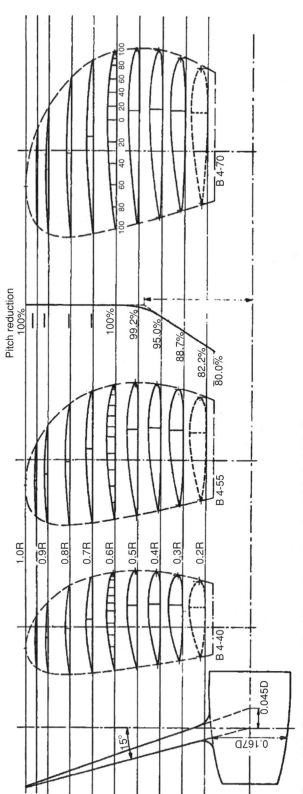

Figure 16.2. Blade outline of the Wageningen B series (4 blades) [16.6].

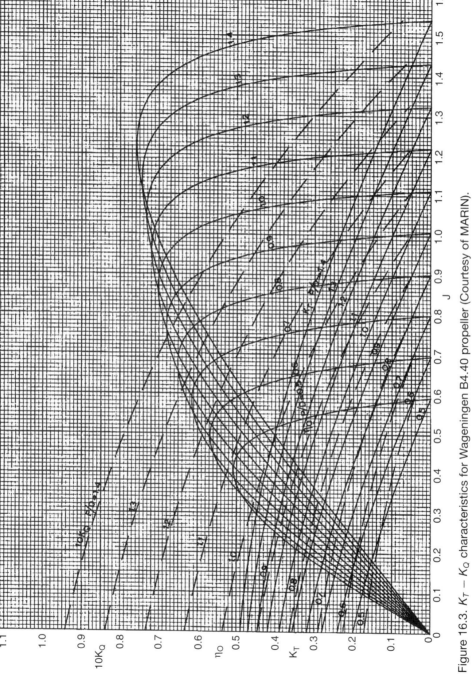

Figure 16.3. $K_T - K_Q$ characteristics for Wageningen B4.40 propeller (Courtesy of MARIN).

373

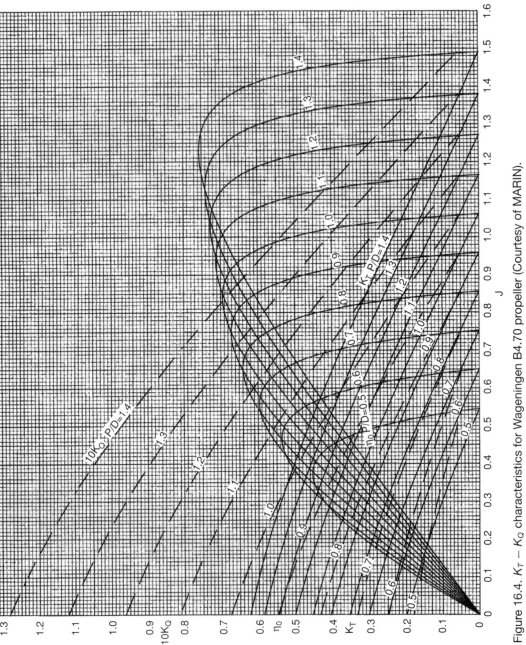

Figure 16.4. $K_T - K_Q$ characteristics for Wageningen B4.70 propeller (Courtesy of MARIN).

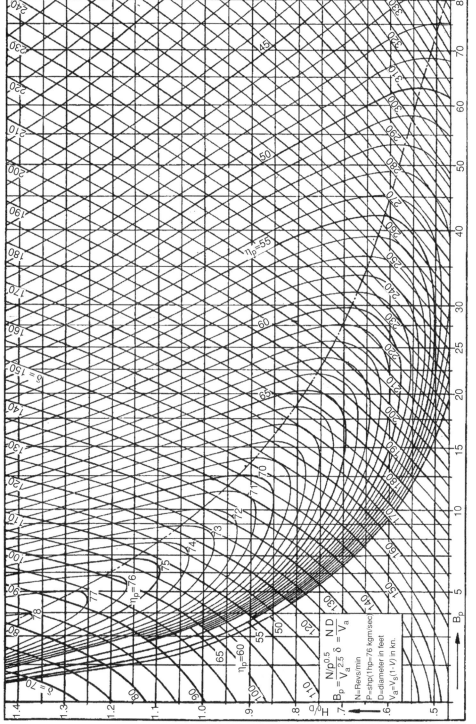

Figure 16.5. *Bp-δ* characteristics for Wageningen B4.40 propeller (Courtesy of MARIN).

375

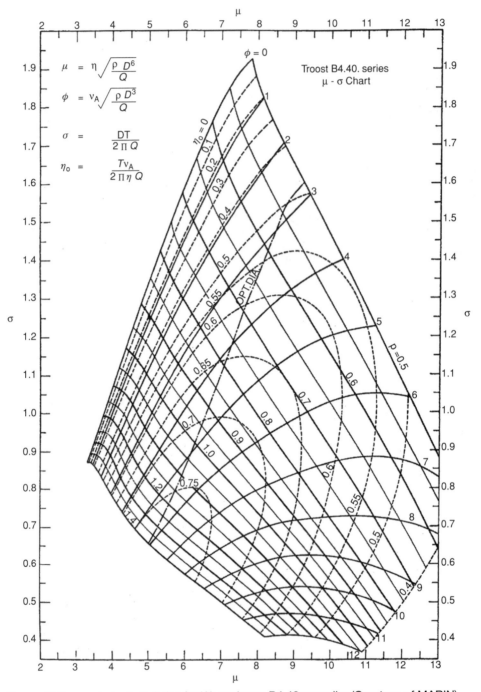

Figure 16.6. μ-σ-ϕ characteristics for Wageningen B4.40 propeller (Courtesy of MARIN).

Table 16.1(a). *Coefficients in K_T and K_Q, Equations (16.9, 16.10), B4.40*

P/D	K_{To}	a	n	K_{Qo}	b	m
0.5	0.200	0.59	1.26	0.0170	0.72	1.50
0.6	0.240	0.70	1.27	0.0225	0.80	1.50
0.7	0.280	0.80	1.29	0.0290	0.90	1.50
0.8	0.320	0.90	1.30	0.0360	0.98	1.60
0.9	0.355	1.01	1.32	0.0445	1.07	1.65
1.0	0.390	1.11	1.35	0.0535	1.18	1.65
1.1	0.420	1.22	1.37	0.0630	1.28	1.67
1.2	0.445	1.32	1.40	0.0730	1.39	1.67

where (A_E/A_0) can be taken as BAR.

$$K_T(Re) = K_T(Re = 2 \times 10^6) + \Delta K_T(Re). \tag{16.5}$$

$$K_Q(Re) = K_Q(Re = 2 \times 10^6) + \Delta K_Q(Re). \tag{16.6}$$

Reynolds number $Re\ (=V_R \cdot c/v)$ is based on the chord length and relative velocity V_R at $0.7R$, as follows:

$$V_R = \sqrt{Va^2 + (0.7\pi nD)^2}. \tag{16.7}$$

An approximation to the chord length at $0.7R$, based on the Wageningen series, Figure 16.2, [16.6] is as follows:

$$\left(\frac{c}{D}\right)_{0.7R} = X_2 \times \text{BAR}, \tag{16.8}$$

where $X_2 = 0.747$ for three blades, 0.520 for four blades and 0.421 for five blades

Approximate Equations for K_T and K_Q. Approximate fits in the form of Equations (16.9) and (16.10) have been made to the Wageningen $K_T - K_Q$ data. These are suitable for approximate estimates of K_T and K_Q at the preliminary design stage.

$$K_T = K_{To}\left[1 - \left(\frac{J}{a}\right)^n\right] \tag{16.9}$$

$$K_Q = K_{Qo}\left[1 - \left(\frac{J}{b}\right)^m\right] \tag{16.10}$$

and

$$\eta_0 = \frac{J\,K_T}{2\pi\,K_Q}, \tag{16.11}$$

where K_{To} and K_{Qo} are values of K_T and K_Q at $J = 0$. K_{To}, K_{Qo} and coefficients a, b, n and m have been derived for a range of P/D for the Wageningen B4.40 and B4.70 propellers and are listed in Tables 16.1(a) and (b).

(II) GAWN SERIES. All the propellers in the Gawn series have three blades, BAR = 0.20–1.10 and P/D = 0.40–2.00. The general blade outline of the Gawn series is shown in Figure 16.7. Typical applications of the Gawn series include ferries,

Table 16.1(b). *Coefficients in K_T and K_Q, Equations (16.9, 16.10), B4.70*

P/D	K_{To}	a	n	K_{Qo}	b	m
0.5	0.200	0.55	1.15	0.0180	0.70	1.18
0.6	0.250	0.65	1.20	0.0250	0.79	1.18
0.7	0.300	0.75	1.20	0.0332	0.86	1.20
0.8	0.352	0.86	1.20	0.0433	0.95	1.23
0.9	0.405	0.96	1.20	0.0545	1.03	1.29
1.0	0.455	1.06	1.20	0.0675	1.12	1.29
1.1	0.500	1.16	1.21	0.0810	1.22	1.30
1.2	0.545	1.27	1.21	0.0960	1.32	1.31

warships, small craft and higher speed craft. Figures 16.8, 16.9 and 16.10, reproduced with permission from [16.7], give examples of Gawn $K_T - K_Q$ charts for propellers with a BAR of 0.35, 0.65 and 0.95.

Polynomials have been fitted to the Gawn $K_T - K_Q$ data [16.16] in the form of Equations (16.3) and (16.4). The results are approximate in places and their application should be limited to $0.8 < P/D < 1.4$. The coefficients of the polynomials are listed in Appendix A4, Table A4.3.

16.2.1.2 Applications of Standard Series $K_T - K_Q$ Charts

Fundamentally, the propeller has to deliver a required thrust (T) at a speed of advance (Va), where $T = R/(1 - t)$, $Va = Vs(1 - w_T)$, R is resistance, t is thrust deduction factor and w_T is the wake fraction.

A typical approach, using $K_T - K_Q$ charts, is shown in Figure 16.11. This shows the approach starting from an assumed diameter, leading to optimum revolutions.

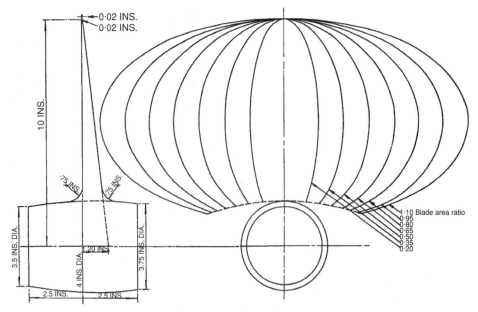

Figure 16.7. Blade outline of the Gawn series [16.10].

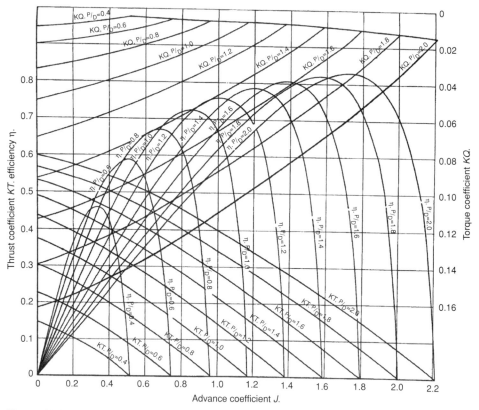

Figure 16.8. $K_T - K_Q$ characteristics for Gawn propeller with BAR = 0.35.

An alternative is to start with assumed revolutions, leading to an optimum diameter.

In theory, there is an infinite range of propeller diameters (D) and pitch ratios (P/D), hence, rpm (N) that meet the design requirement. In practice, there will be an optimum diameter (for maximum efficiency) or diameter limitation due to required clearances, and optimum rpm (or some rpm limit due to engine requirements) that will lead to a suitable solution. The following worked example illustrates the application of the Wageningen series B4.40 $K_T - K_Q$ chart, Figure 16.3, for a typical marine propeller.

WORKED EXAMPLE. Given that P_E, Vs, w_T and t are available for a particular vessel, derive suitable propeller characteristics and efficiency. $P_E = 2800$ kW, $Vs = 14$ knots, $w_T = 0.26$, $t = 0.20$ and assume $\eta_R = 1.0$. $T = R_T/(1-t) = (P_E/V_S)/(1-t) = 2800/(14 \times 0.5144)/(1-0.2) = 486.0$ kN and $Va = 14(1-0.26) \times 0.5144 = 5.329$ m/s.

Case 1: given Diameter and revolutions. Given that $D = 5.2$ m and $N = 120$ rpm ($n = 2.0$ rps) $K_T = T/\rho n^2 D^4 = 486 \times 1000/1025 \times 2^2 \times 5.2^4 = 0.162$,

$$J = Va/nD = 5.329/2.0 \times 5.2 = 0.512.$$

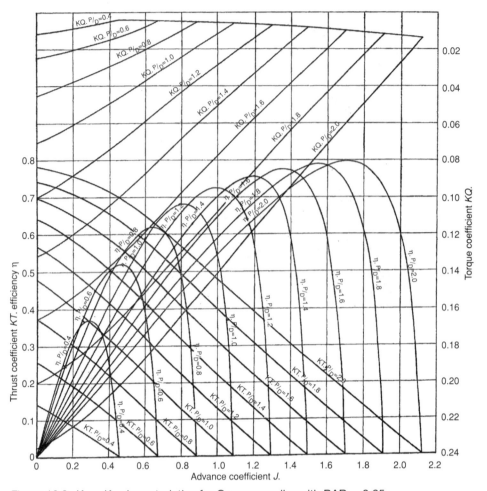

Figure 16.9. $K_T - K_Q$ characteristics for Gawn propeller with BAR $= 0.65$.

The steps are as follows:

(a) Enter chart (B4.40, Figure 16.3) at $J = 0.512$,
(b) Read appropriate P/D from required K_T $(=0.162) = 0.79$,
(c) Read appropriate η_0 from required $P/D = 0.588$,
(d) Read appropriate K_Q from required $P/D = 0.0225$.

From Equation (16.1), $\eta_D = \eta_0 \times \eta_H \times \eta_R = 0.588 \times (1 - 0.2)/(1 - 0.26) \times 1.0 = 0.636$ and $P_D = P_E/\eta_D = 2800/0.636 = 4402.5\,kW$. Or more directly, using the chart K_Q and noting that $P_D = 2\pi nQ/\eta_R$ and $Q = K_Q \times \rho n^2 D^5$

$$\therefore P_D = 2\pi n K_Q \rho n^2 D^5/\eta_R$$
$$= 2\pi 2.0 \times 0.0225 \times 1025 \times 2^2 \times 5.2^5/1000 = 4407.5\,kW$$

(the 5 kW difference is within the accuracy of reading the chart).

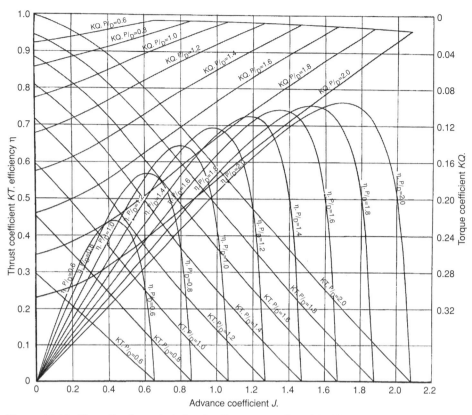

Figure 16.10. $K_T - K_Q$ characteristics for Gawn propeller with BAR $= 0.95$.

Case 2: optimum revolutions. Given that the diameter $D = 5.0$ m, for example, due to tip clearances which might typically be from $0.15D$ to $0.20D$, *find the optimum rpm* for this diameter.

$$K_T = T/\rho n^2 D^4 = 486 \times 1000/1025 \times n^2 \times 5.0^4 = 0.759/n^2.$$
$$J = Va/nD = 5.329/n \times 5.0 = 1.066/n.$$

Assume a range of rpm, deduce η_0 from the chart for each rpm; hence, deduce from say a cross plot the rpm at optimum (maximum) η_0. The steps are shown in Table 16.2. At $J = 0.548$, $P/D = 0.908$ and $K_Q = 0.0301$, $P_D = 2\pi nQ/\eta_R = 2\pi \times 1.945 \times 0.0301 \times 1025 \times 1.945^2 \times 5.0^5/1000 = 4457.3$ kW. {or $\eta_D = \eta_0 \times \eta_H \times \eta_R = 0.583 \times (1 - 0.2)/(1 - 0.26) \times 1.0 = 0.630$} and $P_D = P_E/\eta_D = 2800/0.630 = 4444.4$ kW, the 12.9 kW difference being due to the accuracy of reading from a chart.

Case 3: optimum diameter. Given that the rpm $N = 130$ rpm, for example, due to the requirements of the installed engine, *find the optimum diameter* for these rpm.

$$N = 130 \text{ rpm} = 2.17 \text{ rps}.$$

$$K_T = T/\rho n^2 D^4 = 486 \times 1000/1025 \times 2.17^2 \times D^4 = 100.69/D^4.$$

$$J = Va/nD = 5.329/(2.17 \times D) = 2.456/D.$$

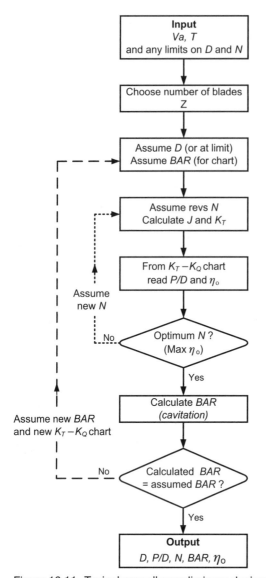

Figure 16.11. Typical propeller preliminary design path–optimum rpm.

Assume a range of D, deduce η_0 from the chart for each D; hence, deduce from say a cross plot the D at optimum (maximum) η_0. The steps are shown in Table 16.3. At $J = 0.483$, $P/D = 0.74$ and $K_Q = 0.0200$. $P_D = 2\pi\, nQ/\eta_R = 2\pi \times 2.17 \times 0.0200 \times 1025 \times 2.17^2 \times 5.09^5/1000 = 4496.8$ kW or $\eta_D = \eta_0 \times \eta_H \times \eta_R = 0.578 \times (1 - 0.2)/(1 - 0.26) \times 1.0 = 0.625$ and $P_D = P_E/\eta_D = 2800/0.625 = 4480.0$ kW, the 16.8 kW difference being due to accuracy of reading from chart.

It should also be noted that the B4.40 chart has been chosen to illustrate the calculation method. A cavitation check, Chapter 12, would indicate which would be the most appropriate chart to use from say the B4.40, B4.55 or B4.70 charts, etc.

Case 4: power approach. Required to design the propeller for a given delivered power at given rpm (for example, having chosen a particular direct drive diesel engine). In this case, speed is also initially not known.

Table 16.2. *Derivation of optimum rpm*

Assume		Calculate		From chart		
N	n	J	K_T	P/D	η_0	From cross plot
102	1.7	0.627	0.263	1.12	0.570	Optimum $J = 0.548$
114	1.9	0.561	0.210	0.94	0.582	$\therefore n = 1.945$ (116.7 rpm)
126	2.1	0.508	0.172	0.81	0.575	$\eta_0 = 0.583$ and $P/D = 0.908$

Approach 1. If the diameter is known, or estimated (e.g. maximum allowing for clearances), assume a range of speeds (covering the expected speed) and repeat Case 1 for each speed. Increase speed until the delivered power P_D matches that required.

Approach 2. Optimum diameter required. Assume a range of speeds and repeat Case 3 for each speed, producing a delivered power and optimum diameter at each speed. Increase speed until delivered power P_D matches that required. A cross plot of P_D and D to a base of speed will yield the speed at the required P_D, together with the optimum diameter at that speed.

It should be noted that $B_P - \delta$ charts (rather than $K_T - K_Q$ charts) may be more suitable for a power approach, see Section 16.2.1.3.

Case 5: bollard pull condition (J = 0). At $J = 0$, $K_{T0} = T/\rho n^2 D^4$, $K_{Q0} = Q/\rho n^2 D^5$, $T = K_{T0} \times \rho n^2 D^4$ and $Q = K_{Q0} \times \rho n^2 D^5$.

From Figure 16.3, for an assumed P/D, K_{T0} and K_{Q0} can be obtained. For an assumed (or limiting) D, a range of rpm may be assumed until a limiting Q or P_D is reached. This point would represent the maximum achievable thrust. This process may be repeated (within any prescribed limits) for a range of D and range of n until the most efficient combination of n and D is achieved. As noted in Section 13.3, the effective thrust $T_E = T \times (1 - t)$, where t is the thrust deduction factor. Values of t for tugs in the bollard pull condition are given in Chapter 8. Worked example application 19, Chapter 17, includes the bollard pull condition for a tug.

Case 6: preliminary screening. A suitable approach at the preliminary design stage is to apply Case 2 a number of times, whereby the sensitivity of efficiency η_0 to diameter D can be checked. The results of such an investigation are shown schematically in Figure 16.12.

16.2.1.3 Applications of Standard Series $B_P - \delta$ Charts

$B_P - \delta$ charts can be usefully employed when it is required to design a propeller for a given delivered power and rpm (for example, having chosen a particular direct drive

Table 16.3. *Derivation of optimum diameter*

Assume	Calculate		From chart		
D	J	K_T	P/D	η_0	From cross plot
4.8	0.512	0.190	0.85	0.571	Optimum $J = 0.483$
5.0	0.491	0.161	0.77	0.577	$D = 5.09$
5.2	0.472	0.138	0.70	0.577	$\eta_0 = 0.578$
5.4	0.455	0.118	0.64	0.562	$P/D = 0.74$

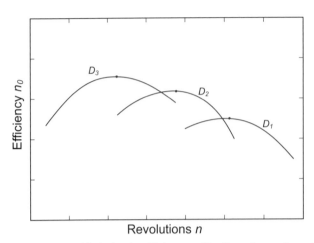

Figure 16.12. Variation in efficiency with diameter and revolutions.

diesel engine). In this case, in the first instance speed has to be assumed as propeller efficiency and delivered propeller thrust are not yet known.

WORKED EXAMPLE. The service delivered power P_D for a ship travelling at an estimated speed of 17 knots is 12,500 kW at 110 rpm.

Calculate the optimum propeller diameter, propeller pitch and efficiency.

From empirical data for this ship type, $w_T = 0.25$, $t = 0.15$ and $\eta_R = 1.01$.

$$\eta_H = (1 - t)/(1 - w_T) = (1 - 0.15)/(1 - 0.25) = 1.133.$$

$$P_D = 12{,}500\,\text{kW} = 12{,}500/0.7457 = 16{,}763\,\text{hp}.$$

$$Vs = 17.0\,\text{knots}.\quad Va = Vs(1 - w_T) = 17(1 - 0.25) = 12.75\,\text{knots}.$$

$$B_P = N \cdot P^{1/2}/Va^{2.5} = 110 \times 16{,}763^{1/2}/12.75^{2.5} = 24.54.$$

Using the Wageningen B4.40 chart shown in Figure 16.5, for $B_P = 24.54$ at the optimum diameter (chain dotted) line, $\eta_0 = 0.625$, $\delta = 200$ and $P/D = 0.730$. $\delta = ND/Va$, diameter $D = \delta Va/N = 200 \times 12.75/110 = 23.18\,\text{ft} = 7.07$ m and $\eta_D = \eta_0 \eta_H \eta_R = 0.625 \times 1.133 \times 1.01 = 0.715.$

If the propeller diameter were limited to say 6.3 m = 20.66 ft, then, in this case, $B_P = 24.54$ as before, but $\delta = ND/Va = 110 \times 20.66/12.75 = 178.24$ and from the B4.40 chart, $\eta_0 = 0.600$ (non-optimum) and $P/D = 0.970$.

$$\eta_D = \eta_0 \eta_H \eta_R = 0.600 \times 1.133 \times 1.01 = 0.687.$$

Considering the optimum diameter case, the 'effective' power output from the engine $P_E' = P_D \times \eta_D = 12{,}500 \times 0.715 = 8937.5$ kW. In practice, if the effective power output P_E' from the installed engine is not equal to P_E for the hull ($R_T \times Vs$) at that speed, then from the $P_E - Vs$ curve for the ship, a new speed for $P_E = 8937.5$ kW can be deduced and calculations can be repeated for the new speed. Hence, the optimum propeller and actual speed can be obtained by iteration.

It should also be noted that the B4.40 chart has been chosen to illustrate the calculation method. A cavitation check, Chapter 12, would indicate which would be the most appropriate chart to use from say the B4.40, B4.55 or B4.70 charts, etc.

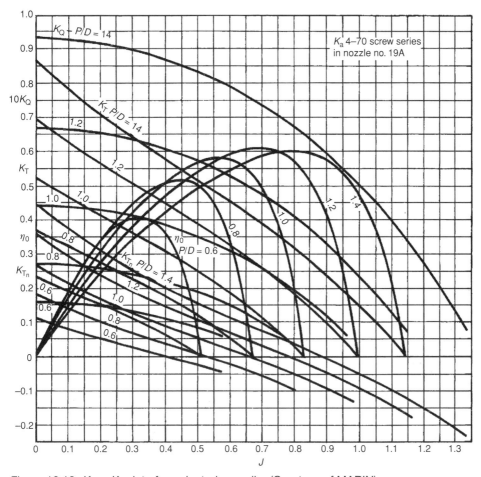

Figure 16.13. $K_T - K_Q$ data for a ducted propeller (Courtesy of MARIN).

16.2.2 Controllable Pitch Propellers

Series $K_T - K_Q$ charts for fixed-pitch propellers, Figures 16.3 to 16.10, can be used, treating pitch as a variable and allowing for a small loss in efficiency due to the increase in boss diameter/propeller diameter ratio. Results in [16.17] would suggest that a 40% increase in boss/diameter ratio (say from 0.18 to 0.25) leads approximately to a 2%–3% decrease in efficiency.

16.2.3 Ducted Propellers

Open water data are available for ducted propellers. An example is given in Figure 16.13 which shows the $K_T - K_Q$ data for the Wageningen Ka4.70 propeller in a 19A nozzle (duct). The use of such a $K_T - K_Q$ chart is the same as for the conventional non-ducted propeller in Section 16.2.1. The contribution of the duct is also shown in Figure 16.13; note that, at lower thrust loadings, higher J values, the duct thrust becomes negative, or a drag force. Further ducted propeller data may be found in [16.18 and 16.19].

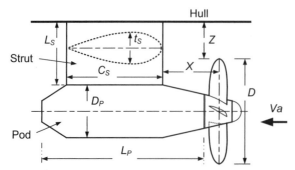

Figure 16.14. Schematic layout of podded unit.

16.2.4 Podded Propellers

The propeller may be assumed to behave in a similar manner to the conventional propeller and the charts in Figures 16.3 to 16.10 provide suitable open water data. The particular differences compared with the conventionally mounted propeller concern the larger propeller boss/diameter ratio, the relatively large diameter of the pod in order to house the electric drive motor, and the drag of the supporting strut. The wake fraction and the thrust deduction factor may be different for the hull forms cut up at the aft end to accommodate the podded unit. The drag of the relatively large pod has to be allowed for and the pod support strut has to be taken into account, either as an appendage drag or a thrust deduction change.

16.2.4.1 Pod Housing Drag

For practical powering predictions, simple semi-empirical methods for estimating the full-scale drag of the pod housing, suggested in International Towing Tank Conference (ITTC) (2008) [16.20], may be applied. The main components of a podded unit are shown schematically in Figure 16.14.

The total pod housing drag (pod plus strut) is assumed to be the following:

$$R_{\text{Pod}} = R_{\text{Body}} + R_{\text{Strut}} + R_{\text{Int}} + R_{\text{Lift}}, \quad (16.12)$$

which are the components of resistance associated with the pod body (nacelle), strut, strut-body interference and lift due to rotating flow downstream of the propeller. For lightly loaded propellers R_{Lift} may be neglected but, if the propeller is heavily loaded (low J), the induced drag can be significant. The effect is similar to that arising from the influence of the propeller slipstream on a rudder, as described by Molland and Turnock [16.21].

A form factor approach is proposed, where $(1 + k_B)$ and $(1 + k_S)$ are the form factors for the pod body and strut, respectively. The semi-empirical formulae, from Hoerner [16.22], are as follows:

16.2.4.2 Pod Body

$$R_{\text{Body}} = (1 + k_B)\left(C_F \frac{1}{2}\rho S_B V_I^2\right) \quad (16.13)$$

and

$$k_B = 1.5 \left(\frac{D_P}{L_P}\right)^{3/2} + 7 \left(\frac{D_P}{L_P}\right)^3, \tag{16.14}$$

where L_P, D_P and S_B are the length, diameter and wetted surface area of the pod body (nacelle) and V_I is the inflow velocity.

STRUT.

$$R_{\text{Strut}} = (1 + k_S) \left(C_F \frac{1}{2} \rho S_S V_I^2\right) \tag{16.15}$$

and

$$k_S = 2 \left(\frac{t_S}{C_S}\right) + 60 \left(\frac{t_S}{C_S}\right)^4, \tag{16.16}$$

where S_S is the wetted surface area of the strut, t_S/C_S is the average thickness ratio of the strut and V_I is the inflow velocity.

STRUT-POD INTERFERENCE.

$$R_{\text{Int}} = \frac{1}{2} \rho t_S^2 V_I^2 f \left(\frac{t_{\text{root}}}{C_{\text{root}}}\right), \tag{16.17}$$

with a fillet joint

$$f \left(\frac{t_{\text{root}}}{C_{\text{root}}}\right) = \left(0.4 \left(\frac{t_{\text{root}}}{C_{\text{root}}}\right) - 0.05\right), \tag{16.18}$$

where t_{root} is maximum thickness at strut root (joint with pod) and C_{root} is the chord length at the root. It is noted that R_{Int} is independent of friction and will not be subject to scale effect.

EFFECT OF PROPELLER SLIPSTREAM. The propeller accelerates the flow over the pod and strut. An equation derived using modified actuator disc theory is proposed [16.21], as follows. The inflow velocity at pod and/or strut

$$V_I = Va + K_R Va \left\{\left[1 + \frac{8K_T}{\pi J^2}\right]^{1/2} - 1\right\}, \tag{16.19}$$

where K_R is a Gutsche-type correction to account for the fluid acceleration between the propeller and strut, based on the distance of strut from the propeller, and can be represented as follows:

$$K_R = 0.5 + \frac{0.5}{[1 + (0.15/(X/D))]}, \tag{16.20}$$

noting that when $X/D = 0$, $K_R = 0.5$ and when $X/D = \infty$, $K_R = 1.0$.

It is recommended that flow velocity over the strut within the propeller slipstream be taken as V_I and that outside the slipstream the flow velocity be taken as Va. Flow over the pod may also be taken as V_I, being an approximate average of the accelerated flow over the pod.

Flikkema et al. [16.23] describe the components of pod resistance. For the strut input velocity, they propose the use of a weighted average of the actuator disc prediction, Equation (16.19) and that for the propeller no-slip condition.

FRICTION COEFFICIENT C_F. It is assumed that, at full scale, the flow over the pod and strut is fully turbulent and that the ITTC1957 line may be used (Equation (4.15)), as follows:

$$C_F = \frac{0.075}{[\log_{10} Re - 2]^2}. \qquad (16.21)$$

It is noted in ITTC2008 [16.20] that, at model scale, the recommendation is to use laminar/transitional flow formulae for the area outside the propeller slipstream.

CORRECTION TO K_{Topen} FOR A CONVENTIONALLY MOUNTED PROPELLER.

$$\text{Correction due to pod drag } K_{Rpod} = \frac{R_{pod}}{\rho n^2 D^4}. \qquad (16.22)$$

Finally, *total* net thrust is propeller thrust minus pod resistance, as follows:

$$K_{Ttotal} = K_{Topen} - K_{Rpod}. \qquad (16.23)$$

The corrected efficiency for podded unit is as follows:

$$\eta_{Opod} = \frac{J_O K_{Ttotal}}{2\pi K_{Qo}}. \qquad (16.24)$$

If the correction is being applied to the open water data for a conventionally mounted propeller with a boss/diameter ratio of say 0.20, then a correction for an increase in the boss/diameter ratio for the podded propeller to say 0.35 should be applied. [16.17] would suggest that this order of increase in boss diameter would lead to a decrease in propeller efficiency of about 3%.

It must be noted that these are first approximations to the likely changes in efficiency when incorporating the results for an open propeller in a podded unit. Any changes in torque and wake (hence, J) have not been accounted for. Some guidance on the likely changes in K_T and K_Q due to a rudder downstream of a propeller (analogous to the pod/strut–propeller arrangement) can be found in Stierman [16.24] and Molland and Turnock [16.21].

APPROXIMATIONS TO POD DIMENSIONS AT INITIAL DESIGN STAGE. The dimensions of the pod will not necessarily be known at the initial design stage. Starting from the propeller diameter D, derived from calculations for a conventionally mounted propeller, Table 16.4 gives approximate average values that have been derived from a survey of existing podded propulsors. This allows suitable pod dimensions to be established for preliminary powering purposes.

APPLICATION OF FORMULAE FOR POD HOUSING DRAG. Apply a podded propulsor to the worked example, Case 1, Section 16.2.1. Given $D = 5.2$ m, $n = 2$ rps, $w_T = 0.26$, $t = 0.20$, $T = 486$ kN, $K_{To} = 0.162$, $Va = (1 - 0.26) \times 14 \times 0.5144 = 5.329$ m/s and $J = 0.512$.

Table 16.4. *Pod dimensions, based on propeller diameter D,*
Figure 16.14

D_P/D	L_P/D	C_S/L_P	L_S/D	t_S/C_S	Z/D	X/D
0.50	1.60	0.60	0.60	0.30	0.35	0.50

From the basic calculations using a B4.40 chart, the following were derived:

$$K_{Qo} = 0.0225, \ P/D = 0.79, \ \eta_o = 0.588.$$

Using Table 16.4, approximate dimensions of podded unit are as follows: with $D = 5.2$ m, $D_{\text{pod}} = 2.6$ m, $L_P = 8.32$ m, $C_S = 4.99$ m, $L_S = 3.12$, $t_S/C_S = 0.30$, $t_S = 1.5$ m, $Z = 1.82$ m, $(L_S - Z) = 1.3$ m.

Applying the correction equations, using Equation (16.19), with $X/D = 0.50$, $Va = 5.329$ m/s, $V_I = 8.177$ m/s, for pod, $L_P = 8.32$, $Re = V_I \times L_P/v = 5.717 \times 10^7$ (with $v = 1.19 \times 10^{-6}$ salt water) and $C_F = 2.263 \times 10^{-3}$. Using Equations (16.13) and (16.14) for pod body, neglecting ends, and assuming the pod to be cylindrical with constant diameter, wetted area $S_B = \pi \times D_P \times L_P = 67.96$ m^2, $k_B = 0.476$ and $R_{\text{Body}} = 7.78$ kN.

For strut, using Equations (16.15) and (16.16), strut in slipstream, $C_S = 4.99$, $Re = V_I \times C_S/v = 3.429 \times 10^7$ (with $v = 1.19 \times 10^{-6}$ salt water), and $C_F = 2.448 \times 10^{-3}$, assuming 10% expansion for girth, girth $= 1.10 \times C_S = 1.1 \times 4.99 = 5.49$ m. The wetted area inside slipstream $S_S = 5.49 \times 1.3 \times 2$ (both sides) $= 14.274$ m^2, $k_S = 1.086$ and in slipstream, $R_{\text{Strut}} = 2.49$ kN. Strut outside slipstream, $C_S = 4.99$, $Re = Va \times C_S/v = 2.235 \times 10^7$ (with $v = 1.19 \times 10^{-6}$ salt water), and $C_F = 2.621 \times 10^{-3}$.

The wetted area outside slipstream $S_S = 5.49 \times 1.82 \times 2$ (both sides) $= 19.98$ m^2. $k_S = 1.086$ and the outside slipstream, $R_{\text{Strut}} = 1.59$ kN.

$$\text{Total strut resistance} \ = R_{\text{Strut}} = 2.49 + 1.59 = 4.08 \text{ kN.}$$

Using Equations (16.17) and (16.18) for strut-pod body interference, assuming t_S/C_S constant across strut, $R_{\text{Int}} = 5.38$ kN.

Total strut-pod resistance $R_{\text{Pod}} = R_{\text{Body}} + R_{\text{Strut}} + R_{\text{Int}} = 7.78 + 4.08 + 5.38 = 17.24$ kN.

$$K_{R\text{pod}} = R_{\text{Pod}}/\rho n^2 D^4 = 17.24 \times 1000/1025 \times 2^2 \times 5.2^4 = 0.0058.$$

$$K_{T\text{total}} = K_{T\text{open}} - K_{R\text{pod}} = 0.162 - 0.0058 = 0.156.$$

The corrected efficiency for podded unit is as follows:

$$\eta_{O\text{pod}} = J \, K_{T\text{total}}/2\pi \, K_{QO} = 0.512 \times 0.156/2\pi \times 0.0225 = 0.565.$$

A further decrease in efficiency of 3% should be included due to the increase in boss diameter, as follows: correction $0.588 \times 0.03 = 0.018$, and the final $\eta_{O\text{pod}} = 0.565 - 0.018 = 0.547(-7\%)$.

It can be noted that this apparent loss in efficiency may be substantially offset by the resistance reduction due to the removal of appendages such as propeller bossing(s) or bracket(s) and rudder(s). These reductions will be seen as a decrease in effective power, P_E.

CORRECTION METHOD ATTRIBUTABLE TO FUNENO. Funeno [16.25] developed a simple design method for podded propellers, using CFD and experimental data, in which account is taken of the differences between when the propeller is operating in open water and when it is operating as part of a podded unit. The design method is based on the use of standard open water series, such as the Wageningen B series, together with suitable corrections for K_Q and K_T due to the presence of the pod and strut.

The propeller is first designed using conventional wake fraction and thrust deduction data and open water propeller data to match the required thrust, leading to the design values of J, K_T, K_Q and a suitable pitch ratio, P/D. Corrections for scale effect on K_T and K_Q (see Chapter 5) should be applied.

The torque is modified according to the following:

$$K_{Q\text{prop}-\text{pod}} = [0.1715 \times J + 1.0019] \times K_{Q\text{open}}. \qquad (16.25)$$

The propeller thrust is modified according to the following:

$$K_{T\text{prop}-\text{pod}} = [0.2491 \times J + 1.0323] \times K_{T\text{open}}. \qquad (16.26)$$

Pod resistance is estimated as follows:

$$K_{R\text{pod}} = [-0.1125 \times J - 0.0625] \times K_{T\text{prop}-\text{pod}}. \qquad (16.27)$$

Finally, *total* net thrust is the sum of the propeller thrust and pod resistance, leading to K_T for the whole unit as follows:

$$K_{T\text{total}} = K_{T\text{prop}-\text{pod}} + K_{R\text{pod}}. \qquad (16.28)$$

Further iterations are applied until $K_{T\text{total}}$ matches the required design K_T. The effect of any change in pitch was not included in Funeno's method because the results of the open water tests indicated that this effect was very small. Funeno's approach is considered suitable for the initial design stage.

APPLICATION OF FUNENO'S METHOD. Apply a podded propulsor to the worked example, Case 1, Section 16.2.1. Given $J = 0.512$, $w_T = 0.26$, $t = 0.20$, $K_{To} = 0.162$.

From the basic calculations using a B4.40 chart the following were derived:

$$K_{Qo} = 0.0225, \ P/D = 0.79, \ \eta_o = 0.588.$$

Apply the correction Equations (16.25) to (16.28).

$$K_{T\text{prop}-\text{pod}} = [0.2491 \times J + 1.0323] \times K_{T\text{open}} = 0.1878\,(+16\%).$$
$$K_{R\text{pod}} = [-0.1125 \times J - 0.0625] \times K_{T\text{prop}-\text{pod}} = -0.0226.$$
$$K_{T\text{total}} = K_{T\text{prop}-\text{pod}} + K_{R\text{pod}} = 0.165\,(+2\%).$$
$$K_{Q\text{prop}-\text{pod}} = [0.1715 \times J + 1.0019] \times K_{Q\text{open}} = 0.0245\,(+9\%).$$

Corrected efficiency is as follows:

$$\eta_{O\text{pod}} = \frac{J_O K_{T\text{total}}}{2\pi K_{Q\text{prop}-\text{pod}}} = 0.512 \times 0.165/2\pi \times 0.0245 = 0.549\,(-6.6\%).$$

It can be noted that this apparent loss in efficiency may be effectively offset by the resistance reduction due to the removal of appendages, such as propeller bossing(s) or bracket(s) and rudder(s). These reductions will be seen as a decrease in effective power, P_E.

Approximations to w_T and t for podded units are discussed in Chapter 8 and an estimate of η_R for podded units is given in Section 16.3.

A number of publications consider the design of podded units. Islam *et al.* [16.26] provide useful information for comparing pushing and pulling podded propulsors and their cavitation characteristics. They concluded that, in general, the puller propellers had higher efficiency than the pusher propellers. The cavitation performance was similar for the two types. Another useful paper by Islam *et al.* [16.27] concerns an investigation into the variation of geometry of podded propulsors. The results indicated that the ratio of the pod body diameter/propeller diameter and the hub angle can have a significant effect on K_T and K_Q. Much of this work is reviewed by Bose [16.28]. Heinke and Heinke [16.29] describe an investigation into the use of podded drives for fast ships. They concluded that the thrust loading and pod diameter/propeller diameter ratio are important parameters in the design; an increase in pod diameter results in increases in thrust and torque and a decrease in efficiency. For the size range considered, it was found that the pod-strut drag of the pushing propeller was lower than that for the pulling propeller and, overall, the efficiency of the pushing propeller was a little higher than the pulling propeller. A small series of podded propellers, suitable for sizes up to 4500 kW, is described by Frolova *et al.* [16.30].

Other useful papers which consider various aspects of podded propulsor design are contained in the proceedings of the T-POD conferences [16.31, 16.32]. Further reviews and information on the design, testing and performance of podded propellers can be found in the proceedings of ITTC conferences [16.20, 16.33, 16.34].

16.2.5 Cavitating Propellers

The influence of cavitation on propeller thrust and torque can be obtained from series tests, such as those reported by Gawn and Burrill [16.10] for the KCA propeller series. Radojcic [16.35] fitted polynomials to these test results, describing K_T and K_Q by Equations (16.29) and (16.30). The influence of cavitation on thrust and torque, ΔK_T and ΔK_Q, for change in cavitation number σ is described by Equations (16.31) and (16.32).

$$K_T = \sum_{n=1}^{16} C_t \cdot 10^e (A_D/A_0)^x (P/D)^y (J)^z. \tag{16.29}$$

$$K_Q = \sum_{n=1}^{17} C_q \cdot 10^e (A_D/A_0)^x (P/D)^y (J)^z. \tag{16.30}$$

$$\Delta K_T = \sum_{n=1}^{20} d_t \cdot 10^e (A_D/A_0)^s (\sigma)^t (K_T)^u (P/D)^v. \tag{16.31}$$

$$\Delta K_Q = \sum_{n=1}^{18} d_q \cdot 10^e (A_D/A_0)^s (\sigma)^t (K_T)^u (P/D)^v, \tag{16.32}$$

where (A_D/A_0) may be taken as BAR.

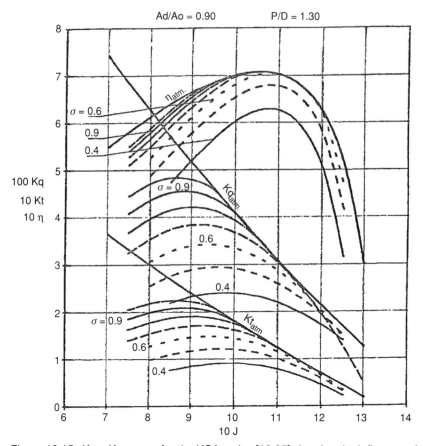

Figure 16.15. $K_T - K_Q$ curves for the KCA series [16.35] showing the influence of cavitation.

The coefficients of the polynomials are listed in Appendix A4, Tables A4.4 and A4.5. The limits of these equations are as follows: $z = 3, 0.5 \leq \text{BAR} \leq 1.1, 0.8 \leq P/D \leq 1.8, J \geq 0.3, K_T \leq [(J/2.5) - 0.1]^{1/2}$.

In this application, cavitation number σ is defined as follows:

$$\sigma = \frac{P - P_V}{0.5\rho Va^2}, \qquad (16.33)$$

where P is the static pressure at the shaft axis ($= \rho gh + P_{AT}$), see Section 12.2, P_V is the vapour pressure and Va is the speed of advance.

Examples of open water $K_T - K_Q$ curves for the KCA series, subject to cavitation, and generated by the polynomials, reproduced with permission, are shown in Figure 16.15. As σ is reduced, K_T, K_Q and η are seen to decrease. Thrust breakdown will not normally start until about 10% back cavitation has developed. Further details of cavitation and estimates of suitable BAR are included in Section 12.2.

16.2.6 Supercavitating Propellers

In the case of the supercavitating propeller, the back (low pressure) side of the section fully cavitates, cavity collapse occurs downstream of the propeller and erosion

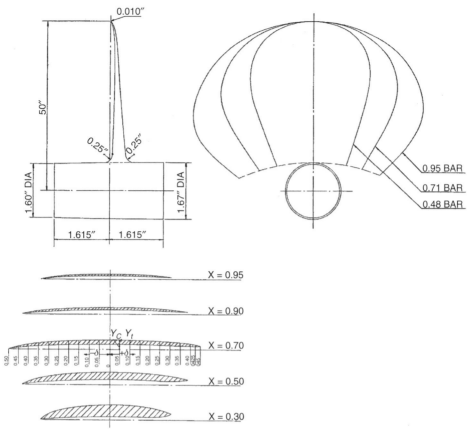

Figure 16.16. Blade outline and section details for the Newton–Rader series of propellers [16.36].

is avoided. Supercavitating propellers suffer a significant loss in thrust when in the supercavitating zone. Figure 16.16 shows the blade outline and section details for the Newton–Rader series of propellers [16.36] and Figure 16.17, reproduced with permission from [16.36], shows examples of K_T, K_Q and η_0 data. These charts are for BAR = 0.71 and P/D = 1.25. Data for other BARs and P/Ds are given in [16.36]. For these tests, cavitation number σ is defined as $\sigma = (P_O - P_V)/\frac{1}{2}\rho Va^2$, where P_O is the static pressure to the centre of the shaft ($= \rho gh + P_{AT}$), see Section 12.2, P_V is the vapour pressure of water and Va is the speed of advance; see the cavitation Section 12.2 for other datum points for pressure and speed. The blade section shape for the Newton–Rader propellers is a slightly concave round back, with modified leading and trailing edges, Figures 16.16 and 16.18(a). It is noted that the use of such $K_T - K_Q$ charts follows the same methodology as for the conventional propellers in Section 16.2.1. In Figures 16.17(a) and 16.17(b), it is seen that at the lowest cavitation number, K_T and K_Q are about half the values at atmospheric pressure. Another significant work on supercavitating foils is that by Tulin [16.37] who developed wedge-shaped sections, Figure 16.18(b). See also [16.14] for further discussion of super-cavitating propellers.

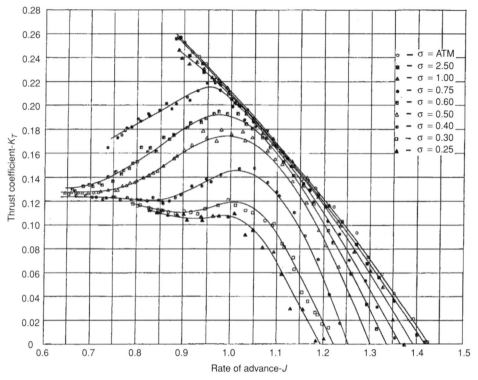

Figure 16.17. (a) K_T data for the Newton–Rader series of propellers [16.36].

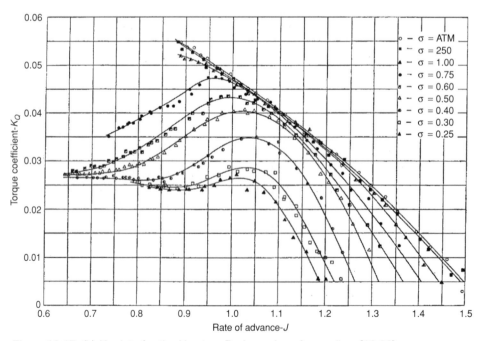

Figure 16.17. (b) K_Q data for the Newton–Rader series of propellers [16.36].

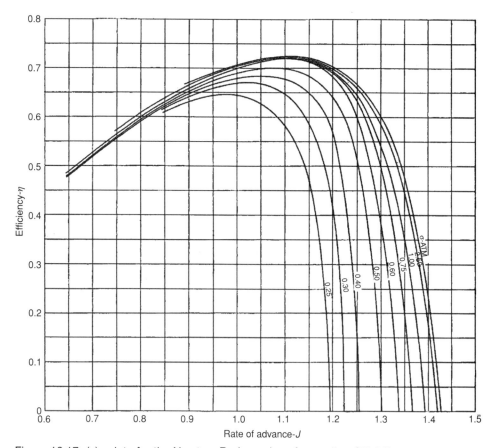

Figure 16.17. (c) η data for the Newton–Rader series of propellers [16.36].

16.2.7 Surface-Piercing Propellers

Data are available for surface-piercing propellers [16.28] and [16.38–16.40]. Figure 16.19 shows typical $K_T - K_Q$ curves for a series of surface-piercing propellers, reproduced with permission from [16.39]. It is noted that the data have a large scatter, which is due in part to the fact that tests were carried out in three different facilities. These charts may be used in a similar way to the conventional fully submerged propellers, Section 16.2.1. In the case of surface-piercing propellers, J is defined as $J_\psi = V \cos \psi / nD$, where ψ is the propeller shaft inclination to the horizontal (typically, $4°–8°$). $K_T' = T/\rho \, n^2 \, D^2 \, A_0'$ and $K_Q' = Q/\rho \, n^2 \, D^3 \, A_0'$, where A_0' is the nominal immersed propeller disc area, Figure 16.20 (typically about $0.5\pi D^2/4$).

Due to the use of A_0' in the denominator, the K_T' and K_Q' values are higher than the K_T and K_Q values for a conventional submerged propeller. Following the tests on the surface piercing propellers [16.39], regression models were developed

(a)   (b)

Figure 16.18. Supercavitating propeller sections: (a) Newton–Rader [16.36], (b) Tulin. [16.37].

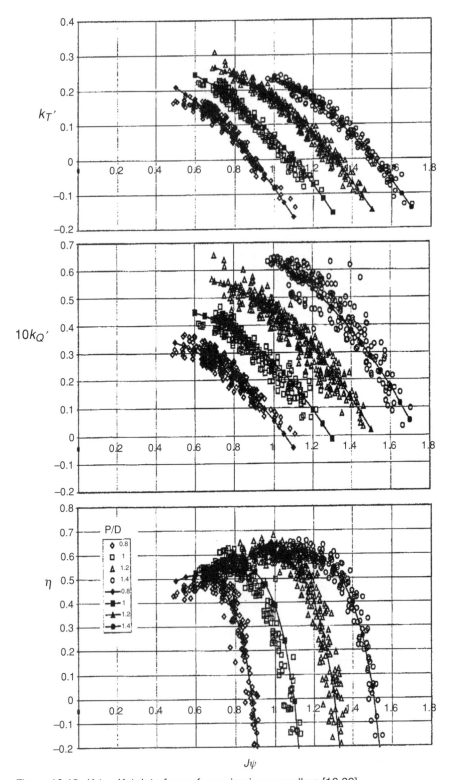

Figure 16.19. $K_T' - K_Q'$ data for surface-piercing propellers [16.39].

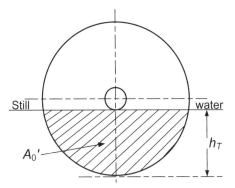

Figure 16.20. Nominal disk-immersed area.

for $K_{T'}$ and $K_{Q'}$ as follows:

$$K_{T'} = -0.691625\,J_\psi + 0.794973(P/D) + 0.870696\,J_\psi(P/D)$$
$$- 0.395012\,J_\psi^2 - 0.515183(P/D)^2. \qquad (16.34)$$

$$10K_{Q'} = -0.300453\,J_\psi + 0.543738(P/D) + 0.877638\,J_\psi(P/D)$$
$$- 0.649314\,J_\psi^2 - 0.208974(P/D)^2. \qquad (16.35)$$

Ferrando et al. [16.39] point out that, as partially submerged propellers operate at the interface between air and water, two additional parameters need to be taken into account. These are the immersion coefficient I_T and the Weber number We which takes account of surface tension effects.

$$I_T = \frac{h_T}{D}, \qquad (16.36)$$

where h_T is the maximum blade tip immersion, Figure 16.20, and I_T is typically about 0.5.

In the context of this work, a special Weber number is defined as follows:

$$We'' = \sqrt{\frac{\rho n^2 D^3 I_T}{\gamma}}, \qquad (16.37)$$

where γ is the surface tension with a value of about 0.073 N/m at the fresh water–air interface, and with a value of about 0.078 N/m for salt water.

In order to produce reliable data for scaling, it is important to know when the transition from fully wetted to fully vented propeller operation takes place. From the experiments it was found that as We is increased, there is an increase in J_ψ at which the transition takes place. The transition was found to be controlled by We and P/D and, from the experimental data, a critical J_ψ value was derived which provides a minimum J_ψ above which the predictions of the propeller performance may be deemed acceptable. The critical J_ψ value is defined as follows:

$$J_{\psi\mathrm{Crit}} = 0.825\frac{P}{D} - 1.25\frac{P}{D}\,e^{-0.018We''}. \qquad (16.38)$$

In general, the influence of Weber number on $J_{\psi\,\mathrm{crit}}$ was found to become negligible at Weber numbers greater than about 260.

WORKED EXAMPLE. Surface-piercing propeller.

Consider a vessel travelling at 50 knots. Propeller diameter $D = 1.1$ m, shaft inclination $\psi = 7°$. Revolutions $= 1400$ rpm ($= 23.33$ rps). $h_T = 0.55$ m, then $I_T = h_T/D = 0.55/1.1 = 0.50$.

Estimate the thrust T and power P_D with a propeller pitch ratio $P/D = 1.2$.

$$A'_o = 0.5\pi D^2/4 = 0.475\,\text{m}^2.$$

Wake fraction w_T is likely to be close to zero, depending on the thickness of the boundary layer. Thrust deduction t for such a hull form with conventional screws is $t = 0.07$ (see Table 8.4). As the surface-piercing propeller will be further from the hull, assume $t = 0.03$.

$$J_\psi = V \cos\psi/nD = 50 \times 0.5144 \times \cos 7°/23.33 \times 1.1 = 0.995.$$

Using Equation(16.34), $\quad K'_T = 0.1725$.

Using Equation(16.35, $\quad K'_Q = 0.04577$.

$$\eta_0 = J_\psi K'_T/2\pi K'_Q = 0.995 \times 0.1725/2\pi \times 0.04577 = 0.597.$$

$$We'' = \sqrt{\frac{\rho n^2 D^3 I_T}{\gamma}} = \sqrt{\frac{1025 \times 23.33^2 \times 1.1^3 \times 0.5}{0.078}} = 2181.7.$$

$$J_{\psi\text{Crit}} = 0.825\frac{P}{D} - 1.25\frac{P}{D}e^{-0.018We''}.$$

$$= 0.825 \times 1.2 - 1.25 \times 1.2 \times e^{-0.018 \times 2181.7} = 0.990.$$

Hence, $J_\psi = 0.995$ is acceptable.

$$\eta_H = (1 - t)/(1 - w_T) = (1 - 0.03)/(1 - 0) = 0.970.$$

Assume $\eta_R = 1.0$.

$$\eta_D = \eta_0 \times \eta_H \times \eta_R = 0.597 \times 0.970 \times 1.0 = 0.579.$$
$$P_D = 2\pi nQ = 2\pi n \times K'_Q \rho n^2 D^3 A'_0$$
$$= 2\pi \times 23.33 \times 0.04577 \times 1025 \times 23.33^2 \times 1.1^3 \times 0.475 = 2366.5\,\text{kW}.$$

Thrust produced is as follows:

$$T = K'_T \rho n^2 D^2 A'_0.$$
$$= 0.1725 \times 1025 \times 23.33^2 \times 1.1^2 \times 0.475 = 55.31\,\text{kN}.$$

[Check: Resistance $R = T(1 - t) = 55.31 \times 0.97 = 53.65$ kN.

$$P_E = R \times Vs = 53.65 \times 50 \times 0.5144 \times \cos 7° = 1369.6\,\text{kW}.$$
$$P_D = P_E/\eta_D = 1369.6/0.579 = 2365.5\,\text{kW}].$$

16.2.8 High-Speed Propellers, Inclined Shaft

Many high-speed propellers are mounted on inclined shafts and the propeller is operating at incidence to the inflow, Figure 16.21. The incidence is typically between $7°$ and $15°$. Two flow components act at the propeller plane, $Va \cos\psi$ and $Va \sin\psi$, where ψ is the propeller shaft inclination to the horizontal. The component of flow $Va \sin\psi$, perpendicular to the propeller shaft, gives rise to an eccentricity of the propeller thrust. On one side, as the propeller blade is upward going, there is a

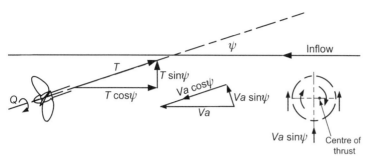

Figure 16.21. Flow and force components with an inclined shaft.

decrease in relative velocity and blade load; on the other side, the propeller blade is downward going and there is an increase in relative velocity and blade load. The net effect is to move the centre of thrust off the centreline. This leads also a significant fluctuation in blade loading as the blade rotates.

The thrust T can be resolved into a horizontal thrust $T \cos \psi$ and a vertical (lift) force $T \sin \psi$, which may make a significant contribution to the balance of masses and forces in the case of a planing craft. It should be noted that, if trim is present, then trim angle should be added to the shaft inclination when resolving for horizontal thrust.

Guidance on the performance of propellers on inclined shafts may be derived from Gutsche [16.41], Hadler [16.42], Peck and Moore [16.43] and Radojcic [16.35].

16.2.9 Small Craft Propellers: Locked, Folding and Self-pitching

The Gawn series propellers, Figures 16.8–16.10, have found many applications in the small craft field. They are all three-bladed and go up to a BAR of 1.10; high BAR is necessary when cavitation is likely to occur, see Chapter 12. The Wageningen B series [16.6] provides one of the few sources of data for two-bladed propellers suitable for small craft.

16.2.9.1 Drag of Locked Propellers

The drag of the propeller can be significant when locked at 0 rpm. There are occasions when propeller thrust is not required, such as in the case of the sailing yacht moving solely under sail power, or a larger merchant ship employing sail assistance or a multiple engine/propeller installation with one engine/propeller shut down. In these situations there are three options, Barnaby [16.44]:

(i) Provide enough power to run the propeller at the point of zero slip, ($K_T = 0$ on the $K_T - K_Q$ chart, Figure 16.3), effectively leading to zero propeller drag. It is, however, generally impractical to carry out this approach.

(ii) Allow the propeller to rotate freely. In this case, with the propeller windmilling, negative torque is developed which must be absorbed by the shaft bearings and clutch. Estimates of drag in this mode are approximate. As pointed out by Barnaby, unless the developed rpm are high enough, requiring free glands, bearings and clutch, etc., it is better to lock the propeller. Both Barnaby [16.44] and Hoerner [16.22] suggest approximate procedures for estimating the drag of a windmilling propeller.

(iii) The propeller is locked. In this case the propeller blades can be treated in a way similar to that of plates in a flow normal to their surface. The drag coefficient will depend on the pitch angle θ. Hoerner [16.22], based on experimental evidence and the blade angle θ at 0.7 radius, suggests the following relationship for the blade drag coefficient C_{DB}:

$$C_{DB} = 0.1 + \cos^2 \theta. \tag{16.39}$$

Using Equation (12.1), θ can be found from

$$\theta = \tan^{-1} \left(\frac{P/D}{0.7\pi} \right) \tag{16.40}$$

and propeller drag

$$D = C_{DB} \times \frac{1}{2} \rho A_B V a^2, \tag{16.41}$$

where blade area (all blades)

$$A_B = \frac{\pi D^2}{4} \times \text{BAR}. \tag{16.42}$$

For example, consider a locked propeller with diameter $D = 400$ mm, blade area ratio BAR $= 0.65$, pitch ratio $P/D = 0.80$ and moving in a wake speed $Va = 8$ knots, then:

$$A_B = (\pi D^2/4) \times \text{BAR} = (\pi \, 0.4^2/4) \times 0.65 = 0.082 \, \text{m}^2$$
$$\theta = \tan^{-1}[(P/D)/0.7\pi] = \tan^{-1}[0.80/0.7\pi] = 19.99°$$
$$C_{DB} = 0.1 + \cos^2 \theta = 0.1 + \cos^2 19.99 = 0.983$$

and propeller drag $D = C_{DB} \times 1/2\rho A_B V a^2 = 0.983 \times 1/2 \times 1025 \times 0.082 \times (8 \times 0.5144)^2 = 699$ N.

It is interesting to note that if the blades are set to $\theta = 90°$, then $C_{DB} = 0.1$ and the propeller drag is only 71 N.

An investigation into the drag of propellers when sailing, with the propeller stationary at 0 rpm, is reported by Mackenzie and Forrester [16.45]. Folding propellers are useful for minimising drag on sail craft when the propeller is not in use. Self-pitching propellers, [16.46], offer a good solution for sailing and motor sailing. The blades are free to rotate through 360°, can fully reverse for astern rpm, and can feather for low drag at 0 rpm. Efficiency is comparable with the conventional propeller, [16.46].

16.2.10 Waterjets

16.2.10.1 Background
An outline of the waterjet and its applications is given in Section 11.3.8. Units are available in the range 50 kW to 38,000 kW. Descriptions of the basic operation of waterjets and the components of efficiency can be found in Allison [16.47] and Carlton [16.14]. A propulsion prediction method for waterjets, using the components of efficiency, losses and hull interaction is given by Van Terwisga [16.48]. Summaries of research into the operation of waterjets and test procedures may be found

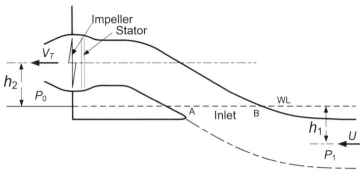

Figure 16.22. Schematic details of waterjet.

in the proceedings of ITTC and conferences [16.49–16.52]. Data suitable for the practical estimation of the performance characteristics and efficiency of waterjets are, however, sparse. This is particularly true for preliminary powering purposes.

Considerable research on waterjets using experimental and CFD methods has been carried out. This has been mainly centred on a suitable shape and slope of the duct at fluid entry, in particular, in areas A and B in Figure 16.22. This has been carried out in order to avoid possible separation, choking and cavitation in the duct and to achieve a clean and uniform flow through the duct to the impeller. Other research and development has concerned the pump, such as [16.53]. Pumps are normally axial or mixed centrifugal/axial flow with one pump stage plus stator blades ahead and/or downstream of the impeller. Increasing the ducting area at the impeller can slow down the flow in order to increase the static pressure and, hence, avoid cavitation on the impeller blades. Care has to be taken as too much diffusion in the ducting can lead to choking.

An outline is given here of the basic principles of operation of the waterjet using a momentum analysis, together with a proposed formula for propulsive efficiency η_D suitable for practical use at the initial design stage.

16.2.10.2 Momentum Analysis of Waterjet Performance

A useful understanding of waterjet performance can be obtained from a momentum analysis of operation. The schematic details of the waterjet are shown in Figure 16.22.

Total pressures are as follows:

at inlet

$$\frac{P_{T1}}{\rho} = \frac{P_1}{\rho} + \frac{1}{2}U^2 - gh_1 \qquad \text{but} \qquad \frac{P_1}{\rho} = \frac{P_0}{\rho} + gh_1$$

and

$$\frac{P_{T1}}{\rho} = \frac{P_0}{\rho} + \frac{1}{2}U^2, \tag{16.43}$$

where U is the free-stream or ship speed.
at tailpipe

$$\frac{P_{T2}}{\rho} = \frac{P_0}{\rho} + \frac{1}{2}V_T^2 + gh_2. \tag{16.44}$$

Equations (16.43) and (16.44) represent the energy per unit volume. Before the inlet, the total pressure is assumed to be constant, but between the inlet and exit nozzle, the pump imparts energy to the fluid. The difference between the two terms represents the useful work (per unit volume) done on the fluid, minus losses in the inlet and grille, ducting, pump and exit nozzle. The mass flow rate $m = \rho A V_T$, where A is the tailpipe or nozzle exit area

and power input is as follows:

$$P_D = m \{P_{T2} - P_{T1} + \text{pressure losses}\}, \tag{16.45}$$

hence,

$$P_D = \rho A V_T \left\{ \frac{1}{2} V_T^2 + gh_2 - \frac{1}{2} U^2 + \text{pressure losses} \right\}. \tag{16.46}$$

Assume that the system pressure losses are $\frac{\Delta P}{\rho} = k \frac{1}{2} V_T^2$, where k is typically 0.10–0.20, then

$$P_D = \rho A V_T \left\{ \frac{1}{2}(1+k) V_T^2 + gh_2 - \frac{1}{2} U^2 \right\}. \tag{16.47}$$

The overall momentum change gives thrust T as the following:

$$T = \rho A V_T \{V_T - U\}. \tag{16.48}$$

The efficiency of the unit $\eta = \dfrac{\text{Useful work}}{\text{Delivered power}} = \dfrac{TU}{P_D} \tag{16.49}$

and

$$\eta = \frac{2U(V_T - U)}{(1+k) V_T^2 + 2gh_2 - U^2}. \tag{16.50}$$

If the exit velocity ratio is written as $x = \frac{V_T}{U}$ and a height factor is written as $H = \frac{2gh_2}{U^2}$, then

$$\eta = \frac{2(x-1)}{(1+k)x^2 + H - 1} \tag{16.51}$$

There is optimum efficiency when $\frac{d\eta}{dx} = 0$ or when

$$x = 1 + \sqrt{\frac{k+H}{1+k}} \tag{16.52}$$

The value of k is typically 0.10–0.20 and its evaluation requires estimates of pressure losses at (i) inlet and grille, (ii) ducting ahead of the impeller, (iii) losses through the pump and (iv) losses in the tailpipe/exit nozzle. The separate assessment of these losses is complicated and, in the present analysis, it is convenient to group them under one heading (k). Van Terwisga [16.48] describes a methodology for estimating these losses separately. Thrust loading may be expressed as follows:

$$C_T = \frac{T}{0.5 \rho A U^2} = 2x(x-1). \tag{16.53}$$

Table 16.5 shows typical optimum values for $H = 0.050$. Figure 16.23 shows the overall influences of losses (k) and thrust loading (C_T) on efficiency, for $H = 0$. It is noted that the effects of the losses are most marked at low values of C_T.

Table 16.5. *Waterjet optimum efficiency*

k	x_{opt}	η_{opt}	C_{Topt}
0.10	1.37	0.66	1.01
0.15	1.42	0.61	1.18
0.20	1.46	0.57	1.33

In assessing overall system performance, account needs to be taken of the wake effect, if the waterjet inlet ingests boundary layer material, when U will be less than ship speed, and the inlet external drag, or thrust deduction. Methods of incorporating these in the design process are described by Van Terwisga [16.48].

Account should also be taken of the weight of water in the unit when it is operational, which adds to the displacement of the craft and, hence drag. Operating at above optimum C_T will reduce unit size and net weight substantially and the overall best unit will be smaller than the 'optimum'. For example, from Equation (16.53), for a given thrust, $A \propto C_T^{-1}$. The volume of the unit is $\propto A^{3/2} \propto C_T^{-3/2}$. Table 16.6 illustrates, for $k = 0.10$ and $H = 0.050$, the influence of C_T and volume on efficiency. It can be seen that increasing C_T from 1.00 to 1.50 almost halves the unit volume for only a 1% loss in unit efficiency. Increasing C_T will ultimately be limited by the possible onset of cavitation.

PUMPS. The pump is normally axial flow or mixed centrifugal-axial flow. A detailed description of pump design is given by Brennen [16.54] and typical research into pumps is described in [16.53]. The pump is required to pass a required flow rate Q at a given rotational speed through a head rise h. This can be expressed as a non-dimensional specific speed N_S, as follows:

$$N_S = \frac{n Q^{1/2}}{g h^{3/4}}. \tag{16.54}$$

Different pump geometries have a typical range of optimum specific speeds for optimum efficiency. For axial flow pumps these are typically 3.0–6.5 and for mixed

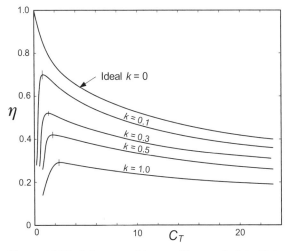

Figure 16.23. Influence of losses (k) on waterjet efficiency.

Table 16.6. *Influence of thrust loading on waterjet volume*

C_T	x	η	Volume factor $= C_T^{-3/2}$
1.00	1.36	0.664	1.00
1.50	1.50	0.656	0.54
2.00	1.62	0.641	0.35
2.50	1.72	0.624	0.25

flow pumps these are 1.5–3.0. The volume, hence weight and cost, of a pump is proportional to the torque required which, for a given speed, varies inversely with speed. Typically, the higher the speed, the lighter and cheaper the pump. The occurrence of cavitation will set an upper limit on viable specific speeds, see Terwisga [16.48]. Finally, the power required to drive the waterjet impeller is given by the following

$$P_i = \rho Q g h. \tag{16.55}$$

16.2.10.3 Practical Estimates of Waterjet Efficiency

Practical design data suitable for estimating the overall thrust, torque and efficiency of waterjets are sparse. Assembling and estimating the various components of efficiency, and losses, is not practical at the early design stage. Information derived from manufacturers' data, including Svensson [16.55], has therefore been used to estimate waterjet efficiency. The data were derived from model and full-scale measurements. An approximate formula for η_D, based on the manufacturers' data and suitable for use at the initial design stage, is given as Equation (16.56).

$$\eta_D = \frac{1}{[1 + 8.64/V]}, \tag{16.56}$$

where V is ship speed in m/s. Equation (16.56) is suitable for speeds up to about 50 knots.

For preliminary design purposes, based on manufacturers' data, the approximate diameter D (m) of the waterjet at the impeller/transom for a given power P_D (kW) may be assumed to be the following:

$$D = \frac{4.05}{1 + [11500/P_D]^{0.9}} \tag{16.57}$$

and the displaced (entrained) volume DV (m^3) may be assumed to be the following:

$$DV = 0.12 \left[\frac{P_D}{1000} \right]^{1.6}. \tag{16.58}$$

16.2.11 Vertical Axis Propellers

There are some limited published data for vertical axis (cycloidal) propellers [16.56, 16.57]. A review of performance prediction and theoretical methods for

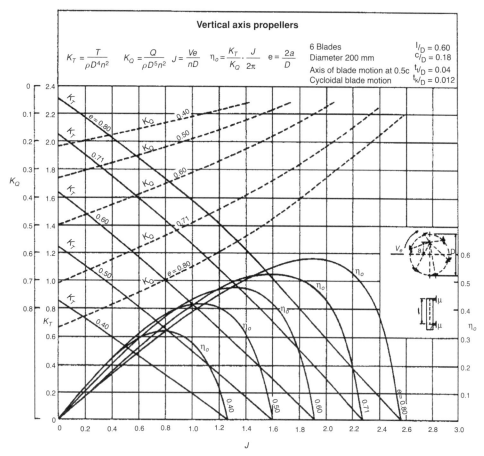

Figure 16.24. $K_T - K_Q$ curves for vertical axis propellers [16.57].

cycloidal propellers is included in Bose [16.28]. An example of such data for K_T and K_Q is that reported by Van Manen [16.57] and reproduced with permission in Figure 16.24. These charts may be used in the same way as for conventional propellers, Section 16.2.1. In Figure 16.24, J $(= Ve/nD)$ is the advance coefficient, Ve is the advance velocity, six-blades with length l, chord c; e is equivalent to pitch and $e = 2a/D$, where a is offset of centre of blade control.

16.2.12 Paddle Wheels

Useful data are available following the tests carried out by Volpich and Bridge [16.58–16.60]. It is found that a paddle employing fully feathering blades can achieve efficiencies reasonably close to the conventional propeller.

16.2.13 Lateral Thrust Units

Data for the design and sizing of lateral thrust units at the initial design stage are derived mainly from manufacturers' information and results from thrust units in service. In the following, simple theory is used to derive the basic equations of operation.

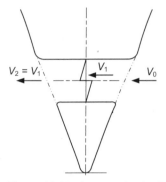

Figure 16.25. Assumed velocities in lateral thrust unit.

Applying actuator disc theory and the velocity changes in Figure 16.25, it can be shown that

$$\text{Thrust } T = \rho A V_1^2 \qquad (16.59)$$

$$\text{Power } P = \frac{1}{2}\rho A V_1^3. \qquad (16.60)$$

Actual power requirements will need to be much higher than this to allow for gearing losses, and drag losses on the impeller and duct walls not taken into account in this simple theory.

In dimensional terms,

$$\text{From Equation (16.59), } T = f(D^2 \cdot V^2)$$
$$\text{and from Equation (16.60), } P = f(D^2 \cdot V^3),$$
$$\text{hence, } T = f(P \cdot D)^{2/3}$$
$$\text{and } T = C[P \cdot D]^{2/3}, \qquad (16.61)$$

where $C = T/(P \cdot D)^{2/3}$ can be seen as a measure of efficiency and may be derived from published data.

Based on published data, and with units T (kN), D (m) and P (kW), a suitable average value is $C = 0.82$.

A suitable approximate value for diameter, for use at the initial design stage is the following:

$$D = 1.9\left[\frac{P}{1000}\right]^{0.45}. \qquad (16.62)$$

Equations (16.61) and (16.62) can be used to determine preliminary dimensions of the thrust unit for a given thrust, power for a required thrust and diameter and the influence of diameter on thrust and efficiency.

A useful background to lateral thrust units and design data are described by English [16.61]. Figure 16.26, reproduced with permission from [16.61], provides typical data for different propulsors and thrust units. The 'C' values in Figure 16.26 for lateral thrust units, defined in the same way as Equation (16.61), lie typically in the range 55–70. It should be noted that the information in Figure 16.26 is in imperial units, is not dimensionless and applies to units operating in salt water. The foregoing

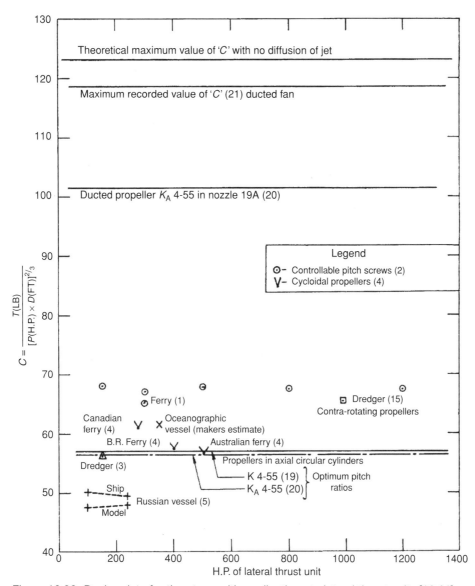

Figure 16.26. Design data for thrusters, with applications to lateral thrust units [16.61].

has considered the ship with zero forward speed. The influence of forward speed on the performance of lateral thrust units is discussed in [16.62, 16.63, 16.64].

Example application of Equations (16.61) and (16.62) follows:

Given an available thruster power of 2000 kW, from Equation (16.62), diameter $D = 1.9 \,(2000/1000)^{0.45} = 2.60$ m, from Equation (16.61), estimated thrust $T = 0.82 \,(2000 \times 2.60)^{2/3} = 246.8$ kN. (In Imperial units, $T = 55.6 \times 10^3$ lb, $P = 2681.0$ hp, $D = 8.53$ ft, and the C factor in Figure 16.26 is 68.8).

16.2.14 Oars

Useful information on the performance of oars, the thrust produced and efficiency is provided in [16.65–16.68].

Table 16.7. *Rig terminology*

I	Height of foretriangle (m)
J	Base of foretriangle (m)
FA	Average freeboard (m)
EHM	Mast height above sheer line (m)
$EMDC$	Average mast diameter (m)
P	Mainsail hoist (m)
BAD	Height of main boom above sheer line (m)
B	Beam (m)

16.2.15 Sails

16.2.15.1 Background

In estimating the performance of a sailing yacht it is necessary to predict the forces generated by the sails. In reality, this depends not only on the apparent wind speed and direction and the sail characteristics, but also on the skill of the sailor in trimming the sails for the prevailing conditions. This aspect of performance is extremely hard to replicate at the design stage and in making predictions of yacht speed.

If wind tunnel tests are undertaken, then it is possible to adjust the trim of the sails in order to maximise the useful drive force from them at each wind speed and direction and for the complete set of sails to be used on the yacht. In this manner, the sail's lift and drag coefficients may be obtained and used for subsequent performance predictions. This process is described, for example, in Marchaj [16.69] and Campbell [16.70].

The treatment of sail performance in most modern velocity prediction programs (VPPs), including that of the Offshore Racing Congress (ORC [16.71]) used for yacht rating, is based on the approach first described by Hazen [16.72]. This produces the force coefficients for the rig as a whole based on the contributions of individual sails. It also includes the facility to consider the effects of 'flattening' and 'reefing' the sails through factors, thus allowing some representation of the manner in which a yacht is sailed in predicting its performance. Since Hazen [16.72], the model has been updated to include the effects of sail–sail interaction and more modern sail types and shapes. These updates are described, for example, in Claughton [16.73] and ORC [16.71]. The presentation included here essentially follows that of Larsson and Eliasson [16.74] and ORC [16.71].

The method is based on the lift and viscous drag coefficients of the individual sails, which are applied to an appropriate vector diagram, such as Figure 11.12.

The rig terminology adopted is that used by the International Measurement System (IMS) in measuring the rig of a yacht. A summary of the parameters is given in Table 16.7.

16.2.15.2 Lift and Drag Coefficients

Typical values of lift and drag coefficient, taken with permission from ORC [16.71], for mainsail and jib for a range of apparent wind angles are given in Tables 16.8 and 16.9 and Figures 16.27 and 16.28.

Table 16.8. *Lift and drag coefficients for mainsail*

β deg	0	7	9	12	28	60	90	120	150	180
C_L	0.000	0.862	1.052	1.164	1.347	1.239	1.125	0.838	0.296	−0.112
C_D	0.043	0.026	0.023	0.023	0.033	0.113	0.383	0.969	1.316	1.345

The total lift and viscous drag coefficients are given by the weighted contribution of each sail as follows:

$$C_L = \frac{C_{Lm} A_m + C_{Lj} A_j}{A_n} \qquad (16.63)$$

and

$$C_{Dp} = \frac{C_{Dm} A_m + C_{Dj} A_j}{A_n}, \qquad (16.64)$$

where the reference (nominal) area is calculated as the sum of the mainsail and foretriangle areas $A_n = A_m + A_F$. The fore triangle area is $A_F = 0.5IJ$.

The total drag coefficient for the rig is given by the following:

$$C_D = C_{Dp} + C_{D0} + C_{DI}, \qquad (16.65)$$

where C_{D0} is the drag of mast and hull topsides and may be estimated as follows:

$$C_{D0} = 1.13 \frac{(FA \cdot B) + (EHM \cdot EMDC)}{A_n}. \qquad (16.66)$$

C_{DI} is the induced drag due to the generation of lift, estimated as follows:

$$C_{DI} = C_L^2 \left(\frac{1}{AR} + 0.005 \right), \qquad (16.67)$$

where the effective rig aspect ratio, AR, for close hauled sailing is as follows:

$$AR = \frac{(1.1 (EHM + FA))^2}{A_n}. \qquad (16.68)$$

The effective AR for other course angles is as follows:

$$AR = \frac{(1.1EHM)^2}{A_n}. \qquad (16.69)$$

The effects of 'flattening' and 'reefing' the sails are given by, $C_L = C_L F \cdot R^2$ and $C_{Dp} = C_{Dp} R^2$, where F and R are factors between 0 and 1, and 1 represents the normal unreefed sail.

The centre of effort of the rig is important, since it determines the overall heeling moment from the rig. The centre of effort is given by weighting the contribution

Table 16.9. *Lift and drag coefficients for jib*

β deg	7	15	20	27	50	60	100	150	180
C_L	0.000	1.000	1.375	1.450	1.430	1.250	0.400	0.000	−0.100
C_D	0.050	0.032	0.031	0.037	0.250	0.350	0.730	0.950	0.900

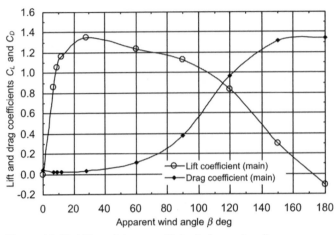

Figure 16.27. Lift and drag coefficients for mainsail.

of each sail's individual centre of effort by its area and contribution to total force (based on lift and viscous drag). Individual sail centres of effort are given as follows:

$$C_{Em} = 0.39P + BAD \tag{16.70}$$

and

$$C_{Ej} = 0.39I. \tag{16.71}$$

In order to account for the effect of heel on the aerodynamic performance of the rig, the apparent wind speed and direction to be used in calculations are those in the heeled mast plane. These are referred to as the 'effective' apparent wind speed and direction and are found as follows:

$$V_{AE} = \sqrt{V_1^2 + V_2^2} \tag{16.72}$$

$$\beta_E = \cos^{-1}\left(\frac{V_1}{V_{AE}}\right), \tag{16.73}$$

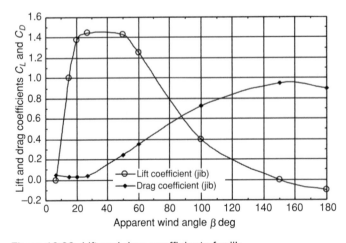

Figure 16.28. Lift and drag coefficients for jib.

with $V_1 = V_S + V_T \cos \gamma$ and $V_2 \approx V_T \sin \gamma \cos \phi$, where γ is the true wind angle (or course angle), V_T is the true wind velocity, V_S is the boat speed and ϕ is the heel angle.

In making performance estimates the total lift and drag values from the rig are used, calculated from the force coefficients as follows:

$$L = \frac{1}{2} \rho_A A_n V_A^2 C_L \qquad (16.74)$$

$$D = \frac{1}{2} \rho_A A_n V_A^2 C_D. \qquad (16.75)$$

These rig forces are resolved into the track axes of the yacht and need to balance the hydrodynamic forces. An example of this procedure is given in example application 17, Chapter 17.

16.3 Hull and Relative Rotative Efficiency Data

16.3.1 Wake Fraction w_T and Thrust Deduction t

Hull efficiency is defined as $\eta_H = (1 - t) / (1 - w_T)$. Wake fractions and thrust deduction factors are dealt with in Chapter 8.

16.3.2 Relative Rotative Efficiency, η_R

The relative rotative efficiency is the ratio of the efficiency of the propeller in a wake behind the hull and the efficiency of the propeller in open water, see Chapter 8. It is obtained from the model self-propulsion test, Section 8.7. η_R ranges typically from 0.95 to 1.05 and is often assumed to be unity for preliminary design purposes. η_R depends on a number of parameters, including propeller diameter and pitch and hull form. In the absence of self-propulsion tests, empirical data for η_R may be used for preliminary powering calculations. Data are available from references such as [16.75–16.84] and are summarised below.

16.3.2.1 Single Screw

BRITISH SHIP RESEARCH ASSOCIATION (BSRA) SERIES [16.75]. Equation (16.76) is the preferred expression. Equation (16.77) is an alternative if pitch ratio (P/D) and blade area ratio (BAR) are not available.

$$\eta_R = 0.8372 + 0.1338 \, C_B + 1.5188 \, D/L_{BP} + 0.1240 \, P/D - 0.1152 \text{BAR}. \quad (16.76)$$

$$\eta_R = 0.5524 + 0.8443 \, C_B - 0.5054 \, C_B^2 + 1.1511 \, D/L_{BP} + 0.4718 D/\nabla^{1/3}, \quad (16.77)$$

where D is the propeller diameter in m, ∇ is displaced volume in m^3, the C_B range is 0.55–0.85, the P/D range is 0.60–1.10, the BAR range is 0.40–0.80 and the D/L_{BP} range is 0.02–0.06.

SERIES 60 [16.76]. A regression equation for η_R was developed, Equation (16.78). The ranges of the hull parameters are as follows: L/B, 5.5–8.5; B/T, 2.5–3.5; C_B, 0.60–0.80 and LCB, −2.48% to +3.51%. The regression 'd' coefficients for a range of speeds are tabulated in Appendix A4, Table A4.6.

$$
\begin{aligned}
\eta_R = {}& d_1 + d_2(L/B) + d_3(B/T) + d_4(C_B) + d_5(LCB) + d_6(L/B)^2 \\
& + d_7(B/T)^2 + d_8(C_B)^2 + d_9(LCB)^2 + d_{10}(L/B)(B/T) \\
& + d_{11}(L/B)(C_B) + d_{12}(L/B)(LCB) + d_{13}(B/T)(C_B) \\
& + d_{14}(B/T)(LCB) + d_{15}(C_B)(LCB).
\end{aligned}
\tag{16.78}
$$

HOLTROP [16.77].

$$
\eta_R = 0.9922 - 0.05908(\mathrm{BAR}) + 0.07424(Cp - 0.0225LCB), \tag{16.79}
$$

where LCB is L_{CB} forward of $0.5L$ as a percentage of L.

16.3.2.2 Twin Screw
HOLTROP [16.77].

$$
\eta_R = 0.9737 + 0.111(Cp - 0.0225LCB) - 0.06325P/D, \tag{16.80}
$$

where LCB is L_{CB} forward of $0.5L$ as a percentage of L.

16.3.2.3 Podded Unit
FLIKKEMA ET AL. [16.23].

$$
\eta_{Rp} = 1.493 - 0.18425\,C_P - 0.4278\,LCB - 0.33804\,P/D \tag{16.81}
$$

where LCB is $(L_{CB}$ forward of $0.5L)/L$.

16.3.2.4 Tug Data
PARKER–DAWSON [16.78] AND MOOR [16.79]. Typical approximate mean values of η_R from [16.78, 16.79] are given in Table 16.10. Further data for changes in propeller diameter and hull form are given in [16.78 and 16.79].

16.3.2.5 Trawler Data
BSRA [16.80, 16.81, 16.82]. Typical approximate values of η_R for trawler forms, from [16.81], are given in Table 16.11. Speed range $Fr = 0.29$–0.33. The influence of $L/\nabla^{1/3}$, B/T, LCB and hull shape variations are given in [16.80 and 16.82].

Table 16.10. *Relative rotative efficiency for tugs*

Source	Fr	η_R	Case
Reference [16.78] $C_B = 0.503$	0.34	1.02	Free-running
	0.21	–	Towing
	0.15	–	Towing
	0.09	–	Towing
	0 (bollard)	–	Bollard
Reference [16.79] $C_B = 0.575$	0.36	1.01	Free-running
	0.21	1.03	Towing
	0.12	1.02	Towing
	0 (bollard)	–	Bollard

Table 16.11. *Relative rotative efficiency for trawlers*

C_B	η_R
0.53	1.048
0.57	1.043
0.60	1.030

Table 16.12. *Relative rotative efficiency for round bilge semi-displacement forms*

C_B range	Fr_∇	η_R
$C_B < 0.45$	0.6	0.95
	1.4	0.95
	2.6	0.95
$C_B > 0.45$	0.6	1.07
	1.4	0.95
	2.2	0.96

16.3.2.6 Round Bilge Semi-displacement Craft

BAILEY [16.83]. Typical approximate mean values of η_R for round bilge forms are given in Table 16.12. The data are generally applicable to round bilge forms in association with twin screws. Speed range $Fr_\nabla = 0.58$–2.76, where $Fr_\nabla = V/\sqrt{g\nabla^{1/3}}$ [V, m/s; ∇, m³] or $Fr_\nabla = 0.165\, V/\Delta^{1/6}$ [V in knots, Δ in tonnes] and the C_B range is 0.37–0.52. Regression equations are derived for η_R in [16.83].

GAMULIN [16.84]. In the round bilge SKLAD series, typical approximate values can be obtained as follows: $C_B \leq 0.45$, speed range $Fr_\nabla = 0.60$–3.0

$$\eta_R = 0.035 Fr_\Delta + 0.90.$$

REFERENCES (CHAPTER 16)

16.1 Van Manen, J.D. The choice of propeller. *Marine Technology*, SNAME, Vol. 3, No. 2, April 1966, pp. 158–171.
16.2 Lewis, E.V. (ed.) *Principles of Naval Architecture.* The Society of Naval Architects and Marine Engineers, New York, 1988.
16.3 Rawson, K.J. and Tupper, E.C. *Basic Ship Theory.* 5th Edition, Combined Volume. Butterworth-Heinemann, Oxford, UK, 2001.
16.4 Molland, A.F. (ed.) *The Maritime Engineering Reference Book.* Butterworth-Heinemann, Oxford, UK, 2008.
16.5 O'Brien, T.P. *The Design of Marine Screw Propellers.* Hutchinson and Co., London, 1969.
16.6 Van Lammeren, W.P.A., Van Manen, J.D. and Oosterveld, M.W.C. The Wageningen B-Screw Series. *Transactions of the Society of Naval Architects and Marine Engineers*, Vol. 77, 1969, pp. 269–317.

16.7 Gawn, R.W.L. Effects of pitch and blade width on propeller performance. *Transactions of the Royal Institution of Naval Architects*, Vol. 95, 1953, pp. 157–193.

16.8 Yazaki, A. Design diagrams of modern four, five, six and seven-bladed propellers developed in Japan. *4th Naval Hydrodynamics Symposium*, National Academy of Sciences, Washington, DC, 1962.

16.9 Lingren, H. Model tests with a family of three and five bladed propellers. SSPA Paper No. 47, Gothenburg, 1961.

16.10 Gawn, R.W.L. and Burrill, L.C. Effect of cavitation on the performance of a series of 16 in. model propellers. *Transactions of the Royal Institution of Naval Architects*, Vol. 99, 1957, pp. 690–728.

16.11 Burrill, L.C. and Emerson, A. Propeller cavitation: some observations from 16 in. propeller tests in the new King's College Cavitation Tunnel. *Transactions of the North East Coast Institution of Engineers and Shipbuilders*, Vol. 70, 1953, pp. 121–150.

16.12 Burrill, L.C. and Emerson, A. Propeller cavitation: further tests on 16 in. propeller models in the King's College Cavitation Tunnel. *Transactions of the North East Coast Institution of Engineers and Shipbuilders*, Vol. 79, 1962–1963, pp. 295–320.

16.13 Emerson, A. and Sinclair, L. Propeller design and model experiments. *Transactions of the North East Coast Institution of Engineers and Shipbuilders*, Vol. 94, No. 6, 1978, pp. 199–234.

16.14 Carlton, J.S. *Marine Propellers and Propulsion*. 2nd Edition Butterworth-Heinemann, Oxford, UK, 2007.

16.15 Oosterveld, M.W.C., van Oossanen, P. Further computer-analysed data of the Wageningen B-screw series. *International Shipbuilding Progress*, Vol. 22, No. 251, July 1975, pp. 251–262.

16.16 Blount, D.L. and Hubble, E.N. Sizing segmental section commercially available propellers for small craft. Propellers 1981. SNAME Symposium, 1981.

16.17 Baker, G.S. The effect of propeller boss diameter upon thrust and efficiency at given revolutions. *Transactions of the Royal Institution of Naval Architects*, Vol. 94, 1952, pp. 92–109.

16.18 Oosterveld, M.W.C. Wake adapted ducted propellers. NSMB Wageningen Publication No. 345, 1970.

16.19 Oosterveld, M.W.C. Ducted propeller characteristics. *RINA Symposium on Ducted Propellers*, RINA, London, 1973.

16.20 ITTC. Report of specialist comitteee on azimuthing podded propulsors. *Proceedings of 25th ITTC*, Vol. II. Fukuoka, 2008.

16.21 Molland, A.F. and Turnock, S.R. *Marine Rudders and Control Surfaces*. Butterworh-Heinemann, Oxford, UK. 2007.

16.22 Hoerner, S.F. *Fluid-Dynamic Drag*. Published by the author, New York, 1965.

16.23 Flikkema, M.B., Holtrop, J. and Van Terwisga, T.J.C. A parametric power prediction model for tractor pods. *Proceedings of the Second International Conference on Advances in Podded Propulsion*, T-POD. University of Brest, France, October 2006.

16.24 Stierman, E.J. The influence of the rudder on the propulsive performance of ships. Parts I and II. *International Shipbuilding Progress*, Vol. 36, No. 407, 1989, pp. 303–334; No. 408, 1989, pp. 405–435.

16.25 Funeno, I. Hydrodynamic development and propeller design method of azimuthing podded propulsion systems. *Ninth Symposium on Practical Design of Ships and Other Floating Structures, PRADS'2004*, Lübeck- Travemünde, Germany, 2004, pp. 888–893.

16.26 Islam, M.F., He, M. and Veitch, B. Cavitation characteristics of some pushing and pulling podded propellers. *Transactions of the Royal Institution of Naval Architects*, Vol. 149, 2007, pp. 41–50.

16.27 Islam, M.F., Molloys, S., He, M., Veitch, B., Bose, N. and Liu, P. Hydro-dynamic study of podded propulsors with systematically varied geometry. *Proceedings of the Second International Conference on Advances in Podded Propulsion*, T-POD. University of Brest, France, October 2006.

16.28 Bose, N. *Marine Powering Predictions and Propulsors*. The Society of Naval Architects and Marine Engineers. New York, 2008.

16.29 Heinke, C. and Heinke, H-J. Investigations about the use of podded drives for fast ships. *Proceedings of Seventh International Conference on Fast Sea Transportation, FAST'2003,* Ischia, Italy, October 2003.

16.30 Frolova, I., Kaprantsev, S., Pustoshny, A. and Veikonheimo, T. Development of the propeller series for azipod compact. *Proceedings of the Second International Conference on Advances in Podded Propulsion, T-POD.* University of Brest, France, October 2006.

16.31 T-POD First International Conference on Advances in Podded Propulsion, Conference Proceedings, University of Newcastle, UK, 2004.

16.32 T-POD Second International Conference on Advances in Podded Propulsion, Conference Proceedings, University of Brest, France, October 2006.

16.33 ITTC. Report of Propulsion Committee. *Proceedings of the 23rd ITTC*, Venice, 2002.

16.34 ITTC. Report of specialist comitteee on azimuthing podded propulsors. *Proceedings of 24th ITTC*, Vol. II. Edinburgh, 2005.

16.35 Radojcic, D. An engineering approach to predicting the hydrodynamic performance of planing craft using computer techniques. *Transactions of the Royal Institution of Naval Architects*, Vol. 133, 1991, pp. 251–267.

16.36 Newton, R.N. and Rader, H.P. Performance data of propellers for high-speed craft. *Transactions of the Royal Institution of Naval Architects*, Vol. 103, 1961, pp. 93–129.

16.37 Tulin, M.P. Supercavitating flow past foils and struts. *Proceedings of NPL Symposium on Cavitation in Hydrodynamics*. Her Majesty's Stationery Office, London, 1956.

16.38 Rose, J.C. and Kruppa, F.L. Surface piercing propellers: Methodical series model test results, *Proceedings of First International Conference on Fast Sea Transportation, FAST'91*, Trondheim, 1991.

16.39 Ferrando, M., Scamardella, A., Bose, N., Liu, P. and Veitch, B. Performance of a family of surface piercing propellers. *Transactions of the Royal Institution of Naval Architects*, Vol. 144, 2002, pp. 63–75.

16.40 Chudley, J., Grieve, D. and Dyson, P.K. Determination of transient loads on surface piercing propellers. *Transactions of the Royal Institution of Naval Architects*, Vol. 144, 2002, pp. 125–141.

16.41 Gutsche, F. The study of ships' propellers in oblique flow. *Schiffbauforschung* 3, 3/4. Translation from German, Defence Research Information Centre DRIC Translation No. 4306, 1964, pp. 97–122.

16.42 Hadler, J.B. The prediction of power performance of planing craft. *Transactions of the Society of Naval Architects and Marine Engineers*, Vol. 74, 1966, pp. 563–610.

16.43 Peck, J.G. and Moore, D.H. Inclined-shaft propeller performance characteristics. Paper G, *SNAME Spring Meeting*, April 1973, pp. G1–G21.

16.44 Barnaby, K.C. *Basic Naval Architecture*. Hutchinson, London, 1963.

16.45 Mackenzie, P.M. and Forrester, M.A. Sailboat propeller drag. *Ocean Engineering*, Vol. 35, No. 1, January 2008, p. 28.

16.46 Miles, A., Wellicome, J.F. and Molland, A.F. The technical and commercial development of self-pitching propellers. *Transactions of the Royal Institution of Naval Architects*, Vol. 135, 1993, pp. 133–147.

16.47 Allison, J.L. Marine waterjet propulsion. *Transactions of the Society of Naval Architects and Marine Engineers*, Vol. 101, 1993, pp. 275–335.

16.48 Van Terwisga, T. A parametric propulsion prediction method for water-jet driven craft. *Proceedings of Fourth International Conference on Fast Sea Transportation, FAST'97*, Sydney, Australia. Baird Publications, South Yarra, Australia, July 1997.

16.49 ITTC. Report of Specialist Committee on Validation of Waterjet Test Procedures. *Proceedings of 23rd ITTC*, Vol. II, Venice, 2002.

16.50 ITTC. Report of Specialist Committee on Validation of Waterjet Test Procedures. *Proceedings of 24th ITTC*, Vol. II, Edinburgh, 2005.

16.51 *Proceedings of International Conference on Waterjet Propulsion III*, RINA, Gothenburg, 2001.

16.52 *Proceedings of International Conference on Waterjet Propulsion IV*, RINA, London, 2004.

16.53 Michael, T.J. and Chesnakas, C.J. Advanced design, analysis and testing of waterjet pumps. *25th Symposium on Naval Hydrodynamics*, St. John's, Newfoundland and Labrador, Canada, August 2004, pp. 104–114.

16.54 Brennen, C.E. *Hydrodynamics of Pumps*. Oxford University Press, 1994.

16.55 Svensson, R. Water jet for naval applications. *RINA International Conference on New Developments in Warship Propulsion*. RINA, London, 1989.

16.56 Van Manen, J.D. Results of systematic tests with vertical axis propellers. *International Shipbuilding Progress*, Vol. 13, 1966.

16.57 Van Manen, J.D. Non-conventional propulsion devices. *International Shipbuilding Progress*, Vol. 20, No. 226, June 1973, pp. 173–193.

16.58 Volpich, H. and Bridge, I.C. Paddle wheels: Part I, Preliminary model experiments. *Transactions Institute of Engineers and Shipbuilders in Scotland IESS*, Vol. 98, 1954–1955, pp. 327–380.

16.59 Volpich, H. and Bridge, I.C. Paddle wheels: Part II, Systematic model experiments. *Transactions Institute of Engineers and Shipbuilders in Scotland IESS*, Vol. 99, 1955–1956, pp. 467–510.

16.60 Volpich, H. and Bridge, I.C. Paddle wheels: Part IIa, III, Further model experiments and ship/model correlation. *Transactions Institute of Engineers and Shipbuilders in Scotland IESS*, Vol. 100, 1956–1957, pp. 505–550.

16.61 English, J.W. The design and performance of lateral thrust units. *Transactions of the Royal Institution of Naval Architects*, Vol. 105, 1963, pp. 251–278.

16.62 Ridley, D.E. Observations on the effect of vessel speed on bow thruster performance. *Marine Technology*, 1971, pp. 93–96.

16.63 Norrby, R.A. and Ridley, D.E. Notes on ship thrusters. *Transactions of the Institute of Marine Engineers*. Vol. 93, Paper 6, 1981, pp. 2–8.

16.64 Brix, J. (ed.) *Manoeuvring Technical Manual*, Seehafen Verlag, Hamburg, 1993.

16.65 Alexander, F.H. The propulsive efficiency of rowing. In *Transactions of the Royal Institution of Naval Architects*, Vol. 69, 1927, pp. 228–244.

16.66 Wellicome, J.F. Some hydrodynamic aspects of rowing. In *Rowing – A Scientific Approach*, ed. J.P.G. Williams and A.C. Scott. A.S. Barnes, New York, 1967.

16.67 Shaw, J.T. Rowing in ships and boats. *Transactions of the Royal Institution of Naval Architects*, Vol. 135, 1993, pp. 211–224.

16.68 Kleshnev, V. Propulsive efficiency of rowing. Proceedings of XVII International Symposium on Biomechanics in Sports, Perth, Australia, 1999, pp. 224–228.

16.69 Marchaj, C.A. Aero-Hydrodynamics of Sailing, Adlard Coles Nautical, London, 1979.

16.70 Campbell, I.M.C. Optimisation of a sailing rig using wind tunnel data, *The 13th Chesapeake Sailing Yacht Symposium, Annapolis*. SNAME. 1997.

16.71 Offshore Racing Congress. VPP Documentation 2009 Offshore Racing Congress. http://www.orc.org/ (last accessed 3 June 2010). 69 pages.

16.72 Hazen, G.S. A model of sail aerodynamics for diverse rig types, *New England Sailing Yacht Symposium*. SNAME, 1980.

16.73 Claughton, A.R. Developments in the IMS VPP formulations, *The 14th Chesapeake Sailing Yacht Symposium, Annapolis*. The Society of Naval Architects and Marine Engineers, 1999, pp. 27–40.

16.74 Larsson, L. and Eliasson, R.E. *Principles of Yacht Design*. 3rd Edition, Adlard Coles Nautical, London, 2009.

16.75 Pattullo, R.N.M. and Wright, B.D.W. Methodical series experiments on single-screw ocean-going merchant ship forms. Extended and revised overall analysis. BSRA Report NS333, 1971.

16.76 Sabit, A.S. An analysis of the Series 60 results: Part II, Regression analysis of the propulsion factors. *International Shipbuilding Progress*, Vol. 19, No. 217, September 1972, pp. 294–301.

16.77 Holtrop, J. and Mennen, G. G. J. An approximate power prediction method. *International Shipbuilding Progress*, Vol. 29, No. 335, July 1982, pp. 166–170.

16.78 Parker, M.N. and Dawson, J. Tug propulsion investigation. The effect of a buttock flow stern on bollard pull, towing and free-running performance. *Transactions of the Royal Institution of Naval Architects*, Vol. 104, 1962, pp. 237–279.

16.79 Moor, D.I. An investigation of tug propulsion. *Transactions of the Royal Institution of Naval Architects*, Vol. 105, 1963, pp. 107–152.

16.80 Pattullo, R.N.M. and Thomson, G.R. The BSRA Trawler Series (Part I). Beam-draught and length-displacement ratio series, resistance and propulsion tests. *Transactions of the Royal Institution of Naval Architects*, Vol. 107, 1965, pp. 215–241.

16.81 Pattulo, R.N.M. The BSRA Trawler Series (Part II). Block coefficient and longitudinal centre of buoyancy series, resistance and propulsion tests. *Transactions of the Royal Institution of Naval Architects*, Vol. 110, 1968, pp. 151–183.

16.82 Thomson, G.R. and Pattulo, R.N.M. The BSRA Trawler Series (Part III). Resistance and propulsion tests with bow and stern variations. *Transactions of the Royal Institution of Naval Architects*, Vol. 111, 1969, pp. 317–342.

16.83 Bailey, D. A statistical analysis of propulsion data obtained from models of high speed round bilge hulls. *Symposium on Small Fast Warships and Security Vessels*. RINA, London, 1982.

16.84 Gamulin, A. A displacement series of ships. *International Shipbuilding Progress*, Vol. 43, No. 434, 1996, pp. 93–107.

17 Applications

17.1 Background

The overall ship powering process is shown in Figure 2.3. A number of worked examples are presented to illustrate typical applications of the resistance and propulsor data and methodologies for estimating ship propulsive power for various ship types and size. The examples are grouped broadly into the estimation of effective power and propeller/propulsor design for large and small displacement ships, semi-displacement ships, planing craft and sailing vessels.

The resistance data are presented in Chapter 10 together with tables of data in Appendix A3. The propeller data are presented in Chapter 16, together with tables of data in Appendix A4.

It should be noted that the example applications use resistance data derived mainly from the results of standard series experiments and regression analyses. As such, they serve a very useful purpose, particularly at the preliminary design stage, but they tend to predict results only at an average level. Many favourable developments in hull form have occurred since the publication of the standard series data, leading to more efficient hull forms. These are, however, generally not in parametric variation form or published. Consequently, improvements in the resistance predictions in some of the worked example applications would be likely with the use of experimental and computational investigations. There have not been such significant improvements in the basic propeller series data and, as such, the series should broadly predict efficiencies likely to be achieved.

17.2 Example Applications

17.2.1 Example Application 1. Tank Test Data: Estimate of Ship Effective Power

The following data relate to the resistance test on a ship model:

Model L: 4.3 m. Ship L: 129 m. (scale $\lambda = 30$).
Model wetted surface area: 3.75 m^2, model test speed: 1.5 m/s.
Model total resistance at 1.5 m/s: 18.0 N.

With the use of empirical data, such as that described in Chapter 4, the form factor $(1 + k)$ is estimated to be 1.15. The ship wetted surface area is as follows:

$$S = 3.75 \times \left(\frac{129}{4.3}\right)^2 = 3375 \, \text{m}^2.$$

The corresponding ship speed (constant Froude number) is as follows:

$$V_S = 1.5 \times \sqrt{\frac{129}{4.3}} = 8.22 \, \text{m/s} = 15.98 \, \text{knots}.$$

From Table A1.2, $\nu_{FW} = 1.140 \times 10^{-6} \, \text{m}^2/\text{s}$ and $\nu_{SW} = 1.190 \times 10^{-6} \, \text{m}^2/\text{s}$.

17.2.1.1 Estimate of Ship P_E Using International Towing Tank Conference (ITTC) Correlation Line, Without a Form Factor

$C_{TM} = R_T/\frac{1}{2}\rho SV^2 = 18.0 \times 2/(1000 \times 3.75 \times 1.5^2) = 4.267 \times 10^{-3}.$

$Re_M = VL/\nu = 1.5 \times 4.3/1.14 \times 10^{-6} = 5.66 \times 10^6.$

$C_{FM} = 0.075/(\log Re - 2)^2 = 0.075/(6.753 - 2)^2 = 3.320 \times 10^{-3}.$

$C_{RM} = C_{TM} - C_{FM} = (4.267 - 3.320) \times 10^{-3} = 0.947 \times 10^{-3}.$

$Re_S = VL/\nu = 8.22 \times 129/1.19 \times 10^{-6} = 8.91 \times 10^8.$

$C_{FS} = 0.075/(\log Re - 2)^2 = 0.075/(8.950 - 2)^2 = 1.553 \times 10^{-3}.$

$C_{RS} = C_{RM} = 0.947 \times 10^{-3}.$

$C_{TS} = C_{FS} + C_{RS} = (1.553 + 0.947) \times 10^{-3} = 2.500 \times 10^{-3}.$

Ship total resistance is as follows:

$$R_{TS} = C_{TS} \times \frac{1}{2}\rho SV_S^2$$
$$= 2.500 \times 10^{-3} \times 1025 \times 3375 \times 8.22^2/(2 \times 1000) = 292.18 \, \text{kN}$$

Ship effective power is as follows:

$$P_E = R_{TS} \times Vs = 292.18 \times 8.22 = 2401.7 \, \text{kW}.$$

17.2.1.2 Estimate of Ship P_E Using the ITTC Correlation Line and a Form Factor (1.15)

$C_{TM} = R_T/\frac{1}{2}\rho SV^2 = 18.0 \times 2/(1000 \times 3.75 \times 1.5^2) = 4.267 \times 10^{-3}.$

$Re_M = VL/\nu = 1.5 \times 4.3/1.14 \times 10^{-6} = 5.66 \times 10^6.$

$C_{FM} = 0.075/(\log Re - 2)^2 = 0.075/(6.753 - 2)^2 = 3.320 \times 10^{-3}.$

$C_{VM} = (1 + k)C_F = 1.15 \times 3.320 \times 10^{-3} = 3.818 \times 10^{-3}.$

$C_{WM} = C_{TM} - C_{VM} = (4.267 - 3.818) \times 10^{-3} = 0.449 \times 10^{-3}.$

$Re_S = VL/\nu = 8.22 \times 129/1.19 \times 10^{-6} = 8.91 \times 10^8.$

$C_{FS} = 0.075/(\log Re - 2)^2 = 0.075/(8.950 - 2)^2 = 1.553 \times 10^{-3}.$

$C_{Vs} = (1 + k)C_F = 1.15 \times 1.553 \times 10^{-3} = 1.786 \times 10^{-3}.$

$C_{WS} = C_{WM} = 0.449 \times 10^{-3}.$

$C_{TS} = C_{VS} + C_{WS} = (1.786 + 0.449) \times 10^{-3} = 2.235 \times 10^{-3}.$

Ship total resistance is as follows:

$$R_{TS} = C_{TS} \times \tfrac{1}{2}\rho S V_S^2$$
$$= 2.235 \times 10^{-3} \times 1025 \times 3375 \times 8.22^2/(2 \times 1000) = 261.2\,\text{kN}.$$

Ship effective power is as follows:

$$P_E = R_{TS} \times V_s = 261.2 \times 8.22 = 2147.1\,\text{kW}.$$

It can be noted that, for this case, the effective power reduces by about 11% when a form factor is employed. Based on modern practice, the use of a form factor would be generally recommended, but see also Chapters 4 and 5.

17.2.2 Example Application 2. Model Self-propulsion Test Analysis

The following data relate to the self-propulsion test on the model of a single-screw ship. The data are for the *ship* self-propulsion point, as described in Chapter 8 and Figure 8.9.

Model speed: 1.33 m/s, propeller diameter $D = 0.19$ m and pitch ratio $P/D = 0.80$
Revolutions at ship self-propulsion point: 10.0 rps
Towed model resistance corresponding to the ship self-propulsion point: 18.5 N
Thrust at self-propulsion point: 22.42 N
Torque at self-propulsion point: 0.58 Nm

It is convenient (for illustrative purposes) to use the $K_T - K_Q$ open water data for the Wageningen propeller B4.40 for $P/D = 0.80$, given in Table 16.1(a). In this case, the following relationships are suitable:

$$K_T = 0.320\left[1 - \left(\frac{J}{0.90}\right)^{1.30}\right]. \quad K_Q = 0.0360\left[1 - \left(\frac{J}{0.98}\right)^{1.60}\right].$$

$$J_b = V_s/n\,D = 1.33/10.0 \times 0.19 = 0.700.$$
$$K_{Tb} = T_b/\rho n^2 D^4 = 22.42/1000 \times 10.0^2 \times 0.19^4 = 0.172.$$
$$K_{Qb} = Q_b/\rho n^2 D^5 = 0.58/1000 \times 10.0^2 \times 0.19^5 = 0.0234.$$

With the use of a thrust identity, $K_{TO} = K_{Tb} = 0.172$. From the open water propeller data at $K_{TO} = 0.172$, the following values are obtained:

$$J_O = 0.497 \quad\text{and}\quad K_{QO} = 0.0239.$$
$$w_T = 1 - (J_O/J_b) = 1 - (0.497/0.700) = 0.290.$$
$$t = 1 - (R/T_b) = 1 - (18.5/22.42) = 0.175.$$
$$\eta_R = \left[\frac{K_{QO}}{K_{Qb}}\right]_{\text{Thrust identity}} = \frac{0.0239}{0.0234} = 1.021.$$
$$\eta_O = \frac{J_O K_{TO}}{2\pi K_{QO}} = \frac{0.497 \times 0.172}{2\pi \times 0.0239} = 0.569.$$
$$\text{QPC} = \eta_D = \eta_O \times \eta_H \times \eta_R = \eta_O \times (1-t)/(1-w_T) \times \eta_R$$
$$= 0.569 \times (1 - 0.175)/(1 - 0.290) \times 1.021$$
$$= 0.675,$$

or directly as

$$\eta_D = (RV)/(2\pi n Q) = 18.5 \times 1.33/2\pi \times 10.0 \times 0.58 = 0.675.$$

K_T and K_Q, hence η_O, can be scaled from model to full size using Equations (5.13) to (5.19), or using the Re correction given with the Wageningen series in Chapter 16. w_T can be scaled from model to full size using Equation (5.22). η_R is assumed the same for model and full scale.

17.2.3 Example Application 3. Wake Analysis from Full-Scale Trials Data

Measurements during a ship trial included ship speed, propeller revolutions and measurement of propeller shaft torque.

The details of the propeller and measurements from the trial are as follows:

Propeller diameter: $D = 6.0\,\mathrm{m}$, pitch ratio $P/D = 0.90$.
Propeller revolutions $N = 115$ rpm.
Delivered power $P_D = 11{,}600\,\mathrm{kW}$ (from measurement of torque and rpm).
Estimated effective power $P_E = 7700\,\mathrm{kW}$ (from model tank tests)
Estimated thrust deduction factor $t = 0.18$ (from model self-propulsion tests).

It is convenient (for illustrative purposes) to use the $K_T - K_Q$ open water data for the Wageningen propeller B4.40 for $P/D = 0.90$, given in Table 16.1(a). In this case, the following relationships are suitable:

$$K_T = 0.355\left[1 - \left(\frac{J}{1.01}\right)^{1.32}\right]. \quad K_Q = 0.0445\left[1 - \left(\frac{J}{1.07}\right)^{1.65}\right].$$

$$N = 115 \text{ rpm and } n = 115/60 = 1.917 \text{ rps.}$$

$$P_D = 2\pi n Q_b,$$

where Q_b is the measured torque in the behind condition.

$$Q_b = P_D/2\pi n = 11600/(2\pi \times 1.917) = 963.1\,\mathrm{kNm}.$$

Behind $\quad K_{Qb} = Q_b/\rho n^2 D^5 = 963.1 \times 1000/(1025 \times 1.917^2 \times 6.0^5) = 0.0329,$
using a torque identity, $K_{QO} = K_{Qb} = 0.0329$, from the open water propeller data at $K_{QO} = 0.0329$,

$$J_0 = 0.472 \text{ and } K_{T0} = 0.225.$$
$$J_b = Vs/nD = 16 \times 0.5144/1.917 \times 6.0 = 0.716$$
$$w_T = 1 - (J_0/J_b) = 1 - (0.472/0.716) = 0.341.$$

Estimated thrust (based on the model experiments) is as follows:

$$T = P_E/[(1 - t)Vs] = 7700/[(1 - 0.18) \times 16 \times 0.5144] = 1140.92\,\mathrm{kN}$$

Table 17.1. *Equivalent 400 ft ship dimensions, cargo ship*

Units	L	B	T
m	140	21.5	8.5
ft	459.2	70.52	27.88
Equivalent 400 ft	400	61.43	24.29
BSRA ft	400	55	26

and

$$K_{Tb} = T/pn^2 D^4 = 1140.92 \times 1000/(1025 \times 1.917^2 \times 6.0^4) = 0.234$$

$$\eta_R = \left[\frac{K_{Tb}}{K_{To}}\right]_{\text{Torque identity}} = \frac{0.234}{0.225} = 1.04.$$

This estimate of η_R must only be seen as approximate, as full-scale measurement of propeller thrust would be required to obtain a fully reliable estimate. Measurement of full-scale propeller thrust is not common practice. The scaling of wake is discussed in Chapter 5.

17.2.4 Example Application 4. 140 m Cargo Ship: Estimate of Effective Power

The dimensions are as follows: $L_{BP} = 140$ m $\times B = 21.5$ m $\times T = 8.5$ m $\times C_B = 0.700 \times LCB = 0.25\% F$. $C_P = 0.722$, $C_W = 0.800$, $S = 4130$ m^2 and propeller diameter D $= 5.0$ m. The service speed is 15 knots. Load displacement: $\nabla = 17,909.5$ m^3, $\Delta = 18357.2$ tonnes, $L/\nabla^{1/3} = 5.35$, $L/B = 6.51$, $B/T = 2.53$.

$$Fr = V/\sqrt{gL} = 15 \times 0.5144/\sqrt{9.81 \times 140} = 0.208.$$

The length, 140 m $= 459.2$ ft, and $V_k/\sqrt{L_f} = 15/\sqrt{459.2} = 0.700$ [$V_{400} = \sqrt{400} \times 0.70 = 14$ knots]. Equivalent 400 ft dimensions are given in Table 17.1.

Using the Moor and Small data, Section 10.3.1.3 and Table A3.6, for $V_k/\sqrt{L_f} = 0.700$ [$V_{400} = 14$ knots], $C_B = 0.700$ and $LCB = +0.25\%$F, $©_{400} = 0.702$ (for standard BSRA dimensions 400 ft $\times$ 55 ft $\times$ 26 ft).

Applying Mumford indices, Equation (10.21),

$$©_1 = ©_2 \times \left(\frac{B_2}{B_1}\right)^{x-\frac{2}{3}} \left(\frac{T_2}{T_1}\right)^{y-\frac{2}{3}}.$$

From Table 10.1, at $V_k/\sqrt{L_f} = 0.70$, $x = 0.90$, $y = 0.60$, and

$$©_{400} = 0.702 \times \left(\frac{61.43}{55}\right)^{0.90-\frac{2}{3}} \left(\frac{24.29}{26}\right)^{0.60-\frac{2}{3}} = 0.724 \text{ (an increase of 3\%)}.$$

The skin friction correction, using Equation (10.2) for $L > 122$ m, is as follows:

$$\delta© = -0.1/[1 + (188/(L-122))],$$

for L $= 140$ m, $\delta© = -0.0087$ (or take from Figure 10.5) and corrected $©_{140} = 0.724 - 0.0087 = 0.715$.

Using Equation (10.3), the effective power is as follows:

$$P_E = \text{©}_s \times \Delta^{2/3} V^3 / 579.8$$
$$= 0.715 \times (18357.2)^{2/3} \times 15^3 / 579.8$$
$$= 2905.8 \text{ kW}.$$

[BSRA series regression estimates 2838 kW (−2.5%)].

Applying a ship correlation/load factor SCF $(1 + x)$, Table 5.1 or Equation (10.4),

$$(1 + x) = 1.2 - \frac{\sqrt{L}}{48} = 0.953.$$

Final naked effective power is as follows:

$$P_{\text{Enaked}} = 2905.8 \times 0.953$$
$$= 2769.2 \text{ kW}.$$

[Holtrop regression estimates 2866(+3.5%) and Hollenbach 2543(−8%)].

BULBOUS BOW. Inspection of Figure 10.6 indicates that, with $C_B = 0.700$ and $V_{400} = 14$, a bulbous bow would not be beneficial. At $C_B = 0.700$, a bulbous bow would be beneficial at speeds greater than about $V_{400} = 15.5$ knots, Figure 10.6 (V_{140} greater than about 16.6 knots).

APPENDAGES ETC. Total effective power would include any resistance due to appendages and still air (see Chapter 3), together with a roughness/correlation allowance, C_A or ΔC_F, if applied to Holtrop or Hollenbach (see Chapter 5).

17.2.5 Example Application 5. Tanker: Estimates of Effective Power in Load and Ballast Conditions

The dimensions are as follows: $L_{BP} = 175 \text{ m} \times B = 32.2 \text{ m} \times T = 11.0 \text{ m} \times C_B = 0.800 \times LCB = 2.0\% F$. $C_P = 0.820$, $S = 7781 \text{ m}^2$. $C_W = 0.880$ and propeller diameter $D = 7.2$ m. The service speed is 14.5 knots. Load displacement

$$\nabla = 49{,}588 \text{ m}^3, \Delta = 50{,}828 \text{ tonnes}, L/\nabla^{1/3} = 4.78, L/B = 5.43, B/T = 2.93.$$

$$Fr = V/\sqrt{gL} = 14.5 \times 0.5144/\sqrt{9.81 \times 175} = 0.180.$$

Length 175 m = 574 ft and $V_k/L_f = 0.605$. Using the regression of the BSRA series, the effective power has been estimated as $P_E = 5051$ kW [Holtrop regression estimates 5012 kW (−2%) and Hollenbach 4899 kW (−3%)]. The displacement in the ballast condition, using all the water ballast, is estimated to be 35,000 tonnes.

Options to estimate the power in the lower (ballast) displacement condition include the BSRA regression analysis in the medium and light conditions and the use of the Moor–O'Connor fractional draught equations, Section 10.3.1.4. This example uses the Moor–O'Connor equations to estimate the effective power in the ballast condition.

Equation (10.22) can be used to estimate the draught in the ballast condition as follows:

$$\frac{\Delta_2}{\Delta_1} = (T)_R^{1.607-0.661C_B}$$

where the draught ratio $(T)_R = (\frac{T_{Ballast}}{T_{Load}})$.

$$\Delta_2/\Delta_1 = 35{,}000/50{,}828 = (T)_R^{1.607-0.661\times0.800} \text{ and } (T)_R = 0.7075.$$

The ballast draught (level) $= 11.0 \times 0.7075 = 7.8$ m.

Equation (10.23) can be used as follows:

$$\frac{P_{Eballast}}{P_{Eload}} = 1 + [(T)_R - 1]\left\{ \left(0.789 - 0.270[(T)_R - 1] + 0.529C_B\left(\frac{L}{10T}\right)^{0.5}\right) \right.$$

$$+ V/\sqrt{L}\left(2.336 + 1.439[(T)_R - 1] - 4.065\,C_B\left(\frac{L}{10T}\right)^{0.5}\right)$$

$$\left. + (V/\sqrt{L})^2\left(-2.056 - 1.485\right)[(T)_R - 1] + 3.798C_B\left(\frac{L}{10T}\right)^{0.5}\right)\right\},$$

where $[(T)_R - 1] = -0.2925$.

$$C_B \times (L/10T)^{0.5} = 0.800 \times (175/(10 \times 11))^{0.5} = 1.009.$$
$$V_k/\sqrt{L_f} = 14.5/\sqrt{(175 \times 3.28)} = 0.605$$

and

$$\frac{P_{Eballast}}{P_{Eload}} = 0.837.$$

The ballast effective power is as follows:

$$P_{Eballast} = 5051 \times 0.837 = 4228 \text{ kW}.$$

It is likely that the ship would operate at more than 14.5 knots in the ballast condition, for example, on the ship measured mile trials and whilst in service. In this case a curve of ballast power would be developed from the curve of power against speed for the loaded condition.

Example application 6 includes a typical propeller design procedure. Example application 18 investigates the performance of the propeller for a tanker working off-design in the ballast condition.

17.2.6 Example Application 6. 8000 TEU Container Ship: Estimates of Effective and Delivered Power

The dimensions are as follows: $L_{BP} = 320$ m $\times B = 43$ m $\times T = 13$ m $\times C_B = 0.650 \times LCB = 1.5\%$ aft. $C_P = 0.663$, $C_W = 0.750$ and $S = 16{,}016$ m^2. The service speed is 25 knots. Using clearance limitations, the propeller diameter is 8.8 m. The load displacement and ratios are as follows:

$$\nabla = 116{,}272 \text{ m}^3, \ \Delta = 119{,}178.8 \text{ tonnes}, \ L/\nabla^{1/3} = 6.58, \ L/B = 7.44, \ B/T = 3.31.$$
$$Fr = V/\sqrt{gL} = 25 \times 0.5144/\sqrt{9.81 \times 320} = 0.230$$

Table 17.2. *Equivalent 400 ft ship dimensions, container ship*

Units	L	B	T
m	320	43.0	13.0
ft	1049.6	141.04	42.64
Equivalent 400 ft	400	53.75	16.25
BSRA ft	400	55	26

The length, 320 m $= 1049.6$ ft, and $V_k/\sqrt{L_f} = 25/\sqrt{1049.6} = 0.772\,[V_{400} = \sqrt{400} \times 0.772 = 15.4\,\text{knots}]$. Equivalent 400 ft dimensions are given in Table 17.2.

17.2.6.1 Effective Power

Using the Moor and Small data, Section 10.3.1.3 and Table A3.6, for $V_k/\sqrt{L_f} = 0.772\,[V_{400} = 15.4\,\text{knots}]$, $C_B = 0.650$ and $LCB = -1.5\%$ A, $©_{400} = 0.691$ (for standard BSRA dimensions 400 ft $\times$ 55 ft $\times$ 26 ft).

Applying Mumford indices, Equation (10.21),

$$©_1 = ©_2 \times \left(\frac{B_2}{B_1}\right)^{x-\frac{2}{3}} \left(\frac{T_2}{T_1}\right)^{y-\frac{2}{3}}$$

From Table 10.1, at $V_k/L_f = 0.772$, $x = 0.90$, $y = 0.63$ and

$$©_{400} = 0.691 \times \left(\frac{53.75}{55}\right)^{0.90-\frac{2}{3}} \left(\frac{16.25}{26}\right)^{0.63-\frac{2}{3}} = 0.700\,(\text{an increase of }1.3\%).$$

The skin friction correction, using Equation (10.2) for $L > 122$ m is

$$\delta© = -0.1/[1 + (188/(L - 122))]$$

for $L = 320$ m, $\delta© = -0.0513$ (or take from Figure 10.5) and corrected $©_{320} = 0.700 - 0.0513 = 0.649$.

Using Equation (10.3), the effective power is:

$$P_E = ©_s \times \Delta^{2/3} V^3/579.8$$
$$= 0.649 \times (119,178.8)^{2/3} \times 25^3/579.8$$
$$= 42,521.8\,\text{kW [using }2/3 = 0.667].$$

[BSRA series regression estimates 43,654 kW $(= +\,2.7\%)$, Holtrop regression estimates 41,422 $(= -2.5\%)$].

Applying a ship correlation/load factor SCF $(1 + x)$, Table 5.1 or Equation (10.4),

$$(1 + x) = 1.2 - \frac{\sqrt{L}}{48} = 0.827.$$

The final naked effective power is as follows:

$$P_{\text{Enaked}} = 42521.8 \times 0.827 = 35165.5\,\text{kW}.$$

[Hollenbach 'mean' estimates 35,352 kW]

BULBOUS BOW. A bulbous bow would normally be fitted to this ship type. Inspection of Figure 10.6 indicates that, with $C_B = 0.650$ and $V_{400} = 15.4$, a bulbous bow would not have significant benefits. If a bulb is fitted, an outline of suitable principal characteristics can be obtained from the Kracht data in Chapter 10.

APPENDAGES FOR THIS SHIP TYPE. The appendages to be added are the rudder, bow thrusters(s) and, possibly, bilge keels. Bilge keels will be omitted in this estimate.

RUDDER. For this type of single-screw aft end arrangement, the propeller diameter will span about 80% of the rudder, see Molland and Turnock [17.1]. The rudder span $= 8.8/0.8 = 11.0$ m, and assuming a geometric aspect ratio $= 1.5$, the rudder chord $= 7.3$ m. (Rudder area $= 11 \times 7.3 = 80.3$ m^2 [$80.3/(L \times T) = 2\%$ $L \times T$], which is suitable).

A practical value for drag coefficient is $C_{D0} = 0.013$, see Section 3.2.1.8(c).

As a first approximation, assume that the decrease in flow speed due to wake is matched by the acceleration due to the propeller, and use ship speed as the flow speed.

$$D_{\text{Rudder}} = C_{D0} \times \tfrac{1}{2}\rho A V^2$$
$$= 0.013 \times 0.5 \times 1025 \times 80.3 \times (25 \times 0.5144)^2/1000 = 88.48 \text{ kN.}$$

$R_T = P_E/V_S = 42{,}521.8/(25 \times 0.5144) = 3306.5$ kN, and rudder drag $= 88.48/3306.5 = 2.7\%$ of total naked resistance.

Check using Equation (3.41) attributable to Holtrop, as follows:

$$D_{\text{Rudder}} = \tfrac{1}{2}\rho V_S^2 \, C_F(1 + k_2)S,$$

where V_S is ship speed, C_F is for ship, S is the wetted area and $(1 + k_2)$ is an appendage form factor which, for rudders, varies from 1.3 to 2.8 depending on rudder type (Table 3.5). Assume for the rudder in this case that $(1 + k_2) = 1.5$. Ship $Re = VL/\nu = 25 \times 0.5144 \times 320/1.19 \times 10^{-6} = 3.458 \times 10^9$, $C_F = 0.075/(\log Re - 2)^2 = 0.00132$ and $D_{\text{Rudder}} = \tfrac{1}{2} \times 1025 \times (25 \times 0.5144)^2 \times 0.00132 \times 1.5 \times (2 \times 80.3)/1000 = 26.95$ kN (about 30% the value of the earlier estimate).

BOW THRUSTERS. Using Equation (3.43) attributable to Holtrop, the resistance due to bow thruster is as follows:

$$R_{BT} = \pi \rho V_S^2 d_T \times C_{BT0},$$

where $d_T =$ thruster diameter. Assuming that $d_T = 2.0$ m and $C_{BT0} = 0.005$, the resistance due to bow thruster is as follows:

$$R_{BT} = \pi \times 1025 \times (25 \times 0.5144)^2 \times 2.0$$
$$\times 0.005 = 5.33 \text{ kN (about 0.15\% of total resistance).}$$

The total drag of appendages $= 88.48 + 5.33 = 93.81$ kN, and the equivalent effective power increase $= 93.81 \times 25 \times 0.5144 = 1206.4$ kW.

Finally the corrected $P_E = (42{,}521.8 + 1206.4) \times 0.827 = 36163.2$ kW.

AIR DRAG. Still air drag can be significant for a container ship, being of the order of up to 6% of the hull resistance for the larger vessels. It is not included in this particular estimating procedure as it is assumed to have been subsumed within the correlation factor.

If air drag is to be estimated and included, for example for use in the ITTC1978 prediction procedure (Chapter 5), then the following approach can be used. A drag coefficient based on the transverse area of hull and superstructure above the water-line is used, see Chapter 3. For this vessel size, the height of the deckhouse above water is approximately 52 m and, allowing for windage of containers and assuming the full breadth of 43 m, the transverse area $A_T = 52 \times 43 = 2236 \, m^2$. Assume a drag coefficient $C_D = 0.80$ for hull and superstructure, see Chapter 3, Section 3.2.2. The still air drag $D_{AIR} = C_D \times \frac{1}{2}\rho_A A_T V^2 = 0.80 \times 0.5 \times 1.23 \times 2236 \times (25 \times 0.5144)^2/1000 = 181.9 \, kN$, and the air drag $= (181.9/3306.5) \times 100 = 5.5\%$ of hull naked resistance.[Check using the ITTC approximate formula, $C_{AA} = 0.001 \, (A_T/S)$, Equation (5.12), where S is the wetted area of the hull and $D_{AIR} = C_{AA} \times \frac{1}{2}\rho_w SV^2 = (0.001 \times (2236/S)) \times \frac{1}{2} \times 1025 \times S \times (25 \times 0.5144)^2/1000 = 189.5 \, kN$].

17.2.6.2 Preliminary Propeller Design
Allowing for propeller tip clearances, for example 20% of diameter, the propeller diameter $D = 8.8 \, m$. The choice of a particular direct drive diesel leads to the requirement for approximately 94 rpm (1.57 rps) at the service power.

WAKE FRACTION. Using Equation (8.14) and Figure 8.12 for the Harvald data for single-screw ships: $w_T = 0.253 +$ correction due to D/L (0.0275) of $+ 0.07 = 0.323$. (Check using BSRA Equation (8.16) for wake fraction, $Fr = 0.230$ and $Dw = 2.079$, gives $w_T = 0.322$).

THRUST DEDUCTION. As a first approximation, using Equation (8.18), $t = 0.60 \times w_T = 0.194$ [check using BSRA Equation (8.19b) for thrust deduction, $Dt = 0.160$ and $L_{CB}/L = -0.015$, gives $t = 0.204$].

RELATIVE ROTATIVE EFFICIENCY. η_R. See Section 16.3. Using Equation (16.77) for BSRA Series, with $\nabla^{1/3} = 48.62$ and $D/L = 8.8/320$, $\eta_R = 1.005$ [check using Holtrop Equation (16.79), assume BAR $= 0.700$ and $C_P = 0.663$, $\eta_R = 1.003$].

Hence, for propeller design purposes, required thrust, T, is as follows:

$$T = R_T/(1-t) = (P_E/Vs)/(1-t)$$
$$= (36163.2/(25 \times 0.5144))/(1 - 0.204) = 3532.75 \, kN.$$

Wake speed, Va, is as follows:

$$Va = Vs(1 - w_T) = 25 \times 0.5144 \times (1 - 0.323) = 8.706 \, m/s.$$
$$J = V/nD = 8.706/(1.57 \times 8.8) = 0.630.$$
$$K_T = T/\rho n^2 D^4 = 3532.75 \times 1000/(1025 \times 1.57^2 \times 8.8^4) = 0.233.$$

Assume a four-bladed propeller and from the $K_T - K_Q$ chart, Figure 16.4, for a Wageningen B4.70 propeller, at $J = 0.630$ and $K_T = 0.233$, $P/D = 1.05$ and $\eta_0 = 0.585$.

Note that, with such a large propeller, there may be scale effects on η_0 derived from the open water propeller charts. K_T and K_Q, hence η_O, can be scaled from model to full size using Equations (5.13) to (5.19), or using the Re correction given with the Wageningen series in Chapter 16.

$$\eta_D = \eta_0 \times \eta_H \times \eta_R \quad \text{where } \eta_H = (1 - t)/(1 - w_T)$$

$$= 0.585 \times (1 - 0.204)/(1 - 0.323) \times 1.005 = 0.691.$$

Delivered power is as follows:

$$P_D = P_E/\eta_D = 36163.2/0.691 = 52334.6 \text{ kW}.$$

Service (or shaft) power is as follows:

$$P_S = P_D/\eta_T,$$

where η_T is the transmission efficiency, and is typically about 0.98 for a direct drive installation. Hence $P_S = 52334.6/0.98 = 53402.7 \text{ kW}$ at 94 rpm.

A margin for roughness, fouling and weather can be established using rigorous techniques described in Chapter 3. In this example, a margin of 15% is assumed. An operator will not normally operate the engine at higher than 90% maximum continuous rating (MCR), hence, the final installed power, P_I can be calculated as follows:

$$P_I = 53402.7 \times 1.15/0.90 = 68236.8 \text{ kW at about 102 rpm, assuming that } P \propto N^3.$$

CAVITATION BLADE AREA CHECK. See Section 12.2.10.

Immersion of the shaft is as follows: $h = 13.0 - 4.40 - 0.20 = 8.4 \text{ m}$.

The required thrust $T = 3532.75 \times 10^3 \text{ N}$, at service speed and power condition.

$$V_R^2 = Va^2 + (0.7\pi nD)^2 = 8.706^2 + (0.7\pi \times 1.57 \times 8.8)^2 = 998.91 \text{ m}^2/\text{s}^2$$
$$\sigma = (\rho gh + P_{AT} - P_V)/\tfrac{1}{2}\rho V_R^2$$
$$= (1025 \times 9.81 \times 8.4 + 101 \times 10^3 - 3 \times 10^3)/\tfrac{1}{2} \times 1025 \times 998.91$$
$$= 0.356$$

Using line (2), Equation (12.18), for upper limit merchant ships, when $\sigma = 0.356$, then $\tau_C = 0.148$. A similar value can be determined directly from Figure 12.31. Then,

$$A_P = T/\tfrac{1}{2}\rho V_R^2 \tau_C = 3532.75 \times 10^3/\tfrac{1}{2} \times 1025 \times 998.91 \times 0.148 = 46.63 \text{ m}^2.$$

$$A_D = A_P/(1.067 - 0.229 \, P/D) = 46.63/(1.067 - 0.229 \times 1.05) = 56.42 \text{ m}^2.$$

and

$$\text{DAR} = (\text{BAR}) = 56.42/(\pi D^2/4) = 56.42/(\pi \times 8.8^2/4) = 0.928.$$

This BAR is reasonably close to 0.70 in order to assume that, for preliminary design purposes, the use of the B4.70 chart is acceptable. For a more precise estimate the calculation would be repeated with the next size chart, and a suitable BAR would be derived by interpolation between the results.

The BAR has been estimated for the service speed, clean hull and calm water. A higher thrust loading may be assumed to account for increases in resistance due to hull fouling and weather. For example, if the thrust required is increased by say 10% at the same speed and revolutions, then the BAR would increase to 1.020.

SUMMARY OF PROPELLER PARTICULARS. $D = 8.8$ m, $P/D = 1.05$, 94 rpm and BAR $= 0.928$.

17.2.6.3 Notes on the Foregoing Calculations

The calculations represent typical procedures carried out to derive preliminary estimates of power at the early design stage. It should be noted that, for this type of direct drive diesel installation, the propeller pitch and revolutions should be matched carefully to the engine, with the initial propeller design curve to the right of the engine N^3 line, see Section 13.2. If the decision is made to build the ship(s), then a full hydrodynamic investigation would normally be carried out. This might include both experimental and computational investigations. The experimental programme might entail hull resistance tests, self-propulsion tests and propeller tests in a cavitation tunnel. Typical investigations would include hull shape, bulbous bow shape and aft end shapes to optimise resistance versus propulsive efficiency. In the case of a large container ship, propeller–rudder cavitation would also be investigated.

Detailed aspects of the propeller design would be investigated. The number of blades might be investigated in terms of propeller excited vibration, taking into account hull natural frequencies. Wake adaption would take place and skew might be introduced to permit further increase in diameter and efficiency.

17.2.7 Example Application 7. 135 m Twin-Screw Ferry, 18 knots: Estimate of Effective Power P_E

The example uses two sets of resistance data for twin-screw vessels, as follows:

(a) The Taylor–Gertler series data for the residuary resistance coefficient C_R, the Schoenherr friction line for C_F and a skin friction allowance $\Delta C_F = 0.0004$
(b) the Zborowski series data for C_{TM}, together with the ITTC friction correlation line for C_F, with and without a form factor

The Holtrop and the Hollenbach regression analyses would also both be suitable for this vessel.

The dimensions of the ferry are as follows: $L_{BP} = 135$ m $\times B = 21$ m $\times T = 6.4$ m $\times C_B = 0.620 \times C_W = 0.720 \times LCB = -1.75\% L$, Propeller diameter $D = 4.5$ m. The service speed is 18 knots.

$$Fr = V/\sqrt{gL} = 0.254, \qquad V_k/\sqrt{L_f} = 0.855.$$
$$\nabla = 11249.3 \text{ m}^3, \Delta = 11530.5 \text{ tonnes}, L/\nabla^{1/3} = 6.04, \nabla/L^3 = 4.57 \times 10^{-3}.$$
$$L/B = 135/21 = 6.43, \quad B/T = 21/6.4 = 3.28.$$

Approximation for C_M using Equation (10.29b), is as follows:

$$C_M = 0.80 + 0.21 C_B = 0.80 + 0.21 \times 0.620 = 0.930.$$
$$C_P = C_B/C_M = 0.667.$$

Estimate the wetted surface area, using Equation (10.86), $S = C_S\sqrt{\nabla L}$ and Table 10.19 for C_S. By interpolation in Table 10.19, for $C_P = 0.667$, $L/\nabla^{1/3} = 6.04$ and $B/T = 3.28$,

$$C_S = 2.573,$$

and the wetted surface area is as follows:

$$S = C_S\sqrt{\nabla L} = 2.573\sqrt{11249.3 \times 135} = 3170.8\,\mathrm{m}^2$$

[check using Denny Mumford, Equation (10.84): $S = 1.7LT + \nabla/T = 1.7 \times 135 \times 6.4 + 11249.3/6.4 = 3226.5\,\mathrm{m}^2 (+1.7\%)$].

17.2.7.1 Using Taylor–Gertler Series

For the Taylor–Gertler series, using the graphs of C_R data in [10.10] or the tabulated C_R data in Tables A3.8 to A3.11 in Appendix A3, and interpolating for $Fr = 0.254$, $C_P = 0.667$, $B/T = 3.28$ and $L/\nabla^{1/3} = 6.04$ ($\nabla/L^3 = 4.57 \times 10^{-3}$),

$$C_R = 1.36 \times 10^{-3} = 0.00136 \text{ (from the tabulated data)}$$
$$Re = VL/\nu = 18.0 \times 0.5144 \times 135/1.19 \times 10^{-6} = 10.50 \times 10^8.$$

Using an approximation to the Schoenherr line, Equation (4.11),

$$C_F = 1/(3.5\log Re - 5.96)^2 = 0.00152$$
$$C_T = C_F + \Delta C_F + C_R = 0.00152 + 0.0004 + 0.00136 = 0.00328,$$

total ship resistance is as follows:

$$\begin{aligned} R_{TS} &= C_T \times \tfrac{1}{2}\rho SV^2 \\ &= 0.00328 \times 0.5 \times 1025 \times 3170.8 \times (18.0 \times 0.5144)^2/1000 \\ &= 457.0\,\mathrm{kN}. \end{aligned}$$

Ship effective power is:

$$P_E = R_{TS} \times V_S = 457.0 \times 18.0 \times 0.5144 \times 4231.5\,\mathrm{kW}$$

(If $\Delta C_F = 0.0004$ is omitted, then $P_E = 3715.1\,\mathrm{kW}$, a reduction of 12.5%) (Holtrop regression estimates 4304 kW, Hollenbach regression estimates 3509 kW).

17.2.7.2 Using Zborowski Series

Interpolating from Table A3.12 for $Fr = 0.254$, $C_B = 0.620$, $L/\nabla^{1/3} = 6.04$ and $B/T = 3.28$,

$$C_{TM} = 5.774 \times 10^{-3}.$$

Model length L_M is 1.90 m, ship length L_S is 135 m, ship speed $Vs = 18 \times 0.5144 = 9.26$ m/s and model speed is as follows:

$$V_M = V_s \times \sqrt{\frac{L_M}{L_S}} = 9.26 \times \sqrt{\frac{1.90}{135}} = 1.099\,\mathrm{m/s}.$$
$$Re_M = VL/\nu = 1.099 \times 1.9/1.14 \times 10^{-6} = 1.832 \times 10^6.$$

$$C_{FM} = 0.075/(\log Re - 2)^2 = 0.075/(6.263 - 2)^2 = 4.130 \times 10^{-3}.$$
$$Re_S = VL/\nu = 9.26 \times 135/1.19 \times 10^{-6} = 10.505 \times 10^8.$$
$$C_{FS} = 0.075/(\log Re - 2)^2 = 0.075/(9.0214 - 2)^2 = 1.521 \times 10^{-3}.$$

WITHOUT A FORM FACTOR

$$C_{TS} = C_{TM} - (C_{FM} - C_{FS})$$
$$= 0.005774 - (0.00413 - 0.001521) = 0.003165.$$

Total ship resistance is as follows:

$$R_{TS} = C_T \times \tfrac{1}{2}\rho SV^2$$
$$= 0.003165 \times 0.5 \times 1025 \times 3170.8 \times (18.0 \times 0.5144)^2/1000$$
$$= 440.94 \, \text{kN}.$$

Ship effective power is:

$$P_E = R_{TS} \times V_S \times 440.94 \times 18.0 \times 0.5144 = 4082.8 \, \text{kW}$$

(3.5% less than the Taylor–Gertler estimate including $\Delta C_F = 0.0004$).

WITH A FORM FACTOR. An estimate of the form factor can be made using the empirical data in Chapter 4; using Watanabe's Equation (4.20) $(1 + k) = 1.12$, Conn and Ferguson's Equation (4.21) $(1 + k) = 1.17$, Holtrop's Equation (4.23) $(1 + k) = 1.19$ and Wright's Equation (4.25) $(1 + k) = 1.20$. Based on these values, $(1 + k) = 1.17$ is assumed.

$$C_{TS} = C_{TM} - (1 + k)(C_{FM} - C_{FS})$$
$$= 0.005774 - (1.17)(0.00413 - 0.001521) = 0.002721.$$

Total ship resistance is as follows:

$$R_{TS} = C_T \times \tfrac{1}{2}\rho SV^2$$
$$= 0.002721 \times 0.5 \times 1025 \times 3170.8 \times (18.0 \times 0.5144)^2/1000$$
$$= 379.1 \, \text{kN}$$

Ship effective power is:

$$P_E = R_{TS} \times V_S = 379.1 \times 18.0 \times 0.5144 = 3510.2 \, \text{kW}$$

(17% less than the Taylor–Gertler estimate, or 5% less than the Taylor–Gertler estimate if the $\Delta C_F = 0.0004$ correction is omitted from the Taylor–Gertler estimate).

Although it is (ITTC) recommended practice to include a form factor and a roughness correction ΔC_F, a realistic comparison is to take the Taylor–Gertler result without ΔC_F (3715.1 kW) and the Zborowski result using a form factor and without ΔC_F (3510.2 kW). This results in a difference of about 5%. Including ΔC_F in both estimates would also lead to a difference of about 5%.

Table 17.3. *NPL interpolated data, passenger ferry*

$V_k/\sqrt{L_f}$	B/T		
	3.63	4.86	Required 4.5
3.2	2.296	2.265	2.274
3.3	2.236	2.206	2.215
Required 3.23	–	–	2.256

17.2.8 Example Application 8. 45.5 m Passenger Ferry, 37 knots, Twin-Screw Monohull: Estimates of Effective and Delivered Power

This 45.5 m ferry can transport 400 passengers. The dimensions are as follows: $L_{OA} = 45.5\,\text{m} \times L_{WL} = 40\,\text{m} \times B = 7.2\,\text{m} \times T = 1.6\,\text{m} \times C_B = 0.400$, and $LCB = -6\%\,L$.

The height of superstructure above base is 8.9 m. The service speed is 37 knots.

$$Fr = V/\sqrt{gL} = 0.961,\ Fr_\nabla = V/\sqrt{g\nabla^{1/3}} = 2.55,\ V_k/\sqrt{L_f} = 3.23,$$
$$\nabla = 184.32\,\text{m}^3,\ \Delta = 188.9\,\text{tonnes},\ L/\nabla^{1/3} = 7.04,$$
$$L/B = 40/7.2 = 5.56,\ B/T = 7.2/1.6 = 4.5.$$

Estimate the wetted surface area, using Equation (10.88), $S = C_S\sqrt{\nabla L}$, Equation 10.89 and from Table 10.20 for the C_S regression coefficients: $C_S = 3.00$. The wetted surface area is as follows:

$$S = C_S\sqrt{\nabla L} = 3.00\sqrt{184.32 \times 40} = 257.60\,\text{m}^2.$$

The above parameters are within the range of the NPL round bilge series and an estimate of C_R is initially made using this series. The data are given in Table A3.16.

At $L/\nabla^{1/3} = 7.10$ (assumed acceptably close enough to 7.04), $B/T = 4.5$, $V_k/L_f = 3.23$ and interpolating from Table A3.16, hence, from Table 17.3,

$$C_R = 2.256 \times 10^{-3} = 0.002256.$$

Check for C_R using the approximate Series 64 data in Tables 10.3 and 10.4. This assumes an extrapolation of the Series 64 lower limit of $L/\nabla^{1/3} = 8.0$ down to 7.04, based on the equation $C_R = a(L/\nabla^{1/3})^n$. B/T at 4.5 is a little outside the Series 64 upper limit of 4.0, although this should be acceptable, see the discussion of the influence of B/T in Section 10.3.2.1. Hence, from Table 17.4,

$$C_R = 2.289 \times 10^{-3} = 0.002289$$

(1.5% higher than NPL estimate).

Table 17.4. *Series 64 interpolated data*

Table	C_B	Fr		
		0.90	1.0	Required 0.961
Table 10.3	0.35	1.575	1.338	1.430
Table 10.4	0.45	3.427	2.968	3.147
–	Required 0.40	–	–	2.289

The NPL model length is 2.54, model speed is as follows:

$$V_M = V_S \times \sqrt{(L_M/L_S)} = 4.796 \, \text{m/s}.$$
$$Re \, \text{model} = VL/\nu = 4.796 \times 2.54/1.14 \times 10^{-6} = 1.069 \times 10^7.$$
$$Re \, \text{ship} = 37 \times 0.5144 \times 40/1.19 \times 10^{-6} = 6.398 \times 10^8.$$
$$C_{FM} = 0.075/(\log Re - 2)^2 = 0.075/(\log 1.069 \times 10^7 - 2)^2$$
$$= 0.075/(7.029 - 2)^2 = 0.00297.$$
$$C_{FS} = 0.075/(\log Re - 2)^2 = 0.075/(\log 6.398 \times 10^8 - 2)^2$$
$$= 0.075/(8.806 - 2)^2 = 0.00162.$$

Using Equation (10.58), form factor $(1 + k) = 2.76 \times (L/\nabla^{1/3})^{-0.4} = 1.264$.
Using Equation (10.60), which includes a form factor, as follows:

$$C_{TS} = C_{FS} + C_R - k(C_{FM} - C_{FS})$$
$$= 0.00162 + 0.002256 - 0.264 \times (0.00297 - 0.00162) = 0.00352.$$
$$R_{TS} = C_{TS} \times \tfrac{1}{2}\rho S V^2$$
$$= 0.00352 \times 0.5 \times 1025 \times 257.6 \times (37 \times 0.5144)^2 = 168.34 \, \text{kN}$$

(if a form factor is not included, $k = 0$ and $R_{TS} = 185.37$ kN).

Air drag is calculated using a suitable drag coefficient based on the transverse area of hull and superstructure. The height of superstructure above water is $8.9 - 1.6 = 7.3$ m, breadth is 7.2 m and transverse area A_T is height $\times$ breadth $= 7.3 \times 7.2 = 52.56 \, \text{m}^2$.

From Figure 3.29, assume that a suitable drag coefficient is $C_D = 0.64$. Then,

$$D_{\text{air}} = C_D \times \tfrac{1}{2}\rho_A A_T V^2$$
$$= 0.64 \times 0.5 \times 1.23 \times 52.56 \times (37 \times 0.5144)^2/1000$$
$$= 7.49 \, \text{kN} \, (4.4\% \, \text{of hull resistance})$$

It is proposed to use waterjets; hence, there will not be any appendage drag of significance. Hence total resistance is $(168.34 + 7.49) = 175.83$ kN and $P_E = R_T \times V_S = 175.83 \times 37 \times 0.5144 = 3346.5$ kW.

Applying an approximate formula for waterjet efficiency η_D, Equation (16.56),

$$\eta_D = 1/[1 + (8.64/Vs)] = 1/[1 + (8.64/37 \times 0.5144)] = 0.688.$$

Delivered power is as follows:

$$P_D = P_E/\eta_D = 3346.5/0.688 = 4864.1 \, \text{kW}.$$

Assuming a transmission efficiency for a geared drive $\eta_T = 0.95$, the service power $P_S = P_D/\eta_T = 4864.1/0.95 = 5120.1$ kW. Allowing a 15% margin for hull fouling and weather, the installed power is:

$$P_I = P_S \times 1.15 = 5120.1 \times 1.15 = 5888 \, \text{kW}.$$

Table 17.5. *Data for range of revs from Gawn chart*

rps	J	K_T	P/D	η_0
13.0	1.273	0.315	1.79	0.68
14.0	1.182	0.272	1.61	0.69
15.5	1.067	0.222	1.41	0.68
17	0.973	0.184	1.25	0.68
20	0.827	0.133	1.05	0.64

17.2.8.1 Outline Propeller Design and Comparison with Waterjet Efficiency

For this twin-screw layout and draught, a propeller with a maximum diameter of 1.15 m is appropriate.

Using Table 8.4 for round bilge twin-screw forms, at $Fr_\nabla = 2.55$ and $C_B = 0.40$, $w_T = 0$ and $t = 0.08$. From Table 16.12, $\eta_R = 0.95$ and from Equation 16.82, $\eta_R = 0.035 \, Fr_\nabla + 0.90 = 0.989$. Assume that $\eta_R = 0.97$. Total resistance R_T is 175.83 kN and total thrust $T = R_T/(1-t) = 175.83/(1-0.08) = 191.12$ kN. Thrust per screw $T = 191.12/2 = 95.56$ kN.

$$Va = Vs(1 - w_T) = 37 \times 0.5144 \times (1 - 0) = 19.03 \, \text{m/s}$$

$$K_T = T/\rho n^2 D^4 = 95.56 \times 1000/1025 \times n^2 \times 1.15^4 = 53.30/n^2$$

$$J = Va/nD = 19.03/(n \times 1.15) = 16.548/n.$$

The Gawn series of propellers is appropriate for this vessel type. Using the Gawn G3.95 chart, Figure 16.10 and assuming a range of rpm, from a cross plot of data, take optimum rpm as 14.5 rps [870 rpm], (although η_0 is not very sensitive to rps), $P/D = 1.54$ and $\eta_0 = 0.690$. See Table 17.5.

$$\eta_H = (1 - t)/(1 - w_T) = (1 - 0.08)/(1 - 0) = 0.920, \, \eta_R = 0.97$$
$$\eta_D = \eta_0 \times \eta_H \times \eta_R = 0.690 \times 0.920 \times 0.97 = 0.616$$

CAVITATION BLADE AREA CHECK. See Section 12.2.10.

The immersion of the shaft is as follows:

$$h = 1.6 - (1.15/2) = 1.025 \, \text{m}.$$

The required thrust per screw $T = 95.56 \times 10^3$ N, at service speed. $D = 1.15$ m, $P/D = 1.54$, and $n = 14.5$ rps.

$$V_R^2 = V_a^2 + (0.7\pi nD)^2 = 19.03^2 + (0.7\pi \times 14.5 \times 1.15)^2 = 1706.85$$
$$\sigma = (\rho gh + P_{AT} - P_V)/\tfrac{1}{2}\rho V_R^2$$
$$= (1025 \times 9.81 \times 1.025 + 101 \times 10^3 - 3 \times 10^3)/\tfrac{1}{2} \times 1025 \times 1706.85$$
$$= 0.124,$$

using line (2), Equation (12.18), for upper limit merchant ships, when $\sigma = 0.124$, $\tau_C = 0.28 (\sigma - 0.03)^{0.57} = 0.0728$. A similar value can be determined directly from Figure 12.31. Then,

$$A_P = T/\tfrac{1}{2}\rho V_R^2 \tau_C = 95.56 \times 10^3 / \tfrac{1}{2} \times 1025 \times 1706.85 \times 0.0728 = 1.50 \text{ m}^2.$$

$$A_D = A_P/(1.067 - 0.229 P/D) = 1.50/(1.067 - 0.229 \times 1.54) = 2.09 \text{ m}^2.$$

and

$$\text{DAR} = (\text{BAR}) = 2.09/(\pi D^2/4) = 2.09/(\pi \times 1.15^2/4) = 2.01.$$

This BAR is clearly too large for practical application. If the rps are increased to 17 rps and $P/D = 1.25$, without much loss in efficiency, then BAR is reduced to about 1.75, which is still too large. If the higher cavitation limit line is used, say line (3), Equation (12.19), the BAR reduces to about 1.5.

One option is to use three screws, with the penalties of extra first and running costs, when the BAR is reduced to about 1.1 which is acceptable. Other options would be to use two supercavitating screws, Section 16.2.6, or tandem propellers, Section 11.3.5.

Notwithstanding the problems with blade area ratio, it is seen that, at this speed, the overall efficiency of the conventional propeller is much lower than the waterjet efficiency $\eta_D = 0.688$. The effects of the resistance of the propeller shaft brackets and rudders also have to be added. This supports the comments in Section 11.3.8 that, above about 30 knots, a waterjet can become more efficient than a comparable conventional propeller.

Surface-piercing propellers might also be considered, but inspection of Figure 16.19 would suggest that the maximum η_0 that could be achieved is about 0.620. This would agree with comments in Sections 11.3.1.and 11.3.8 indicating that surface-piercing propellers are only likely to be superior at speeds greater than about 40–45 knots.

17.2.9 Example Application 9. 98 m Passenger/Car Ferry, 38 knots, Monohull: Estimates of Effective and Delivered Power

The passenger/car ferry can transport 650 passengers and 150 cars. The dimensions are as follows: $L_{OA} = 98 \text{ m} \times L_{WL} = 86.5 \text{ m} \times B = 14.5 \text{ m} \times T = 2.35 \text{ m} \times C_B = 0.36$. Height of the superstructure above base is 13 m. The service speed is 38 knots.

$$Fr = V/\sqrt{gL} = 0.671, \; V_k/\sqrt{L_f} = 2.26,$$

$$\nabla = 1061.10 \text{ m}^3, \Delta = 1087.6 \text{ tonnes}, L/\nabla^{1/3} = 8.50,$$

$$L/B = 86.5/14.5 = 5.97, B/T = 14.5/2.35 = 6.17.$$

A possible hull form and series would be the National Technical University of Athens (NTUA) double-chine series. The hull form parameters are inside the series limits of $L/\nabla^{1/3} = 6.2–8.5$ and $B/T = 3.2–6.2$.

The regression equation for C_R is given as Equation (10.39), and the regression coefficients are given in Table A3.17. Repetitive calculations are best carried out

Table 17.6. *NPL interpolated data, passenger/car ferry*

$V_k/\sqrt{L_f}$	B/T		
	4.90	5.80	Required 6.17
2.2	2.375	2.226	2.165
2.3	2.246	2.108	2.051
Required 2.26	–	–	2.097

using a computer program or spreadsheet. For the above hull parameters, the C_R value is found to be $C_R = 0.001748$.

These parameters are within the range of the NPL round bilge series, with the exception of C_B which is fixed at 0.40 for the NPL series. An outline check estimate of C_R is made using this series. The data are given in Table A3.16, although the required parameters do not occur in the same combinations and some extrapolation is required.

At $L/\nabla^{1/3} = 8.3$ (closest to 8.5), $B/T = 6.17$, $V_k/L_f = 2.26$, the interpolated results from Table A3.16 are shown in Table 17.6. Hence, $C_R = 2.097 \times 10^{-3} = 0.002097$ (20% higher than NTUA). The data had to be extrapolated to $B/T = 6.17$, and estimating the effects of extrapolating from $L/\nabla^{1/3} = 8.3$–8.5 would indicate a decrease in C_R of about 8%. The effects of using data for $C_B = 0.4$ (rather than for 0.36) are not clear. Hence, as C_R is about 60% of total C_T (see later equation for C_{TS}), use of the NPL series result would lead to an increase in the predicted P_E of about 6%.

The estimate of wetted surface area, using Equation (10.92) and coefficients in Table 10.22, yields wetted surface area as: $S = 1022.0 \, \text{m}^2$. The approximate NTUA model length is 2.35 m. Model speed is as follows:

$$V_M = V_S \times \sqrt{(L_M/LS)} = 3.222 \text{ m/s}$$
$$Re \, \text{model} = VL/\nu = 3.222 \times 2.35/1.14 \times 10^{-6} = 6.642 \times 10^6.$$
$$Re \, \text{ship} = 38 \times 0.5144 \times 86.5/1.19 \times 10^{-6} = 1.421 \times 10^9.$$
$$C_{FM} = 0.075/(\log Re - 2)^2 = 0.075/(\log 6.642 \times 10^6 - 2)^2$$
$$= 0.075/(6.822 - 2)^2 = 0.00323.$$
$$C_{FS} = 0.075/(\log Re - 2)^2 = 0.075/(\log 1.421 \times 10^9 - 2)^2$$
$$= 0.075/(9.153 - 2)^2 = 0.00147.$$

Using Equation (10.58), form factor $(1 + k) = 2.76 \times (L/\nabla^{1/3})^{-0.4} = 1.173$. Using Equation (10.60), which includes a form factor,

$$C_{TS} = C_{FS} + C_{Rmono} - k(C_{FM} - C_{FS})$$
$$= 0.00147 + 0.001748 - 0.173 \times (0.00323 - 0.00147)$$
$$= 0.00291.$$
$$R_{TS} = C_{TS} \times \tfrac{1}{2}\rho SV^2$$
$$= 0.00291 \times 0.5 \times 1025 \times 1022.0 \times (38 \times 0.5144)^2/1000$$
$$= 582.38 \, \text{kN}.$$

(if a form factor is not included, $k = 0$, and $R_{TS} = 644.02$ kN).

Air drag is calculated using a suitable drag coefficient based on the transverse area of hull and superstructure. The height of superstructure above water $= 13.0 - 2.35 = 10.65$ m, breadth is 14.5 m and the transverse area $A_T =$ height $\times$ breadth $= 10.65 \times 14.5 = 154.4 \, \text{m}^2$.

From Figure 3.29, assume a suitable drag coefficient to be $C_D = 0.55$. Then,

$$D_{\text{air}} = C_D \times \tfrac{1}{2} \rho_A A_T V^2$$
$$= 0.55 \times 0.5 \times 1.23 \times 154.4 \times (38 \times 0.5144)^2 / 1000$$
$$= 19.96 \, \text{kN} \; (3.5\% \text{ of hull resistance}).$$

It is proposed to use waterjets; hence there will not be any appendage drag of significance. Hence, total resistance $= (582.38 + 19.96) = 602.34$ kN and $P_E = R_T \times Vs = 602.34 \times 38 \times 0.5144 = 11,774.1$ kW.

Applying an approximate formula for waterjet efficiency η_D, Equation (16.56), η_D is as follows:

$$\eta_D = 1/[1 + (8.64/Vs)] = 1/[1 + (8.64/38 \times 0.5144)] = 0.693$$

Delivered power is:

$$P_D = P_E/\eta_D = 11774.1/0.693 = 16990.0 \, \text{kW}.$$

Assuming a transmission efficiency for a geared drive $\eta_T = 0.95$, the service power is as follows:

$$P_S = P_D/\eta_T = 16990.0/0.95 = 17,884.2 \, \text{kW}.$$

Allowing a 15% margin for hull fouling and weather, the installed power is:

$$P_I = P_S \times 1.15 = 17884.2 \times 1.15 = 20,567 \, \text{kW}.$$

Alternatively, allowances for fouling and weather can be estimated in some detail. They will depend on operational patterns and expected weather in the areas of operation, see Chapter 3.

17.2.10 Example Application 10. 82 m Passenger/Car Catamaran Ferry, 36 knots: Estimates of Effective and Delivered Power

The catamaran passenger/car ferry can transport 650 passengers and 150 cars. The dimensions are as follows: $L_{OA} = 82$ m $\times L_{WL} = 72$ m $\times B = 21.5$ m $\times$ b $= 6.7$ m $\times T = 2.80$ m $\times C_B = 0.420$. The height of superstructure above base is 13.25 m. The service speed is 36 knots.

$$Fr = V/\sqrt{gL} = 0.697, \; V_k/\sqrt{L_f} = 2.34.$$
$$\nabla = 567.3 \, \text{m}^3 \text{ per hull, total } \nabla = 1134.6 \, \text{m}^3$$
$$\Delta = 1163.0 \text{ tonnes} = 581.5 \text{ tonnes per hull.}$$
$$L/\nabla^{1/3} = 8.72 \, (\text{for one hull}).$$
$$L/b = 72/6.7 = 10.75, \; b/T = 6.7/2.80 = 2.39,$$
$$S/L = (21.5 - 2 \times (6.7/2))/72 = 0.206.$$

Table 17.7. *Southampton catamaran data*

		Fr	
	0.60	0.70	Required 0.697
Table			
Table 10.10	2.799	2.272	2.288

Estimating the wetted surface area, using Equation (10.88), $S = C_S\sqrt{\nabla L}$, Equation 10.89 and Table 10.20 for the C_S regression coefficients, $C_S = 2.67$. The wetted surface area per hull is as follows: $S = C_S\sqrt{\nabla L} = 2.67\sqrt{567.3 \times 72} = 539.61$ m². For the total wetted area, both hulls $S = 1079.23$ m².

Use Equation (10.64) for catamarans:

$$C_{TS} = C_{FS} + \tau_R C_R - \beta k(C_{FM} - C_{FS}),$$

where C_R is for a monohull and τ_R is the residuary resistance interference factor. τ_R is given in Table 10.11, and C_R can be obtained from the Southampton extended NPL catamaran series, data or the Series 64 data.

Estimate using the Southampton extended NPL catamaran series as follows. This series is strictly for $C_B = 0.40$, but as the resistance for these semi-displacement hull types is dominated by $L/\nabla^{1/3}$, it is reasonable to assume that the series is applicable to $C_B = 0.42$. The approximate C_R data are given in Table 10.10, given by the equation $C_R = a\,(L/\nabla^{1/3})^n$ and interpolation can be made for $Fr = 0.697$, shown in Table 17.7. Hence, $C_R = 2.288 \times 10^{-3} = 0.002288$. This can be checked as acceptable by inspection of the actual values of C_R for the extended NPL series, contained in Table A3.26. The NPL extended series model length is 1.60 m. Model speed is as follows:

$$V_M = V_S \times \sqrt{(L_M/L_S)} = 2.761 \text{ m/s}.$$

$$Re\,\text{model} = VL/\nu = 2.761 \times 1.6/1.14 \times 10^{-6} = 3.875 \times 10^6.$$

$$Re\,\text{ship} = 36 \times 0.5144 \times 72/1.19 \times 10^{-6} = 1.120 \times 10^9.$$

$$C_{FM} = 0.075/(\log Re - 2)^2 = 0.075/(\log 3.875 \times 10^6 - 2)^2$$
$$= 0.075/(6.588 - 2)^2$$
$$= 0.00356.$$
$$C_{FS} = 0.075/(\log Re - 2)^2 = 0.075/(\log 1.120 \times 10^9 - 2)^2$$
$$= 0.075/(9.049 - 2)^2$$
$$= 0.00151.$$

Using Equation (10.59) for catamarans, form factor $(1 + \beta k) = 3.03 \times (L/\nabla^{1/3})^{-0.4} = 1.274$.

The residuary resistance interference factor τ_R can be obtained from Table 10.11 and checked, if required, from Table A3.26. Using Table 10.11, the results are shown in Table 17.8. Hence $\tau_R = 1.328$.

Table 17.8. *Catamaran interference factors*

S/L	Fr		
	0.60	0.70	Required 0.697
0.20	1.430	1.329	1.332
0.30	1.266	1.268	1.268
Required 0.206	–	–	1.328

Using Equation (10.64), which includes the form factor and the residuary resistance interference factor τ_R,

$$C_{TS} = C_{FS} + \tau_R C_R - \beta k(C_{FM} - C_{FS})$$
$$= 0.00151 + 1.328 \times 0.002288 - 0.274 \times (0.00356 - 0.00151)$$
$$= 0.00399$$
$$R_{TS} = C_{TS} \times \tfrac{1}{2}\rho S V^2$$
$$= 0.00399 \times 0.5 \times 1025 \times 1079.23 \times (36 \times 0.5144)^2/1000$$
$$= 756.8 \text{ kN}$$

(if a form factor is not included, $k = 0$, and $R_{TS} = 862.7$ kN).

Air drag is calculated using a suitable drag coefficient based on the transverse area of hulls and superstructure. The height of superstructure above water is $13.25 - 2.8 = 10.45$ m, breadth is 21.5 m and the transverse area A_T = height × breadth = $10.45 \times 21.5 = 224.7$ m^2 (neglecting the gap between the hulls).

From Figure 3.29, assume a suitable drag coefficient to be $C_D = 0.55$. Then,

$$D_{air} = C_D \times \tfrac{1}{2}\rho_A A_T V^2$$
$$= 0.55 \times 0.5 \times 1.23 \times 224.7 \times (36 \times 0.5144)^2/1000$$
$$= 26.06 \text{ kN (3.4\% of hull resistance)}.$$

It is proposed to use waterjets, hence there will not be any appendage drag of significance. Hence, total resistance is $(756.8 + 26.06) = 782.86$ kN and $P_E = R_T \times Vs = 782.86 \times 36 \times 0.5144 = 14,497.3$ kW.

Applying an approximate formula for waterjet efficiency η_D, Equation (16.56), η_D is found as follows:

$$\eta_D = 1/[1 + (8.64/Vs)] = 1/[1 + (8.64/36 \times 0.5144)] = 0.682.$$

Delivered power is:

$$P_D = P_E/\eta_D = 14,497.3/0.682 = 21,257.0 \text{ kW}.$$

Assuming a transmission efficiency $\eta_T = 0.95$, the service power $P_S = P_D/\eta_T = 21257.0/0.95 = 22,375.8$. Allowing a 15% margin for hull fouling and weather, the installed power $P_I = P_S \times 1.15 = 25732$ kW.

Alternatively, allowances for fouling and weather can be estimated in some detail. They will depend on operational patterns and expected weather in the areas of operation, see Chapter 3.

17.2.11 Example Application 11. 130 m Twin-Screw Warship, 28 knots, Monohull: Estimates of Effective and Delivered Power

The dimensions are as follows: $L_{WL} = 130\,\text{m} \times B = 16.0\,\text{m} \times T = 4.7\,\text{m} \times C_B = 0.390 \times C_P = 0.530 \times C_W = 0.526 \times LCB = 6\% L$ Aft. Full speed is 28 knots. Cruise speed is 15 knots. Based on a twin-screw layout and clearance limitations, propeller diameter $D = 4.0\,\text{m}$.

$\nabla = 3812.6\,\text{m}^3$, $\Delta = 3908.0$ tonnes, $L/\nabla^{1/3} = 8.34$, $L/B = 8.13$, $B/T = 3.40$ and $\nabla/L^3 = 1.735 \times 10^{-3}$.

$$Fr = V/\sqrt{gL} = 28 \times 0.5144/\sqrt{9.81 \times 130} = 0.403.$$

$$Fr_\nabla = V/\sqrt{g\nabla^{1/3}} = 28 \times \sqrt{9.81 \times 3812.6^{1/3}} = 1.17.$$

The length 130 m = 426.4 ft, and $V_k/\sqrt{L_f} = 28/\sqrt{426.4} = 1.36$.

Estimating the wetted surface area, using Equation (10.86), $S = C_S\sqrt{\nabla L}$, and Table 10.19 for C_S, by interpolating in Table 10.19, for $C_P = 0.530$, $L/\nabla^{1/3} = 8.34$ and $B/T = 3.40$,

$$C_S = 2.534.$$

For wetted surface area,

$$S = C_S\sqrt{\nabla L} = 2.534\sqrt{3812.6 \times 130} = 1784\,\text{m}^2.$$

Using Equation (10.85),

$$S = 3.4\nabla^{2/3} + 0.485 L.\nabla^{1/3} = 1814\,\text{m}^2.$$

Take $S = (1784 + 1814)/2 = 1799\,\text{m}^2$ [check using Denny Mumford, Equation (10.84): $S = 1.7\,LT + \nabla/T = 1.7 \times 130 \times 4.7 + 3812.6/4.7 = 1850\,\text{m}^2 (+2.8\%)$].

17.2.11.1 Estimate of Effective Power

The Taylor–Gertler series data are used for C_R, together with C_F Schoenherr and a roughness allowance $\Delta C_F = 0.0004$. It can be noted that the LCB for all the Taylor–Gertler models was fixed at amidships.

For the Taylor–Gertler series, using the graphs of C_R data in [10.10] or the tabulated C_R data in Tables A3.8 to A3.11 in Appendix A3, and interpolating for $Fr = 0.403$ ($V_k/\sqrt{L_f} = 1.36$), $C_P = 0.530$, $B/T = 3.40$ and $L/\nabla^{1/3} = 8.34$ ($\nabla/L^3 = 1.735 \times 10^{-3}$),

$$C_R = 3.16 \times 10^{-3} = 0.00316. \quad \text{(from Graphs).}$$

$$Re = VL/\nu = 28.0 \times 0.5144 \times 130/1.19 \times 10^{-6} = 1.573 \times 10^9.$$

Using an approximation to the Schoenherr line, Equation (4.11),

$$C_F = 1/(3.5\,Log\,Re - 5.96)^2 = 0.001454.$$

$$C_T = C_F + \Delta C_F + C_R = 0.001454 + 0.0004 + 0.00316 = 0.005014.$$

Total ship resistance is as follows:

$$R_{TS} = C_T \times \tfrac{1}{2}\rho SV^2$$
$$= 0.005014 \times 0.5 \times 1025 \times 1799 \times (28.0 \times 0.5144)^2/1000$$
$$= 959.0\,\text{kN}.$$

Without ΔC_F, $R_{TS} = 882$ kN. An estimate using the Holtrop regression gives 774 kN (-12%). An estimate using Series 64 data gives 758 kN (-14%).

The Taylor series has LCB fixed at amidships, which is non-optimum for this ship. It also has a cruiser-type stern, rather than the transom used in modern warships. Both of these factors are likely to contribute to the relatively high estimate using the Taylor series. (The most useful aspects of the Taylor series are its very wide range of parameters and the facility to carry out relative parametric studies).

17.2.11.2 Estimates of Wake Fraction w_T and Wake Speed for Use with Appendages

Extrapolation of Harvald data, Figure 8.13, would suggest a value of $w_T = 0.08$. Semi-displacement data, Table 8.4 would suggest $w_T = 0$.

As the aft end is closer to a conventional form, assume a value of $w_T = 0.05$. Hence, wake speed $Va = Vs\,(1 - w_T) = 28 \times 0.5144\,(1 - 0.05) = 13.68$ m/s.

17.2.11.3 Appendage Drag

APPENDAGES FOR THIS SHIP TYPE. The appendages to be added are the two rudders, propeller A-brackets and shafting.

RUDDER. For this size and type of vessel, typical rudder dimensions will be rudder span = 3.5 m and rudder chord = 2.5 m. [Rudder area = $3.5 \times 2.5 = 8.75$ m^2[$8.75 \times 2/(L \times T) = 2.9\% L \times T$, which is suitable.]

A practical value for the drag coefficient is $C_{D0} = 0.013$, see section 3.2.1.8 (c).

As a first approximation, assume a 20% acceleration due to the propeller, and rudder inflow speed is $Va \times 1.2 = 13.68 \times 1.2 = 16.42$ m/s.

$$
\begin{aligned}
D_{\text{Rudder}} &= C_{D0} \times \tfrac{1}{2}\rho A V^2 \\
&= 0.013 \times 0.5 \times 1025 \times 8.75 \times (16.42)^2/1000 \\
&= 15.72\,\text{kN} \times \text{two rudders} = 31.44\,\text{kN},
\end{aligned}
$$

i.e. rudder drag = $31.44/959.0 = 3.3\%$ of naked hull resistance.

Check using Equation (3.48) attributable to Holtrop, as follows:

$$
D_{\text{Rudder}} = \tfrac{1}{2}\rho V_S^2 C_F(1 + k_2)S,
$$

where Vs is ship speed, C_F is for ship, S is the wetted area and $(1 + k_2)$ is an appendage resistance factor which, for rudders, varies from 1.3 to 2.8 depending on rudder type (Table 3.5). For twin-screw balanced rudders, $(1 + k_2) = 2.8$.

Ship $Re = VL/\nu = 28 \times 0.5144 \times 130/1.19 \times 10^{-6} = 1.573 \times 10^9$, and $C_F = 0.075/(\log Re - 2)^2 = 0.001448$. The total wetted area for both rudders is $(8.75 \times 2) \times 2 = 35$ m^2, and $D_{\text{Rudder}} = \tfrac{1}{2} \times 1025 \times (28 \times 0.5144)^2 \times 0.001448 \times 2.8 \times 35/1000 = 15.09$ kN (about 48% of the value of the earlier estimate).

PROPELLER A BRACKETS: ONE PORT, ONE STARBOARD. There are two struts per A bracket. The approximate length of each strut is 3.0 m, t/c is 0.25, chord is 700 mm and thickness is 175 mm. The wake speed $Va = Vs(1 - w_T) = 28 \times 0.5144(1 - 0.05) = 13.68$ m/s. Based on strut chord,

$$Re = VL/\nu = 13.68 \times 0.700/1.19 \times 10^{-6} = 8.047 \times 10^6.$$

$$C_F = 0.075/(\log Re - 2)^2 = 0.00311.$$

Using Equation (3.42),

$$C_D = C_F \left[1 + 2 \left(\frac{t}{c}\right) + 60 \left(\frac{t}{c}\right)^4\right]$$

$$C_D = 0.00311 \times (1 + 2 \times 0.25 + 60 \times 0.25^4) = 0.00539.$$

Assume that area = span × chord = $3.0 \times 0.700 = 2.1$ m^2, then total wetted area = $2 \times 2.1 = 4.2$ m^2 per strut, and

$$D_{strut} = C_D \times \tfrac{1}{2}\rho A V^2$$
$$= 0.00539 \times 0.5 \times 1025 \times 4.2 \times 13.68^2/1000 = 2.17 \text{ kN per strut.}$$

With two struts per A bracket, and two A brackets, the total drag is $4 \times 2.17 = 8.68$ kN. [Check using Equation (3.41), as follows:

$$D_{CS} = \frac{1}{2}\rho S V^2 C_F \left[1.25\frac{Cm}{Cf} + \frac{S}{A} + 40 \left(\frac{t}{Ca}\right)^3\right] \times 10^{-1}.$$

In Figure 3.23, Cm is 0.700, Cf is 0.200, Ca is 0.500, wetted area S is 4.2 m^2, t is 0.17 m, A is frontal area = $t \times$ span = $0.175 \times 3.0 = 0.525$ m^2 and

$$D_{CS} = 0.5 \times 1025 \times 4.2 \times (13.68)^2 \times 0.00311 \times [1.25 \times 0.700/0.200$$
$$+ 4.2/0.525 + 40 \times (0.175/0.500)^3] \times 10^{-1}/1000$$
$$= 1.765 \text{ kN per strut} \times 4 \text{ struts} = 7.06 \text{ kN}].$$

INTERFERENCE BETWEEN STRUTS AND HULL. Using Equation (3.45), with $t/c = 0.25$, $t = 0.175$ m and $V = 13.68$ m/s,

$$D_{INT} = \tfrac{1}{2}\rho V^2 t^2 \left[0.75\frac{t}{c} - \frac{0.0003}{(t/c)^2}\right]$$
$$D_{INT} = 0.5 \times 1025 \times 13.68^2 \times 0.175^2 \times [0.75 \times 0.25 - 0.0003/(0.25)^2]$$
$$= 0.54 \text{ kN per strut.}$$

For four struts, $D_{INT} = 4 \times 0.54 = 2.16$ kN.

Total drag of struts, including interference is $8.68 + 2.16 = 10.84$ kN.

DRAG OF SHAFTS. Using Equation (3.46),

$$D_{SH} = \tfrac{1}{2}\rho L_{SH} Ds V^2 (1.1 \sin^3 \alpha + \pi C_F).$$

The details of the shaft follows: length $L_{SH} = 12$ m, diameter $Ds = 0.450$ m, angle is 10° and $V = 13.68$ m/s. The layout will be broadly as shown in Figure 14.28. A P-bracket may also be employed at the hull end of the shaft, but this has not been included in the present analysis.

Re based on shaft diameter $= VDs/\nu = 13.68 \times 0.450/1.19 \times 10^{-6} = 5.173 \times 10^6$.

$$C_F = 0.075/(\log Re - 2)^2 = 0.00338.$$

$D_{SH} = 0.5 \times 1025 \times 12 \times 0.450 \times 13.68^2 \times (1.1 \times \sin^3 10 + \pi \times 0.00338) = 8.49\,\text{kN}$.

For two shafts, $D_{SH} = 2 \times 8.49 = 16.98$ kN. Allowing say 5% for shaft-strut interference, total drag of shafts, $D_{SH} = 1.05 \times 16.98 = 17.83$ kN.

SUMMARY OF APPENDAGE DRAG.

Rudders $\times$ 2: 31.44 kN.
Struts $\times$ 4, including interference: 10.84 kN.
Shafts $\times$ 2, including interference: 17.83 kN.
Total appendage drag $= 60.11$ kN.

Appendage drag is $60.11/959 = 6.3\%$ of naked hull resistance.

17.2.11.4 Air Drag
A drag coefficient based on the transverse area of hull and superstructure above the waterline is used, see Chapter 3. For this vessel size, the height of the deckhouse above water is approximately 17.3 m and, assuming the full breadth of 16 m, the transverse area A_T is $17.3 \times 16 = 276.8\,\text{m}^2$.

Assume a drag coefficient $C_D = 0.80$ for hull and superstructure, see Chapter 3, Section 3.2.2.

$$
\begin{aligned}
\text{Still air drag } D_{\text{AIR}} &= C_D \times \tfrac{1}{2}\rho_A A_T V^2 \\
&= 0.80 \times 0.5 \times 1.23 \times 276.8 \times (28 \times 0.5144)^2/1000 \\
&= 28.25\,\text{kN}
\end{aligned}
$$

(and air drag $= (28.25/959) \times 100 = 3\%$ of hull naked resistance). [Check using the ITTC approximate formula, $C_{AA} = 0.001(A_T/S)$, Equation (5.12), where S is the wetted area of the hull and $D_{\text{AIR}} = C_{AA} \times \tfrac{1}{2}\rho_w SV^2 = (0.001 \times (276.8/S)) \times \tfrac{1}{2} \times 1025 \times S \times (28 \times 0.5144)^2/1000 = 29.4\,\text{kN}$].

17.2.11.5 Summary
Total resistance $R_T =$ naked hull $+$ appendages $+$ air drag $= 959.0 + 60.11 + 28.25 = 1047.4$ kN. The effective power $P_E = R_T \times Vs = 1047.4 \times 28 \times 0.5144 = 15085.9\,\text{kW}$.

17.2.11.6 Outline Propeller Design
For this twin-screw layout and draught, and allowing for hull clearances and some projection of the propeller below the keel line, a propeller with a diameter of 4.0 m is appropriate.

Estimate the thrust deduction fraction t, as follows. As a first approximation for twin screw, $t = w_T = 0.050$ [Equation (8.25)]. Using the Holtrop formula, Equation (8.26), for twin-screw vessels,

$$t = 0.325C_B - 0.1885\frac{D}{\sqrt{BT}},$$

and $t = 0.325 \times 0.390 - 0.1885 \times 4.0/(16 \times 4.7)^{0.5} = 0.040$.

Table 17.9. *Gawn data for range of revs*

rpm	rps	J	K_T	P/D	η_0
150	2.5	1.368	0.333	1.92	0.695
180	3.0	1.140	0.231	1.51	0.700
210	3.5	0.977	0.170	1.21	0.690

Assume that $t = 0.04$. The total resistance $R_T = 1047.4$ kN. The total required thrust $T = R_T/(1 - t) = 1047.4/(1 - 0.04) = 1091.0$ kN. Required thrust per screw $T = 1091.0/2 = 545.5$ kN.

$$V_a = Vs(1 - w_T) = 28 \times 0.5144 \times (1 - 0.05) = 13.68 \text{ m/s}.$$

$$J = Va/nD = 13.68/(n \times 4.0) = 3.42/n.$$

$$K_T = T/\rho n^2 D^4 = 545.5 \times 1000/1025 \times n^2 \times 4.0^4 = 2.079/n^2.$$

For preliminary design purposes, the Gawn series of propellers is appropriate for this vessel type. Using the Gawn G3.95 chart, Figure 16.10, and assuming a range of rpm, the results are shown in Table 17.9.

From a cross plot of data, take optimum revs as 3.0 rps (180 rpm), (although η_0 is not very sensitive to rps), $P/D = 1.51$ and $\eta_0 = 0.700$. Note that the values of P/D and η_0 were read manually from the chart (Figure 16.10) and, consequently, are only approximate. Use of a larger chart or the regression of the Gawn series (Chapter 16) should lead to more accurate answers.

$$\eta_H = (1 - t)/(1 - w_T) = (1 - 0.04)/(1 - 0.05) = 1.011.$$

Estimate η_R for $C_P = 0.530$, $LCB = -0.06$.

Using Equation (16.80), $\eta_R = 0.952$; from Table 16.12, $\eta_R = 0.950$; take $\eta_R = 0.95$. Then, $\eta_D = \eta_0 \times \eta_H \times \eta_R = 0.700 \times 1.011 \times 0.95 = 0.672$. The delivered power $P_D = P_E/\eta_D = 15085.9/0.672 = 22449.3$ kW. The service (or shaft) power $P_S = P_D/\eta_T$, where η_T is transmission efficiency, and is typically about 0.95 for a geared drive installation. Hence, $P_S = 22449.3/0.95 = 23630.8$ kW.

Allowing say 15% margin for fouling and weather, the installed power $P_I = P_S \times 1.15 = 27175$ kW.

CAVITATION BLADE AREA CHECK. See Section 12.2.10. From the propeller and shafting layout, the immersion of shaft h is 3.7 m. Required thrust per screw $T = 545.5 \times 10^3$ N, at service speed. $D = 4.0$m, $P/D = 1.51$ and $n = 3.0$ rps.

$$V_R^2 = V_a^2 + (0.7\pi nD)^2 = 13.68^2 + (0.7\pi \times 3.0 \times 4.0)^2 = 883.54.$$
$$\sigma = (\rho gh + P_{AT} - P_V)/\tfrac{1}{2}\rho V_R^2$$
$$= (1025 \times 9.81 \times 3.7 + 101 \times 10^3 - 3 \times 10^3)/\tfrac{1}{2} \times 1025 \times 883.54$$
$$= 0.299.$$

using line (3), Equation (12.19), for naval vessels and fast craft, when $\sigma = 0.299$, $\tau_C = 0.43\,(\sigma - 0.02)^{0.71} = 0.174$. A similar value can be determined directly from Figure 12.31. Then,

$$A_P = T/\tfrac{1}{2}\rho V_R^2 \tau_C = 545.5 \times 10^3/\tfrac{1}{2} \times 1025 \times 883.54 \times 0.174 = 6.923\,\text{m}^2.$$
$$A_D = A_P/(1.067 - 0.229\,P/D) = 6.923/(1.067 - 0.229 \times 1.51) = 9.599\,\text{m}^2.$$
$$\text{DAR} = (\text{BAR}) = 9.599/(\pi D^2/4) = 9.599/(\pi \times 4.0^2/4) = 0.764.$$

The calculations could be repeated using the Gawn chart G3.65, Figure 16.9, and interpolation carried out between the BAR $= 0.95$ and 0.65 charts. The use of the smaller BAR chart would lead to a small improvement in efficiency.

SUMMARY OF PROPELLER PARTICULARS

$$D = 4.0\,\text{m}, \quad 180\,\text{rpm}, \quad P/D = 1.51, \quad \text{BAR} = 0.764.$$

It should be noted that, from vibration considerations, the propeller is likely to be five-bladed rather than the three-bladed propeller used in this analysis. There is likely to be only a very small decrease in efficiency, see Section 16.1.2 concerning choice of number of blades.

17.2.11.7 Outline Power Estimate for Cruise Speed of 15 knots

Assume the same percentage addition for appendages and still air drag as for 28 knots and the same wake fraction, thrust deduction factor and η_R. Assume the propeller is fixed pitch at $P/D = 1.51$.

$$Fr = V/\sqrt{gL} = 15 \times 0.5144/\sqrt{9.81 \times 130} = 0.216.$$

The length $130\,\text{m} = 426.4\,\text{ft}$, and $V_k/\sqrt{L_f} = 15/\sqrt{426.4} = 0.726$.

For the Taylor–Gertler series, using the graphs of C_R data in [10.10] or the tabulated C_R data in Tables A3.8 to A3.11 in Appendix A3, and interpolating for $Fr = 0.216$ ($V_k/\sqrt{L_f} = 0.726$), $C_P = 0.530$, $B/T = 3.40$ and $L/\nabla^{1/3} = 8.34(\nabla/L^3 = 1.735 \times 10^{-3})$,

$$C_R = 0.40 \times 10^{-3} = 0.00040$$

$$Re = V\,L/\nu = 15.0 \times 0.5144 \times 130/1.19 \times 10^{-6} = 8.423 \times 10^8.$$

Using an approximation to the Schoenherr line, Equation (4.11),

$$C_F = 1/(3.5\,\text{Log}\,Re - 5.96)^2 = 0.00157$$

$$C_T = C_F + \Delta C_F + C_R = 0.00157 + 0.0004 + 0.00040 = 0.00237$$

Total ship resistance is as follows:

$$R_{TS} = C_T \times \tfrac{1}{2}\rho S V^2$$
$$= 0.00237 \times 0.5 \times 1025 \times 1799 \times (15.0 \times 0.5144)^2/1000$$
$$= 130.10\,\text{kN}.$$

Assuming the same 6% resistance increase due to appendages and 3% for still air drag, total $R_T = 130.10 \times 1.09 = 141.8$ kN. The effective power $P_E = R_T \times Vs = 141.8 \times 15 \times 0.5144 = 1094.1$ kW. The required thrust per screw $T = (R_T/2)/(1 - t) = (141.8/2)/(1 - 0.04) = 73.85$ kN

$$Va = Vs(1 - w_T) = 15 \times 0.5144 \times (1 - 0.05) = 7.33 \text{ m/s}.$$

$$J = Va/nD = 7.33/(n \times 4.0) = 1.833/n.$$

$$K_T = T/\rho n^2 D^4 = 73.85 \times 1000/(1025 \times n^2 \times 4.0)^4 = 0.2814/n^2.$$

Inspection of the Gawn B3.95 propeller chart, Figure 16.10, indicates that with the (fixed) pitch ratio of 1.51, the propeller will run at 84 rpm (1.40 rps) with $J = 1.31$, $K_T = 0.144$ and $\eta_0 = 0.705$. Then, $\eta_D = \eta_0 \times \eta_H \times \eta_R = 0.705 \times 1.011 \times 0.95 = 0.677$. The delivered power $P_D = P_E/\eta_D = 1094.1/0.677 = 1616.1$ kW. The service (or shaft) power $P_S = P_D/\eta_T = 1616.1/0.95 = 1701.2$ kW.

Published data would suggest that this power is low. A check using Series 64 (with transom stern) would indicate an approximate $C_R = 0.0017$ and a total shaft power of about 2650 kW.

A reasonable margin will be added to allow for the development of fouling and the maintenance of 15 knots in rough weather.

17.2.12 Example Application 12. 35 m Patrol Boat, Monohull: Estimate of Effective Power

The dimensions are as follows: $L_{OA} = 41$ m $\times L_{WL} = 35$ m $\times B = 7.0$ m $\times T = 2.4$ m $\times C_B = 0.400 \times LCB = 6\%L$ Aft. Speed = 25 knots.

$$Fr = V/\sqrt{gL} = 0.694, \qquad Fr_\nabla = V/\sqrt{g\nabla^{1/3}} = 1.654, \qquad V_k/\sqrt{L_f} = 2.33,$$

$$\nabla = 235.2 \text{m}^3, \qquad \Delta = 241.1 \text{ tonnes}, \qquad L/\nabla^{1/3} = 5.68,$$

$$L/B = 35/7 = 5.0, \qquad B/T = 7/2.4 = 2.92.$$

The above parameters are within the range of the NPL round bilge series and an estimate of C_R is initially made using this series. The data are given in Table A3.16.

At $L/\nabla^{1/3} = 5.76$ (assumed acceptably close enough to 5.68), $B/T = 2.92$, $V_k/L_f = 2.33$ and, interpolating from Table A3.16, the results are shown in Table 17.10. Hence,

$$C_R = 7.454 \times 10^{-3} = 0.007454.$$

Table 17.10. *Interpolated NPL data, patrol boat*

	B/T		
$V_k/\sqrt{L_f}$	2.59	3.67	Required 2.92
2.3	7.903	7.333	7.573
2.4	7.321	6.845	7.176
Required 2.33	–	–	7.454

Table 17.11. *Results from WUMTIA regression, patrol boat*

Fr_∇	Vs knots	P_E
1.50	22.7	2464.3
1.75	26.45	3204.33
–	Required 25.0	2918

Estimating the wetted surface area, using Equation (10.88), $S = C_S\sqrt{\nabla L}$, and Equation (10.90) for C_S, $C_S = 2.794$. And the wetted surface area is as follows:

$$S = C_S\sqrt{\nabla L} = 2.794 \times \sqrt{235.2 \times 35} = 253.5 \text{ m}^2.$$

NPL model length $= 2.54$.

$$\text{Model speed } V_M = V_S \times \sqrt{(L_M/L_S)} = 3.464 \text{ m/s.}$$

$$Re \text{ model} = V\,L/\nu = 3.464 \times 2.54/1.14 \times 10^{-6} = 7.718 \times 10^6.$$
$$Re \text{ ship} = 25 \times 0.5144 \times 35/1.19 \times 10^{-6} = 3.782 \times 10^8.$$
$$C_{FM} = 0.075/(\log Re - 2)^2 = 0.075/(\log 7.718 \times 10^6 - 2)^2$$
$$= 0.075/(6.888 - 2)^2 = 0.00314.$$
$$C_{FS} = 0.075/(\log Re - 2)^2 = 0.075/(\log 3.782 \times 10^8 - 2)^2$$
$$= 0.075/(8.577 - 2)^2 = 0.00173.$$

If a form factor is used, using Equation (10.58), the form factor $(1 + k) = 2.76 \times (L/\nabla^{1/3})^{-0.4} = 1.378$. Using Equation (10.60), which includes a form factor,

$$C_{TS} = C_{FS} + C_R - k(C_{FM} - C_{FS})$$
$$= 0.00173 + 0.007454 - 0.378 \times (0.00314 - 0.00173) = 0.00865.$$
$$R_{TS} = C_{TS} \times \tfrac{1}{2}\rho SV^2 = 0.00865 \times 0.5 \times 1025 \times 253.5 \times (25 \times 0.5144)^2/1000$$
$$= 185.85 \text{ kN.}$$

In general, it is not recommended, for example by the ITTC, to use a form factor with such a hull form with a low $L/\nabla^{1/3}$. In addition, it can be noted that the form factor Equation (10.58) was not derived for hulls with such a low $L/\nabla^{1/3}$.

If a form factor is *not* included, $k = 0$ and $R_{TS} = 197.33$ kN. Then effective power $P_E = R_T \times Vs = 197.33 \times 25 \times 0.5144 = 2537.7$ kW.

Check using the Wolfson Unit for Marine Technology and Industrial Aerodynamics (WUMTIA) regression for round bilge forms, Section 10.3.4.2 and Table A3.24. The results are shown in Table 17.11. $P_E = 2918$ kW and the wetted surface area $S = 263.01$ m^2. Applying the correction factor for round bilge forms, described in Section 10.3.4.2, the corrected $P_E = 2918 \times 0.96 = 2801$ kW (about 10% higher than the NPL series estimate).

17.2.13 Example Application 13. 37 m Ocean-Going Tug: Estimate
of Effective Power

The dimensions are as follows: $L_{OA} = 37\,\text{m} \times L_{BP} = 35\,\text{m} \times B = 13.6\,\text{m} \times T = 6.1\,\text{m} \times C_B = 0.580 \times C_M = 0.830 \times C_P = 0.700 \times LCB = 2\%A \times \frac{1}{2}\alpha_E = 18°$. The free-running speed is 14 knots.

Using the Oortmerssen regression (Section 10.3.4.1),

$$L_D = (L_{OA} + L_{BP})/2 = 36\,\text{m}.$$

$$\nabla = 1732.2\text{m}^3, \quad \Delta = 1775.5\,\text{tonnes}, \quad L/\nabla^{1/3} = 3.00, \quad L/B = 2.65, \quad B/T = 2.23,$$

$$Fr = V/\sqrt{gL} = 14 \times 0.5144/\sqrt{9.81 \times 36} = 0.383.$$

$$\text{At} \quad Fr = 0.383, \quad R_R/\Delta = 0.02062, \quad \text{wetted surface area S} = 699.0\,\text{m}^2$$

and

$R_R = (R_R/\Delta) \times (\nabla \times \rho \times g) = 0.02062 \times 1732.2 \times 1025 \times 9.81/1000 = 359.2\,\text{kN}.$
$C_R = R_R/\frac{1}{2}\rho S V^2 = 359.2 \times 1000/0.5 \times 1025 \times 699.0 \times (14 \times 0.5144)^2 = 0.01933.$
$Re = VL/\nu = 14 \times 0.5144 \times 36/1.19 \times 10^{-6} = 2.179 \times 10^8.$
$C_F = 0.075/(\log Re - 2)^2 = 0.001867.$
$\Delta C_F = 0.00051$ (using all items in Table 10.7.)

Then

$$C_T = C_F + \Delta C_F + C_R = 0.001867 + 0.00051 + 0.01933 = 0.02171.$$
$$R_T = C_T \times \frac{1}{2}\rho S V^2$$
$$= 0.02171 \times 0.5 \times 1025 \times 699.0 \times (14 \times 0.5144)^2/1000$$
$$= 403.35\,\text{kN}.$$
Effective power $\quad P_E = R_T \times Vs = 403.35 \times 14 \times 0.5144 = 2904.8\,\text{kW}.$

The performance of a tug propeller, working off design, is described in Example Application 19.

17.2.14 Example Application 14. 14 m Harbour Work Boat, Monohull:
Estimate of Effective Power

The dimensions are as follows: $L_{OA} = 14.5\,\text{m} \times L_{WL} = 14\,\text{m} \times B = 3.3\,\text{m} \times T = 0.90\,\text{m} \times C_B = 0.410 \times LCB = 4\%L$ Aft. The speed range is 5–15 knots.

$$\nabla = 17.05\,\text{m}^3, \quad \Delta = 17.48\,\text{tonnes}, \quad L/\nabla^{1/3} = 5.44,$$
$$L/B = 14/3.3 = 4.24, \quad B/T = 3.3/0.90 = 3.67,$$
$$Fr = V/\sqrt{gL}, \quad Fr_\nabla = V/\sqrt{g\nabla^{1/3}}.$$

The above parameters are within the range of the WUMTIA regression for round bilge forms, Section 10.3.4.2 and Table A3.24. An estimate of effective power P_E is made using these data. The results are shown in Table 17.12.

A correction factor of 4% has been applied, see Chapter 10.

Table 17.12. *Results from WUMTIA regression, work boat*

Fr_∇	Fr	$V_k/\sqrt{L_f}$	Vs knots	P_E	Corrected P_E
0.50	0.21	0.72	4.89	3.43	3.3
0.75	0.32	1.08	7.33	12.13	11.6
1.00	0.43	1.45	9.80	27.36	26.3
1.25	0.54	1.80	12.21	74.72	71.7
1.50	0.64	2.16	14.66	122.52	117.6
1.75	0.75	2.52	17.09	133.07	127.7

The above parameters are within the range of the NPL round bilge series and an estimate of C_R is made using this series. (C_B is in fact just outside $C_B = 0.400$ for the NPL Series). The data are given in Table A3.16.

At $B/T = 3.67$, C_R values were interpolated from Table A3.16 and values at the required $L/\nabla^{1/3} = 5.44$ were further interpolated as shown in Table 17.13. An example calculation is as follows: At 13.55 knots,

$$Re = VL/\nu = 13.55 \times 0.5144 \times 14/1.19 \times 10^{-6} = 8.200 \times 10^7.$$
$$C_F = 0.075/(\log Re - 2)^2 = 0.075/(\log 8.200 \times 10^7 - 2)^2 = 0.00214.$$
$$C_T = C_F + C_R = 0.00214 + 0.01131 = 0.01345.$$

Estimating the wetted surface area, using Equation (10.88), $S = C_S\sqrt{\nabla L}$, and Equation (10.90) for C_S, $C_S = 2.895$. The wetted surface area is as follows:

$$S = C_S\sqrt{\nabla L} = 2.895 \times \sqrt{17.05 \times 14.0} = 44.73 \text{ m}^2.$$
$$R_T = C_T \times \tfrac{1}{2}\rho S V^2$$
$$= 0.01345 \times 0.5 \times 1025 \times 44.73 \times (13.55 \times 0.5144)^2/1000 = 14.979 \text{ kN}.$$
$$P_E = R_T \times Vs = 14.979 \times 13.55 \times 0.5144 = 104.4 \text{ kW}.$$

C_R was also estimated using the NTUA double-chine series. The $L/\nabla^{1/3}$ at 5.44 is below the lower recommended limit of 6.2 for the NTUA series and caution should be exercised in such an extrapolation of regression data. The results are shown in Table 17.14. Wetted surface area using the NTUA data gives $S = 42.4 \text{ m}^2$.

The results are plotted in Figure 17.1. It is seen that between 12 and 14 knots the three predictions are within about 8% of each other. Note that a skeg will often

Table 17.13. *Interpolated NPL data, work boat*

V_K	V_K/L_f	$L/\nabla^{1/3}$			C_F	C_T	P_E
		5.23	5.76	5.44			
6.78	1.0	5.247	4.547	4.970×10^{-3}	0.00238	0.00735	7.2
10.16	1.5	11.092	9.278	10.374×10^{-3}	0.00224	0.01261	41.3
13.55	2.0	12.832	8.991	11.311×10^{-3}	0.00214	0.01345	104.4
16.94	2.5	8.532	6.383	7.681×10^{-3}	0.00208	0.00976	148.0

Table 17.14. *NTUA regression data*

V_K	C_R	C_F	C_T	P_E
6.78	0.010364	0.00238	0.01274	11.7
10.16	0.01322	0.00224	0.01546	48.0
13.55	0.01101	0.00214	0.01315	96.8
16.94	0.00735	0.00208	0.00943	135.5

be incorporated in the design of such craft. In this case, the wetted area of the skeg should be added to that of the hull.

Some form of roughness allowance might also be made. Oortmerssen, Table 10.7, would suggest a ΔC_F correction of the order of 0.0004 for a small craft. From the above calculations, this would suggest an increase in P_E of about 3%.

17.2.15 Example Application 15. 18 m Planing Craft, Single-Chine Hull: Estimates of Effective Power Preplaning and Planing

The example uses resistance regression data for Series 62, the WUMTIA regression for hard chine forms and the Savitsky equations. The displacement of the example craft has been chosen as equivalent to 100,000 lb so that, for the purposes of comparison, the Series 62 predictions do not require a skin friction correction (see Chapter 10). The dimensions are as follows: $L_{WL} = 18\,\text{m} \times B = 4.6\,\text{m} \times T = 1.3\,\text{m} \times C_B = 0.411$.

$$\nabla = 44.2\,\text{m}^3, \quad \Delta = 45.4\,\text{tonnes} = 444.8\,\text{kN}[\approx 100,000\,\text{lb}].$$

$$L/B = 3.91; \quad L/\nabla^{1/3} = 5.1.$$

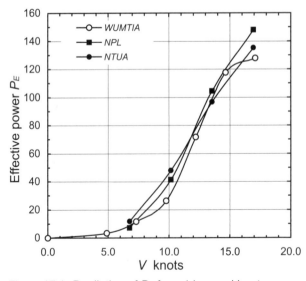

Figure 17.1. Prediction of P_E for a 14 m workboat.

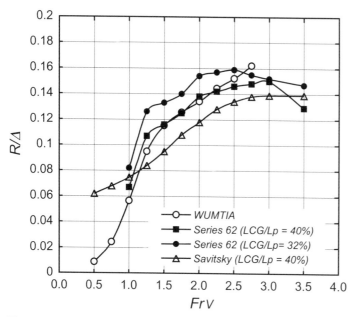

Figure 17.2. Prediction of R_T/Δ for an 18 m planing craft.

For the Series 62 regression, $l_p = 18 \times 1.02 = 18.36\,\text{m}$, $B_{px} = B = 4.6\,\text{m}$ and $B_{pa} = B_{px}/1.21 = 3.80\,\text{m}$.

$$l_p/B_{pa} = 18.36/3.80 = 4.83.$$

$$A_p = l_p \times B_{pa} = 18.36 \times 3.80 = 69.77 \quad \text{and} \quad A_p/\nabla^{2/3} = 5.57.$$

The speed range of 6 knots to 40 knots is considered.

The results for R_T/Δ are shown in Figure 17.2. The results, including trim, wetted area and wetted length are shown in Table 17.15.

An example calculation for the WUMTIA regression at $Fr_\nabla = 2.5$, $Fr = 1.11$, 28.6 knots is as follows:

Output from regression, $P_E = 1027.28\,\text{kW}$. This is reduced by 3% (see Chapter 10), hence, $P_E = 1027.28 \times 0.97 = 996.5\,\text{kW}$.

$$R_T = P_E/(Vs \times 0.5144)$$
$$= 996.5/(28.6 \times 0.5144) = 67.73\,\text{kN}$$
$$R_T/\Delta = 67.73/444.8 = 0.152.$$

The Series 62 regression was run for $A_p/\nabla^{2/3} = 5.57$, $l_p/B_{pa} = 4.83$ and deadrise $\beta = 20°$. Two positions of longitudinal centre of gravity (LCG) were investigated: LCG/l_p = 40% and 32%.

Figure 17.2 shows that the WUMTIA R_T/Δ values are in good agreement with the Series 62 data up to $Fr_\nabla = 2.5$ for the case of LCG/l_p = 40%. For the LCG/l_p = 32% case, the WUMTIA values are about 10% to 15% lower than Series 62. It is also seen in Table 17.15 that there is a marked increase in resistance and running trim when LCG/l_p is moved aft from 40% l_p to 32% l_p.

Figure 17.2 and Table 17.15 show that the Savitsky R_T/Δ values for LCG/l_p = 40% over the range $Fr_\nabla = 1.5–3.0$ are about 5% to 15% lower than the

Table 17.15. *Planing craft resistance data*

FrV	Fr	Vk	WUMTIA			Series 62 (LCG/lp = 40%)				Series 62 (LCG/lp = 32%)				Savitsky 40%		Savitsky 32%	
			P_E	R_T	R_T/Δ	R_T/Δ	Trim	$S\nabla^{2/3}$	L_W/L_P	R_T/Δ	Trim	$S\nabla^{2/3}$	L_W/L_P	R_T/Δ	Trim	R_T/Δ	Trim
0.50	0.22	5.72	11.4	3.9	0.0088	–	–	–	–	–	–	–	–	0.062	3.4	0.102	5.78
0.75	0.33	8.58	48.1	10.9	0.0245	–	–	–	–	–	–	–	–	0.068	3.7	0.111	6.24
1.00	0.44	11.44	148.0	25.1	0.0564	0.067	1.4	6.61	0.99	0.082	3.2	6.80	0.86	0.075	3.9	0.124	6.87
1.25	0.55	14.30	318.7	42.3	0.0951	0.107	3.3	6.41	0.92	0.126	5.1	6.57	0.82	0.084	4.3	0.140	7.68
1.50	0.66	17.16	451.5	51.2	0.1151	0.116	3.7	6.42	0.87	0.133	5.5	6.51	0.79	0.095	4.7	0.158	8.60
1.75	0.77	20.02	575.5	55.9	0.1257	0.125	4.3	6.21	0.84	0.140	6.3	5.95	0.74	0.100	5.2	0.173	9.39
2.00	0.89	22.88	703.1	59.7	0.1342	0.138	5.0	6.02	0.78	0.154	7.2	5.51	0.66	0.118	5.7	0.182	9.79
2.25	1.00	25.74	852.2	64.3	0.1446	(0.142)	–	–	–	(0.157)	–	–	–	0.128	6.0	0.184	9.75
2.50	1.11	28.60	996.5	67.7	0.1522	0.146	5.5	4.93	0.67	0.159	7.0	3.99	0.52	0.134	6.2	0.181	9.38
2.75	1.22	31.47	1167.8	72.2	0.1623	(0.148)	–	–	–	(0.155)	–	–	–	0.138	6.2	0.175	8.84
3.00	1.33	34.33	–	–	–	0.150	4.8	4.60	0.61	0.152	5.6	3.48	0.44	0.131	6.1	0.168	8.23
3.50	1.55	40.05	–	–	–	0.129	4.7	4.21	0.57	0.147	5.2	2.68	0.41	0.139	5.5	0.156	7.1

Numbers in parentheses are estimated values.

Series 62 40% data, and the trim values are higher. For the Savitsky $LCG/l_p = 32\%$ case, it is seen in Table 17.15 that the R_T/Δ and trim values are significantly higher than the Series 62 32% values.

The above calculations show that the R_T/Δ and trim values are very sensitive to the position of LCG. Whilst the position of the LCG forms part of the basic ship design process regarding the balance of masses and moments, it is clear that, for planing craft, hydrodynamic performance will have an important influence on its choice.

17.2.16 Example Application 16. 25 m Planing Craft, 35 knots, Single-Chine Hull: Estimate of Effective Power

The example uses resistance regression data for Series 62, the WUMTIA regression for hard chine forms and the Savitsky equations. The displacement of the example craft is larger than the 100,000 lb for the Series 62 basis displacement, hence the Series 62 prediction requires a skin friction correction (see Chapter 10).

The dimensions are as follows: $L_{WL} = 25$ m $\times$ $B = 7.0$ m $\times$ $T = 1.9$ m $\times$ $C_B = 0.380$

$$\nabla = 126.4\,\text{m}^3, \quad \Delta = 129.6\,\text{tonnes} = 1271.0\,\text{kN}\,[\approx 285{,}743\,\text{lb}]$$
$$L/B = 3.57, \quad L/\nabla^{1/3} = 4.99.$$

The assumed deadrise angle β is $20°$. Speed is 35 knots, $Fr = 1.15$, $Fr_\nabla = 2.57$

For the Series 62 regression, $l_p = 25 \times 1.02 = 25.5$ m, $B_{px} = B = 7.0$ m, and $B_{pa} = B_{px}/1.21 = 5.79$ m.

$$l_p/B_{pa} = 25.5/5.79 = 4.40.$$

$$A_p = l_p \times B_{pa} = 25.5 \times 5.79 = 147.6\,\text{m}^2 \quad \text{and} \quad A_p/\nabla^{2/3} = 5.85.$$

The Series 62 regression was run for $A_p/\nabla^{2/3} = 5.85$, $l_p/B_{pa} = 4.40$ and deadrise $\beta = 20°$. LCG was assumed as follows: $LCG/l_p = 44\%$. The results are shown in Table 17.16.

$$R_T = R/\Delta \times \Delta = 0.1528 \times 1271.0 = 194.21\,\text{kN}$$

The skin friction correction is as follows:

$$S = 4.805 \times \nabla^{2/3} = 4.805 \times 126.4^{2/3} = 121.22\,\text{m}^2 \text{ and } L_W/L_p = 0.717.$$
$$\lambda = 3\sqrt{\frac{129.6}{45.3}} = 1.419.$$
$$L_{pnew} = L_{WL} \times 1.02 = 25 \times 1.02 = 25.5\,\text{m}.$$

Table 17.16. *Series 62 regression data*

Fr_∇	R_T/Δ	Trim (deg.)	$S/\nabla^{2/3}$	Lw/Lp
2.5	0.1524	4.93	4.774	0.721
3.0	0.1549	4.30	4.995	0.690
2.57	0.1528	4.81	4.805	0.717

Table 17.17. *Results from WUMTIA*
regression, planing craft

Fr_∇	P_E
2.5	3600.1
2.75	4123.6
2.57	3746.7

$$L_{pbasis} = L_{pnew}/1.419 = 25.5/1.419 = 17.97 \, \text{m}.$$

$$L_{basis} = (L_W/L_p) \times L_{pbasis} = 0.717 \times 17.97 = 12.88 \, \text{m}.$$

$$L_{new} = (L_W/L_p) \times L_{pnew} = 0.717 \times 25.5 = 18.28 \, \text{m}.$$

$$Re_{basis} = VL/\nu = 35 \times 0.5144 \times 12.88/1.19 \times 10^{-6} = 1.949 \times 10^8.$$

$$C_{Fbasis} = 0.075/(\log Re - 2)^2 = 0.075/(8.290 - 2)^2 = 0.001896.$$

$$Re_{new} = VL/\nu = 35 \times 0.5144 \times 18.28/1.19 \times 10^{-6} = 2.766 \times 10^8.$$

$$C_{Fnew} = 0.075/(\log Re - 2)^2 = 0.075/(8.442 - 2)^2 = 0.001807.$$

$$
\begin{aligned}
\text{Skin friction correction} \quad &= (C_{Fbasis} - C_{Fnew}) \times \tfrac{1}{2}\rho \times S \times V^2/1000 \\
&= (0.001896 - 0.001807) \times \tfrac{1}{2} \times 1025 \times 121.22 \\
&\quad \times (35 \times 0.5144)^2/1000 \\
&= 1.79 \, \text{kN}.
\end{aligned}
$$

$$\text{Uncorrected, } R_T = 194.21 \, \text{kN}.$$

$$\text{Corrected, } R_T = 194.21 - 1.79 = 192.42 \, \text{kN}.$$

$$P_E = R_T \times Vs = 192.42 \times 35 \times 0.5144 = 3464.3 \, \text{kW}.$$

Using the WUMTIA regression for 35 knots, $Fr_\nabla = 2.57$, $Fr = 1.15$. The output from the regression is shown in Table 17.17. $P_E = 3746.7 \, \text{kW}$. This is reduced by 3% (see Chapter 10), hence, $P_E = 3746.7 \times 0.97 = 3634.3 \, \text{kW}$ (5% higher than Series 62).

Using the Savitsky equations,

For $LCG/l_p = 44\%$, $\beta = 20°$.
$T = 153.4 \, \text{kN}$ and $\tau = 5.20°$.
$P_E = R_T \times Vs = 153.4 \times \cos 5.2 \times 35 \times 0.5144 = 2750.4 \, \text{kW}$.
(21% less than Series 62).

It is interesting to note the sensitivity of the Savitsky prediction to change in LCG position. If LCG/l_p is moved to 40%, then P_E increases to 3133.3 kW, and when moved to 32%, P_E increases to 4069.4 kW (Series 62 increases to 3561 kW).

It can be noted that the effects of appendages, propulsion forces and air drag can also be incorporated in the resistance estimating procedure, Hadler [17.2].

17.2.17 Example Application 17. 10 m Yacht: Estimate of Performance

The example uses the hull resistance regression data for the Delft yacht series, Chapter 10, and the sail force data described in Chapter 16.

The yacht has the following particulars:

Hull particulars	Rig/sail particulars
$L_{WL} = 10$ m, $B_{WL} = 3.0$ m	Mainsail luff length (P) = 12.4 m
$T_C = 0.50$ m, $T = 1.63$ m	Foot of mainsail (E) = 4.25 m
$\nabla_C = 6.90$ m^3, $\nabla_K = 0.26$ m^3	Height of fore triangle (I) = 13.85 m
$\nabla_T = 7.16$ m^3 (7340 kg)	Base of fore triangle (J) = 4.0 m
$C_M = 0.750$, $C_P = 0.55$	Boom height above sheer (BAD) = 2.00 m
LCB (aft FP) = 5.36 m, LCF (aft FP) = 5.65 m	Mast height above sheer (EHM) = 13.85 m
Waterplane area = 21.0 m^2	Effective mast diameter (EMDC) = 0.15 m
Keel wetted surface area = 3.4 m^2	Average freeboard (FA) = 1.5 m
Keel chord = 1.5 m	Mainsail area = 29.0 m^2
Keel $VCB = 1.0$ m (below hull)	Genoa (jib) area = 40.2 m^2
Rudder wetted surface area = 1.3 m^2	
Rudder chord = 0.5 m	
GM (upright) = 1.1 m	

An estimate is required of the yacht speed V_S for given wind angle γ and true wind strength V_T. The Offshore Racing Congress (ORC) rig model is used, as described in Chapter 16.

The wind velocity vector diagram is shown in Figure 17.3 and the procedure follows that in the velocity prediction program (VPP) flow chart, Figure 17.4.

17.2.17.1 Sail Force

Take the case of wind angle $\gamma = 60°$ and $V_T = 10$ knots ($= 5.144$ m/s). Estimate the likely boat speed, V_S, say $V_S = 6.0$ knots ($= 3.086$ m/s), then

$$\beta = \tan^{-1}\left(\frac{\sin\gamma}{\cos\gamma + \dfrac{V_S}{V_T}}\right) = \tan^{-1}(0.787) = 38.22°$$

and

$$V_{App} = \frac{\sin\gamma}{\sin\beta}V_T = 7.20 \text{ m/s.}$$

The apparent wind strength and direction need to be found in the plane of the heeled yacht, accounting for heel effects on the aerodynamic forces. A heel angle thus needs to be estimated initially. Say heel angle $\phi = 15°$. Resolving velocities in the heeled plane as follows:

$$V_{Ae} = \sqrt{V_1^2 + V_2^2} \quad \text{and} \quad \beta_{Ae} = \cos^{-1}\left(\frac{V_1}{V_{Ae}}\right),$$

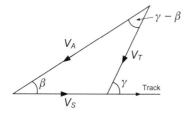

Figure 17.3. Velocity vector diagram.

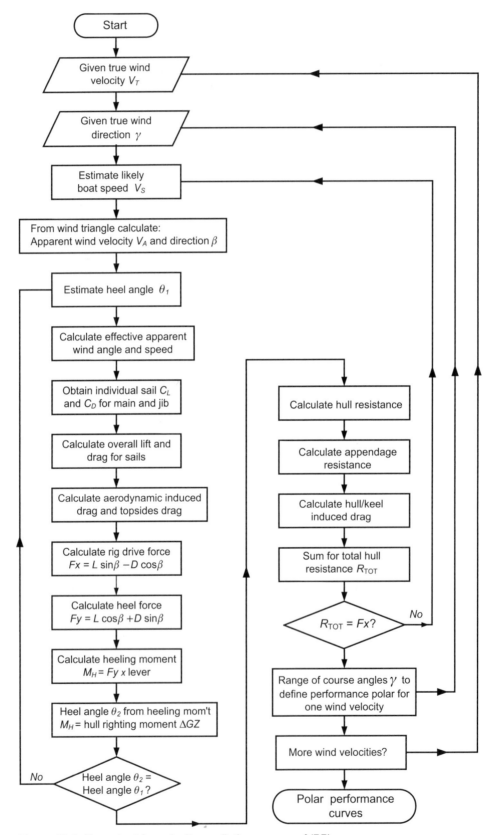

Figure 17.4. Flow chart for velocity prediction program (VPP).

where V_{Ae} is the effective apparent wind speed and β_{Ae} is the effective apparent wind angle.

$$V_1 = V_S + V_T \cos\gamma \quad \text{and} \quad V_2 \approx V_T \sin\gamma \cos\gamma.$$

Hence, $V_1 = 5.658$ m/s and $V_2 = 4.303$ m/s. Then, $V_{Ae} = 7.108$ m/s and $\beta_{Ae} = 37.25°$. For these V_{Ae} and β_{Ae}, calculate the sail forces. For $\beta_{Ae} = 37.25°$, using the ORC rig model, Tables 16.8 and 16.9 or Figures 16.27 and 16.28 give the following:

$$C_{Lm} = 1.3519, \quad C_{Dm} = 0.0454.$$

$$C_{Lj} = 1.4594, \quad C_{Dj} = 0.1083.$$

Given an area of mainsail $A_m = 29.0$ m^2 and area of genoa (jib) $A_j = 40.2$ m^2, then the reference sail area $A_n = A_m + \frac{1}{2} IJ = 56.70$ m^2.

$$C_L = \frac{C_{Lm}A_m + C_{Lj}A_j}{A_n} = 1.7261.$$

$$C_{Dp} = \frac{C_{Dm}A_m + C_{Dj}A_j}{A_n} = 0.100.$$

For β close to the wind, the aspect ratio of the rig is given by Equation (16.68) as follows:

$$AR = \frac{(1.1(EHM + FA))^2}{A_n} = 5.028.$$

$$C_{DI} = C_L^2 \left(\frac{1}{\pi AR} + 0.005\right) = 0.2035.$$

Drag of mast and hull above water using Equation (16.66) are as follows:

$$C_{D0} = 1.13\frac{(FA \cdot B) + (EHM \cdot EMDC)}{A_n} = 0.1311$$
$$C_D = C_{Dp} + C_{DI} + C_{D0} = 0.4346.$$

Sail lift $L = \frac{1}{2}\rho_a A_n V_{Ae}^2 C_L = \frac{1}{2} \times 1.23 \times 56.7 \times 7.108^2 \times 1.7261 = 3041.4$ N.

Sail drag $D = \frac{1}{2}\rho a A_n V_{Ae}^2 C_D = \frac{1}{2} \times 1.23 \times 56.7 \times 7.108^2 \times 0.4346 = 765.72$ N.

This acts at a centre of effort given by weighting individual sail centres of effort by area and a partial force contribution.

Partial force coefficient (main)

$$F_m = \frac{\sqrt{C_{Lm}^2 + C_{Dm}^2}}{\sqrt{C_{Lm}^2 + C_{Dm}^2} + \sqrt{C_{Lj}^2 + C_{Dj}^2}} = 0.4803.$$

Partial force coefficient (jib),

$$F_j = \frac{\sqrt{C_{Lj}^2 + C_{Dj}^2}}{\sqrt{C_{Lm}^2 + C_{Dm}^2} + \sqrt{C_{Lj}^2 + C_{Dj}^2}} = 0.5197.$$

$$C_{Em} = 0.39 \times P + BAD = 0.39 \times 12.4 + 2.00 = 6.836 \text{ m}.$$

$$C_{Ej} = 0.39 \times I = 0.39 \times 13.85 = 5.402 \text{ m}.$$

Combined $C_E = (29.0 \times 6.836 \times 0.4803 + 40.2$

$\times 5.402 \times 0.5197)/(29.0 \times 0.4803 + 40.2 \times 0.5197)$

$= 5.975\,\text{m (above sheer line)}.$

Force in direction of yacht motion is $Fx = L\sin\beta - D\cos\beta = 1279.8\,\text{N}$. Force perpendicular to yacht motion is $Fy = L\cos\beta + D\sin\beta = 2863.3\,\text{N}$.

The sail heeling moment $= Fy \times (C_E + C_{HE} + FA) = 2863.3 \times (5.975 + 1.5 + 1.5) = 25.699\,\text{kNm}$ (about waterline), where C_{HE} is the centre of hydrodynamic side force (in this case, 1.5 m below the waterline).

Sail heeling moment $=$ hull righting moment $= \Delta\,GZ = \Delta\,GM\sin\phi$ (approximate, or use $GZ - \phi$ curve if available), i.e. $25.699 \times 1000 = 7340 \times 9.81 \times 1.1 \times \sin\phi$ and $\phi' = 18.94°$, which can be compared with the assumed $\phi = 15°$. Further iterations would yield $\phi = 18.63°$.

With $\phi = 18.63°$, $Fx = 1267.5\,\text{N}$ and $Fy = 2818.7\,\text{N}$. $Fx = 1267.5\,\text{N}$ is the component of sail drive that has to balance the total hull resistance, R_{Total}.

17.2.17.2 Hull Resistance

Estimate of hull resistance using the Delft series at a speed of 3.086 m/s and $\phi = 18.63°$. $Fr = V/\sqrt{gL} = 0.312$ and from Equation (10.94), upright wetted surface area, $S = 23.72\,\text{m}^2$.

Using Equation (10.68) for R_{Total},

$$R_{\text{Total}} = R_{Fh} + R_{Rh} + R_{VK} + R_{VR} + R_{RK} + \Delta R_{Rh} + \Delta R_{RK} + R_{\text{Ind}}.$$

Using Equation (10.72), and interpolating between calculated R_{Rh} at different Fr, $R_{Rh} = 205.40\,\text{N}$.

Using Equation (10.95), heeled wetted surface area $= 22.62\,\text{m}^2$.

$$Re = VL/\nu = 3.086 \times 7.0/1.19 \times 10^{-6} = 1.815 \times 10^7, \text{ using } L = 0.7 L_{WL}.$$

$$C_F\,(\text{ITTC}) = 2.712 \times 10^{-3} \quad \text{and } R_{Fh} = 299.36\,\text{N}.$$

Using Equation (10.73), change in hull residuary resistance with heel of $\phi = 18.63°$, $\Delta R_{Rh} = 8.99\,\text{N}$ and $R_{\text{Hull}} = 205.4 + 299.36 + 8.99 = 513.75\,\text{N}$.

17.2.17.3 Appendage Resistance

From Equation (10.77), form factor keel $= (1 + k)_K = 1.252$, assuming $t/c = 0.12$, and form factor rudder $= (1 + k)_R = 1.252$, assuming $t/c = 0.12$.

$$Re_K = 3.890 \times 10^6, C_{FK} = 3.560 \times 10^{-3} \text{ and } R_{FK} = 59.076\,\text{N}.$$

$$Re_R = 1.297 \times 10^6, C_{FR} = 4.434 \times 10^{-3} \text{ and } R_{FR} = 28.13\,\text{N}.$$

Then, $R_{VK} = 73.989\,\text{N}$ and $R_{VR} = 35.23\,\text{N}$.

Using Equation (10.78), and interpolating between calculated R_{RK} at different Fr, $R_{RK} = 10.705\,\text{N}$.

Change in keel residuary resistance with heel $\phi = 18.63°$, $\Delta R_{RK} = 63.21\,\text{N}$.
Total $R_{RK} = 10.705 + 63.21 = 73.92\,\text{N}$
Hull and keel induced resistance, R_{Ind} is calculated as follows:
Calculation of effective draught, T_E, using Equation (10.82), $T_E = 1.748\,\text{m}$.

Table 17.18. *Balance of sail and hull forces*

V_s, m/s	Fx (calculated), N	R_{Total} (calculated), N
3.086	1267.5	866.5
3.50	1298.3	1318.4
3.45	1294.7	1238.8
3.487	1297.4	1296.9

Using Equation (10.81),

$$R_{Ind} = \frac{Fh^2}{\pi\,T_E^2 \dfrac{1}{2}\rho V^2} = 169.58\,\text{N}.$$

where Fh is the required sail heeling force $= Fy = 2818.7$ N, hence, at speed $V_S = 3.086$ m/s, total resistance $R_{Total} = 513.75 + 73.99 + 35.23 + 73.92 + 169.58 = 866.47$ N, but the sail drive force $Fx = 1267.5$ N; hence, the yacht would travel faster than 3.086 m/s.

The process is repeated until equilibrium (sail force Fx = hull resistance R_{Total}) is reached, as shown in Table 17.18. Balance occurs at an interpolated $V_S = 3.49$ m/s = 6.8 knots.

The process can be repeated for different course angles γ to give a complete performance polar for the yacht. Examples of the performance polars for this yacht are shown in Figure 17.5.

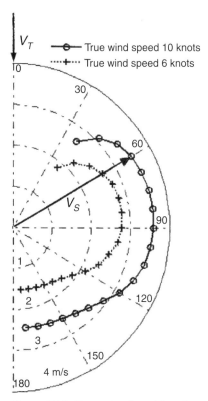

Figure 17.5. Example of yacht performance polars.

Table 17.19. *Data for tanker*

Vs, knots	P_E, kW
14	3055
15	4070
16	5830

17.2.18 Example Application 18. Tanker: Propeller Off-Design Calculations

The propeller has been designed for the loaded condition and it is required to estimate the performance in the ballast condition. This involves estimating the delivered power and propeller revolutions in the ballast condition at 14 knots and the speed attainable and corresponding propeller revolutions with a delivered power of 5300 kW. It is also required to estimate the maximum speed attainable if the propeller revolutions are not to exceed 108 rpm, a typical requirement for a ballast trials estimate where the revolutions are limited by the engine, see Figures 13.2 and 13.3.

The preliminary propeller design, based on the loaded condition, resulted in a propeller with a diameter $D = 5.8$ m and a pitch ratio $P/D = 0.80$. The effective power in the ballast condition is given in Table 17.19.

The hull interaction factors in the ballast condition are estimated to be $w_T = 0.41$, $t = 0.24$ and $\eta_R = 1.0$ (see Chapter 8 for estimates of w_T and t at fractional draughts).

It is convenient (for illustrative purposes) to use the $K_T - K_Q$ data for the Wageningen propeller B4.40 for $P/D = 0.80$, given in Table 16.1(a). In this case, the following relationships are suitable:

$$K_T = 0.320 \left[1 - \left(\frac{J}{0.90} \right)^{1.3} \right].$$

$$K_Q = 0.036 \left[1 - \left(\frac{J}{0.98} \right)^{1.6} \right].$$

17.2.18.1 Power and rpm at 14 knots

$$\text{Required } T = R/(1 - t) = (P_E/Vs)/(1 - t) = (3055/14 \times 0.5144)/(1 - 0.24)$$
$$= 558.17 \text{ kN.}$$
$$J = Va/nD = 14(1 - 0.41) \times 0.5144/(n \times 5.8) = 0.7326/n.$$

Table 17.20. *Assumed range of rps, tanker at 14 knots*

n, rps	J	K_{Topen}	T'
1.55	0.4726	0.1815	505.80
1.65	0.4440	0.1923	607.27
Check 1.60	0.4579	0.1871	555.58

Table 17.21. *Assumed range of rps, tanker at 15 knots*

n, rps	J	K_{Topen}	T'
1.7	0.4617	0.1856	622.17
1.8	0.4361	0.1952	733.60
Check 1.765	0.4447	0.1920	693.80

Thrust produced by the propeller is as follows:

$$T' = \rho n^2 D^4 \times K_T = 1025 \times 5.8^4 \times n^2 \times K_T = 1159.94\, n^2 K_T \text{ kN}.$$

Using graphical or linear interpolation, $T' = T$ when $n = 1.60$ rps ($N = 96.0$ rpm), Table 17.20.

At $J = 0.4579$,

$K_Q = 0.0253$.

$$P_D = 2\pi n Q_O/\eta_R = 2\pi n K_Q \rho n^2 D^5/\eta_R$$
$$= 2\pi \times 1.60 \times 0.0253 \times 1025 \times 1.60^2 \times 5.8^5/(1.0 \times 1000) = 4380.51 \text{ kW}.$$

17.2.18.2 Speed and Revolutions for $P_D = 5300$ kW

Consider 15 knots

$$\text{Required } T = R/(1-t) = (P_E/Vs)/(1-t)$$
$$= (4070/15 \times 0.5144)/(1 - 0.24) = 694.05 \text{ kN}.$$
$$J = Va/nD = 15(1 - 0.41) \times 0.5144/(n \times 5.8) = 0.7849/n.$$

Thrust produced by the propeller is as follows:

$$T' = \rho n^2 D^4 \times K_T = 1025 \times 5.8^4 \times n^2 \times K_T/1000 = 1159.94\, n^2 K_T \text{ kN}.$$

Using graphical or linear interpolation, $T' = T$ when $n = 1.765$ rps ($N = 105.90$ rpm), Table 17.21.

At $J = 0.4447$,

$K_Q = 0.0258$.

$$P_D = 2\pi n Q_O/\eta_R = 2\pi n K_Q \rho n^2 D^5/\eta_R$$
$$= 2\pi \times 1.765 \times 0.0258 \times 1025 \times 1.765^2 \times 5.8^5/(1.0 \times 1000) = 5996.50 \text{ kW}.$$

Hence, at 14 knots, $P_D = 4380.51$ kW and at 15 knots, $P_D = 5996.50$ and speed to absorb 5300 kW $= 14 + 1.0 \times (5300 - 4380.51)/(5996.50 - 4380.51) = 14.57$ knots and revs $N = 96.0 + (105.9) - 96.0) \times (5300 - 4380.51)/(5996.50 - 4380.51) = 101.6$ rpm.

17.2.18.3 Speed for 108 rpm (1.8 rps)

At 16 knots,

$$\text{Required } T = R/(1-t) = (P_E/Vs)/(1-t) = (5830/16 \times 0.5144)/(1 - 0.24)$$
$$= 932.0 \text{ kN}.$$
$$J = Va/nD = Vs(1 - 0.41) \times 0.5144/(1.8 \times 5.8) = 0.02907\, Vs.$$

Table 17.22 *Assumed range of Vs*

Vs, knots	J	K_{Topen}	T'	Required T
15	0.4361	0.1952	733.60	694.05
16	0.4651	0.1843	692.64	932.00

Thrust produced by the propeller is as follows:

$$T' = \rho n^2 D^4 \times K_T = 1025 \times 1.8^2 \times 5.8^4 \times K_T/1000 = 3758.21 K_T \text{ kN.}$$

Using graphical or linear interpolation, $T' = T$ when $Vs = 15.14$ knots, Table 17.22. (An alternative approach is to calculate the revolutions at 16 knots and interpolate between 15 and 16 knots to obtain the speed when $N = 108$).

17.2.19 Example Application 19. Twin-Screw Ocean-Going Tug: Propeller Off-Design Calculations

The propellers have been designed for the free-running condition at 14 knots and it is required to estimate the power, rpm and available tow rope pull when towing at 6 knots and for the bollard pull ($J = 0$) condition. The propeller torque in the towing and bollard pull conditions is not to exceed that in the free-running condition. Typical estimation of the effective power for an ocean-going tug is described in Example 13.

The preliminary design of the fixed-pitch propellers, based on the free-running condition at 14 knots, resulted in propellers with a diameter $D = 4.0$ m and a pitch ratio $P/D = 1.0$. The effective power and wake estimates for free-running at 14 knots and towing at 6 knots are given in Table 17.23.

It is convenient (for illustrative purposes) to use the $K_T - K_Q$ data for the Wageningen propeller B4.70 for $P/D = 1.0$ given in Table 16.1(b). In this case, the following relationships are suitable:

$$K_T = 0.455 \left[1 - \left(\frac{J}{1.06} \right)^{1.20} \right] \qquad K_Q = 0.0675 \left[1 - \left(\frac{J}{1.12} \right)^{1.29} \right].$$

17.2.19.1 Free-Running at 14 knots

$$\text{Required } T = R/(1 - t) = (P_E/Vs)/(1 - t) = (1650/14 \times 0.5144)/(1 - 0.22)$$
$$= 293.74 \text{ kN per prop}$$
$$J = Va/nD = 14(1 - 0.21) \times 0.5144/(n \times 4.0) = 1.422/n.$$

Table 17.23. *Effective power and wake estimates*

Vs, knots	P_E, kW	P_E (kW) per prop	w_T	t	η_R
14	3300	1650	0.21	0.22	1.02
6	480	240	0.21	0.13	1.02

Table 17.24. *Assumed range of rps, tug at 14 knots*

n, rps	J	K_{Topen}	T'
2.2	0.6464	0.2037	258.70
2.4	0.5925	0.2286	345.51
Check 2.283	0.6229	0.2146	293.50

Thrust produced by the propeller is as follows:

$$T' = \rho n^2 D^4 \times K_T = 1025 \times 4.0^4 \times n^2 \times K_T/1000 = 262.40 \, n^2 K_T, \text{kN}.$$

Using graphical or linear interpolation, $T' = T$ when $n = 2.283$ rps ($N = 137.0$ rpm), Table 17.24.

At $J = 0.6229$,

$$K_Q = 0.0358$$
$$P_D = 2\pi n Q_O/\eta_R = 2\pi n K_Q \rho n^2 D^5/\eta_R$$
$$= 2\pi \times 2.283 \times 0.0358 \times 1025 \times 2.283^2 \times 4.0^5/1.02 \times 1000$$
$$= 2754.3 \text{ kW per prop.}$$

Total $P_D = 2 \times 2754.3 = 5508.6$ kW.

Torque (maximum) $Q_O = K_Q \rho n^2 D^5 = 0.0358 \times 1025 \times 2.283^2 \times 4.0^5/1000$
$$= 195.85 \text{ kNm.}$$

17.2.19.2 Towing at 6 knots

$$J = Va/nD = 6(1 - 0.21) \times 0.5144/(n \times 4.0) = 0.6096/n.$$

Torque produced by the propeller is as follows:

$$Q_O' = K_Q \rho n^2 D^5 = 1049.6 n^2 K_Q.$$

Assume a range of rps until propeller torque is at the limit of 195.85 kNm.

Using graphical or linear interpolation, $Q_O' = Q_O$ when $n = 1.864$ rps ($N = 111.8$ rpm), Table 17.25.

$$P_D = 2\pi n Q_O/\eta_R = 2\pi n K_Q \rho n^2 D^5/\eta_R$$
$$= 2\pi \times 1.864 \times 0.05371 \times 1025 \times 1.864^2 \times 4.0^5/1.02 \times 1000$$
$$= 2249.0 \text{ kW per prop.}$$

(or, for constant Q, $P_D \propto n$ and $P_D = 2754.3 \times 1.864/2.283 = 224.8$ kW).

Total $P_D = 2 \times 2249.0 \times 4498$ kW.

Table 17.25. *Assumed range of rps, tug at 6 knots*

n, rps	J	K_{Qopen}	Q_O'
1.8	0.3387	0.05307	180.48
1.9	0.3208	0.05405	204.80
Check 1.864	0.3270	0.05371	195.87

17.2.19.3 Available Tow Rope Pull at 6 Knots

$$n = 1.864 \text{ rps}, \quad J = 0.3270 \quad \text{and} \quad K_T = 0.3441.$$

Thrust produced by one propeller is as follows:

$$T' = \rho n^2 D^4 \times K_T$$
$$= 1025 \times 1.864^2 \times 4.0^4 \times 0.3441/1000 = 313.72 \text{ kN}.$$

Allowing for thrust deduction, the effective thrust per prop, $T_E = T (1 - t) =$ 313.72 $(1 - 0.13) = 272.9$ kN and allowing for hull resistance, the hull resistance $R = P_E/Vs = 480 / (6 \times 0.5144) = 155.52$ kN.

Available tow rope pull $= T_E - R = (272.9 \times 2) - 155.52 = 390.3$ kN (38.5 tonnes).

17.2.19.4 Bollard Pull (J = 0)

Maximum torque is 195.85 kNm. From Table 16.1(b), at $J = 0$, $K_{TO} = 0.455$ and $K_{QO} = 0.0675$.

$$\text{Maximum torque} = Q_O = K_{QO} \times \rho n^2 D^5 = 0.0675 \times 1025 \, n^2 \, 4.0^5$$
$$= 195.85 \text{ kNm per prop,}$$

whence $n = 1.663$ rps $= 99.8$ rpm. Thrust (or bollard pull) is as follows:

$$T' = K_{TO} \times \rho n^2 D^4$$
$$= 0.455 \times 1025 \times 1.663^2 \times 4.0^4 = 330.19 \text{ kN per prop.}$$

$$\text{Total bollard pull} = 2 \times 330.19 = 660.4 \text{ kN} (67.3 \text{ tonnes}).$$

$$\text{Total delivered power} \quad P_D = 2 \times (2\pi n Q)/\eta_R = 2 \times 2006.3 = 4012.6 \text{ kW}.$$

It is seen that, with the torque limitation, the delivered power at 2006.3 kW per propeller is much less than that available in the free-running condition at 2754.3 kW per propeller. This problem, which is frequently encountered with dual-role craft such as tugs, is often overcome by using a controllable pitch propeller. For example, if for the bollard condition the pitch is reduced to $P/D = 0.73$. (see also Figure 13.4), then from Table 16.1(b), $K_{TO} = 0.316$ and $K_{QO} = 0.0362$.

With $P/D = 0.73$, then at the torque limit, the rps rise to 2.27 (136.2 rpm), P_D rises to 2738.6 kW per propeller (close to the free-running case) and the total bollard pull increases to $2 \times 427.3 = 854.6$ kN (87.1 tonnes).

A two-speed gearbox may also be considered, if available for this power range. If a fixed propeller is to be used, then the propeller design point may be taken closer to the towing condition, depending on the proportions of time the tug may spend free-running or towing.

The relevant discussion of propeller–engine matching and the need for torque limits is contained in Chapter 13, Section 13.2.

17.2.20 Example Application 20. Ship Speed Trials: Correction for Natural Wind

Assume a head wind and natural wind velocity gradient. The correction is based on the BSRA recommendation, including a velocity gradient allowance, Section 5.3.5.

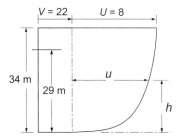

Figure 17.6. Wind velocity gradient.

The measured mile trial on a passenger ship was conducted in a head wind. The recorded delivered power was $P_D = 19200$ kW at 22 knots. It is required to correct this power to an equivalent still air value. Further information is as follows:

Relative wind speed at 34 m above sea level: 30 knots.
Hull and superstructure width: 30 m.
Superstructure height above sea level: 29 m.

It is estimated that $\eta_D = 0.700$. The drag coefficient for hull and superstructure C_D may be taken as 0.80, Section 3.2.2. The natural wind velocity may be assumed to be proportional to $h^{1/6}$, where h is the height above sea level.

Relative velocity $V_R = V + U = 30$ knots, Figure 17.6
Ship speed $V = 22$ knots $= 11.33$ m/s.
$U = 8$ knots $= 4.12$ m/s (with a natural wind gradient)

$$\frac{u}{U} = \left(\frac{h}{H}\right)^{1/6} = \frac{1}{34^{1/6}}h^{1/6} \quad \text{and} \quad u = \frac{4.12}{34^{1/6}}h^{1/6} = 2.289h^{1/6}.$$

Wind resistance due to *relative* wind velocity:

$$R_W = 0.5\rho BC_D \int_0^{29} (V+u)^2 dh$$

$$= \frac{1.23}{2} \times 30 \times 0.80 \int_0^{29} (11.33 + 2.289h^{1/6})^2 dh$$

$$= 14.76 \int_0^{29} (128.369 + 51.869h^{1/6} + 5.240h^{2/6}) dh$$

$$= 14.76 \left[128.369h + 51.869h^{7/6} \times \frac{6}{7} + 5.240h^{8/6} \times \frac{6}{8} \right]_0^{29}$$

$$= 93.48 \text{ kN}$$

$$= \text{total deduction for vacuum.}$$

Addition for still air $= \frac{1}{2} \rho A_T C_D V^2 = (1.23/2) \times [30 \times 29] \times 0.80 \times$
  $11.33^2 /1000 = + 54.95$ kN.
Hence net deduction, leading to still air $= -93.48 + 54.95 = -38.53$ kN
Net power deduction $= R \times Vs/\eta_D = 38.53 \times 22 \times 0.5144/0.700 = -622.91$ kW
  (3.2% of recorded power).

Corrected delivered power $P_D = 19200 - 622.91 = 18577.1$ kW.

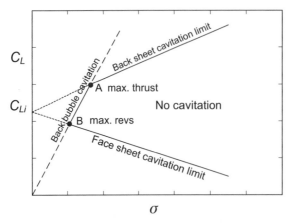

Figure 17.7. Cavitation limits.

17.2.21 Example Application 21. Detailed Cavitation Check on Propeller Blade Section

The propeller of a fast ferry is to operate at an advance speed Va of 37 knots. An elemental blade section of the propeller under consideration has a radius of 0.35 m and a minimum immersion below the water surface of 1.0 m.

Details of the section are as follows:

Thickness/chord ratio $t/c = 0.030$.
Nose radius/chord ratio $r/c = 0.00045$.
Ideal lift coefficient $C_{Li} = 0.104$.

Cavitation limits using Equations (12.10) and (12.11) are as follows:

Back bubble cavitation limit $\sigma = \frac{2}{3}C_L + \frac{5}{2}(t/c)$.
Sheet cavitation limits $\sigma = \frac{0.06(C_L - C_{Li})^2}{(r/c)}$.
Atmospheric pressure $= 101 \times 10^3$ N/m^2.
Vapour pressure for water $= 3.0 \times 10^3$ N/m^2.

17.2.21.1 Point of Maximum Thrust

Along the back bubble limit, the cavitation number σ varies slowly and blade forces increase as C_L increases. Along the back sheet limit C_L varies slowly with $V_R^2 \propto 1/\sigma$ and blade forces increase as σ decreases. For a given craft speed and propeller immersion, σ falls as rpm increase; hence the maximum thrust is at A in Figure 17.7.

17.2.21.2 Cavitation Number at Maximum Thrust Condition, and Maximum Lift for Cavitation-Free Operation

$$\text{Sheet cavitation limit} \quad \sigma = \frac{0.06(C_L - C_{Li})^2}{(r/c)},$$

Table 17.26 *Assumed range of σ (sheet)*

σ	C_{Lmax}	C_{Lmin}
0.1	0.1284	0.07961
0.2	0.1427	0.06527
0.3	0.2652	0.05710

with $r/c = 0.00045$ and $C_{Li} = 0.104$.

$$\text{Sheet cavitation limits } |C_L - 0.104| = \sqrt{\frac{0.00045 \times \sigma}{0.06}} = 0.0866\sqrt{\sigma}.$$

$$\text{Back bubble limit } \sigma = \frac{2}{3}C_L + \frac{5}{2}(t/c) = \frac{2}{3}C_L + 0.0750.$$

From diagram or numerical evaluation, $\sigma = 0.1 + 0.1(0.1284 - 0.0375)/[(0.1284 - 0.0375) + (0.1875 - 0.1427)] = 0.167$, Tables 17.26 and 17.27.

Giving maximum thrust point at $\sigma = 0.167$ and $C_L = 0.138$.

17.2.21.3 Maximum Propeller Revolutions at the Maximum Thrust Condition

At 1.0 m immersion,

$$\sigma = (\rho g h + P_{AT} - P_V)/\tfrac{1}{2}\rho V_R^2$$

$$0.167 = [(1025 \times 9.81 \times 1.0) + 101 \times 10^3 - 3 \times 10^3]/\tfrac{1}{2} \times 1025 \times V_R^2$$

and $V_R^2 = 1262.57 (\text{m/s})^2$.

$$Va = 37 \text{ knots} = 37 \times 0.5144 = 19.033 \text{ m/s}.$$

From Figure 17.8,

$$V_R^2 = Va^2 + (2\pi nr)^2$$

$$1262.57 = 19.033^2 + (2\pi \times 0.35)^2 \times n^2$$

and $n = 13.64$ rps, $N = 818.7$ rpm.

17.2.22 Example Application 22. Estimate of Propeller Blade Root Stresses

Details of the propeller are as follows:

Four blades, $D = 4.7$ m, $P/D = 1.1$, BAR $= 0.60$, $N = 120$ rpm ($n = 2$ rps), $Va = 14$ knots.

Table 17.27. *Assumed range of σ (bubble)*

σ	C_L
0.1	0.0375
0.2	0.1875

Figure 17.8. Velocity vectors.

$K_T = 0.190$, $K_Q = 0.035$, Blade rake $\mu = 0.40$ m aft at tip, and linear to zero at the centreline. Density ρ of propeller material (manganese bronze) $= 8300$ kg/m^3.

At 0.2R, for four blades, from Table 12.4, $t/D = 0.036$ and $t = 0.036 \times 4.7 = 0.169$ m.

From Equation (12.35), $(c/D)_{0.2R} = 0.416 \times 0.60 \times 4/4 = 0.250$ and chord length $c = 0.250 \times 4.7 = 1.18$ m $(t/c = 0.169/1.18 = 0.143 = 14.3\%)$. The blade section pitch at 0.2R $\theta = \tan^{-1}((P/D)/\pi x) = \tan^{-1}(1.1/\pi \times 0.2) = 60.27°$.

$$\Omega = 2\pi n = 2\pi \, 120/60 = 12.57 \text{ rads/sec.}$$

Assume that $\bar{r}/R = 0.68$, hence, $\bar{r} = 0.68 \times 4.7/2 = 1.60$ m. From Equation (12.32), $I/y = 0.095ct^2 = 0.095 \times 1.18 \times 0.169^2 = 0.00320$ m^3.

From Equation (12.34), at 0.2R, area $A = 0.70ct = 0.70 \times 1.18 \times 0.169 = 0.139$ m^2 and assume that the area varies linearly to zero at the tip.

$$T = K_T \times n^2 \times D^4$$
$$= 0.190 \times 1025 \times (120/60)^2 \times 4.7^4/1000 = 380.1 \text{ kN (for four blades).}$$
$$Q = K_Q \times n^2 \times D^5$$
$$= 0.035 \times 1025 \times (120/60)^2 \times 4.7^5/1000 = 329.1 \text{ kNm (for four blades).}$$

Using Equation (12.22), $(M_T) = T_{0.2R}(\bar{r} - r_{0.2R}) = (380.1/4) \times (1.60 - 0.47)/1000 = 0.1074$ MNm

Using Equation (12.23), $(M_Q) = Q_{0.2R}(1 - r/\bar{r}) = (329.1/4) \times (1 - 0.47/1.60)/1000 = 0.0581$ MNm.

From Equation (12.42),

$$m(r) = \frac{M}{0.32R}\left(1 - \frac{r}{R}\right)$$

and from Equation (12.41), mass $M = 0.139 \times 8300 \times 0.8 \times 2.35/2 = 1084.5$ kg.

From Equation (12.43), $Z'(r) = 0.40\,(r/R - 0.2)$.

Substitute for $m(r)$ and $Z'(r)$ in Equation (12.29) as follows:

$$M_{R_{0.2}} = \int_{0.2R}^{R} m(r) \cdot r \cdot \Omega^2 Z'(r) \, dr = \int_{0.2R}^{R} \frac{\Omega^2 \, m\,(1 - r/R)}{0.32R} \cdot \mu\left(\frac{r}{R} - 0.2\right) r\, dr$$

$$= \frac{\mu M \Omega^2}{0.32R} \int_{0.2R}^{R}\left(1 - \frac{r}{R}\right)\left(\frac{r}{R} - 0.2\right) r\, dr = \frac{\mu M \Omega^2}{0.32R} \int_{0.2R}^{R}\left(1.2\frac{r^2}{R} - \frac{r^3}{R^2} - 0.2r\right) dr$$

$$= \frac{\mu M \Omega^2}{0.32R}\left[1.2\frac{r^3}{3R} - \frac{r^4}{4R^2} - \frac{0.2r^2}{2}\right]_{0.2R}^{R} = 0.160\,\mu M R \Omega^2,$$

i.e. with the above assumptions for mass, mass distribution and rake,

$$M_{R_{0.2}} = 0.160 \mu\, M R \Omega^2$$
$$= 0.160 \times 0.40 \times 1084.5 \times 2.35 \times 12.57^2/1000^2 = 0.02577 \text{ MNm}.$$

Using Equation (12.31),

$$M_N = (M_T + M_R)\cos\theta + M_Q \sin\theta$$
$$= (0.1074 + 0.02577)\cos 60.27 + 0.0581 \sin 60.27 = 0.1165 \text{ MNm}.$$

Substitute for $m(r)$ in Equation (12.30) for F_C, as follows:

$$F_C = \int_{0.2R}^{R} m(r)\cdot r \cdot \Omega^2.dr = \Omega^2 \int_{0.2R}^{R} \frac{M}{0.32R}\left(1 - \frac{r}{R}\right) r\, dr = \frac{M\Omega^2}{0.32R}\int\left(r - \frac{r^2}{R}\right) dr$$
$$= \frac{M\Omega^2}{0.32R}\left[\frac{r^2}{2} - \frac{r^3}{3R}\right]_{0.2R}^{R} = 0.4667\, R M \Omega^2,$$

i.e. with the above assumption for mass distribution,

$$F_C = 0.4777\, R M \Omega^2 = 0.4667 \times 2.35 \times 1084.5 \times 12.57^2/1000^2 = 0.1879 \text{ MN}.$$

Using Equation (12.36), root stress $\sigma = \text{direct stress} + \text{bending stress} = F_C/A + M_N/I/y = 0.1879/0.139 + 0.1165/0.0032 = 1.35 + 36.41 = 37.76 \text{ MN/m}^2$. This is a little below the recommended allowable design stress for manganese bronze in Table 12.3.

17.2.23 Example Application 23. Propeller Performance Estimates Using Blade Element–Momentum Theory

The propeller has four blades and is to operate at $J = 0.58$. At a radius fraction $x = 0.60$ the section chord ratio $c/D = 0.41$, pitch ratio $P/D = 0.80$, design angle of attack $\alpha = 0.80°$, Goldstein factor $\kappa = 0.96$ (Figure 15.4) and drag coefficient $C_D = 0.008$.

(i) Calculate dK_T/dx and C_L for this section. Given the data in Table 17.28, already calculated for other radii, estimate the overall K_T for the propeller.
(i) Use the Ludweig–Ginzel method to estimate the camber required for the section at $x = 0.60$. $k_1 \cdot k_2$ is estimated to be 1.58 (Figure 15.14). Assume $dC_L/dx = 0.10$ per degree and $C_L = 12 \times m/c$ at $\alpha = 0°$.
(ii) Assuming the $x = 0.60$ section to be representative of the whole propeller, calculate the approximate loss in propeller efficiency if the drag coefficient is increased by 20% due to roughness and fouling.

Table 17.28. *Data at other radii*

x	0.20	0.40	0.80	1.0
dK_T/dx	0.031	0.114	0.264	0

Table 17.29. *Integration of dK_T/dx*

x	0.2	0.4	0.6	0.8	1.0
dK_T/dx	0.031	0.114	0.2121	0.264	0
SM	1	4	2	4	1

(i) The calculations follow part of the blade element-momentum theory flow chart, Figure 15.7, as follows:

$\tan(\phi + \alpha) = ((P/D)/\pi x) = 0.8/\pi \times 0.60 = 0.4244.$
$(\phi + \alpha) = 23.0°, \quad \alpha = 0.80°$ (given/assumed)
and $\quad \phi = 22.20°$ and $\quad \tan \phi = 0.408.$
$\tan \psi = J/\pi x = 0.58/\pi \times 0.60 = 0.3077.$
Ideal efficiency $\quad \eta_i = \tan \psi / \tan \phi = 0.3077/0.408 = 0.754.$
Inflow factor $\quad a = (1 - \eta_i)/[\eta_i + \tan^2 \psi/\eta].$

For the first iteration, assume that the blade drag is zero i.e $\gamma = 0$ and $\eta = \eta_i$, then $a = (1 - 0.754)/[0.754 + 0.3077^2/0.754] = 0.2797 \qquad \{0.2736\}$
$\tan \psi = 0.408, \quad \lambda_i = x \tan \phi = 0.245$ and $1/\lambda_i = 4.08$, from Figure 15.4(b), for four blades, $K = 0.96$,

$$\frac{dK_T}{dx} = \pi J^2 x Ka(1 + a) = 0.2179 \quad \{0.2121\}$$

also

$$\frac{dK_T}{dx} = \frac{\pi^2}{4} \left(\frac{Zc}{D} \right) C_L x^2 (1 - a')^2 \sec \phi (1 - \tan \phi \tan \gamma).$$

For the first iteration, assuming no drag, $\gamma = 0$ and $(1 - \tan \phi \tan \gamma) = 1.0.$

$$(1 - a') = \eta_i (1 + a) = 0.9649 \qquad \{0.9603\}.$$
$$Zc/D = 4 \times 0.41 = 1.64.$$
$$C_L = \frac{dK_T}{dx} / \frac{\pi^2}{4} \frac{Zc}{D} x^2 (1 - a')^2 \sec \phi (1 - \tan \phi \tan \gamma) = 0.1488 \quad \{0.1494\}.$$
$$\tan \gamma = C_D/C_L = 0.0080/0.1488 = 0.05376 \qquad \{0.05355\}.$$
$$\gamma = 3.077° \qquad \{3.065°\}.$$
$$\text{Efficiency including drag } \eta = \frac{\tan \psi}{\tan(\phi + \gamma)} = 0.652 \qquad \{0.652\}.$$

Return to new estimate of inflow factor a using the new estimate of efficiency. The updated values from the second iteration are shown in {braces}. Two cycles are adequate for this particular case resulting in $\eta = 0.652$ and $dK_T/dx = 0.2121$. The numerical integration (Simpson's first rule) for estimating total K_T is shown in Table 17.29, where SM is the Simpson multiplier.

$$\text{Total overall } K_T = (0.2/3) \times 1.967 = 0.1311.$$

(ii) Two-dimensional camber at $\alpha = 0°$ is as follows: $m_0/c = C_L/12 = 0.1494/12 = 0.01245$ and required section camber at $\alpha = 0° = k_1 \cdot k_2 \, m_0/c = 1.58 \times 0.01245 = 0.01967$

Table 17.30. *Distribution of axial wake*

x	0.2	0.4	0.6	0.8	1.0
w_T	0.45	0.34	0.27	0.24	0.22

$dC_L/dx = 0.10$ and at $\alpha = 0.80°$, $C_L = 0.10 \times 0.8 = 0.08$ and equivalent camber reduction $\delta(m/c) = 0.08/12 = 0.0067$; hence, camber required for operation at $\alpha = 0.80°$ is $m/c = 0.01967 - 0.0067 = 0.01297$.

(iii) Assuming the drag coefficient C_D is increased by 20%, the new $C_D = 0.008 \times 1.2 = 0.0096$.

$$\tan \gamma = C_D/C_L = 0.0096/0.1494 = 0.06426 \quad \text{and} \quad \gamma = 3.677°$$
$$\phi = 22.20°$$
$$\text{New overall } \eta = \frac{\tan \psi}{\tan(\phi + \gamma)} = 0.3077/0.4851 = 0.634$$

and approximate loss in efficiency is $((0.652 - 0.634)/0.652) \times 100 = 2.8\%$.

17.2.24 Example Application 24. Wake-Adapted Propeller

The radial distribution of axial wake for a particular ship is given in Table 17.30. The propeller is designed to operate at an advance coefficient $Js = 1.1$ based on ship speed. The average ideal efficiency is estimated to be $\eta_i = 0.82$. The blade sections are to operate at an angle of attack $\alpha = 0.5°$.

The Lerbs criterion is used to calculate the optimum radial distribution of efficiency and hence the optimum radial distribution of pitch ratio for the propeller.

The Lerbs criterion assumes $\eta \propto \sqrt{1 - w_T}$,

$$Ja = Js(1 - w_T) = 1.1(1 - w_T).$$
$$\tan \psi = \frac{Ja}{\pi x}, \quad \tan \phi = \frac{\tan \psi}{\eta_i}, \quad \tan(\phi + \alpha) = \frac{P/D}{\pi x} \quad \text{and} \quad \frac{P}{D} = \pi x \tan(\phi + \alpha).$$
$$\eta_i = 0.82 \quad \text{and} \quad \text{Optimum } \eta_i = 0.82 \times \frac{\sqrt{1 - w_T}}{0.8326}.$$

The results are shown in Table 17.31.

Table 17.31. *Distribution of pitch ratio*

x	w_T	Ja	$\tan \psi$	$\sqrt{1 - w_T}$	η_i opt	$\tan \phi$	$\tan(\phi + \alpha)$	P/D
0.2	0.45	0.605	0.9629	0.742	0.731	1.3172	1.3413	0.843
0.4	0.34	0.726	0.5777	0.812	0.800	0.7221	0.7355	0.924
0.6	0.27	0.803	0.4260	0.854	0.841	0.5065	0.5175	0.975
0.8	0.24	0.836	0.3326	0.872	0.859	0.3872	0.3973	0.998
1.0	0.22	0.858	0.2731	0.883	0.870	0.3139	0.3235	1.016
				$\Sigma/5 = 0.8326$	$\Sigma/5 = 0.82$			

It is noted that the radial distribution of a and a' could, if required, be obtained as follows:

$$a = \frac{1 - \eta_i}{\eta_i + \dfrac{1}{\eta}\tan^2\psi} \qquad a' = \frac{a}{\eta}\tan^2\psi,$$

with $\eta = \eta_i$ if viscous losses are neglected.

REFERENCES (CHAPTER 17)

17.1 Molland, A.F. and Turnock, S.R. *Marine Rudders and Control Surfaces.* Butterworth-Heinemann, Oxford, UK, 2007.

17.2 Hadler, J.B. The prediction of power performance of planing craft. *Transactions of the Society of Naval Architects and Marine Engineers.* Vol. 74, 1966, pp. 563–610.

Background Physics

A1.1 Background

This appendix provides a background to basic fluid flow patterns, terminology and definitions, together with the basic laws governing fluid flow. The depth of description is intended to provide the background necessary to understand the basic fluid flows relating to ship resistance and propulsion. Some topics have been taken, with permission, from Molland and Turnock [A1.1]. Other topics, such as skin friction drag, effects of surface roughness, pressure drag and cavitation are included within the main body of the text. Descriptions of fluid mechanics to a greater depth can be found in standard texts such as Massey and Ward-Smith [A1.2] and Duncan *et al.* [A1.3].

A1.2 Basic Fluid Properties and Flow

Fluid Properties

From an engineering perspective, it is sufficient to consider a fluid to be a continuous medium which will deform continuously to take up the shape of its container, being incapable of remaining in a fixed shape of its own accord.

Fluids are of two kinds: *liquids*, which are only slightly compressible and which naturally occupy a fixed volume in the lowest available space within a container, and *gases*, which are easily compressed and expand to fill the whole space available within a container.

For flows at low speeds it is frequently unnecessary to distinguish between these two types of fluid as the changes of pressure within the fluid are not large enough to cause a significant density change, even within a gas.

As with a solid material, the material within the fluid is in a state of stress involving two kinds of stress component:

(i) Direct stress: Direct stresses act normal to the surface of an element of material and the local stress is defined as the normal force per unit area of surface. In a fluid at rest or in motion, the average direct stress acting over a small element of fluid is called the fluid *pressure* acting at that point in the fluid.

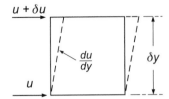

Figure A1.1. Shear stress.

(ii) Shear stress: Shear stresses act tangentially to the surface of an element of material and the local shear stress is defined as the tangential force per unit area of surface. In a fluid at rest there are no shear stresses. In a solid material the shear stress is a function of the shear strain. In a fluid in motion, the shear stress is a function of the rate at which shear strain is occurring, Figure A1.1, that is of the velocity gradient within the flow.

For most engineering fluids the relation is a linear one:

$$\tau = \mu \left(\frac{\partial u}{\partial y} \right),$$ (A1.1)

where τ is shear stress and μ is a constant for that fluid.

Fluids that generate a shear stress due to shear flow are said to be viscous and the viscosity of the fluid is measured by μ, the coefficient of viscosity (or coefficient of *dynamic* viscosity) or $v = \frac{\mu}{\rho}$, the coefficient of *kinematic* viscosity, where ρ is the fluid mass density. The most common fluids, for example air and water, are only slightly viscous.

Values of density and kinematic viscosity for fresh water (FW), salt water (SW) and air, suitable for practical engineering design applications are given in Tables A1.1 and A1.2.

Steady Flow

In steady flow the various parameters such as velocity, pressure and density *at any point* in the flow do not change with time. In practice, this tends to be the exception rather than the rule. Velocity and pressure may vary from *point to point*.

Uniform Flow

If the various parameters such as velocity, pressure and density do not change *from point to point* over a specified region, at a particular instant, then the flow is said to be uniform over that region. For example, in a constant section pipe (and neglecting

Table A1.1. *Density of fresh water, salt water and air*

Temperature, °C		10	15	20
Density kg/m^3	FW	1000	1000	998
	SW	1025	1025	1025
[Pressure = 1 atm]	Air	1.26	1.23	1.21

Table A1.2. *Viscosity of fresh water, salt water and air*

Temperature, °C		10	15	20
Kinematic viscosity m²/s	$FW \times 10^6$	1.30	1.14	1.00
	$SW \times 10^6$	1.35	1.19	1.05
[Pressure = 1 atm]	$Air \times 10^5$	1.42	1.46	1.50

the region close to the walls) the flow is steady and uniform. In a tapering pipe, the flow is steady and non-uniform. If the flow is accelerating in the constant section pipe, then the flow will be non-steady and uniform, and if the flow is accelerating in the tapering pipe, then it will be non-steady and non-uniform.

Streamline

A streamline is an imaginary curve in the fluid across which, at that instant, no fluid is flowing. At that instant, the velocity of every particle on the streamline is in a direction tangential to the line, for example line a–a in Figure A1.2. This gives a good indication of the flow, but only with steady flow is the pattern unchanging. The pattern should therefore be considered as instantaneous. Boundaries are always streamlines as there is no flow across them. If an indicator, such as a dye, is injected into the fluid, then in steady flow the streamlines can be identified. A bundle of streamlines is termed a streamtube.

A1.3 Continuity of Flow

Continuity exists on the basis that what flows in must flow out. For example, consider the flow between (1) and (2) in Figure A1.3, in a streamtube (bundle of streamlines).

For no flow through the walls and a constant flow rate, then for continuity,

$$\text{Mass flow rate} = \rho_1 A_1 V_1 = \rho_2 A_2 V_2 \text{ kg/s},$$

and if the fluid is incompressible, $\rho_1 = \rho_2$ and $A_1 V_1 = A_2 V_2 =$ volume flow rate m³/s. If Q is the volume rate, then

$$Q = A_1 V_1 = A_2 V_2 = \text{constant.} \tag{A1.2}$$

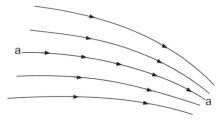

Figure A1.2. Streamlines.

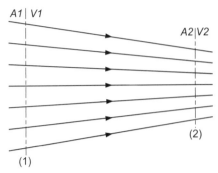

Figure A1.3. Continuity of flow.

A1.4 Forces Due to Fluids in Motion

Forces occur on fluids due to accelerations in the flow. Applying Newton's Second Law:

$$\text{Force} = \text{mass} \times \text{acceleration}$$

or

$$\text{Force} = \text{Rate of change of momentum.}$$

A typical application is a propeller where thrust (T) is produced by accelerating the fluid from velocity from V_1 to V_2, and

$$T = \dot{m}(V_2 - V_1), \tag{A1.3}$$

where $\dot{m}$ is the mass flow rate.

A1.5 Pressure and Velocity Changes in a Moving Fluid

The changes are described by Bernoulli's equation as follows:

$$\frac{P}{\rho g} + \frac{u^2}{2g} + z = H = \text{constant (units of } m\text{),} \tag{A1.4}$$

which is strictly valid when the flow is frictionless, termed inviscid, steady and of constant density. H represents the total head, or total energy and, under these conditions, is constant for any one fluid particle throughout its motion along any one streamline. In Equation (A1.4), $P/\rho g$ represents the pressure head, $u^2/2g$ represents the velocity head (kinetic energy) and z represents the position or potential head (energy) due to gravity. An alternative presentation of Bernoulli's equation in terms of pressure is as follows:

$$P + \frac{1}{2}\rho u^2 + \rho g z = P_T = \text{constant (units of pressure, N/m}^2\text{),} \tag{A1.5}$$

where P_T is total pressure.

Figure A1.4. Pressure and velocity changes.

As an example, consider the flow between two points on a streamline, Figure A1.4, then,

$$P_0 + \frac{1}{2}\rho u_0^2 + \rho g z_0 = P_L + \frac{1}{2}\rho u_L^2 + \rho g z_L, \tag{A1.6}$$

where P_0 and u_0 are in the undisturbed flow upstream and P_L and u_L are local to the body.

Similarly, from Figure A1.4,

$$P_0 + \frac{1}{2}\rho u_0^2 + \rho g z_0 = P_S + \frac{1}{2}\rho u_S^2 + \rho g z_S. \tag{A1.7}$$

In the case of air, its density is small relative to other quantities. Hence, the $\rho g z$ term becomes small and is often neglected.

Bernoulli's equation is strictly applicable to inviscid fluids. It can also be noted that whilst, in reality, frictionless or inviscid fluids do not exist, it is a useful assumption that is often made in the description of fluid flows, in particular, in the field of computational fluid dynamics (CFD). If, however, Bernoulli's equation is applied to real fluids (with viscosity) it does not necessarily lead to significant errors, since the influence of viscosity in steady flow is usually confined to the immediate vicinity of solid boundaries and wakes behind solid bodies. The remainder of the flow, well clear of a solid body and termed the outer flow, behaves effectively as if it were inviscid, even if it is not so. The outer flow is discussed in more detail in Section A1.6.

A1.6 Boundary Layer

Origins

When a slightly viscous fluid flows past a body, shear stresses are large only within a thin layer close to the body, called the boundary layer, and in the viscous wake formed by fluid within the boundary layer being swept downstream of the body, Figure A1.5. The boundary layer increases in thickness along the body length.

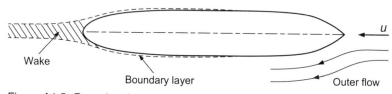

Figure A1.5. Boundary layer and outer flow.

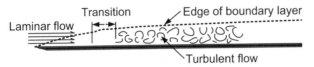

Figure A1.6. Boundary layer development.

Outer Flow

Outside the boundary layer, in the so-called outer flow in Figure A1.5, shear stresses are negligibly small and the fluid behaves as if it were totally inviscid, that is, non-viscous or frictionless. In an inviscid fluid, the fluid elements are moving under the influence of pressure alone. Consideration of a spherical element of fluid shows that such pressures act through the centre of the sphere to produce a net force causing a translation motion. There is, however, no mechanism for producing a moment that can change the angular momentum of the element. Consequently, the angular momentum remains constant for all time and if flow initially started from rest, the angular momentum of all fluid elements is zero for all time. Thus, the outer flow has no rotation and is termed irrotational.

Flow Within the Boundary Layer

Flows within a boundary layer are unstable and a flow that is smooth and steady at the forward end of the boundary layer will break up into a highly unsteady flow which can extend over most of the boundary layer.

Three regions can be distinguished, Figure A1.6, as follows:

1. Laminar flow region: In this region, the flow within the boundary layer is smooth, orderly and steady, or varies only slowly with time.
2. Transition region: In this region, the smooth flow breaks down.
3. Turbulent flow region: In this region the flow becomes erratic with a random motion and the boundary layer thickens. Within the turbulent region, the flow can be described by superimposing turbulence velocity components, having a zero mean averaged over a period of time, on top of a steady or slowly varying mean flow. The randomly distributed turbulence velocity components are typically ±20% of the mean velocity. The turbulent boundary layer also has a thin laminar sublayer close to the body surface. It should be noted that flow outside the turbulent boundary layer can still be smooth and steady and turbulent flow is not due to poor body streamlining as it can happen on a flat plate. Figure A1.7 shows typical velocity distributions for laminar and turbulent boundary layers. At the surface of the solid body, the fluid is at rest relative to the body. At the outer edge of the boundary layer, distance δ, the fluid effectively has the full free-stream velocity relative to the body.

The onset of the transition from laminar to turbulent flow will depend on the fluid velocity (v), the distance (l) it has travelled along the body and the fluid kinematic viscosity (ν). This is characterised by the Reynolds number (Re) of the flow, defined as:

$$Re = \frac{vl}{\nu}.$$

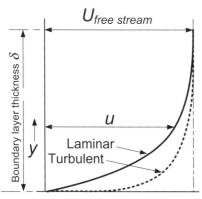

Figure A1.7. Boundary layer velocity profiles.

It is found that when Re exceeds about 0.5×10^6 then, even for a smooth body, the flow will become turbulent. At the same time, the surface finish of the body, for example, its level of roughness, will influence transition from laminar to turbulent flow.

Transition will also depend on the amount of turbulence already in the fluid through which the body travels. Due to the actions of ocean waves, currents, shallow water and other local disruptions, ships will be operating mainly in water with relatively high levels of turbulence. Consequently, their boundary layer will normally be turbulent.

Displacement Thickness

The boundary layer causes a reduction in flow, shown by the shaded area in Figure A1.8. The flow of an inviscid or frictionless fluid may be reduced by the same amount if the surface is displaced outwards by the distance δ^*, where δ^* is termed the displacement thickness. The displacement thickness δ^* may be employed to reduce the effective span and effective aspect ratio of a control surface whose root area is operating in a boundary layer. Similarly, in theoretical simulations of fluid flow with assumed inviscid flow, and hence no boundary layer present, the surface of the body may be displaced outwards by δ^* to produce a body shape equivalent to that with

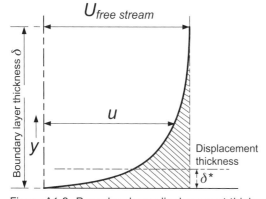

Figure A1.8. Boundary layer displacement thickness.

no boundary layer. Approximate estimates of displacement thickness may be made as follows.

Laminar Flow

$$\frac{\delta^*}{x} = 1.721 \, Re_x^{-1/2}. \tag{A1.8}$$

Turbulent Flow, using a 1/7 power law velocity distribution:

$$\frac{\delta^*}{x} = 0.0463 \, Re_x^{-1/5}. \tag{A1.9}$$

A1.7 Flow Separation

For flow along a flat surface, with constant pressure in the direction of flow, the boundary layer grows in thickness with distance, but the flow will not separate from the surface. If the pressure is falling in the direction of the flow, termed a favourable pressure gradient, then the flow is not likely to separate. If, however, the pressure is increasing along the direction of flow, known as an adverse pressure gradient, then there is a relative loss of speed within the boundary layer. This process can reduce the velocity in the inner layers of the boundary layer to zero at some point along the body length, such as point S, Figure A1.9. At such a point, the characteristic mean flow within the boundary layer changes dramatically and the boundary layer starts to become much thicker. The flow is reversed on the body surface, the main boundary layer detaches from the body surface and a series of large vortices or eddies form behind the separation point S. Separated flows are usually unsteady, the vortices periodically breaking away into the wake downstream.

It should be noted that separation can occur in a laminar boundary layer as well as a turbulent one and, indeed, is more likely to occur in the laminar case. Inspection of the boundary layer velocity profiles in Figure A1.7 indicates that the laminar layer has less momentum near the surface than the turbulent layer does and is thus likely to separate earlier, Figure A1.10. Thus, turbulent boundary layers are much more resistant to separation than laminar boundary layers. This leads to the result that drag due to separation is higher in laminar flow than in turbulent flow. This also explains why golf balls with dimples that promote turbulent flow have less drag and travel further than the original smooth golf balls. It is also worth noting that a thick wake following separation should not be confused with the thickening

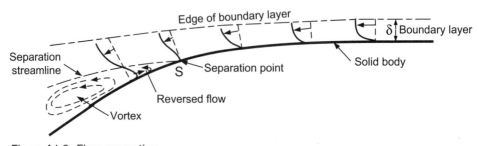

Figure A1.9. Flow separation.

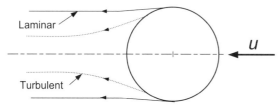

Figure A1.10. Separation drag.

of the boundary layer following transition from laminar to turbulent flow, described earlier.

A1.8 Wave Properties

Winds create natural waves on the oceans and ships create waves during their passage through water. A wave system is created by the passage of a disturbance, not the bulk movement of water. For example, a small floating object will simply rise and fall with the passage of a wave beneath it. The water particles move in orbital paths which are approximately circular. These orbital paths decrease exponentially with increasing depth, Figure A1.11. The orbital motion is generally not of concern for large displacement ships. There may be some influence on the wake of twin-screw ships, depending on whether the propellers are near a crest or trough when the wake will be increased or decreased by the orbital motion. Hydrofoil craft may experience a change in effective inflow velocity to the foils. Smaller craft may experience problems with control with, for example, the effects of the orbital motion of a following sea.

The wave contour is given by a trochoid function, which is a path traced out by a point on the radius of a rolling circle, Figure A1.11. The theory of the trochoid can be found in [A1.4] and standard texts such as [A1.5] and [A1.6]. The trochoid is generally applied to ship hydrostatic calculations for the hull in a longitudinal wave. Other applications tend to use a sine wave which can include the orbital motion of

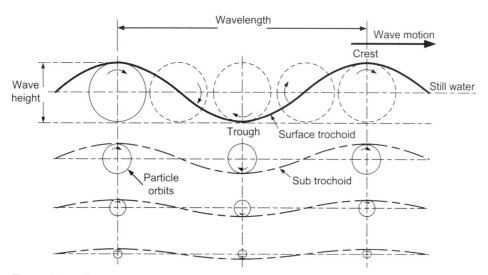

Figure A1.11. Deep water wave.

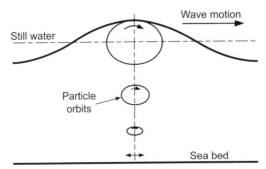

Figure A1.12. Shallow water wave.

the water particles and is easier to manipulate mathematically. The actual difference between a trochoid and a sine wave is small [A1.6].

Wave Speed

Wave theory [A1.7] yields the wave velocity c as follows:

$$c = \left[g \frac{\lambda}{2\pi} . \tanh \left(\frac{2\pi h}{\lambda} \right) \right]^{1/2}, \tag{A1.10}$$

where h is the water depth from the still water level and λ is the wavelength, crest to crest.

Deep Water
When h/λ is large,

$$\left(\tanh \frac{2\pi h}{\lambda} \right) \rightarrow 1.0$$

and

$$c^2 = \frac{g\lambda}{2\pi} \quad \text{or} \quad c = \sqrt{\frac{g\lambda}{2\pi}}. \tag{A1.11}$$

This deep water relationship is suitable for approximately $\frac{h}{\lambda} \geq \frac{1}{2}$. Also, $c = \frac{\lambda}{T_W}$, where T_W is the wave period.

If ω is the angular velocity of the orbital motion, then $\omega = \frac{2\pi}{T_W}$ rads/sec and wave frequency is as follows:

$$f_W = \frac{1}{T_W} = \frac{\omega}{2\pi}. \tag{A1.12}$$

Shallow Water
With a ground effect, the particle paths take up an elliptical motion, Figure A1.12. When h/λ is small,

$$\left(\tanh \frac{2\pi h}{\lambda} \right) \rightarrow \frac{2\pi h}{\lambda}$$

and

$$c^2 = g\,h \quad \text{or} \quad c = \sqrt{g\,h}. \tag{A1.13}$$

The velocity now depends only on the water depth and waves of different wavelength propagate at the same speed. This shallow water relationship is suitable for approximately $\frac{h}{\lambda} \leq \frac{1}{20} \cdot c = \sqrt{gh}$ is known as the critical speed.

REFERENCES (APPENDIX A1)

A1.1 Molland, A.F. and Turnock, S.R. *Marine Rudders and Control Surfaces.* Butterworth-Heinemann, Oxford, UK, 2007.

A1.2 Massey, B.S. and Ward-Smith J. *Mechanics of Fluids.* 8th Edition. Taylor and Francis, London, 2006.

A1.3 Duncan, W.J., Thom, A.S. and Young, A.D. *Mechanics of Fluids.* Edward Arnold, Port Melbourne, Australia, 1974.

A1.4 Froude, W. On the rolling of ships. *Transactions of the Royal Institution of Naval Architects*, Vol. 2, 1861, pp. 180–229.

A1.5 Rossell, H.E. and Chapman, L.B. *Principles of Naval Architecture.* The Society of Naval Architects and Marine Engineers, New York, 1939.

A1.6 Molland, A.F. (ed.) *The Maritime Engineering Reference Book.* Butterworth-Heinemann, Oxford, 2008.

A1.7 Lamb, H. *Hydrodynamics.* Cambridge University Press, Cambridge, 1962.

APPENDIX A2

Derivation of Eggers Formula for Wave Resistance

A summary derivation of the Eggers formula for wave resistance was given in Chapter 7. The following provides a more detailed account of the derivation.

The Eggers series for the far field wave pattern was derived in Chapter 7 as:

$$\zeta = \sum_{n=0}^{\infty} \left[\xi_n \cos\left(x \gamma_n \cos\theta_n\right) + \eta_n \sin\left(x \gamma_n \cos\theta_n\right) \right] \cos\left(\frac{2\pi n y}{b}\right).$$

In terms of velocity potential the wave elevation is given by the following:

$$\zeta = -\frac{c}{g} \left.\frac{\partial \theta}{\delta x}\right|_{z=0}.$$

Using this result it can be shown that the velocity potential,

$$\phi = \frac{g}{c} \sum_{n=0}^{\infty} \frac{\cosh \gamma_n (z+h)}{\lambda_n \cosh (\gamma_n h)} \left[\eta_n \cos \lambda_n x - \xi_n \sin \lambda_n x \right] \cos \frac{2\pi n y}{b},$$

corresponds to the Eggers wave pattern, where $\lambda_n = \gamma_n \cos\theta_n$ (Note, in this instance, that λ_n is a constant, not wavelength).

It satisfies the Laplace equation $\frac{\partial^2 \phi}{\partial x^2} + \frac{\partial^2 \phi}{\partial y^2} + \frac{\partial^2 \phi}{\partial z^2} = 0$ as required and the boundary conditions $\frac{\partial \phi}{\partial z} = 0$ on $z = -h$ and $\frac{\partial \phi}{\partial y} = 0$ on $y = \pm\frac{b}{2}$, i.e. condition of no flow through tank walls.

From the momentum analysis of the flow around a hull, see Chapter 3, Equation (3.10), the wave pattern resistance is:

$$R_w = \left\{ \frac{1}{2}\rho g \int_{-b/2}^{b/2} \zeta_B^2 dy + \frac{1}{2}\rho \int_{-b/2}^{b/2} \int_{-h}^{0} \left(v^2 + w^2 - u^2\right) dz dy \right\}$$

since the wave elevation is assumed to be small.

Each term can be evaluated in this expression from the following:

$$u = \frac{\partial \phi}{\partial x} \qquad v = \frac{\partial \phi}{\partial y} \qquad w = \frac{\partial \phi}{\partial z}$$

e.g.

$$u = \frac{\partial \phi}{\partial x} = \frac{-g}{c} \sum_{n=0}^{\infty} \frac{\cosh \gamma_n (z+h)}{\cosh (\gamma_n h)} [\eta_n \sin \lambda_n x + \xi_n \cos \lambda_n x] \cos \frac{2\pi n y}{b}$$

$$u^2 \text{ leads to } \left(\cos \frac{2\pi n y}{b} \right)^2, \text{ hence to solve, } \text{let } n^2 = n \times m$$

now

$$\int_{-b/2}^{b/2} \cos \left(\frac{2\pi n y}{b} \right) \cos \left(\frac{2\pi m y}{b} \right) dy = \frac{1}{2} \int_{-b/2}^{b/2} \left[\cos \left(\frac{2\pi y (n+m)}{b} \right) \right.$$
$$\left. + \cos \left(\frac{2\pi y (n-m)}{b} \right) \right] dy$$
$$= \begin{cases} 0 & \text{for } n \neq m \\ \frac{b}{2} & \text{for } n = m \neq 0 \\ b & \text{for } n = m = 0. \end{cases}$$

Using this result,

$$\int_{-b/2}^{b/2} \left(\frac{\partial \phi}{\partial x} \right)^2 dy = \frac{g^2 b}{2c^2} \sum_{n=0}^{\infty} {}' \left(\frac{\cosh \gamma_n (z+h)}{\cosh (\gamma_n h)} \right)^2 [\eta_n \sin \lambda_n x + \xi_n \cos \lambda_n x]^2$$

(note: $\sum'$ denotes that the $n = 0$ term is doubled). As $[\cosh \gamma_n (z+h)]^2 = \frac{1}{2} [\cosh 2\gamma_n (z+h) + 1]$, we can substitute and then integrate to get the following:

$$\int_{-b/2}^{b/2} \int_{-h}^{0} \left(\frac{\partial \phi}{\partial x} \right)^2 dy = \frac{g^2 b}{4c^2} \sum_{n=0}^{\infty} {}' [\eta_n \sin \lambda_n x + \xi_n \cos \lambda_n x]^2 \frac{\left[z + \frac{\sinh 2\gamma_n (z+h)}{2\gamma_n} \right]_{z=-h}^{0}}{\cosh^2 (\gamma_n h)},$$

then, substituting the integration in the z direction, we get the following:

$$\int_{-b/2}^{b/2} \int_{-h}^{0} u^2 dy = \frac{g^2 b}{4c^2} \sum_{n=0}^{\infty} {}' [\eta_n \sin \lambda_n x + \xi_n \cos \lambda_n x]^2 \left(\frac{\sinh 2\gamma_n h + 2\gamma_n h}{2\gamma_n \cosh^2 \gamma_n h} \right).$$

From wave speed relation,

$$\gamma_n \cos^2 \theta_n = \frac{g}{c^2} \tanh \gamma_n h$$

$$\frac{g}{c^2} = \frac{\gamma_n \cos^2 \theta_n}{\tanh \gamma_n h}$$

$$\int_{-b/2}^{b/2} \int_{-h}^{0} u^2 dy = \frac{gb}{4} \sum_{n=0}^{\infty} {}' [\eta_n \sin \lambda_n x + \xi_n \cos \lambda_n x]^2 \cos^2 \theta_n \left(1 + \frac{2\gamma_n h}{\sinh 2\gamma_n h} \right).$$

A similar analysis can be carried through for the other terms in R_w and it is found that the total contribution to R_w for the nth term $(n \neq 0)$

finally becomes the following:

$$\delta R_w = \tfrac{1}{4}\rho g b \left\{ (\xi_n \cos \lambda_n x + \eta_n \sin \lambda_n x)^2 \right.$$

$$- \tfrac{1}{2} [\eta_n \sin \lambda_n x + \xi_n \cos \lambda_n x]^2 \cos^2 \theta_n \left(1 + \frac{2\gamma_n h}{\sinh 2\gamma_n h}\right)$$

$$+ \tfrac{1}{2} [\eta_n \cos \lambda_n x - \xi_n \sin \lambda_n x]^2 \left(1 - \frac{2\gamma_n h}{\sinh 2\gamma_n h}\right)$$

$$\left. + \tfrac{1}{2} [\eta_n \cos \lambda_n x - \xi_n \sin \lambda_n x]^2 \sin^2 \theta_n \left(1 + \frac{2\gamma_n h}{\sinh 2\gamma_n h}\right) \right\}.$$

On expanding the $(\ )^2$ terms and noting $\cos^2 + \sin^2 = 1$, this gives the following:

$$\delta R_w = \tfrac{1}{4}\rho g b \left(\xi_n^2 + \eta_n^2\right) \left\{ \left(1 - \tfrac{1}{2}\cos^2 \theta_n \left[1 + \frac{2\gamma_n h}{\sinh 2\gamma_n h}\right]\right) \right\}.$$

The term for $n = 0$ is special and the complete result, valid at all speeds is, as follows:

$$R_w = \tfrac{1}{4}\rho g b \left\{ \left(\xi_0^2 + \eta_0^2\right) \left(1 - \frac{2\gamma_0 h}{\sinh 2\gamma_0 h}\right) \right.$$

$$\left. + \sum_{n=1}^{\infty} \left(\xi_n^2 + \eta_n^2\right) \left[1 - \tfrac{1}{2}\cos^2 \theta_n \left(1 + \frac{2\gamma_n h}{\sinh 2\gamma_n h}\right)\right] \right\}, \qquad \text{(A2.1)}$$

i.e. Eggers formula for wave resistance.

For the special case of deep water, noting $\sinh \to \infty$, this simplifies to the following:

$$R_w = \tfrac{1}{4}\rho g b \left\{ \left(\xi_0^2 + \eta_0^2\right) + \sum_{n=1}^{\infty} \left(\xi_n^2 + \eta_n^2\right) \left[1 - \tfrac{1}{2}\cos^2 \theta_n\right] \right\}. \qquad \text{(A2.2)}$$

The nth term in this series represents the contribution δR_w (or $\delta C_w = \delta R_w / \tfrac{1}{2}\rho S V^2$) to the wave resistance due to the nth component of the wave pattern. It depends only on the square of the amplitude of the wave components and its phase $\tan^{-1} \eta_n \xi_n$ is not relevant. Hence, if the coefficients γ_n and θ_n have been determined, the wave resistance may readily be found once the coefficients ξ_n and η_n have been determined. As described in Chapter 7, ξ_n and η_n can be found by measuring the wave pattern elevation. Chapter 9 describes how ξ_n and η_n can be found theoretically.

APPENDIX A3

Tabulations of Resistance Design Data

Table A3.1 Isherwood wind resistance coefficients
Table A3.2 Sabit: BSRA resistance regression coefficients: load draught
Table A3.3 Sabit: BSRA resistance regression coefficients: medium draught
Table A3.4 Sabit: BSRA resistance regression coefficients: light draught
Table A3.5 Sabit: Series 60 resistance regression coefficients
Table A3.6 Moor and Small: average © values
Table A3.7 Hollenbach: resistance regression coefficients
Table A3.8 Taylor–Gertler: series resistance data $C_P = 0.50$
Table A3.9 Taylor–Gertler: series resistance data $C_P = 0.60$
Table A3.10 Taylor–Gertler: series resistance data $C_P = 0.70$
Table A3.11 Taylor–Gertler: series resistance data $C_P = 0.80$
Table A3.12 Zborowski: twin-screw series resistance data
Table A3.13 Yeh: Series 64 resistance data $C_P = 0.35$
Table A3.14 Yeh: Series 64 resistance data $C_P = 0.45$
Table A3.15 Yeh: Series 64 resistance data $C_P = 0.55$
Table A3.16 Bailey: NPL series resistance data
Table A3.17 Radojcic et al.: NTUA double-chine series resistance regression coefficients
Table A3.18 Radojcic et al.: NTUA double-chine series trim regression coefficients
Table A3.19 Radojcic: Series 62 planing craft: resistance regression coefficients
Table A3.20 Radojcic: Series 62 planing craft: trim regression coefficients
Table A3.21 Radojcic: Series 62 planing craft: wetted area regression coefficients
Table A3.22 Radojcic: Series 62 planing craft: wetted length regression coefficients
Table A3.23 Oortsmerssen: small ships: resistance regression coefficients
Table A3.24 WUMTIA resistance regression coefficients for C-factor: round bilge
Table A3.25 WUMTIA resistance regression coefficients for C-factor: hard chine
Table A3.26 Molland et al.: Southampton catamaran series resistance data
Table A3.27 Müller-Graf and Zips: VWS catamaran series resistance regression coefficients

It should be noted that some data, such as the coefficients for the Holtrop and Mennen regressions, Savitsky's equations for planing craft and resistance data for yachts, are contained fully within the text in Chapter 10.

Table A3.1. *Isherwood wind resistance coefficients*

				Isherwood wind force coefficients			
γ_R (deg.)	A0	A1	A2	A3	A4	A5	A6
0	2.152	−5.00	0.243	−0.164	−	−	−
10	1.714	−3.33	0.145	−0.121	−	−	−
20	1.818	−3.97	0.211	−0.143	−	−	0.033
30	1.965	−4.81	0.243	−0.154	−	−	0.041
40	2.333	−5.99	0.247	−0.190	−	−	0.042
50	1.726	−6.54	0.189	−0.173	0.348	−	0.048
60	0.913	−4.68	−	−0.104	0.482	−	0.052
70	0.457	−2.88	−	−0.068	0.346	−	0.043
80	0.341	−0.91	−	−0.031	−	−	0.032
90	0.355	−	−	−	−0.247	−	0.018
100	0.601	−	−	−	−0.372	−	−0.020
110	0.651	1.29	−	−	−0.582	−	−0.031
120	0.564	2.54	−	−	−0.748	−	−0.024
130	−0.142	3.58	−	0.047	−0.700	−	−0.028
140	−0.677	3.64	−	0.069	−0.529	−	−0.032
150	−0.723	3.14	−	0.064	−0.475	−	−0.032
160	−2.148	2.56	−	0.081	−	1.27	−0.027
170	−2.707	3.97	−0.175	0.126	−	1.81	−
180	−2.529	3.76	−0.174	0.128	−	1.55	−

Table A3.2. *Sabit: BSRA resistance regression coefficients: load draught*

BSRA series regression coefficients				Load draught condition			
$V_K/\sqrt{L_f}$	0.50	0.55	0.60	0.65	0.70	0.75	0.80
a1	−0.7750	−0.7612	−0.7336	−0.6836	−0.5760	−0.3290	−0.0384
a2	0.2107	0.2223	0.2399	0.2765	0.3161	0.3562	0.4550
a3	0.0872	0.0911	0.0964	0.0995	0.1108	0.1134	0.0661
a4	0.0900	0.0768	0.0701	0.0856	0.1563	0.4449	1.0124
a5	0.0116	0.0354	0.0210	0.0496	0.2020	0.3557	0.2985
a6	0.0883	0.0842	0.0939	0.1270	0.1790	0.1272	0.0930
a7	0.0081	0.0151	0.0177	0.0175	0.0170	0.0066	0.0118
a8	0.0631	0.0644	0.0656	0.0957	0.1193	0.1415	0.5080
a9	0.0429	0.0650	0.1062	0.1463	0.1706	0.1238	0.2203
a10	−0.0249	−0.0187	−0.0270	−0.0502	−0.0699	−0.0051	−0.0514
a11	−0.0124	0.0292	0.0647	0.1629	0.3574	0.2882	0.2110
a12	0.0236	−0.0245	−0.0776	−0.1313	−0.3034	−0.2508	0.0486
a13	−0.0301	−0.0442	−0.0537	−0.0863	−0.0944	−0.0115	0.0046
a14	0.0877	0.1124	0.1151	0.1133	0.0839	−0.0156	−0.1433
a15	−0.1243	−0.1341	−0.0775	0.0355	0.1715	0.2569	0.2680
a16	−0.0269	−0.0006	0.1145	0.2255	0.2006	0.0138	0.2283

Table A3.3. *Sabit: BSRA resistance regression coefficients: medium draught*

BSRA series regression coefficients				Medium draught condition			
$V_K/\sqrt{L_f}$	0.50	0.55	0.60	0.65	0.70	0.75	0.80
b1	−0.6979	−0.6799	−0.6443	−0.5943	−0.4908	−0.2874	−0.0317
b2	0.2191	0.2297	0.2598	0.2870	0.3273	0.3701	0.4490
b3	0.1180	0.1231	0.1224	0.1381	0.1399	0.1389	0.1033
b4	0.1568	0.1510	0.2079	0.1913	0.2681	0.5093	0.9315
b5	−0.0948	−0.0736	−0.1715	−0.0707	0.0233	0.1098	0.1370
b6	0.0671	0.0647	0.0788	0.1072	0.1392	0.1079	0.0825
b7	−0.0030	0.0063	−0.0057	0.0171	0.0104	−0.0108	−0.0043
b8	0.0463	0.0359	0.0442	0.0691	0.0858	0.1353	0.3402
b9	0.0501	0.0688	0.1181	0.1431	0.1817	0.1973	0.2402
b10	−0.0169	−0.0186	−0.0153	−0.0332	−0.0466	−0.0154	−0.0481
b11	0.0094	0.0057	0.0061	0.0998	0.2114	0.1805	0.1123
b12	0.0255	0.0542	0.0759	0.0084	−0.0491	0.0308	0.3415
b13	0.0448	0.0306	0.0649	0.0041	0.0006	0.0485	0.0619
b14	−0.0468	−0.0163	−0.0996	−0.0180	−0.0467	−0.1125	−0.2319
b15	−0.1128	−0.1075	−0.0748	0.0127	0.1538	0.1922	0.1293
b16	−0.0494	−0.0198	0.0622	0.1492	0.1591	0.0687	−0.0222

Table A3.4. *Sabit: BSRA resistance regression coefficients: light draught*

BSRA series regression coefficients				Light draught condition			
$V_K/\sqrt{L_f}$	0.50	0.55	0.60	0.65	0.70	0.75	0.80
c1	−0.6318	−0.6068	−0.5584	−0.4868	−0.3846	−0.2050	0.0098
c2	0.2760	0.2876	0.3126	0.3476	0.3678	0.4012	0.4901
c3	0.1179	0.1242	0.1340	0.1356	0.1392	0.1566	0.1583
c4	0.1572	0.1636	0.1683	0.1797	0.2718	0.4848	0.8435
c5	−0.0918	−0.0662	−0.0387	0.0145	0.0722	0.1451	0.1736
c6	0.0231	0.0231	0.0498	0.0458	0.0480	0.0316	0.0232
c7	−0.0048	−0.0001	−0.0053	−0.0041	−0.0144	−0.0210	−0.0141
c8	0.0045	0.0129	0.0184	0.0118	0.0211	0.0728	0.2510
c9	0.0106	0.0262	0.0686	0.0849	0.1237	0.1174	0.1483
c10	0.0065	0.0010	−0.0089	−0.0089	−0.0176	0.0089	0.0100
c11	−0.0313	0.0066	0.0376	0.0513	0.0471	0.0212	0.0122
c12	0.1180	0.1091	0.1128	0.1692	0.2713	0.3626	0.5950
c13	0.0484	0.0359	0.0061	−0.0255	−0.0170	0.0312	0.0697
c14	−0.0770	−0.0507	−0.0173	0.0286	−0.0189	−0.0471	−0.1412
c15	−0.0479	−0.0568	−0.0407	0.0650	0.2150	0.2722	0.3212
c16	−0.0599	−0.0439	0.0088	0.0891	0.0918	0.0497	0.0637

Table A3.5. *Sabit: Series 60 resistance regression coefficients*

| Series 60 resistance regression coefficients | | | | Load draught condition | | | | |
$V_K/\sqrt{L_f}$	0.50	0.55	0.60	0.65	0.70	0.75	0.80	0.85	0.90
a1	−0.8244	−0.8249	−0.8278	−0.7970	−0.7562	−0.6619	−0.5200	−0.3570	−0.0267
a2	0.1906	0.1865	0.2050	0.2332	0.2496	0.2607	0.3185	0.3528	0.1333
a3	0.1164	0.1133	0.1042	0.1116	0.1221	0.1298	0.1302	0.1533	0.1015
a4	−0.0519	0.0060	0.0832	0.1075	0.1494	0.2603	0.5236	0.5455	0.4568
a5	0.0057	−0.0109	−0.0451	−0.0165	0.0472	0.1491	0.2289	0.4001	0.4677
a6	0.0072	0.0198	0.0211	0.0172	0.0216	0.0361	−0.0017	−0.0027	0.0181
a7	−0.0052	−0.0036	0.0067	0.0068	0.0064	0.0033	−0.0023	−0.0025	0.0175
a8	0.1134	0.1109	0.0933	0.1041	0.1585	0.1859	0.2930	0.2579	0.0506
a9	0.0670	0.0917	0.0708	0.0826	0.1428	0.1562	0.1742	0.1861	0.1558
a10	0.0483	0.0510	0.0400	0.0409	0.0414	0.0403	0.0368	0.0118	0.0279
a11	−0.1276	−0.0745	−0.0729	−0.0879	−0.0744	−0.0636	−0.1171	−0.1500	−0.0988
a12	0.1125	0.0971	0.1269	0.1882	0.2115	0.2289	0.3315	0.4253	0.0834
a13	−0.0481	−0.0213	0.0232	0.0265	0.0188	0.0103	0.0132	−0.0068	−0.0151
a14	0.0372	0.0206	−0.0105	−0.0049	0.0135	0.0378	0.0190	0.0789	0.0582
a15	−0.0954	−0.1924	−0.0855	−0.0189	0.0018	0.0793	0.1247	0.2562	0.3376
a16	−0.0629	0.0108	0.0036	0.0581	0.0884	0.0671	−0.0053	0.0601	0.1429

Table A3.6. *Moor and Small: average ⊙ values*

LCB	2.00 A	1.75 A	1.50 A	1.25 A	1.00 A	0.75 A	0.50 A	0.25 A	Amidships	0.25 F	0.50 F	0.75 F	1.00 F	1.25 F	1.50 F	1.75 F	2.00 F	2.25 F	2.50 F
C_B									10 knots $V/\sqrt{L} = 0.50$										
0.625	0.643	0.644	0.645	0.646	0.647	0.647	0.647	0.649	0.649	0.650	0.650	0.650	—	—	—	—	—	—	—
0.650	0.645	0.645	0.645	0.645	0.645	0.645	0.645	0.645	0.645	0.645	0.645	0.645	0.645	0.645	0.645	0.645	—	—	—
0.675	0.650	0.649	0.648	0.648	0.648	0.647	0.646	0.646	0.646	0.646	0.645	0.644	0.642	0.642	0.641	0.640	—	—	—
0.700	—	—	—	0.657	0.656	0.654	0.652	0.650	0.649	0.648	0.646	0.645	0.643	0.642	0.640	0.640	0.638	0.637	0.636
0.725	—	—	—	—	—	0.668	0.665	0.661	0.659	0.656	0.653	0.650	0.648	0.646	0.644	0.642	0.640	0.638	0.636
0.750	—	—	—	—	—	—	—	0.683	0.678	0.671	0.667	0.664	0.660	0.656	0.653	0.650	0.647	0.644	0.642
0.775	—	—	—	—	—	—	—	—	—	0.700	0.695	0.688	0.681	0.675	0.669	0.665	0.660	0.657	0.655
0.800	—	—	—	—	—	—	—	—	—	0.746	0.738	0.729	0.719	0.710	0.703	0.697	0.691	0.685	0.680
0.825	—	—	—	—	—	—	—	—	—	—	—	—	0.790	0.784	0.777	0.771	0.768	—	—
									11 knots $V/\sqrt{L} = 0.55$										
0.625	0.640	0.643	0.645	0.646	0.646	0.647	0.647	0.650	0.652	0.653	0.656	0.657	—	—	—	—	—	—	—
0.650	0.642	0.643	0.645	0.645	0.645	0.646	0.646	0.647	0.650	0.650	0.651	0.652	0.652	0.653	0.655	0.656	—	—	—
0.675	0.646	0.647	0.648	0.648	0.648	0.648	0.648	0.648	0.649	0.649	0.650	0.650	0.650	0.650	0.651	0.652	—	—	—
0.700	—	—	—	0.655	0.655	0.654	0.653	0.653	0.653	0.653	0.653	0.653	0.652	0.652	0.651	0.651	0.651	0.651	0.650
0.725	—	—	—	—	—	0.666	0.665	0.664	0.663	0.661	0.660	0.659	0.658	0.657	0.654	0.654	0.652	0.652	0.650
0.750	—	—	—	—	—	—	—	0.680	0.679	0.676	0.676	0.674	0.670	0.667	0.664	0.662	0.660	0.658	0.654
0.775	—	—	—	—	—	—	—	—	—	0.701	0.699	0.694	0.690	0.686	0.681	0.678	0.675	0.671	0.666
0.800	—	—	—	—	—	—	—	—	—	0.737	0.733	0.729	0.727	0.722	0.718	0.713	0.710	0.704	0.700
0.825	—	—	—	—	—	—	—	—	—	—	—	—	0.791	0.790	0.790	0.789	0.789	—	—
									12 knots $V/\sqrt{L} = 0.60$										
6.625	0.658	0.659	0.659	0.659	0.659	0.659	0.659	0.660	0.660	0.661	0.662	0.662	—	—	—	—	—	—	—
0.650	0.653	0.655	0.655	0.655	0.655	0.655	0.655	0.658	0.658	0.659	0.659	0.659	0.660	0.660	0.660	0.661	—	—	—
0.675	0.655	0.657	0.657	0.657	0.657	0.657	0.657	0.659	0.660	0.660	0.660	0.660	0.660	0.661	0.661	0.661	—	—	—
0.700	—	—	—	0.665	0.664	0.663	0.663	0.665	0.665	0.666	0.666	0.666	0.665	0.665	0.664	0.664	0.664	0.664	0.664
0.725	—	—	—	—	—	0.680	0.680	0.680	0.680	0.679	0.677	0.676	0.673	0.672	0.670	0.670	0.668	0.668	0.667
0.750	—	—	—	—	—	—	—	0.712	0.707	0.702	0.698	0.692	0.687	0.683	0.680	0.679	0.680	0.680	0.680
0.775	—	—	—	—	—	—	—	—	—	0.738	0.725	0.715	0.708	0.703	0.700	0.700	0.702	0.706	0.712
0.800	—	—	—	—	—	—	—	—	—	0.784	0.760	0.745	0.735	0.730	0.731	0.738	0.751	0.774	0.808
0.825	—	—	—	—	—	—	—	—	—	—	—	—	0.775	0.776	0.796	0.836	0.890	—	—

(continued)

Table A3.6. *Moor and Small: average ⊙ values (continued)*

LCB	2.00 A	1.75 A	1.50 A	1.25 A	1.00 A	0.75 A	0.50 A	0.25 A	Amidships	0.25 F	0.50 F	0.75 F	1.00 F	1.25 F	1.50 F	1.75 F	2.00 F	2.25 F	2.50 F
C_B																			
13 knots $V/\sqrt{L} = 0.65$																			
0.625	0.680	0.680	0.680	0.680	0.680	0.680	0.680	0.679	0.679	0.678	0.678	0.678	–	–	–	–	–	–	–
0.650	0.676	0.676	0.676	0.676	0.676	0.676	0.676	0.676	0.676	0.676	0.676	0.676	0.676	0.676	0.676	0.676	–	–	–
0.675	0.674	0.676	0.676	0.677	0.677	0.677	0.677	0.678	0.678	0.678	0.678	0.678	0.679	0.679	0.679	0.679	–	–	–
0.700	–	–	–	0.684	0.684	0.684	0.684	0.684	0.684	0.684	0.684	0.684	0.683	0.683	0.683	0.683	0.682	0.682	0.681
0.725	–	–	–	–	–	0.703	0.703	0.701	0.700	0.699	0.697	0.696	0.695	0.693	0.693	0.693	0.689	0.689	0.689
0.750	–	–	–	–	–	–	–	0.744	0.735	0.727	0.720	0.716	0.712	0.710	0.710	0.710	0.709	0.710	0.712
0.775	–	–	–	–	–	–	–	–	–	0.773	0.759	0.751	0.745	0.740	0.739	0.739	0.746	0.756	0.770
0.800	–	–	–	–	–	–	–	–	–	0.841	0.820	0.806	0.799	0.795	0.795	0.803	0.818	0.843	0.885
0.825	–	–	–	–	–	–	–	–	–	–	–	–	0.885	0.888	0.902	0.929	0.973	–	–
14 knots $V/\sqrt{L} = 0.70$																			
0.625	0.684	0.684	0.684	0.683	0.682	0.682	0.682	0.680	0.681	0.680	0.680	0.678	–	–	–	–	–	–	–
0.650	0.684	0.684	0.684	0.684	0.684	0.684	0.684	0.684	0.684	0.684	0.684	0.684	0.684	0.684	0.684	0.684	–	–	–
0.675	0.684	0.684	0.684	0.685	0.686	0.687	0.688	0.689	0.690	0.691	0.694	0.695	0.697	0.698	0.700	0.702	–	–	–
0.700	–	–	–	0.687	0.689	0.692	0.695	0.697	0.698	0.702	0.705	0.707	0.711	0.714	0.720	0.721	0.727	0.730	0.734
0.725	–	–	–	–	–	0.704	0.708	0.712	0.717	0.720	0.724	0.729	0.733	0.738	0.745	0.748	0.754	0.760	0.766
0.750	–	–	–	–	–	–	–	0.754	0.754	0.756	0.757	0.761	0.765	0.769	0.777	0.784	0.792	0.802	0.813
0.775	–	–	–	–	–	–	–	–	–	0.810	0.810	0.810	0.812	0.815	0.822	0.833	0.850	0.871	0.898
0.800	–	–	–	–	–	–	–	–	–	0.891	0.883	0.880	0.880	0.889	0.903	0.927	0.957	0.998	1.052
0.825	–	–	–	–	–	–	–	–	–	–	–	–	1.005	1.020	1.044	1.080	1.134	–	–
15 knots $V/\sqrt{L} = 0.75$																			
0.625	0.689	0.690	0.688	0.688	0.688	0.688	0.688	0.686	0.687	0.688	0.688	0.688	–	–	–	–	–	–	–
0.650	0.687	0.688	0.689	0.689	0.689	0.690	0.690	0.691	0.692	0.693	0.693	0.694	0.694	0.696	0.697	0.698	–	–	–
0.675	0.687	0.691	0.693	0.696	0.698	0.701	0.702	0.706	0.707	0.711	0.714	0.716	0.719	0.722	0.724	0.725	–	–	–
0.700	–	–	–	0.714	0.720	0.723	0.728	0.732	0.737	0.741	0.746	0.751	0.757	0.762	0.766	0.771	0.776	0.781	0.787
0.725	–	–	–	–	–	0.771	0.775	0.781	0.787	0.791	0.797	0.801	0.808	0.813	0.822	0.829	0.838	0.852	0.863
0.750	–	–	–	–	–	–	–	–	–	0.856	0.856	0.859	0.866	0.877	0.891	0.908	0.928	0.951	0.976
0.775	–	–	–	–	–	–	–	–	–	–	0.932	0.931	0.937	0.952	0.974	1.002	1.037	1.076	1.120
0.800	–	–	–	–	–	–	–	–	–	–	1.020	1.021	1.029	1.047	1.077	1.116	1.164	1.220	1.275

Table A3.6. *Moor and Small: average ⓒ values (continued)*

LCB	2.00 A	1.75 A	1.50 A	1.25 A	1.00 A	0.75 A	0.50 A	0.25 A	Amidships	0.25 F	0.50 F	0.75 F	1.00 F	1.25 F	1.50 F	1.75 F
C_B																
16 knots $V/\sqrt{L} = 0.80$																
0.625	0.686	0.689	0.691	0.694	0.696	0.700	0.701	0.704	0.708	0.710	0.719	0.731	–	–	–	–
0.650	0.688	0.690	0.693	0.696	0.699	0.704	0.706	0.712	0.718	0.725	0.734	0.746	0.759	0.775	0.796	0.820
0.675	0.690	0.696	0.700	0.706	0.712	0.718	0.725	0.732	0.739	0.749	0.760	0.772	0.788	0.804	0.824	0.848
0.700	–	–	–	0.742	0.750	0.760	0.770	0.780	0.792	0.805	0.818	0.832	0.849	0.866	0.883	0.905
0.725	–	–	–	–	–	0.860	0.870	0.881	0.892	0.902	0.914	0.928	0.941	0.957	0.972	0.998
0.750	–	–	–	–	–	–	–	–	–	1.053	1.052	1.056	1.061	1.072	1.087	1.119
0.775	–	–	–	–	–	–	–	–	–	–	–	–	–	1.197	1.215	1.268
17 knots $V/\sqrt{L} = 0.85$																
0.625	0.716	0.717	0.719	0.723	0.725	0.729	0.737	0.748	0.759	0.773	0.790	0.813	–	–	–	–
0.650	0.737	0.738	0.740	0.742	0.746	0.749	0.756	0.766	0.778	0.790	0.808	0.829	0.855	0.887	–	–
0.675	0.765	0.768	0.771	0.775	0.778	0.784	0.792	0.800	0.810	0.825	0.842	0.867	0.894	0.932	–	–
0.700	–	–	–	–	–	0.855	0.859	0.865	0.875	0.890	0.909	0.933	0.962	1.009	–	–
0.725	–	–	–	–	–	–	–	–	–	1.000	1.011	1.029	1.058	1.107	–	–
18 knots $V/\sqrt{L} = 0.90$																
0.625	0.824	0.834	0.847	0.862	0.879	0.898	0.919	0.942	0.967	0.992	–	–	–	–	–	–
0.650	0.862	0.866	0.876	0.887	0.902	0.920	0.940	0.960	0.981	1.004	–	–	–	–	–	–
0.675	0.897	0.900	0.907	0.918	0.932	0.950	0.968	0.988	1.010	1.032	–	–	–	–	–	–
0.700	–	–	–	–	–	0.989	1.010	1.033	1.059	1.085	–	–	–	–	–	–

Table A3.7. *Hollenbach: resistance regression coefficients*

Hollenbach: Resistance regression coefficients

	Mean			Minimum	
	Single-screw			Single-screw	
	Design draught	Ballast draught	Twin-screw	Design draught	Twin-screw
$a1$	-0.3382	-0.7139	-0.2748	-0.3382	-0.2748
$a2$	0.8086	0.2558	0.5747	0.8086	0.5747
$a3$	-6.0258	1.1606	-6.7610	-6.0258	-6.7610
$a4$	-3.5632	0.4534	-4.3834	-3.5632	-4.3834
$a5$	9.4405	11.222	8.8158	0	0
$a6$	0.0146	0.4524	-0.1418	0	0
$a7$	0	0	-0.1258	0	0
$a8$	0	0	0.0481	0	0
$a9$	0	0	0.1699	0	0
$a10$	0	0	0.0728	0	0
b_{11}	-0.57424	-1.50162	-5.34750	-0.91424	3.27279
b_{12}	13.3893	12.9678	55.6532	13.38930	-44.1138
b_{13}	90.5960	-36.7985	-114.905	90.59600	171.692
b_{21}	4.6614	5.55536	19.2714	4.6614	-11.5012
b_{22}	-39.721	-45.8815	-192.388	-39.7210	166.559
b_{23}	-351.483	121.820	388.333	-351.483	-644.456
b_{31}	-1.14215	-4.33571	-14.35710	-1.14215	12.4626
b_{32}	-12.3296	36.0782	142.73800	-12.3296	-179.505
b_{33}	459.254	-85.3741	-254.76200	459.25400	680.921
c_1	*Fr/Fr·krit*	$10C_B(Fr/Fr\cdot krit\ -1)$	*Fr/Fr·krit*	–	–
d_1	0.854	0.032	0.8970	–	–
d_2	-1.228	0.803	-1.4570	–	–
d_3	0.497	-0.739	0.7670	–	–
e_1	2.1701	1.9994	1.8319	–	–
e_2	-0.1602	-0.1446	-0.1237	–	–
f_1	0.17	0.15	0.16	0.17	0.14
f_2	0.20	0.10	0.24	0.20	0
f_3	0.60	0.50	0.60	0.60	0
g_1	0.642	0.42	0.50	0.614	0.952
g_2	-0.635	-0.20	0.66	-0.717	-1.406
g_3	0.150	0	0.50	0.261	0.643
h_1	1.204	1.194	1.206	–	–
Ship length $L(m)$	42.0–205.0	50.2–224.8	30.6–206.8	42.0–205.0	30.6–206.8
$L/\nabla^{1/3}$	4.49–6.01	5.45–7.05	4.41–7.27	4.49–6.01	4.41–7.27
C_B	0.60–0.83	0.56–0.79	0.51–0.78	0.60–0.83	0.51–0.78
L/B	4.71–7.11	4.95–6.62	3.96–7.13	4.71–7.11	3.96–7.13
B/T	1.99–4.00	2.97–6.12	2.31–6.11	1.99–4.00	2.31–6.11
L_{OS}/L_{WL}	1.00–1.05	1.00–1.05	1.00–1.05	1.00–1.05	1.00–1.05
L_{WL}/L	1.00–1.06	0.95–1.00	1.0–1.07	1.00–1.06	1.00–1.07
D_P/T	0.43–0.84	0.66–1.05	0.50–0.86	0.43–0.84	0.50–0.86

Table A3.8. *Taylor–Gertler: series resistance data* $C_P = 0.50$

Taylor–Gertler series data: $C_R \times 1000$ $C_P = 0.50$

$L/\nabla^{1/3}$	5.5			6.0			7.0			8.0			9.0			10.0		
B/T	2.25	3.00	3.75	2.25	3.00	3.75	2.25	3.00	3.75	2.25	3.00	3.75	2.25	3.00	3.75	2.25	3.00	3.75
Fr																		
0.16	0.45	0.54	0.65	0.38	0.46	0.55	0.30	0.33	0.44	0.24	0.28	0.37	0.22	0.26	0.34	0.20	0.25	0.33
0.18	0.48	0.55	0.67	0.39	0.47	0.57	0.30	0.34	0.45	0.26	0.28	0.38	0.22	0.27	0.34	0.20	0.25	0.33
0.20	0.50	0.60	0.71	0.41	0.48	0.59	0.31	0.35	0.47	0.28	0.29	0.39	0.23	0.27	0.37	0.20	0.25	0.33
0.22	0.55	0.73	0.84	0.47	0.58	0.72	0.35	0.42	0.52	0.29	0.36	0.43	0.24	0.30	0.40	0.20	0.27	0.37
0.24	0.68	0.97	1.12	0.58	0.79	0.90	0.43	0.62	0.65	0.30	0.51	0.52	0.25	0.40	0.48	0.20	0.33	0.43
0.26	0.83	1.11	1.32	0.67	0.91	1.05	0.49	0.72	0.75	0.37	0.59	0.60	0.27	0.48	0.52	0.20	0.39	0.48
0.28	0.88	1.11	1.40	0.70	0.92	1.11	0.52	0.75	0.80	0.39	0.58	0.60	0.28	0.47	0.51	0.20	0.38	0.46
0.30	–	1.40	1.63	0.85	1.16	1.32	0.59	0.81	0.87	0.41	0.60	0.61	0.32	0.50	0.55	0.22	0.44	0.46
0.32	–	2.20	2.55	1.42	1.71	1.91	0.98	1.15	1.18	0.68	0.85	0.83	0.51	0.70	0.70	0.40	0.60	0.59
0.34	–	–	–	2.71	3.10	3.20	1.78	1.90	1.96	1.20	1.30	1.30	0.95	1.05	1.06	0.68	0.86	0.81
0.36	–	–	–	–	4.91	5.00	2.96	3.00	3.00	1.97	1.97	2.00	1.45	1.50	1.51	1.00	1.16	1.14
0.38					–	–	4.32	4.32	4.58	2.83	2.82	3.00	2.03	2.05	2.10	1.40	1.50	1.51
0.40							–	–	–	3.71	3.80	3.95	2.58	2.63	2.64	1.75	1.82	1.90
0.42										4.40	4.50	4.65	2.93	3.08	3.10	1.98	2.10	2.20
0.44										4.83	5.02	5.17	3.20	3.40	3.45	2.10	2.26	2.40
0.46										5.05	–	–	3.32	3.58	3.63	2.18	2.34	2.52
0.48										5.16			3.38	3.59	3.71	2.20	2.36	2.60
0.50										5.15			3.38	3.60	3.72	2.19	2.35	2.61
0.52										5.08			3.35	3.59	3.71	2.15	2.30	2.60
0.54										4.93			3.28	3.50	3.65	2.10	2.25	2.58
0.56										4.72			3.17	3.41	3.51	2.05	2.18	2.51
0.58										4.45			3.00	3.30	3.41	1.99	2.15	2.47

Table A3.9. *Taylor–Gertler: series resistance data $C_P = 0.60$*

Taylor–Gertler series data: $C_R \times 1000$ — $C_P = 0.60$

$L/\nabla^{1/3}$	B/T 5.5			6.0			7.0			8.0			9.0			10.0		
Fr	2.25	3.00	3.75	2.25	3.00	3.75	2.25	3.00	3.75	2.25	3.00	3.75	2.25	3.00	3.75	2.25	3.00	3.75
0.16	0.50	0.60	0.75	0.42	0.51	0.65	0.35	0.42	0.51	0.29	0.38	0.48	0.22	0.32	0.44	0.23	0.30	0.42
0.18	0.50	0.60	0.77	0.42	0.51	0.65	0.35	0.42	0.53	0.29	0.38	0.49	0.22	0.32	0.44	0.23	0.30	0.42
0.20	0.52	0.64	0.80	0.43	0.54	0.68	0.34	0.42	0.55	0.29	0.39	0.49	0.22	0.33	0.44	0.23	0.30	0.42
0.22	0.60	0.74	0.90	0.51	0.61	0.75	0.38	0.49	0.59	0.30	0.45	0.51	0.25	0.38	0.45	0.23	0.35	0.43
0.24	0.71	0.85	0.96	0.61	0.74	0.82	0.47	0.64	0.64	0.37	0.58	0.54	0.30	0.50	0.49	0.25	0.46	0.45
0.26	1.01	1.06	1.21	0.87	0.98	1.02	0.66	0.87	0.85	0.49	0.76	0.69	0.37	0.64	0.57	0.28	0.56	0.49
0.28	1.67	1.64	1.80	1.43	1.37	1.50	1.00	1.20	1.12	0.71	0.98	0.87	0.51	0.77	0.70	0.38	0.64	0.57
0.30	–	2.45	2.50	2.08	2.11	2.08	1.40	1.51	1.45	1.00	1.18	1.09	0.55	0.91	0.80	0.46	0.70	0.61
0.32	–	2.90	3.04	2.29	2.45	2.57	1.48	1.68	1.75	1.05	1.29	1.26	0.56	0.95	0.92	0.46	0.75	0.70
0.34	–	–	–	2.50	2.83	3.08	1.60	2.00	2.05	1.10	1.47	1.41	0.63	1.07	1.01	0.52	0.80	0.78
0.36	–	–	–	3.30	3.62	3.84	2.12	2.58	2.59	1.41	1.74	1.75	0.86	1.20	1.19	0.69	0.92	0.82
0.38	–	–	–	5.10	5.10	5.31	3.12	3.40	3.50	1.95	2.28	2.30	1.19	1.47	1.60	0.92	1.12	1.08
0.40	–	–	–	–	–	–	–	–	–	2.62	2.98	3.01	1.50	1.90	2.07	1.20	1.41	1.38
0.42										3.20	3.58	3.60	1.79	2.32	2.40	1.40	1.60	1.58
0.44										3.60	4.03	4.07	2.03	2.64	2.65	1.52	1.74	1.70
0.46										3.82	4.31	4.38	2.19	2.80	2.87	1.61	1.80	1.81
0.48										3.95	4.42	4.58	2.19	2.89	3.00	1.68	1.85	1.89
0.50										4.02	–	–	2.20	2.92	3.10	1.70	1.86	1.91
0.52										4.00	–	–	2.20	2.94	3.11	1.70	1.86	1.92
0.54										3.92			2.19	2.93	3.09	1.67	1.85	1.90
0.56										3.81			2.10	2.85	2.95	1.62	1.81	1.88
0.58										3.66			2.00	2.75	2.85	1.60	1.78	1.80

Table A3.10. *Taylor–Gertler: series resistance data* $C_P = 0.70$

Taylor–Gertler series data: $C_R \times 1000$ $C_P = 0.70$

$L/\nabla^{1/3}$	5.5			6.0			7.0			8.0			9.0			10.0		
B/T Fr	2.25	3.00	3.75	2.25	3.00	3.75	2.25	3.00	3.75	2.25	3.00	3.75	2.25	3.00	3.75	2.25	3.00	3.75
0.16	0.52	0.68	0.83	0.43	0.58	0.74	0.33	0.47	0.62	0.29	0.38	0.56	0.25	–	–	0.23	–	–
0.18	0.57	0.72	0.84	0.46	0.60	0.75	0.35	0.48	0.62	0.29	0.38	0.56	0.25	–	–	0.23		
0.20	0.68	0.87	0.97	0.57	0.73	0.85	0.40	0.57	0.69	0.30	0.45	0.58	0.26	–		0.23		
0.22	0.89	1.10	1.19	0.77	1.00	1.03	0.58	0.79	0.85	0.44	0.60	0.70	0.34			0.29		
0.24	1.20	1.40	1.49	1.07	1.30	1.32	0.86	1.07	1.09	0.68	0.78	0.87	0.52			0.40		
0.26	1.68	1.84	2.00	1.50	1.67	1.75	1.20	1.37	1.39	0.95	1.03	1.10	0.73			0.56		
0.28	3.24	3.13	3.24	2.70	2.71	2.82	1.90	2.09	2.14	1.42	1.52	1.57	1.02			0.80		
0.30	–	–	–	4.95	–	–	3.17	3.21	3.28	2.25	2.30	2.30	1.62			1.14		
0.32				5.71			3.70	3.90	3.90	2.40	2.60	2.60	1.71			1.24		
0.34				5.69			3.52	4.04	4.20	2.29	2.60	2.69	1.62			1.15		
0.36				5.58			3.60	4.13	4.30	2.23	2.63	2.80	1.56			1.14		
0.38				6.12			3.88	4.42	4.65	2.43	2.78	3.00	1.72			1.24		
0.40				–			–	–	–	2.82	3.08	3.40	2.00			1.37		
0.42										3.20	3.50	3.65	2.19			1.45		
0.44										3.50	3.86	4.00	2.33			1.55		
0.46										3.67	4.02	4.28	2.45			1.60		
0.48										3.78	–	–	2.50			1.65		
0.50										3.80			2.52			1.66		
0.52										3.80			2.52			1.65		
0.54										3.79			2.51			1.65		
0.56										3.73			2.45			1.62		
0.58										3.60			2.40			1.59		

Table A3.11. *Taylor–Gertler: series resistance data* $C_P = 0.80$

	Taylor–Gertler series data: $C_R \times 1000$												$C_P = 0.80$						
$L/\nabla^{1/3}$	5.5			6.0			7.0			8.0			9.0			10.0		
B/T	2.25	3.00	3.75	2.25	3.00	3.75	2.25	3.00	3.75	2.25	3.00	3.75	2.25	3.00	3.75	2.25	3.00	3.75
Fr																		
0.16	0.62	0.82	0.92	0.53	0.75	0.84	0.41	0.62	0.71	0.35	0.43	0.65	0.33	–	–	0.32	–	–
0.18	0.83	1.03	1.05	0.73	0.97	0.95	0.58	0.82	0.82	0.51	0.61	0.74	0.49	–	–	0.46	–	–
0.20	1.32	1.34	1.38	1.20	1.28	1.27	0.96	1.12	1.13	0.78	0.89	1.05	0.69	–	–	0.62	–	–
0.22	2.10	1.92	1.98	1.90	1.82	1.85	1.55	1.62	1.63	1.26	1.35	1.52	1.07	–	–	0.91	–	–
0.24	2.90	3.05	2.90	2.67	2.88	2.73	2.27	2.50	2.48	1.90	1.99	2.15	1.55	–	–	1.30	–	–
0.26	–	–	–	3.60	–	–	2.97	3.40	3.18	2.28	2.57	2.58	1.77	–	–	1.45	–	–
0.28	–	–	–	–	–	–	3.48	–	–	2.58	2.99	3.02	2.06	–	–	1.70	–	–
0.30							5.20	–	–	3.70	4.00	3.87	2.77	–	–	2.20	–	–
0.32							6.51	–	–	4.23	4.85	4.55	3.05	–	–	2.18	–	–
0.34							6.71	–	–	4.25	4.73	4.63	2.97	–	–	2.06	–	–
0.36							6.35	–	–	4.10	4.50	4.66	2.77	–	–	1.96	–	–
0.38							6.01	–	–	3.92	4.38	4.52	2.68	–	–	1.90	–	–
0.40										3.93	–	–	2.65	–	–	1.85	–	–
0.42										4.07	–	–	2.69	–	–	1.80	–	–
0.44										4.18	–	–	2.74	–	–	1.79	–	–
0.46										4.26	–	–	2.80	–	–	1.77	–	–
0.48										4.28	–	–	2.80	–	–	1.75	–	–
0.50										4.27	–	–	2.80	–	–	1.78	–	–
0.52										4.19	–	–	2.79	–	–	1.80	–	–
0.54										4.12	–	–	2.76	–	–	1.80	–	–
0.56										4.03	–	–	2.70	–	–	1.79	–	–
0.58										3.92	–	–	2.65	–	–	1.78	–	–

Table A3.12. *Zborowski: twin-screw series resistance data*

<table>
<tr><td colspan="15" align="center">Zborowski: twin-screw series resistance data $C_{TM} \times 1000$</td></tr>
</table>

a: Series with B/T and $L/\nabla^{1/3}$ variation [$C_B = 0.518$]

B/T	$L/\nabla^{1/3}$	Fr	0.25	0.26	0.27	0.28	0.29	0.30	0.31	0.32	0.33	0.34	0.35	Model symbol
2.25	6.00		5.169	5.168	5.183	5.193	5.227	5.296	5.355	5.520	5.794	6.225	6.829	A−1
	6.50		5.300	5.279	5.262	5.253	5.245	5.270	5.333	5.469	5.657	6.013	6.527	A−2
	7.00		5.028	5.022	5.095	5.112	5.089	5.132	5.158	5.237	5.431	5.728	6.084	A−3
2.80	6.00		5.427	5.430	5.420	5.420	5.440	5.500	5.580	5.740	6.060	6.490	7.030	B−1
	6.50		5.374	5.336	5.338	5.327	5.342	5.379	5.444	5.569	5.823	6.198	6.651	B−2
	7.00		5.202	5.217	5.220	5.220	5.203	5.220	5.271	5.385	5.558	5.819	6.185	B−3
3.35	6.00		5.351	5.356	5.383	5.379	5.400	5.450	5.530	5.677	5.987	6.428	6.963	C−1
	6.50		5.354	5.306	5.311	5.313	5.319	5.340	5.395	5.538	5.790	6.170	6.620	C−2
	7.00		5.104	5.103	5.104	5.091	5.098	5.120	5.162	5.269	5.435	5.699	6.074	C−3

b: Series with C_B variations [$L/\nabla^{1/3} = 6.5$]

B/T	C_B	Fr	0.25	0.26	0.27	0.28	0.29	0.30	0.31	0.32	0.33	0.34	0.35	
3.35	0.518		5.354	5.306	5.311	5.313	5.319	5.340	5.395	5.538	5.790	6.170	6.620	C−2
	0.564		5.298	5.310	5.395	5.461	5.534	5.625	5.270	5.752	5.940	6.166	6.378	R−2
	0.605		5.458	5.500	5.600	5.720	5.858	6.097	6.236	6.370	6.470	6.580	6.717	R−1
	0.645		6.055	6.083	6.179	6.360	6.598	6.923	7.228	7.405	7.500	7.606	7.773	R−2

Table A3.13. *Yeh: Series 64 resistance data $C_B = 0.35$*

| Yeh: Series 64 resistance data $C_B = 0.35$ $C_R \times 1000$ | | | | | | | | | |
| $L/\nabla^{1/3}$ | 9.3 | | | 10.5 | | | 12.4 | | |
B/T $V_K/\sqrt{L_f}$	2.0	3.0	4.0	2.0	3.0	4.0	2.0	3.0	4.0
0.2	1.533	1.500	1.403	1.869	1.830	1.718	1.100	1.075	1.003
0.4	0.950	0.929	0.868	0.889	0.869	0.813	0.854	0.835	0.782
0.6	1.477	1.348	1.172	1.053	1.149	1.194	0.936	1.070	0.862
0.8	1.362	1.448	1.310	1.093	1.005	1.135	0.923	1.076	0.932
1.0	1.539	1.617	1.422	1.191	1.083	1.182	0.954	0.990	0.828
1.2	1.633	1.701	1.463	1.227	1.115	1.192	0.913	0.933	0.771
1.4	1.687	1.747	1.509	1.194	1.106	1.167	0.882	0.865	0.709
1.6	1.708	1.732	1.502	1.139	1.067	1.117	0.794	0.778	0.693
1.8	1.690	1.636	1.501	1.076	1.003	1.071	0.720	0.706	0.665
2.0	1.561	1.477	1.462	1.014	0.931	0.990	0.672	0.645	0.660
2.2	1.433	1.332	1.390	0.956	0.850	0.918	0.652	0.639	0.667
2.4	1.300	1.200	1.310	0.872	0.789	0.863	0.619	0.626	0.662
2.6	1.183	1.091	1.216	0.801	0.742	0.825	0.588	0.625	0.643
2.8	1.076	1.014	1.126	0.752	0.710	0.794	0.580	0.632	0.603
3.0	0.998	0.959	1.052	0.718	0.690	0.777	0.570	0.644	0.579
3.2	0.937	0.919	1.000	0.686	0.690	0.765	0.567	0.642	0.576
3.4	0.899	0.901	0.979	0.679	0.700	0.747	0.576	0.646	0.583
3.6	0.870	0.892	0.967	0.686	0.709	0.769	0.580	0.658	0.593
3.8	0.849	0.887	0.962	0.700	0.722	0.774	0.591	0.671	0.601
4.0	0.836	0.885	0.952	0.715	0.736	0.776	0.604	0.678	0.611
4.2	0.833	0.889	0.947	0.728	0.747	0.778	0.615	0.688	0.626
4.4	0.830	0.894	0.939	0.741	0.756	0.777	0.625	0.695	0.632
4.6	0.832	0.898	0.932	0.750	0.766	0.775	0.633	0.704	0.639
4.8	0.836	0.898	0.926	0.759	0.772	0.775	0.643	0.709	0.646
5.0	0.836	0.895	0.918	0.763	0.776	0.772	0.649	0.713	0.651

Table A3.14. *Yeh: Series 64 resistance data $C_B = 0.45$*

	Yeh: Series 64 resistance data $C_B = 0.45$ $C_R \times 1000$								
$L/\nabla^{1/3}$		8.6			9.6			11.3	
B/T	2.0	3.0	4.0	2.0	3.0	4.0	2.0	3.0	4.0
$V_K/\sqrt{L_f}$									
0.2	3.187	3.120	2.968	2.778	2.781	2.585	3.693	3.584	3.377
0.4	2.201	2.155	2.051	2.112	1.816	1.487	1.766	1.407	1.643
0.6	2.076	2.127	1.936	1.900	1.637	1.237	1.592	1.273	1.482
0.8	2.161	2.331	2.069	1.541	1.445	1.135	1.263	0.996	1.252
1.0	2.157	2.455	1.851	1.326	1.259	1.236	1.041	0.865	1.116
1.2	2.832	2.775	2.375	1.556	1.551	1.478	0.992	0.829	0.788
1.4	3.042	2.981	2.695	1.732	1.697	1.618	0.952	0.901	0.762
1.6	2.915	2.924	2.587	1.682	1.648	1.572	0.880	0.761	0.743
1.8	2.723	2.700	2,346	1.572	1.540	1.468	0.781	0.686	0.727
2.0	2.536	2.418	2.111	1.469	1.439	1.372	0.689	0.649	0.690
2.2	2.254	2.145	1.926	1.341	1.314	1.253	0.615	0.613	0.634
2.4	1.984	1.909	1.776	1.221	1.189	1.127	0.547	0.562	0.569
2.6	1.766	1.707	1.643	1.107	1.072	1.028	0.483	0.503	0.507
2.8	1.545	1.541	1.524	1.007	0.971	0.945	0.438	0.455	0.464
3.0	1.381	1.414	1.425	0.919	0.891	0.892	0.401	0.415	0.438
3.2	1.298	1.318	1.334	0.839	0.839	0.851	0.379	0.411	0.424
3.4	1.248	1.265	1.270	0.789	0.815	0.826	0.373	0.408	0.422
3.6	1.218	1.228	1.224	0.777	0.804	0.811	0.385	0.413	0.427
3.8	1.206	1.206	1.200	0.806	0.806	0.816	0.405	0.432	0.441
4.0	1.194	1.188	1.182	0.777	0.814	0.822	0.424	0.454	0.459
4.2	1.184	1.180	1.168	0.784	0.828	0.833	0.451	0.479	0.481
4.4	1.175	1.176	1.163	0.797	0.841	0.842	0.478	0.506	0.508
4.6	1.172	1.173	1.160	0.808	0.852	0.851	0.507	0.533	0.537
4.8	1.166	1.172	1.159	0.819	0.862	0.862	0.539	0.564	0.563
5.0	1.162	1.173	1.161	0.825	0.871	0.871	0.566	0.592	0.587

Table A3.15. *Yeh: Series 64 resistance data $C_B = 0.55$*

$L/\nabla^{1/3}$	8.0			8.9			10.5		
B/T $V_K/\sqrt{L_f}$	2.0	3.0	4.0	2.0	3.0	4.0	2.0	3.0	4.0
0.2	2.888	2.856	2.750	2.504	2.476	2.383	2.065	2.041	1.963
0.4	2.593	2.565	2.473	1.206	1.192	0.923	1.282	1.267	1.218
0.6	2.504	2.478	2.389	1.095	1.082	1.041	1.162	1.149	1.105
0.8	2.612	2.585	2.494	1.450	1.435	0.824	1.203	1.190	1.075
1.0	2.440	2.415	2.330	1.984	1.963	1.357	1.403	1.388	1.157
1.2	2.846	2.817	2.718	2.472	2.446	1.864	1.610	1.560	1.253
1.4	3.163	3.131	3.023	2.734	2.536	2.246	1.552	1.512	1.273
1.6	3.169	3.208	3.052	2.489	2.521	2.266	1.471	1.455	1.245
1.8	2.975	2.993	2.946	2.302	2.335	2.044	1.385	1.356	1.168
2.0	2.711	2.754	2.797	2.042	2.058	1.915	1.278	1.229	1.084
2.2	2.397	2.477	2.580	1.788	1.815	1.685	1.165	1.114	0.981
2.4	2.086	2.206	2.352	1.567	1.609	1.503	1.063	1.019	0.881
2.6	1.856	1.967	2.134	1.390	1.452	1.364	0.970	0.925	0.778
2.8	1.648	1.764	1.928	1.251	1.332	1.263	0.903	0.852	0.706
3.0	1.477	1.595	1.757	1.144	1.247	1.163	0.831	0.801	0.656
3.2	1.347	1.478	1.643	1.067	1.186	1.109	0.818	0.777	0.648
3.4	1.261	1.393	1.561	1.012	1.149	1.062	0.802	0.766	0.657
3.6	1.205	1.329	1.497	0.963	1.133	1.032	0.804	0.774	0.666
3.8	1.170	1.289	1.446	0.932	1.124	1.006	0.820	0.786	0.676
4.0	1.152	1.258	1.415	0.908	1.115	0.991	0.831	0.799	0.689
4.2	1.140	1.257	1.378	0.900	1.121	0.986	0.848	0.808	0.731
4.4	1.152	1.303	1.425	0.904	1.120	0.998	0.863	0.813	0.751
4.6	1.212	1.382	1.491	0.932	1.117	1.012	0.877	0.819	0.764
4.8	1.321	1.487	1.576	0.998	1.169	1.049	0.886	0.822	0.779
5.0	1.441	1.592	1.660	1.097	1.222	1.094	0.898	0.825	0.791

Table caption: Yeh: Series 64 resistance data $C_B = 0.55$ $C_R \times 1000$

Table A3.16. *Bailey: NPL series resistance data*

NPL series resistance data $C_R \times 1000$

B/T	$L/\nabla^{1/3}$ 4.47			4.86			5.23			5.76			
$V_K/\sqrt{L_f}$	1.72	2.43	3.19	2.19	3.16	4.08	1.94	2.75	5.10	1.93	2.59	3.67	6.80
0.8	4.800	5.167	5.533	3.524	4.088	4.651	3.251	4.250	4.172	3.373	3.953	3.900	3.768
0.9	5.163	5.758	6.352	4.062	4.691	5.319	3.846	4.563	4.789	3.777	4.209	4.187	4.311
1.0	5.673	6.472	7.270	4.722	5.391	6.060	4.449	5.098	5.480	4.240	4.499	4.547	4.900
1.1	6.429	7.292	8.155	5.453	6.100	6.746	5.125	5.603	6.040	4.687	4.899	5.002	5.485
1.2	7.382	8.491	9.599	6.456	7.119	7.782	5.834	6.406	6.900	5.244	5.470	5.580	6.068
1.3	8.833	10.087	11.340	8.007	8.559	9.110	6.955	7.557	7.908	6.134	6.316	6.506	6.749
1.4	11.510	12.580	13.650	10.334	10.782	11.230	8.526	9.132	9.290	7.572	7.498	7.805	7.428
1.5	15.711	16.005	16.298	13.133	13.481	13.828	11.111	11.081	11.108	9.383	9.205	9.278	8.055
1.6	20.312	19.854	19.395	16.482	16.398	16.315	13.558	13.278	13.225	10.568	11.012	10.525	8.835
1.7	23.559	23.440	23.320	18.603	18.482	18.360	15.336	14.974	14.040	11.103	11.963	10.818	9.380
1.8	24.854	25.445	26.035	18.947	18.671	18.395	15.749	15.144	13.885	11.114	12.013	10.461	9.095
1.9	25.171	25.160	25.148	18.237	17.873	17.508	15.068	14.332	13.118	10.561	11.143	9.726	8.509
2.0	24.636	24.119	23.601	16.851	16.636	16.421	13.950	13.122	12.381	9.622	10.087	8.991	7.895
2.1	23.400	22.627	21.853	15.690	15.417	15.143	12.991	12.236	11.493	8.881	9.267	8.431	7.357
2.2	21.461	20.834	20.204	14.452	14.228	14.004	12.046	11.248	10.624	8.177	8.548	7.869	6.792

Table A3.16. *Bailey: NPL series resistance data (continued)*

NPL series resistance data $C_R \times 1000$

$L/\nabla^{1/3}$	4.47			4.86			5.23			5.76			
B/T — $V_K/\sqrt{L_f}$	1.72	2.43	3.19	2.19	3.16	4.08	1.94	2.75	5.10	1.93	2.59	3.67	6.80
2.3	19.723	19.095	18.475	13.340	13.087	12.835	11.126	10.387	9.755	7.535	7.903	7.333	6.328
2.4	18.134	17.365	16.955	12.327	12.071	11.815	10.305	9.549	8.865	7.004	7.321	6.845	5.888
2.5	16.571	16.113	15.655	11.314	11.104	10.895	9.533	8.761	8.175	6.498	6.850	6.383	5.499
2.6	15.306	14.954	14.584	10.500	10.302	10.104	8.812	8.147	7.634	6.104	6.429	6.019	5.184
2.7	13.982	13.788	13.603	9.661	9.507	9.353	8.140	7.584	7.163	5.760	6.033	5.656	4.869
2.8	12.876	12.799	12.721	8.921	8.816	8.711	7.555	7.044	6.772	5.376	5.711	5.342	4.604
2.9	11.836	11.738	11.999	8.307	8.252	8.196	7.007	6.605	6.431	5.081	5.378	5.028	4.364
3.0	10.896	11.142	11.387	7.742	7.774	7.805	6.610	6.265	6.191	4.810	5.130	4.789	4.174
3.1	10.055	10.460	10.864	7.402	7.458	7.513	6.287	5.975	5.950	4.551	4.895	4.548	4.059
3.2	9.139	9.820	10.501	7.112	7.167	7.221	6.027	5.710	5.734	4.367	4.698	4.358	3.919
3.3	—	—		6.972	6.989	7.005	5.792	5.545	5.594	4.195	4.513	4.218	3.804
3.4				6.966	6.954	6.941	5.633	5.429	5.504	4.048	4.354	4.053	3.714
3.5				7.005	6.940	6.875	5.501	5.389	5.363	3.900	4.268	3.912	3.674
3.6				7.094	7.015	6.935	5.413	5.348	5.309	3.802	4.233	3.846	3.708
3.7				7.209	7.115	7.021	5.376	5.432	5.256	3.704	4.224	3.755	3.693
3.8				7.299	7.192	7.084	5.377	5.469	5.269	3.644	4.249	3.739	3.731
3.9				7.413	7.292	7.171	5.402	5.608	5.281	3.558	4.275	3.748	3.743
4.0				7.552	7.418	7.284	5.464	5.722	5.292	3.522	4.374	3.706	3.729
4.1				7.716	7.594	7.472	5.527	5.912	5.455	3.475	4.500	3.740	3.766

Table A3.16. *Bailey: NPL series resistance data (continued)*

NPL series resistance data $C_R \times 1000$

$L/\nabla^{1/3}$	6.59				7.10				8.30		
B/T	2.01	2.90	3.88	5.49	2.51	3.63	4.86	6.87	4.02	4.90	5.80
$V_K/\sqrt{L_f}$											
0.8	3.584	2.899	2.889	3.149	3.308	2.449	3.426	2.347	2.864	2.389	1.913
0.9	3.785	3.165	3.081	3.436	3.504	2.654	3.454	2.485	2.921	2.432	1.943
1.0	3.973	3.479	3.446	3.846	3.754	2.893	3.496	2.670	2.977	2.492	2.007
1.1	4.176	3.839	3.808	4.251	4.018	3.140	3.632	2.999	3.017	2.536	2.055
1.2	4.388	4.346	4.166	4.729	4.381	3.460	3.828	3.302	3.073	2.593	2.112
1.3	4.734	5.086	4.661	5.304	4.904	3.932	4.159	3.627	3.150	2.690	2.229
1.4	5.550	6.038	5.417	5.778	5.631	4.503	4.514	3.926	3.315	2.830	2.344
1.5	7.033	6.550	6.121	6.124	6.169	4.879	4.691	4.147	3.471	2.996	2.520
1.6	7.577	6.700	6.397	6.319	6.255	5.016	4.704	4.342	3.438	3.054	2.670
1.7	7.661	6.636	6.396	6.313	6.177	4.964	4.627	4.360	3.303	3.024	2.744
1.8	7.520	6.398	6.194	6.081	5.954	4.763	4.514	4.228	3.149	2.933	2.717
1.9	7.113	6.095	5.927	5.771	5.567	4.393	4.322	3.969	2.987	2.788	2.589
2.0	6.538	5.604	5.523	5.412	5.179	4.095	4.056	3.685	2.825	2.624	2.423
2.1	6.068	5.263	5.130	5.102	4.859	3.903	3.888	3.525	2.668	2.506	2.344
2.2	5.647	4.883	4.786	4.740	4.539	3.673	3.670	3.339	2.523	2.375	2.226

Table A3.16. *Bailey: NPL series resistance data (continued)*

NPL series resistance data $C_R \times 1000$

B/T	$L/\nabla^{1/3}$ 6.59				7.10				8.30		
$V_K/\sqrt{L_f}$	2.01	2.90	3.88	5.49	2.51	3.63	4.86	6.87	4.02	4.90	5.80
2.3	5.245	4.515	4.505	4.504	4.243	3.430	3.477	3.178	2.383	2.246	2.108
2.4	4.898	4.246	4.185	4.292	3.982	3.249	3.283	2.965	2.250	2.132	2.013
2.5	4.589	4.016	3.953	4.005	3.744	3.080	3.101	2.854	2.141	2.000	1.919
2.6	4.342	3.771	3.733	3.817	3.547	2.960	2.930	2.740	2.044	1.953	1.862
2.7	4.094	3.539	3.533	3.654	3.343	2.803	2.785	2.627	1.947	1.876	1.805
2.8	3.896	3.355	3.341	3.490	3.182	2.692	2.639	2.538	1.861	1.816	1.771
2.9	3.660	3.209	3.158	3.326	3.009	2.536	2.531	2.450	1.781	1.753	1.725
3.0	3.493	3.038	3.024	3.187	2.860	2.440	2.422	2.335	1.727	1.715	1.703
3.1	3.319	2.917	2.914	2.996	2.724	2.355	2.312	2.296	1.665	1.667	1.669
3.2	3.182	2.795	2.830	2.906	2.624	2.296	2.265	2.205	1.623	1.641	1.659
3.3	3.058	2.685	2.771	2.817	2.519	2.236	2.206	2.141	1.580	1.628	1.675
3.4	2.971	2.625	2.711	2.777	2.439	2.202	2.196	2.101	1.543	1.604	1.665
3.5	2.877	2.553	2.689	2.661	2.370	2.166	2.199	2.060	1.512	1.608	1.704
3.6	2.833	2.530	2.665	2.569	2.338	2.180	2.214	2.044	1.487	1.615	1.743
3.7	2.789	2.494	2.693	2.529	2.307	2.189	2.241	1.978	1.468	1.626	1.783
3.8	2.807	2.483	2.720	2.438	2.288	2.209	2.256	1.962	1.474	1.644	1.847
3.9	2.777	2.497	2.760	2.422	2.282	2.248	2.321	1.921	1.493	1.696	1.899
4.0	2.794	2.523	2.862	2.354	2.287	2.311	2.411	1.904	1.530	1.765	1.999
4.1	2.813	2.576	2.977	2.339	2.319	2.376	2.551	1.888	1.580	1.835	2.089

Table A3.17. *Radojcic et al.: NTUA double-chine*
series resistance regression coefficients

NTUA double-chine series resistance regression
coefficients: $C_R = \Sigma a_i \cdot x_i$

Variables	Coefficients
x_i	a_i
1	81.947561
$Fr(L/\nabla^{1/3})^2$	-18.238522
$(L/\nabla^{1/3})$	-44.283380
(B/T)	-7.775629
$(L/B)(B/T)$	1.731934
$Fr^2(L/\nabla^{1/3})^2$	17.075124
$Fr^3(B/T)^2$	12.079902
$Fr^5(B/T)$	273.294648
$Fr^5\,(B/T)^2$	-16.121701
$(B/T)^2$	1.294730
$Fr(B/T)^3$	-0.187700
$Fr^7(B/T)^3$	-1.459234
$Fr^6(B/T)^3$	3.399356
$(L/B)(B/T)^2$	-0.235111
$(L/\nabla^{1/3})^2$	5.323100
$(L/B)^2(L/\nabla^{1/3})$	-0.021188
$Fr\,(L/\nabla^{1/3})$	108.448244
$Fr^2(L/\nabla^{1/3})$	-92.667206
$Fr^7(B/T)(L/\nabla^{1/3})^2$	0.176635
$Fr^6(B/T)$	-216.312999
$Fr^6(B/T)\,(L/\nabla^{1/3})$	-3.354160
$Fr(L/\nabla^{1/3})^4$	0.070018
$Fr(B/T)\,(L/\nabla^{1/3})$	1.145943
$(L/\nabla^{1/3})^4$	-0.017590
$Fr^2(L/\nabla^{1/3})^4$	-0.062264
$Fr^7(B/T)(L/B)$	-0.264296
$Fr^3(B/T)$	-105.059107
$Fr^7\,(B/T)$	55.703462
$Fr^5(B/T)^3$	-1.810860
$Fr^7(B/T)^2$	4.310164
$Fr^3(L/\nabla^{1/3})^2$	-1.240887

Table A3.18. *Radojcic et al.: NTUA double-chine series trim regression coefficients*

NTUA double-chine series trim regression
coefficients: $\tau = \Sigma b_i \cdot x_i$

Variables	Coefficients
x_i	b_i
1	1.444311
$Fr(L/B)$	-33.570197
$Fr(L/B)(L/\nabla^{1/3})$	0.246174
$Fr^3(L/B)$	-158.702695
$Fr^4(L/\nabla^{1/3})^2$	-4.977716
$Fr^3(L/\nabla^{1/3})^2$	7.433941
$Fr^3(L/B)^2$	4.833377
$Fr^4(L/\nabla^{1/3})$	8.684395
$Fr^2(L/B)^2(L/\nabla^{1/3})$	0.095234
$Fr(L/B)^2$	1.661479
$Fr^5(L/B)$	-18.319249
$Fr^2(L/\nabla^{1/3})$	-9.214203
$(L/B)(L/\nabla^{1/3})^2$	-0.006817
$(L/B)^2$	-0.123263
$Fr^5(B/T)(L/\nabla^{1/3})^2$	0.038370
$Fr^4(L/B)$	82.474887
$Fr^5(L/B)^3$	-0.066572
$Fr^2(L/B)(L/\nabla^{1/3})$	-1.861194
$Fr(B/T)$	1.562730
$Fr^2(L/\nabla^{1/3})^2$	-3.780954
$(B/T)(L/\nabla^{1/3})$	-0.091927
$Fr^5(L/\nabla^{1/3})$	-2.133727
(L/B)	2.371311
$Fr^2(L/B)$	128.709296
$Fr^2(L/B)^2$	-5.739856
$Fr^4(L/\nabla^{1/3})^2 (B/T)$	-0.067986
$Fr(L/\nabla^{1/3})^2$	1.120683
$Fr^5(L/\nabla^{1/3})^2$	1.040239

Table A3.19. *Radojcic: Series 62 planing craft: resistance regression coefficients*

	Series 62			Resistance data				
				Fr_∇				
Coeff	1.0	1.25	1.5	1.75	2.0	2.5	3.0	3.5
b0	0.061666	0.092185	0.102614	0.109973	0.123424	0.133448	0.13738	0.1271
b1	−0.02069	−0.04135	−0.03582	−0.03369	−0.0304	−0.02033	−0.00487	0
b2	−0.01159	−0.01133	−0.00928	−0.00722	−0.00997	0	0.015175	0.013447
b3	−0.03102	−0.05631	−0.04982	−0.04533	−0.04212	−0.02675	−0.01721	0.025926
b4	0	0	0	0	0	0	0.014291	0.029002
b5	0	0	0	0	0	0.010282	0.018113	0.040553
b6	0.00286	0.012382	0.009146	0.009664	0.011089	0	0	0
b7	0	0	0	0	0	0	0	0
b8	0	0	0	0	0	0	0	0
b9	0	0	0	0	0	0	0	0
b10	0	0	0	0	0	−0.00839	−0.01831	−0.01585
b11	0.009057	0.014487	0.012373	0.018814	0.021142	0.016951	0.014405	0.037687
b12	0.0063	0.016982	0.016704	0.016669	0.014986	0.014537	0.009484	0
b13	0.021537	0.035708	0.024975	0.016109	0.004506	−0.00454	0	0.016351
b14	0	0	0	0	0	0	0.007653	0.024701
b15	0	0	0	0	0	0	0	0
b16	0	0	0	0	0	0	0	0
b17	0	0	0	−0.00585	−0.01213	−0.01539	−0.02448	−0.03109
b18	0	0	0	0	0.010251	0.011794	0	−0.03996
b19	−0.01066	−0.01427	−0.0097	−0.00777	0	0	−0.00778	−0.02575
b20	0	0	0	0	−0.00928	−0.02265	−0.0239	0
b21	0	0	0	0	0	0	0	0
b22	0	0	0	0	0	0	0	0
b23	0	0.004174	0.00535	0.007986	0.011154	0.011167	0	−0.04083
b24	0	0	0	0	0	0	0	0
b25	0	0	0	0	0	0	0	0
b26	0	0	0	0	0	0	0	0

Table A3.20 *Radojcic: Series 62 planing craft: trim regression coefficients*

| | Series 62 | | | Trim data | | | | |
| | | | | Fr_∇ | | | | |
Coeff	1.0	1.25	1.5	1.75	2.0	2.5	3.0	3.5
a0	1.33336	2.75105	3.20994	3.463689	3.933821	4.339123	3.840853	4.02473
a1	−1.03703	−1.86276	−1.79649	−2.26844	−2.7588	−2.48144	−1.9728	−1.38826
a2	−1.83433	−1.53407	−1.55056	−1.72094	−1.84238	−1.53977	−0.9005	−0.83942
a3	−1.15781	−2.57566	−2.73104	−2.87897	−2.67157	−1.63892	−0.75973	1.021179
a4	−0.44248	−0.36603	−0.38527	−0.48195	−0.42807	−0.73198	−0.74778	0
a5	0	0	0	0	0	0	0	0
a6	0.48222	1.149628	1.323168	1.300026	1.135659	0.30212	−0.45416	−0.41309
a7	0	0	0	0	0	0	0	0
a8	1.560221	0.783089	0.771597	0.744954	0.606753	0	−0.48899	0
a9	0	0	0	0	0	−0.53871	−0.65202	−1.02696
a10	0	0	0	0	0	0	0	0
a11	0.231295	0.72789	0.781132	1.087874	1.485322	0.835769	0.815711	0.570996
a12	0.817052	0.990262	0.933988	0.891951	0.820895	0.673848	0.508176	−0.53188
a13	0.777217	1.637322	1.661001	1.45161	0.619359	−0.73224	−0.6305	−0.24371
a14	0.420958	0.480769	0.611616	0.722073	0.773467	0.329024	0	0
a15	0	0	0	0	0	0	0	0
a16	0	−0.76619	−1.02118	−0.45775	0	0.460668	0	0
a17	0	0	0	0	0	0	0	0
a18	0	0	0	0	0	0	0	−1.01829
a19	−1.31998	−1.06647	−0.75931	−0.46489	0	0.748834	0.686297	0
a20	0	0	0	0	0	0	0	0
a21	0	0	0	0	0	0	0	0
a22	0	0	0	0	0	0	0	0
a23	0	0	0	0	0	0.242369	0.778294	−1.01221
a24	0	0	0	0	0	0	0	0
a25	1.776541	0.985106	0.827274	0.700908	0.315896	1.059782	0.898469	0
a26	0	0	0	0	0	0	0	0

Table A3.21. *Radojcic: Series 62 planing craft: wetted area regression coefficients*

	Series 62			Wetted area				
				Fr_∇				
Coeff	1.0	1.25	1.5	1.75	2.0	2.5	3.0	3.5
c0	7.359729	7.282876	7.208292	7.15098	7.018296	5.95292	5.432009	5.362447
c1	1.67294	1.903142	1.88377	1.971286	2.035661	2.115156	2.009723	2.02278
c2	−0.07193	−0.17774	−0.17757	0.014623	0.160293	0.628621	0.83802	1.347096
c3	0.428002	0.725644	0.994301	1.292172	1.47999	1.477329	1.269205	0.722797
c4	−0.06056	−0.11251	−0.02734	0.130288	0.298054	0.728084	0.858963	1.619465
c5	0	0	0	0	0	0	0	0
c6	0	0	0	0	0	0	0	0
c7	0	0	0	0	0	0	0	0
c8	0	0	0	0	0	0	0	0
c9	0	0	0	0	0	0	0	0
c10	0	0	0	0	0	0	0	0
c11	0	0	0	0	0	0	0	0
c12	0.274736	−0.03968	−0.22836	−0.58284	−0.81368	−0.63394	−0.53871	−0.16625
c13	−0.38915	−0.49279	−0.60403	−0.63793	−0.71262	0.098037	−0.00353	−0.46934
c14	0	0	0	0	0	0	0	0
c15	0	0	0	0	0	0	0	0
c16	0	0	0	0	0	0	0	0
c17	0	0	0	0	0	0	0	0
c18	0	0	0	0	0	0	0	0
c19	0.449769	0.584691	0.545012	0.286942	0.247044	−0.26484	−0.38301	0.295878
c20	0	0	0	0	0	0	0	0
c21	0	0	0	0	0	0	0	0
c22	0	0	0	0	0	0	0	0
c23	0	0	0	0	0	0	0	0
c24	0	0	0	0	0	0	0	0
c25	0	0	0	0	0	0	0	0
c26	0	0	0	0	0	0	0	0

Table A3.22. *Radojcic: Series 62 planing craft: wetted length regression coefficients*

	Series 62			Wetted length				
				Fr_∇				
Coeff	1.0	1.25	1.5	1.75	2.0	2.5	3.0	3.5
d0	0.77418	0.765015	0.752199	0.736158	0.701285	0.612134	0.562071	0.535208
d1	−0.12184	−0.10418	−0.09482	−0.07337	−0.05133	−0.01425	0	0
d2	0.184347	0.131686	0.103753	0.109122	0.117522	0.146182	0.154206	0.158171
d3	0.023393	0.047758	0.072623	0.097962	0.116865	0.116218	0.092023	0.0546
d4	−0.2557	−0.1944	−0.14539	−0.10503	−0.05034	0	0.023845	0.034613
d5	0	0	0	0	0	0	0	0
d6	0	0	0	0	0	0	0	0
d7	−0.01291	−0.09672	−0.11006	−0.11534	−0.11284	−0.07581	−0.03929	−0.03873
d8	0	0	0	0	0	0	0	0
d9	0	0	0	0	0	0	0	0
d10	0	0	0	0	0	0	0	0
d11	0	0	0	0	0	0	0	0
d12	0	0	0	0	0	0	0	0
d13	0	−0.02083	−0.04265	−0.0649	−0.05404	0	0	0
d14	0	0	0	0	0	0	0	0
d15	0	0	0	0	0	0	0	0
d16	0	0	0	0	0	0	0	0
d17	0	0	0	0	0	0	0	0
d18	0	0	0	0	0	0	0	0
d19	0	0	0	0	0	0	0	0
d20	−0.36473	−0.239	−0.15159	−0.12144	−0.06682	−0.04427	−0.01844	−0.02929
d21	0	0	0	0	0	0	0	0
d22	0	0	0	0	0	0	0	0
d23	0	0	0	0	0	0	0	0
d24	0	0	0	0	0	0	0	0
d25	0	0	0	0	0	0	0	0
d26	0	0	0	0	0	0	0	0

Table A3.23. *Oortmerssen: small ships: resistance regression coefficients*

Oortmerssen: Resistance regression coefficients

$i =$	1	2	3	4
di, 0	79.32134	6714.88397	− 908.44371	3012.14549
di, 1	− 0.09287	19.83000	2.52704	2.71437
di, 2	− 0.00209	2.66997	− 0.35794	0.25521
di, 3	−246.45896	−19662.02400	755.18660	−9198.80840
di, 4	187.13664	14099.90400	− 48.93952	6886.60416
di, 5	− 1.42893	137.33613	9.86873	− 159.92694
di, 6	0.11898	− 13.36938	− 0.77652	16.23621
di, 7	0.15727	− 4.49852	3.79020	− 0.82014
di, 8	− 0.00064	0.02100	− 0.01879	0.00225
di, 9	− 2.52862	216.44923	− 9.24399	236.37970
di, 10	0.50619	− 35.07602	1.28571	− 44.17820
di, 11	1.62851	− 128.72535	250.64910	207.25580

Table A3.24. *WUMTIA resistance regression coefficients for C-factor: round bilge*

WUMTIA Resistance regression coefficients: round bilge

Parameter	Fr_∇	0.50	0.75	1.00	1.25	1.50	1.75	2.00	2.25	2.50	2.75
	a_0	1136.829	−4276.159	−921.0902	−449.8701	−605.9794	−437.3817	351.1909	813.1732	−622.9426	−1219.095
$(L/\nabla^{1/3})$	a_1	−54.50337	859.2251	460.6681	243.5577	3.073361	40.51505	−183.7483	−194.1047	200.5628	346.1326
(L/B)	a_2	−261.8232	98.15745	2.604913	59.9026	−32.77933	−87.85154	−101.2289	−63.92188	−138.7268	−139.0729
$(S/L^2)^{0.5}$	a_3	−2695.885	16369.41	737.8893	−223.2636	4097.999	3101.983	956.9388	−1884.341	2745.177	4659.579
$(L/\nabla^{1/3})^2$	a_4	5.365086	−153.5496	−75.42524	−40.36861	−1.682758	−4.308722	35.62357	36.56844	−31.93601	−56.785
$(L/B)^2$	a_5	59.31649	−24.77183	−0.706952	−12.58654	6.486023	20.96359	25.14769	17.01779	36.50832	36.71361
(S/L^2)	a_6	5300.271	−32787.07	−2325.398	7.616481	−7823.835	−6339.599	−2061.44	3417.534	−5770.126	−9650.592
$(L/\nabla^{1/3})^3$	a_7	−0.136343	9.031855	4.114508	2.222081	0.200794	0.104035	−2.254183	−2.264704	1.682685	3.096224
$(L/B)^3$	a_8	−4.207338	1.970939	0.112712	0.901679	−0.341222	−1.586765	−2.004468	−1.429349	−3.082187	−3.124286
$(S/L^2)^{3/2}$	a_9	−3592.034	21484.47	2095.625	274.9351	4961.028	4302.659	1473.702	−2022.774	4046.07	6655.716

Table A3.25. *WUMTIA resistance regression coefficients for C-factor: hard chine*

WUMTIA resistance regression coefficients: hard chine

Parameter	Fr_∇	0.50	0.75	1.00	1.25	1.50	1.75	2.00	2.25	2.50	2.75
	a_0	89.22488	−113.6004	45.69113	17.73174	73.06773	113.3937	158.9419	214.5219	237.5662	298.0709
$(L/\nabla^{1/3})$	a_1	−121.3817	83.82636	−27.37823	−16.27857	−30.75841	−29.30795	−36.04947	−62.45653	−83.58921	−104.4391
(L/B)	a_2	201.9296	10.56977	32.59633	32.93866	18.25139	−12.03086	−31.44896	−28.86708	−10.46636	−16.86174
$(L/\nabla^{1/3})^2$	a_3	19.60451	−12.83540	5.164941	3.565837	5.486712	5.134334	5.980537	10.57904	14.01806	16.13167
$(L/B)^2$	a_4	−49.30790	−2.686623	−5.827831	−6.978205	−3.283823	4.229486	8.982328	7.123322	2.163294	5.264803
$(L/\nabla^{1/3})^3$	a_5	−1.058321	0.630652	−0.285075	−0.186828	−0.273739	−0.257980	−0.298946	−0.566186	−0.753518	−0.791785
$(L/B)^3$	a_6	4.033562	0.348702	0.399769	0.504987	0.196993	−0.395147	−0.771390	−0.528270	−0.099098	−0.524063

Table A3.26. *Molland et al.: Southampton catamaran series resistance data $C_R \times 1000$*

	Model 3b residuary resistance ($C_T - C_{FITTC}$)						Model 5a residuary resistance ($C_T - C_{FITTC}$)				
Fr	Monohull	$S/L = 0.2$	$S/L = 0.3$	$S/L = 0.4$	$S/L = 0.5$	Fr	Monohull	$S/L = 0.2$	$S/L = 0.3$	$S/L = 0.4$	$S/L = 0.5$
0.200	2.971	3.192	3.214	2.642	2.555	0.200	1.862	2.565	2.565	2.381	2.592
0.250	3.510	4.540	3.726	4.019	3.299	0.250	2.485	3.074	2.991	3.031	3.123
0.300	3.808	5.303	4.750	4.464	3.938	0.300	3.009	3.959	3.589	3.686	3.473
0.350	4.800	6.771	5.943	5.472	4.803	0.350	3.260	4.018	3.756	3.589	3.716
0.400	5.621	8.972	7.648	7.085	6.589	0.400	3.677	4.472	4.604	4.616	4.403
0.450	8.036	12.393	12.569	10.934	9.064	0.450	4.103	6.068	5.563	5.099	4.929
0.500	9.038	14.874	14.237	12.027	10.112	0.500	3.884	5.805	4.950	4.581	4.501
0.550	8.543	15.417	12.275	10.538	9.394	0.550	3.442	4.914	4.221	4.015	3.966
0.600	7.626	12.818	10.089	8.962	8.361	0.600	3.063	4.065	3.596	3.516	3.499
0.650	6.736	8.371	8.123	7.592	7.488	0.650	2.736	3.429	3.138	3.126	3.140
0.700	5.954		6.852	6.642	6.726	0.700	2.461	3.004	2.827	2.845	2.882
0.750	5.383		5.934	5.921	6.078	0.750	2.278	2.705	2.615	2.658	2.699
0.800	4.911		5.289	5.373	5.537	0.800	2.138	2.494	2.465	2.519	2.559
0.850	4.484		4.814	4.949	5.046	0.850	2.038	2.342	2.351	2.406	2.453
0.900	4.102		4.452	4.543	4.624	0.900	1.931	2.231	2.260	2.308	2.354
0.950	3.785		4.172	4.236	4.335	0.950	1.871	2.153	2.183	2.238	2.272
1.000	3.579		3.936	3.996	4.099	1.000	1.818	2.100	2.124	2.179	2.201

Table A3.26. Molland et al.: Southampton catamaran series resistance data (continued) $C_R \times 1000$

	Mode 4a residuary resistance ($C_T - C_{FITTC}$)						Model 5b residuary resistance ($C_T - C_{FITTC}$)				
Fr	Monohull	$S/L = 0.2$	$S/L = 0.3$	$S/L = 0.4$	$S/L = 0.5$	Fr	Monohull	$S/L = 0.2$	$S/L = 0.3$	$S/L = 0.4$	$S/L = 0.5$
0.200	1.909	2.327	2.564	2.495	2.719	0.200	1.406	2.288	2.849	2.538	3.006
0.250	2.465	3.148	3.315	2.937	3.484	0.250	2.362	2.843	3.200	3.260	3.093
0.300	3.273	3.954	4.283	4.396	3.875	0.300	2.632	3.643	3.539	3.693	3.330
0.350	3.585	5.073	4.576	4.064	4.173	0.350	2.890	4.194	3.952	3.711	3.437
0.400	4.100	4.874	5.871	5.900	5.109	0.400	3.514	4.520	4.687	4.622	4.303
0.450	5.305	8.111	7.953	7.220	6.299	0.450	3.691	5.506	5.218	4.960	4.648
0.500	5.526	8.365	7.150	6.650	6.140	0.500	3.518	5.581	4.903	4.632	4.324
0.550	5.086	7.138	5.990	5.692	5.615	0.550	3.125	4.927	4.323	4.057	3.804
0.600	4.431	5.878	5.090	4.880	4.981	0.600	2.851	4.177	3.783	3.504	3.286
0.650	3.924	4.815	4.392	4.269	4.387	0.650	2.599	3.555	3.302	3.090	2.872
0.700	3.477	4.047	3.949	3.834	3.911	0.700	2.285	3.051	2.989	2.759	2.576
0.750	3.128	3.556	3.594	3.512	3.570	0.750	2.155	2.744	2.752	2.515	2.396
0.800	2.904	3.224	3.187	3.252	3.296	0.800	2.010	2.529	2.584	2.327	2.310
0.850	2.706	2.923	2.966	3.054	3.070	0.850	1.938	2.383	2.462	2.163	2.322
0.900	2.544	2.729	2.839	2.881	2.873	0.900	1.830	2.298	2.375	2.111	2.382
0.950	2.398	2.550	2.657	2.767	2.707	0.950	1.852	2.221	2.324	2.128	1.852
1.000	2.272	2.433	2.437	2.687	2.558	1.000	1.803	2.186	2.279	2.145	1.803

517

Table A3.26. Molland et al.: Southampton catamaran series resistance data (continued) $C_R \times 1000$

Fr	Model 4b residuary resistance ($C_T - C_{FITTC}$)					Fr	Model 5c residuaty resistance ($C_T - C_{FITTC}$)				
	Monohull	$S/L = 0.2$	$S/L = 0.3$	$S/L = 0.4$	$S/L = 0.5$		Monohull	$S/L = 0.2$	$S/L = 0.3$	$S/L = 0.4$	$S/L = 0.5$
0.200	2.613	2.929	2.841	2.721	2.820	0.200	2.517	2.731	2.801	2.718	2.983
0.250	2.629	3.686	3.374	3.365	3.396	0.250	2.756	3.256	3.199	3.203	3.290
0.300	3.532	4.311	4.113	4.150	3.902	0.300	3.010	3.445	3.599	3.386	3.371
0.350	3.763	3.483	4.816	4.557	4.329	0.350	3.273	3.937	3.779	3.623	3.625
0.400	4.520	5.897	5.934	5.940	5.716	0.400	3.687	4.635	4.813	4.731	4.519
0.450	5.402	7.748	7.777	7.078	6.741	0.450	3.891	5.908	5.543	4.969	4.644
0.500	5.389	8.420	7.669	6.922	6.581	0.500	3.621	5.864	5.016	4.513	4.340
0.550	4.865	8.099	6.639	6.145	5.921	0.550	3.232	5.095	4.274	3.945	3.855
0.600	4.276	7.159	5.471	5.315	5.209	0.600	3.048	4.231	3.703	3.495	3.512
0.650	3.787	6.008	4.620	4.605	4.593	0.650	2.685	3.576	3.267	3.183	3.187
0.700	3.394	4.769	4.061	4.098	4.125	0.700	2.417	3.074	2.930	2.920	2.936
0.750	3.098	4.041	3.641	3.718	3.786	0.750	2.205	2.771	2.741	2.717	2.779
0.800	2.848	3.605	3.326	3.440	3.520	0.800	2.076	2.558	2.632	2.564	2.594
0.850	2.647		3.153	3.247	3.319	0.850	1.903	2.434	2.607	2.476	2.514
0.900	2.476		2.917	3.078	3.131	0.900	1.863	2.346	2.599	2.404	2.454
0.950	2.361		2.834	2.968	2.988	0.950	1.915	2.259	2.550	2.341	2.358
1.000	2.347			2.882	2.870	1.000	1.785	2.213	2.481	2.256	2.281

Table A3.26. Molland et al.: Southampton catamaran series resistance data (continued) $C_R \times 1000$

	Model 4c residuary resistance $(C_T - C_{FITTC})$						Model 6a residuary resistance $(C_T - C_{FITTC})$				
Fr	Monohull	$S/L = 0.2$	$S/L = 0.3$	$S/L = 0.4$	$S/L = 0.5$	Fr	Monohull	$S/L = 0.2$	$S/L = 0.3$	$S/L = 0.4$	$S/L = 0.5$
0.200	2.169	2.983	2.830	2.801	2.690	0.200	1.916	2.727	2.660	2.807	2.484
0.250	2.506	3.718	3.459	3.412	3.336	0.250	2.257	3.379	3.244	3.595	3.515
0.300	2.987	4.401	4.110	4.067	3.960	0.300	2.443	3.792	3.548	3.761	3.665
0.350	3.349	5.336	4.777	4.321	4.275	0.350	2.527	3.665	3.381	3.754	3.566
0.400	4.371	5.905	5.850	5.919	5.722	0.400	2.723	4.377	4.403	4.257	4.009
0.450	5.525	8.567	8.454	7.605	7.061	0.450	2.796	4.703	4.593	4.339	3.998
0.500	5.512	9.474	7.892	7.013	6.633	0.500	2.658	4.592	3.974	3.855	3.635
0.550	5.021	8.316	6.625	6.087	5.907	0.550	2.434	3.799	3.382	3.338	3.243
0.600	4.473	6.845	5.522	5.249	5.204	0.600	2.246	3.193	2.994	2.955	2.916
0.650	3.995	5.584	4.720	4.617	4.637	0.650	2.111	2.812	2.703	2.689	2.651
0.700	3.632	4.718	4.167	4.165	4.203	0.700	1.917	2.534	2.496	2.505	2.475
0.750	3.360	4.216	3.785	3.845	3.871	0.750	1.781	2.367	2.348	2.379	2.336
0.800	3.119	3.784	3.503	3.587	3.608	0.800	1.633	2.253	2.261	2.304	2.243
0.850	2.922	3.459	3.276	3.364	3.387	0.850	1.544	2.176	2.194	2.230	2.171
0.900	2.743	3.276	3.089	3.165	3.190	0.900	1.478	2.110	2.155	2.146	2.093
0.950	2.603	3.076	2.934	3.003	3.017	0.950	1.528	2.062	2.110	2.047	2.021
1.000	2.481	2.904	2.821	2.875	2.875	1.000	1.521	2.027	2.064	1.976	1.962

Table A3.26. *Molland et al.: Southampton catamaran series resistance data (continued)* $C_R \times 1000$

Fr	Model 6b residuary resistance $(C_T - C_{FITTC})$					Fr	Model 6c residuary resistance $(C_T - C_{FITTC})$				
	Monohull	$S/L = 0.2$	$S/L = 0.3$	$S/L = 0.4$	$S/L = 0.5$		Monohull	$S/L = 0.2$	$S/L = 0.3$	$S/L = 0.4$	$S/L = 0.5$
0.200	1.755	2.864	2.297	2.933	2.353	0.200	1.882	2.979	1.909	2.608	2.515
0.250	2.136	3.217	3.235	3.203	2.335	0.250	2.395	3.169	3.328	3.056	2.911
0.300	2.255	3.769	3.162	3.251	2.833	0.300	2.581	3.539	3.401	3.252	3.191
0.350	2.150	3.667	3.299	3.502	3.158	0.350	2.666	3.531	3.309	3.385	3.366
0.400	2.639	4.007	3.721	3.913	3.479	0.400	2.785	3.684	3.774	3.813	3.629
0.450	2.696	4.534	4.092	3.950	3.570	0.450	2.816	4.229	3.932	3.813	3.676
0.500	2.510	4.379	3.771	3.592	3.393	0.500	2.626	4.154	3.719	3.527	3.446
0.550	2.338	3.734	3.202	3.196	3.085	0.550	2.394	3.573	3.256	3.187	3.145
0.600	2.084	3.144	2.762	2.866	2.662	0.600	2.177	3.080	2.855	2.866	2.851
0.650	1.900	2.738	2.507	2.635	2.565	0.650	2.006	2.809	2.595	2.609	2.608
0.700	1.747	2.477	2.355	2.468	2.378	0.700	1.866	2.504	2.437	2.432	2.487
0.750	1.656	2.311	2.249	2.339	2.268	0.750	1.754	2.305	2.331	2.345	2.358
0.800	1.575	2.184	2.158	2.241	2.214	0.800	1.682	2.165	2.199	2.232	2.297
0.850	1.527	2.093	2.068	2.172	2.112	0.850	1.633	2.138	2.167	2.210	2.249
0.900	1.523	2.052	2.056	2.129	2.064	0.900	1.568	2.108	2.120	2.174	2.227
0.950	1.482	2.020	2.046	2.089	2.048	0.950	1.628	2.078	2.121	2.149	2.227
1.000	1.426	2.001	2.001	2.063	2.036	1.000	1.672	2.067	2.134	2.157	2.193

Table A3.27. *Müller-Graf and Zips: VWS catamaran series resistance coefficients*

VWS catamaran series resistance coefficients $R_R/\Delta = \Sigma (XRi \times CRi)/100$

Parameter XRi	Fr_∇	CRi							
		1.0	1.25	1.50	1.75	2.00	2.50	3.0	3.5
X0 = 1		2.348312	4.629531	5.635988	5.627470	5.690865	6.209794	7.243674	7.555179
X1 = $f(L/b)$		−0.706875	−2.708625	−2.371713	−2.266895	−2.500808	−2.900769	−3.246017	−2.647421
X2 = $f(\beta_M)$		−0.272668	−0.447266	−0.328047	−0.428999	−0.422339	−0.391296	0	0.453125
X3 = $f(\delta_W)$		0.558673	0	0	0	−0.288437	−0.447655	0	0
X4 = $X1^2$		0.256967	0.701719	0.349687	0.416250	0.571875	0.832031	0.554213	0.332042
X5 = $X1^3$		0	0	0.165938	0.342187	0.496875	0.658719	1.196250	1.884844
X6 = $X2^2$		0	0.148359	0	0	0	0	0	−0.276875
X7 = $X3^{1/2}$		0	0	0	0	0	0	−1.87743	0
X8 = $X3^{1/3}$		−0.152163	0	−0.251026	−0.429128	−0.450245	−0.866917	0	−1.036289
X9 = $X3^{1/4}$		0	0	0	0	0	0	0	−0.767250
X10 = X1×X2		0	0.149062	0.090188	0.089625	0.076125	0	−0.332250	−0.767250
X11 = X1 × X6		−0.151312	−0.090188	−0.135563	−0.194250	−0.190125	−0.225938	−0.211125	0
X12 = X4 × X6		−0.0592	−0.322734	0	0	0	0	0	0
X13 = X4 × X2		0	−0.148359	−0.096328	0	0	0	0	0
X14 = X1 × X3		0	0	0.484800	0	0.817200	1.189350	1.007700	0
X15 = X4 × X7		0	0.409500	0	0	0	0	0.588758	0.683505
X16 = X1 × X9		0	0	0	0	0	0	0	−0.241426
X17 = X1 × $X7^3$		0	0	0	0.704463	0	0	0	0
X18 = X2 × X8		−0.083789	0	0	0.120516	0.137585	0.257507	0	0

Tabulations of Propulsor Design Data

It should be noted that some data, such as the design charts for the Wageningen series, Gawn series, ducted propellers, supercavitating and surface-piercing propellers, are contained fully within the text in Chapter 16.

Table A4.1. *Wageningen propeller series polynomial coefficients*

	Wageningen propeller series polynomial coefficients									
Thrust K_T						Torque K_Q				
n	C_n	s	t	u	v	n	C_n	s	t	u	v
1	0.00880496	0	0	0	0	1	0.00379368	0	0	0	0
2	−0.20455400	1	0	0	0	2	0.00886523	2	0	0	0
3	0.16635100	0	1	0	0	3	−0.032241	1	1	0	0
4	0.15811400	0	2	0	0	4	0.00344778	0	2	0	0
5	−0.14758100	2	0	1	0	5	−0.0408811	0	1	1	0
6	−0.48149700	1	1	1	0	6	−0.108009	1	1	1	0
7	0.41543700	0	2	1	0	7	−0.0885381	2	1	1	0
8	0.01440430	0	0	0	1	8	0.188561	0	2	1	0
9	−0.05300540	2	0	0	1	9	−0.00370871	1	0	0	1
10	0.01434810	0	1	0	1	10	0.00513696	0	1	0	1
11	0.06068260	1	1	0	1	11	0.0209449	1	1	0	1
12	−0.01258940	0	0	1	1	12	0.00474319	2	1	0	1
13	0.01096890	1	0	1	1	13	−0.00723408	2	0	1	1
14	−0.13369800	0	3	0	0	14	0.00438388	1	1	1	1
15	0.00638407	0	6	0	0	15	−0.0269403	0	2	1	1
16	−0.00132718	2	6	0	0	16	0.0558082	3	0	1	0
17	0.16849600	3	0	1	0	17	0.0161886	0	3	1	0
18	−0.05072140	0	0	2	0	18	0.00318086	1	3	1	0
19	0.08545590	2	0	2	0	19	0.015896	0	0	2	0
20	−0.05044750	3	0	2	0	20	0.0471729	1	0	2	0
21	0.01046500	1	6	2	0	21	0.0196283	3	0	2	0
22	−0.00648272	2	6	2	0	22	−0.0502782	0	1	2	0
23	−0.00841728	0	3	0	1	23	−0.030055	3	1	2	0
24	0.01684240	1	3	0	1	24	0.0417122	2	2	2	0
25	−0.00102296	3	3	0	1	25	−0.0397722	0	3	2	0
26	−0.03177910	0	3	1	1	26	−0.00350024	0	6	2	0
27	0.01860400	1	0	2	1	27	−0.0106854	3	0	0	1
28	−0.00410798	0	2	2	1	28	0.00110903	3	3	0	1
29	−0.000606848	0	0	0	2	29	−0.000313912	0	6	0	1
30	−0.004981900	1	0	0	2	30	0.0035985	3	0	1	1
31	0.002598300	2	0	0	2	31	−0.00142121	0	6	1	1
32	−0.000560528	3	0	0	2	32	−0.00383637	1	0	2	1
33	−0.001636520	1	2	0	2	33	0.0126803	0	2	2	1
34	−0.000328787	1	6	0	2	34	−0.00318278	2	3	2	1
35	0.000116502	2	6	0	2	35	0.00334268	0	6	2	1
36	0.000690904	0	0	1	2	36	−0.00183491	1	1	0	2
37	0.004217490	0	3	1	2	37	0.000112451	3	2	0	2
38	0.0000565229	3	6	1	2	38	−0.0000297228	3	6	0	2
39	−0.001465640	0	3	2	2	39	0.000269551	1	0	1	2
						40	0.00083265	2	0	1	2
						41	0.00155334	0	2	1	2
						42	0.000302683	0	6	1	2
						43	−0.0001843	0	0	2	2
						44	−0.000425399	0	3	2	2
						45	0.0000869243	3	3	2	2
						46	−0.0004659	0	6	2	2
						47	0.0000554194	1	6	2	2

Table A4.2. *Wageningen propeller series polynomial coefficients: influence of Re*

Wageningen propeller series polynomial coefficients: influence of Re			
$\Delta K_T = \Sigma\, a_i \cdot P_i$		$\Delta K_Q = \Sigma\, b_i\, P_i$	
P_i	a_i	P_i	b_i
1	0.0003534850	1	−0.00059141200
$(A_E/A_0)J^2$	−0.0033375800	(P/D)	0.00696898000
$(A_E/A_0)(P/D)J$	−0.0047812500	$Z \cdot (P/D)^6$	−0.00006666540
$(\log Re - 0.301)^2 \cdot (A_E/A_0) \cdot J^2$	0.0002577920	$(A_E/A_0)^2$	0.01608180000
$(\log Re - 0.301)\,(P/D)^6 J^2$	0.0000643192	$(\log Re - 0.301) \cdot (P/D)$	−0.00093809000
$(\log Re - 0.301)^2 \cdot (P/D)^6 \cdot J^2$	−0.0000110636	$(\log Re - 0.301) \cdot (P/D)^2$	−0.00059593000
$(\log Re - 0.301)^2 \cdot Z \cdot (A_E/A_0) \cdot J^2$	−0.0000276305	$(\log Re - 0.301)^2 \cdot (P/D)^2$	0.00007820990
$(\log Re - 0.301) \cdot Z \cdot (A_E/A_0) \cdot (P/D) \cdot J$	0.0000954000	$(\log Re - 0.301) \cdot Z \cdot (A_E/A_0) \cdot J^2$	0.00000521990
$(\log Re - 0.301) \cdot Z^2 \cdot (A_E/A_0) (P/D)^3 \cdot J$	0.0000032049	$(\log Re - 0.301)^2 \cdot Z \cdot (A_E/A_0) \cdot (P/D) \cdot J$	−0.00000088528
		$(\log Re - 0.301) \cdot Z \cdot (P/D)^6$	0.00002301710
		$(\log Re - 0.301)^2 \cdot Z \cdot (P/D)^6$	−0.00000184341
		$(\log Re - 0.301)\,(A_E/A_0)^2$	−0.00400252000
		$(\log Re - 0.301)^2 (A_E/A_0)^2$	0.00022091500

Table A4.3. *Gawn propeller series polynomial coefficients*

Gawn propeller series polynomial coefficients											
Thrust K_T						Torque K_Q					
n	C_n	s	t	u	v	n	C_n	s	t	u	v
1	−0.0558636300	0	0	0	0	1	0.0051589800	0	0	0	0
2	−0.2173010900	1	0	0	0	2	0.0160666800	2	0	0	0
3	0.260531400	0	1	0	0	3	−0.044115300	1	1	0	0
4	0.158114000	0	2	0	0	4	0.0068222300	0	2	0	0
5	−0.147581000	2	0	1	0	5	−0.040881100	0	1	1	0
6	−0.481497000	1	1	1	0	6	−0.077329670	1	1	1	0
7	0.3781227800	0	2	1	0	7	−0.088538100	2	1	1	0
8	0.0144043000	0	0	0	1	8	0.1693750200	0	2	1	0
9	−0.0530054000	2	0	0	1	9	−0.003708710	1	0	0	1
10	0.0143481000	0	1	0	1	10	0.0051369600	0	1	0	1
11	0.0606826000	1	1	0	1	11	0.0209449000	1	1	0	1
12	−0.0125894000	0	0	1	1	12	0.0047431900	2	1	0	1
13	0.0109689000	1	0	1	1	13	−0.007234080	2	0	1	1
14	−0.1336980000	0	3	0	0	14	0.0043838800	1	1	1	1
15	0.0024115700	0	6	0	0	15	−0.026940300	0	2	1	1
16	−0.0005300200	2	6	0	0	16	0.0558082000	3	0	1	0
17	0.1684960000	3	0	1	0	17	0.0161886000	0	3	1	0
18	0.0263454200	0	0	2	0	18	0.0031808600	1	3	1	0
19	0.0436013600	2	0	2	0	19	0.0129043500	0	0	2	0
20	−0.0311849300	3	0	2	0	20	0.024450840	1	0	2	0
21	0.0124921500	1	6	2	0	21	0.0070064300	3	0	2	0
22	−0.0064827200	2	6	2	0	22	−0.027190460	0	1	2	0
23	−0.0084172800	0	3	0	1	23	−0.016645860	3	1	2	0
24	0.0168424000	1	3	0	1	24	0.0300449000	2	2	2	0
25	−0.0010229600	3	3	0	1	25	−0.033697490	0	3	2	0
26	−0.0317791000	0	3	1	1	26	−0.003500240	0	6	2	0
27	0.018604000	1	0	2	1	27	−0.010685400	3	0	0	1
28	−0.0041079800	0	2	2	1	28	0.0011090300	3	3	0	1
29	−0.0006068480	0	0	0	2	29	−0.000313912	0	6	0	1
30	−0.0049819000	1	0	0	2	30	0.0035895000	3	0	1	1
31	0.0025963000	2	0	0	2	31	−0.001421210	0	6	1	1
32	−0.0005605280	3	0	0	2	32	−0.003836370	1	0	2	1
33	−0.0016365200	1	2	0	2	33	0.0126803000	0	2	2	1
34	−0.0003287870	1	6	0	2	34	−0.003182780	2	3	2	1
35	0.0001165020	2	6	0	2	35	0.0033426800	0	6	2	1
36	0.0006909040	0	0	1	2	36	−0.001834910	1	1	0	2
37	0.0042174900	0	3	1	2	37	0.0001124510	3	2	0	2
38	0.0000565229	3	6	1	2	38	−0.0000297228	3	6	0	2
39	−0.0014656400	0	3	2	2	39	0.000269551	1	0	1	2
						40	0.0008326500	2	0	1	2
						41	0.0015533400	0	2	1	2
						42	0.0003026830	0	6	1	2
						43	−0.000184300	0	0	2	2
						44	−0.000425399	0	3	2	2
						45	0.0000869243	3	3	2	2
						46	−0.0004659000	0	6	2	2
						47	0.0000554194	1	6	2	2

Table A4.4. *KCA propeller series polynomial coefficients*

	KCA propeller series polynomial coefficients									
	C_t	e	x	y	z	C_q	e	x	y	z
1	0.1193852	0	0	0	0	1.5411660	−3	0	0	0
2	−0.6574682	0	0	0	1	0.1091688	0	0	0	1
3	0.3493294	0	0	1	0	−0.3102420	0	0	0	2
4	0.4119366	0	0	1	1	0.1547428	0	0	0	3
5	−0.1991927	0	0	2	1	−4.3706150	−2	0	1	0
6	5.8630510	−2	0	2	2	0.2490295	0	0	1	2
7	−1.1077350	−2	0	2	3	−0.1594602	0	0	1	3
8	−0.1341679	0	1	0	0	8.5367470	−2	0	2	0
9	0.2628839	0	1	0	1	−9.5121630	−2	0	2	1
10	−0.5217023	0	1	1	1	−9.3203070	−3	0	2	2
11	0.2970728	0	1	2	0	3.2878050	−2	0	2	3
12	6.1525800	−2	2	1	3	5.4960340	−2	1	0	1
13	−2.4708400	−2	2	2	3	−4.8650630	−2	1	1	0
14	−4.0801660	−3	1	6	0	−0.1062500	0	1	1	1
15	4.1542010	−3	1	6	1	8.5299550	−2	1	2	0
16	−1.1364520	−3	2	6	0	1.1010230	−2	2	0	3
17	−	−	−	−	−	−3.1517560	−3	2	2	2

Table A4.5. *KCA propeller series polynomial coefficients: influence of cavitation*

	KCA propeller series polynomial coefficients: influence of cavitation										
	d_t	e	s	t	u	v	d_q	e	s	t	u	v
1	6.688144	−2	0	0	0	0	4.024475	−3	0	0	0	0
2	3.579195	0	0	0	2	0	1.202447	−1	0	0	2	0
3	−5.700350	0	0	0	3	0	−9.836070	−2	1	1	0	0
4	−1.359994	0	1	1	0	0	−8.318840	−1	1	1	1	0
5	−8.111903	0	1	1	1	0	5.098177	0	1	1	3	0
6	4.770548	1	1	1	3	0	−5.192839	−1	2	1	1	0
7	−2.313208	−1	2	1	0	0	2.641109	0	2	2	0	0
8	−1.387858	1	2	1	2	0	−1.688934	1	2	2	3	0
9	4.992201	1	2	1	3	0	4.928417	−2	0	0	1	1
10	−7.161204	1	2	1	4	0	1.024274	−2	0	0	0	2
11	1.721436	1	2	2	0	0	−1.194521	−1	0	1	1	1
12	2.322218	1	2	2	1	0	5.498736	−2	1	0	1	1
13	−1.156897	2	2	2	2	0	−2.488235	−1	1	1	0	1
14	5.014178	−2	0	0	0	2	−5.832879	−1	0	0	5	0
15	−6.555364	−2	0	0	1	2	1.503955	−1	0	3	0	0
16	2.852867	−1	1	0	1	1	−3.316121	0	3	3	0	0
17	−8.081759	−1	1	1	0	1	3.890792	0	3	3	1	0
18	8.671852	1	3	2	5	0	1.682032	1	3	3	3	0
19	−3.727835	1	3	3	0	0	−	−	−	−	−	−
20	8.043970	1	3	3	1	0	−	−	−	−	−	−

Table A4.6. *Series 60 relative rotative efficiency regression coefficients*

$V_K/\sqrt{L_f}$	Series 60 relative rotative efficiency regression coefficients, load draught condition								
	0.50	0.55	0.60	0.65	0.70	0.75	0.80	0.85	0.90
d1	0.1825	0.2300	0.2734	0.3271	0.3661	0.3748	0.3533	0.3356	−0.4972
d2	−0.2491	−0.2721	−0.2895	−0.3076	−0.2667	−0.2155	−0.2139	−0.1158	−1.3337
d3	−0.1100	−0.1046	−0.1061	−0.1006	−0.0884	−0.0582	−0.0616	0.0056	−0.5433
d4	0.5135	0.4472	−0.3708	0.2956	0.1927	0.1226	0.1211	0.1683	−1.3800
d5	−0.4949	−0.4104	−0.3446	−0.3111	−0.2275	−0.1495	−0.2094	−0.1753	−0.5107
d6	0.1810	0.1604	0.1147	0.0382	0.0376	0.0935	0.1504	0.1477	0.0631
d7	−0.0579	−0.0631	−0.0879	−0.1190	−0.1130	−0.0921	−0.0954	−0.0966	−0.0752
d8	−0.0242	0.0237	0.0518	0.0405	−0.0591	−0.1327	−0.0936	0.0029	−1.1904
d9	−0.2544	−0.2958	−0.2720	−0.2688	−0.3438	−0.3352	−0.2724	−0.1821	−0.3510
d10	0.1006	0.0513	0.0101	−0.0310	−0.0279	−0.0006	0.0215	−0.0150	−0.0601
d11	0.2174	0.2904	0.2999	0.2666	0.1486	0.1591	0.2869	0.3187	0.2347
d12	0.0571	−0.0811	−0.2046	−0.3065	−0.1286	−0.0518	−0.1748	−0.1018	−1.9488
d13	0.4328	0.3663	0.2863	0.1974	0.1296	0.0855	0.2038	0.2525	0.3341
d14	−0.4880	−0.4161	−0.3420	−0.2699	−0.1713	−0.0513	−0.1365	−0.0933	−1.0238
d15	0.1961	0.1269	0.0683	0.0334	0.1639	0.2359	0.1411	0.0616	0.5313

Index